012309211

OF LI
NOT TO
REMOVED FROM
THE LI RARY

Elsevier Materials Selector

Vol 3

Elsevier Materials Selector

Vol 3

Edited by

Norman A. Waterman, B.Sc., Ph.D.
Chief Executive, Quo-Tec Ltd,
Amersham, Buckinghamshire, UK

and

Michael F. Ashby, F.R.S.
Professor of Engineering,
University of Cambridge, UK

UNIVERSITY
LIBRARY

Elsevier Applied Science
London

ELSEVIER SCIENCE PUBLISHERS LTD
Crown House, Linton Road, Barking, Essex, IG11 8JU, England, UK

Published in the USA and Canada as *CRC–Elsevier Materials Selector* by CRC Press, Inc., 2000 Corporate Blvd., N.W., Boca Raton, Florida 33431, USA

WITH 777 TABLES AND 441 ILLUSTRATIONS

© 1991 ELSEVIER SCIENCE PUBLISHERS LTD
© 1991 M. F. ASHBY, Figs 1.1.5–1.1.16, 1.1.18, 1.1.19, Vol.1, Chap.1.1.
(Reproduced with the permission of Pergamon Press)
Extracts from British Standards are reproduced with the permission of BSI.
Complete copies can be obtained from BSI Sales, Linford Wood, Milton Keynes, MK14 6LE, UK

British Library Cataloguing in Publication Data
Elsevier materials selector.
1. Materials. Selection
I. Waterman, N. A. (Norman Allan) *1941–* II. Ashby, Michael F.
620.11
ISBN 1 85166 605 2

ISBN 1 85166 606 0 (Vol. 1)
ISBN 1 85166 607 9 (Vol. 2)
ISBN 1 85166 608 7 (Vol. 3)
ISBN 1 85166 605 2 (Set)

Library of Congress Cataloging-in-Publication Data
CRC–Elsevier materials selector / edited by Norman A. Waterman and Michael F. Ashby.
p. cm.
ISBN 0-8493-7790-0 (set)
1. Materials. I. Waterman, Norman A. II. Ashby, M. F.
TA403.C733 1991 91-9391
620.1'1—dc20 CIP

ISBN 0 8493 7791 9 (Vol. 1)
ISBN 0 8493 7792 7 (Vol. 2)
ISBN 0 8493 7793 5 (Vol. 3)
ISBN 0 8493 7790 0 (Set)

No responsibility is assumed by the publisher for any injury and/or damage to persons or property as a matter of products liability, negligence or otherwise, or from any use or operation of any methods, products, instructions or ideas contained in the material herein.

Special regulations for readers in the USA
This publication has been registered with the Copyright Clearance Centre Inc. (CCC), Salem, Massachusetts. Information can be obtained from the CCC about conditions under which photocopies of parts of this publication may be made in the USA. All other copyright questions, including photocopying outside of the USA, should be referred to the publisher.

All rights reserved. No part of this publication may be reproduced, stored in a retrieval system or transmitted in any form or by any means, electronic, mechanical, photocopying, recording, or otherwise, without the prior written permission of the publisher.

Indexed by Paul Nash MSc BTech

Typeset in Great Britain by Variorum Publishing Limited, Northampton

Printed in Great Britain at the Alden Press Ltd, Oxford

Preface

The main aim of these three volumes is to provide a system and the necessary information for the selection and specification of engineering materials and related component manufacturing processes.

These volumes are intended for use by designers, materials engineers and production engineers who are seeking to identify the most suitable materials and manufacturing methods for a specific application.

Throughout the volumes, extensive use has been made of tabular and graphical information to facilitate the comparison of candidate materials for a specific application. Every material and process is described in terms of what it will do for the component designer and product maker and no detailed knowledge of metallurgy, polymer chemistry or materials science is assumed or necessary.

The information is arranged in order of increasing detail with the aim that no section need be read unless it is likely to be of direct relevance to the reader and his quest; whether this is the selection of material for a new product or the search for a substitute material.

Disclaimer

Whilst every effort has been made to check the accuracy of the information contained in these volumes, no material should ever be selected and specified for a component or product on a paper exercise alone. The purpose of these volumes is to provide enough information for a short list of candidates for testing and to reduce the number of fruitless tests. No liability can be accepted for loss or damage resulting from the use of information contained herein.

How to use the Elsevier Materials Selector

There will be two main reasons for using this information system.

(a) To select and specify materials and manufacturing routes for a new product.
(b) To evaluate alternative materials or manufacturing routes for an existing product.

The method of using the *Elsevier Materials Selector* for each of these purposes is as follows:

(a) Selection and Specification of Materials and Manufacturing Routes for a New Product

1. Define the function of the product and translate into materials requirements of strength, stiffness, corrosion and wear resistance, etc. (see Volume 1, Chapters 1.1–1.7).
2. Define the production requirements in terms of number required, tolerances, surface finish, etc.
3. Search for possible combinations of materials and production routes using Volume 1 and compile a short list according to performance/cost relationship.
4. Investigate candidate materials in more detail using Volume 2 for metallic materials and ceramics, and Volume 3 for plastics, elastomers and composites.
5. Specify optimum materials and processing routes.

(b) Evaluation of Alternative Materials and Manufacturing Routes for an Existing Product

1. Characterise currently used materials in terms of performance (see Volume 2 for metallic materials and ceramics, and Volume 3 for plastics, elastomers and composites), manufacturing requirements and cost (from in-house data).

2. Evaluate which characteristics are necessary for product function (see Volume 1, Chapter 1.1).
3. Search for alternative materials and, if permissible, alternative manufacturing routes (using Volume 1).
4. Compile short list of materials and manufacturing routes and estimate costs.
5. Compare existing materials and production routes with alternatives.

Acknowledgements and history of the publication

The origins of the *Elsevier Materials Selector* may be traced back to the early 1970s. At that time I was employed as a materials engineer in industry and perceived the need for a selection system which would compare the performance and cost of materials in terms which could be understood by designers and production engineers without the benefit of degrees in metallurgy, polymer chemistry, materials science, etc. I am grateful to my employers at that time, Danfoss A/S, for providing a stimulating working environment in which the idea was born.

The Fulmer Research Institute, my next employer, provided the financial and technical resources to convert the idea into reality; the *Fulmer Materials Optimizer* which was published in 1976. I am particularly grateful for the support and encouragement of Dr W. E. Duckworth, Managing Director and Mr M. A. P. Dewey (then Assistant Director) of Fulmer. As the first editor of the *Optimizer*, I am acutely aware that the original production and successful launch would have been impossible without the insight and hard work of the contributors of individual sections, in particular:

Dr T. J. Baker	—	Steels
Mr G. B. Brook	—	Aluminium alloys
Mr J. N. Cheetham	—	Polyurethanes
Mr D. G. S. Davies	—	Ceramics and unit conversion tables
Dr H. Deighton	—	Mechanical properties
Mr D. W. Mason	—	Copper alloys
Mr V. Micuksi	—	Surface coatings
Mr M. J. Neale	—	Wear
Mr R. Newnham	—	Ceramics
Mr G. Sanderson	—	Corrosion
Mr J. A. Shelton	—	Nylons & Polyacetals
Mr W. Titov	—	PVC

who produced work of exceptional quality.

It is also a pleasure to acknowledge the painstaking and thorough work of Mr A.M. Pye who undertook, in 1979, the first major review and updating of the *Optimizer*.

Between 1979 and 1987, my only relationship with the *Optimizer* was that of user of the system in support of my activities as a consultant on materials selection and specification. In the intervening period, the *Optimizer* was edited by Dr M.A. Moore, Dr U. Lenel and Mr L. Wyatt at Fulmer. A particularly valuable section on adhesive bonding was added by Mr W. A. Lees at this time.

In 1988, Elsevier purchased the rights of the *Fulmer Materials Optimizer* and invited me to undertake the task of converting the information in the 1987 edition of the *Optimizer* into a new three-volume materials selection to be known as the *Elsevier Materials Selector*.

I am very grateful to the publishers for this opportunity and also wish to thank the following sub-editors for their efforts in updating and checking individual sections.

Volume 2, Chapter 2.1 Wrought steels—Dr T.J. Baker (*Imperial College*)
Volume 3—Dr James Maxwell (*Formerly ICI Advanced Materials*)
Dr David Wright (*Technical Director, RAPRA*)
Also for Volumes 1–3 Amanda White,
Mathew Poole,
Martin Smith,
and Michael Weston
(*all of Quo-Tec*)

It is a special pleasure to acknowledge the help and inspiration of my associate editor, Professor M. F. Ashby, whose work in recent years has created the ideal introduction to the *Elsevier Materials Selector.*

Last, but not least, I wish to thank my wife, Margaret, for patience, sacrifice and support, without which I could not have started, let alone finished.

N. A. Waterman

Contents

Vol 3

3–1

Introduction to polymers and general information

Contents

List of tables

List of figures

3.1.1 Introduction to polymers, fillers and fibre reinforcement

3.1.1.1 Characteristics of polymers

Polymeric materials are customarily divided into three categories—thermoplastics, elastomers and thermosetting plastics

Historically, thermosets have been distinguished from thermoplastics (otherwise known as linear polymers, in common with elastomers) by their structure, containing a sufficient number of chemical cross-links to render the mass rigid at room temperature once moulded and incapable of further softening under heat. But means exist of creating chemical cross-links between individual polymer chains in thermoplastics (these means being chemical agents or short-wave irradiation) with loss of softening under heat and an increase in useful temperature range.

In more recent times the distinctions have become blurred with the development of thermoplastic elastomers (see Chapter 3.5) and thermosetting materials which can be injection moulded. The main characteristics of thermosetting plastics and elastomers are summarised in:

Table 3.1.1 *Classification of polymeric materials*

The materials are described and their important variables analysed in:

Table 3.1.2 *Thermoplastics and elastomers* and

Table 3.1.3 *Thermosetting resins (or plastics)*

Provided that the molecular chains in a linear polymer are (on average) above a critical length, their flexibility leads to the possibility of there being entanglements between chains which provide a degree of structure (affecting both the 'solid' properties and 'melt' properties) and which act effectively as temporary (physical) cross-links. This is why average molecular weight and the form and extent of molecular weight distribution on a linear polymer mass have an important effect on processing behaviour and product performance.

In certain block copolymers in which one component of the block chain is flexible and the other rigid at ambient temperature, a thermoplastic elastomer is achieved: the rigid sections of several molecules mutually form physical cross-links, the flexible component in the chains providing the overall flexibility. At the processing temperatures, the physical cross-links loosen and deformation and flow are possible as in a normal thermoplastic.

Thermoplastics and elastomers consist of agglomerations of large numbers of long chain-like molecules which are capable of relative movement above a characteristic temperature (known as Tg), differing from one chemical class to another. In the case of elastomers their characteristic temperatures are considerably below normal ambient temperature. In the case of non-crystalline (i.e amorphous) thermoplastics their characteristic temperatures are above ambient. In the case of the partially crystalline thermoplastics the Tg values are usually below ambient, but flexibility is inhibited by the presence of the crystalline structure, the Tg being relatable only to non-crystalline areas.

Polymer characteristics can be modified during the manufacturing process. For example, 'cross-linking' may transform a thermoplastic polymer to a thermoset and may therefore combine the advantages of both types.

'Orientation' may enhance properties in one or more directions.

'Expansion' may greatly enhance flexibility, heat insulation and lightness and the useful properties of two materials may be combined during manufacture.

The characteristics and relative advantages of these processes are set out in:

Table 3.1.4 *Characteristics of resin modification and combination*

3.1.1.2 Additives and reinforcement

Polymers are, however, seldom used 'neat'. Even when a polymer appears to have the ideal properties for a specific application, manufacturing considerations almost always necessitate the use of an additive. Other additives are employed to improve mechanical or physical properties, or to enhance appearance. The advantages and limitations of the use of fillers is summarised in:

Table 3.1.5 *General summary of the effect of fillers on the performance, properties and processing characteristics of thermoplastics*

Mechanical properties can be greatly enhanced and, where appropriate, given directionality by reinforcement. Fibres and wire, which may be continuous or in short lengths, can be aligned in the direction of principle tensile stress and greatly increase the engineering applications of polymers. A general listing of additives and reinforcement is provided in:

Table 3.1.6 *Polymer additives and reinforcement*

Fillers are dealt with comprehensively in:

Table 3.1.7 *Common fillers for plastics*

Preliminary information on reinforcement (which is treated in detail in Chapter 3.8) is given in:

Table 3.1.8 *Fibre reinforcement of plastics*

TABLE 3.1.1 Classification of polymeric materials

Category	*Principal characteristics*		*Structural characteristics*	*Factors which affect characteristics*	*Typical examples*
	Mechanical	*Thermal*			
Thermoplastics	Rigid at room temperature.	Reversible softening on heating and flow.	Either rigid chain (a)	Composition of polymer chain, molecular weight and distribution, presence of branching.	Polystyrene, ABS, SAN Polycarbonate PPO Polysulphones
			or crystallisation present (b)	As above, plus degree of crystallisation governed by presence of branching nature of chain constituents.	Polyethylene[a] Polypropylene Polyamides[a] Polyacetals Mouldable polyesters[a] PPS
Elastomers	Flexible at room temperature.	Increasing tendency to flow on heating.	Flexible chains and on crystallisation, chains present in coiled form.	Composition of the polymer chain, molecular weight, molecular weight distribution, degree of cross-linking.	Natural rubber Various synthetic rubbers
Thermosetting plastics	Rigid at room temperaure.	No softening on heating after initial moulding.	Chemically cross-linked.	Reacting constituents fillers and reinforcements.	Phenol formaldehyde Urea formaldehyde Melamine formaldehyde Unsaturated polyesters Epoxy resins

[a] Transparent types are amorphous

Table 3.1.2

TABLE 3.1.2 Thermoplastics and elastomers

Molecular feature	Variations		Structure (description)	Structure (diagrammatic)	Main effects	Examples
Structure of the polymer chain	Homopolymer	All	From a single monomer all repeating units alike except at each end of chain.		Chain may be flexible or relatively stiff according to nature of links between repeating units.	Polystyrene
		Symmetrical	Without sizable sub groups or branches		Likely to form partly crystalline mass.	High-density Polyethylene
		Non-symmetrical with sizeable side groups	Atactic: Irregular side group relationship.		Waxy, properties not useful when chain is flexible. If chain is stiffer an amorphous polymer results. Can crystallise only when groups attached to the asymmetric carbon atom are of like size.	Atactic Polypropylene (not used commercially)
			Syndiotactic: regular alternating arrangement.			Syndiotactic polypropylene
			Isotactic: consistent arrangement (all groups are one side of chain)		Partially crystalline mass.	Isotactic polypropylene
		Containing variably sized branches	Chains with reduced linear form.		Reduced amount of crystallinity.	Low-density Polyethylene
	Copolymer	All	Produced by copolymerising two or more selected monomers			*Note:* Copolymers are more often known by trade names
		Random	Units A & B in random arrangement along chain	—AABAAABBABABBAAAB—	A tendency to 'average' properties and processing behaviour. Reduction in crystallinity—this may result in much greater flexibility.	EVA
		Alternating	Units A & B alternating along chain	—ABABABABAB—		
		Block	Repeating blocks of A and blocks of B along chain	—AAAAAAAAABBBBBBBAAA—		
		Graft	The structure is something like a comb (with A as back, and B as teeth)	—AAAAAAAAAAAA— BBBBBB	If A is rigid (e.g. polystyrene) and B is flexible (isoprene) a physically 'cross-linked' elastomer results, which can be processed conventionally because links will melt.	EP rubber Thermoplastic elastomers

TABLE 3.1.2 Thermoplastics and elastomers—*continued*

Blending of two (or more) polymers	Blended polymers may be	Compatible	Intimate mixture of chains (homogeneous alloy)		A tendency to average properties but different in detail from copolymerisation. Adjustment of processing behaviour.	Mod. PPO (Noryl) PVC alloys. α-methyl styrene–styrene–acrylonitrile
		Incompatible	Often discrete elastomer particles in a polymer matrix, formed by alloying a graft or block copolymer, of which one constituent is compatible, with a third, matrix, polymer		The toughness is markedly increased.	Toughened polystyrene ABS PVC (impact modified)
			Alloys of incompatible polymers can also be formed if an interfacial agent can be found			Addition of chlorinated polyethylene to PVC and polyethylene
Variation of molecular size and size distribution	Increase average length of chain		Fewer end groups and increased entanglement as average length is increased		Increase results in an improvement in stability and an improvement in mechanical properties, but increases difficulty of processing.	All thermoplastics and elastomers.
	Increase breadth distribution of molecular sizes Presence of low molecular weight fraction should be avoided, and free monomer content should be very low.				In processing, a broader range of molecular weights makes melt more shear sensitive, but too high a proportion of shorter chains is undesirable for good mechanical properties.	

Table 3.1.2—*continued*

TABLE 3.1.3 Thermosetting resins (or plastics)

Characteristics of setting reaction	***Examples***	***Variables***	***Processing characteristics***
Condensation	Phenol formaldehyde Melamine formaldehyde Urea formaldehyde Some polyurethanes	Powders or pellets available as B-stage solid condensates incorporating fillers of various kinds and catalysts to promote cross-linking on heating. Properties depend on initial reactants, filler types and amounts.	Water is evolved during cure. Processes limited to high pressure techniques because of this. Hot cure required. Volatile by-products.
Addition polymerisation	Silicones	Unreacted resins, which may be liquid or solid, are blended with high percentage of a hardener and, possibly, fillers. Hardeners and curing schedules varied to adjust properties of cured resin. Can be used with various kinds of reinforcement.	Since there are no volatile products, low-pressure or no-pressure processes are feasible. Cold or hot cure possible depending on hardener. Encapsulating and moulding possible. No volatile by-products.
Ring opening polymerisation	Epoxides		
Free radical polymerisation (monofunctional)	Unsaturated polyesters Unsaturated vinyl esters	These resins are usually used as matrices for heavy reinforcement with various forms of glass-fibre. Other reinforcements can be used. Resin composition affects chemical and fire resistance properties.	Low pressure and high pressure processes can be used by lay-up and matched-mould techniques. No volatile by-products.
Free radical polymerisation (difunctional)	Cross-linked polystyrene for casting Di-allyl phthalate (DAP) Allyl diglycol carbonate (CR39)	For transparent casting since fillers are nor necessary; useful for optical and electrical application. DAP and CR 39 can be filled and coloured if required.	For casting, encapsulating and moulding. No volatile by-products.
Pre-polymer rearrangement polymerisation	Polyimides and related materials	Variations occur in the nature and constitution of the pre-polymer.	The required form is produced in the pre-polymer which is then raised in temperature to form imide groups possibly with loss of water or other simple molecules in rearrangement. Wire coatings, films and solid articles made by specially developed procedures.

TABLE 3.1.4 Characteristics of resin modification and combination

Material modification	*Types*	*Function*	*Aspects of incorporation*	*Aspects of converting (moulding extrusion, etc.)*
Cross-linking	A process involving input of energy in one form or another. The most common is chemical. Chemical cross-linking-agent with or without heat. Chemical action (usually with the elimination of water) gives rigid three-dimensional thermoset structure, e.g. phenol/ formaldehyde urea/ formaldehyde. An alternative process is irradiation by the application of high-energy radiation.	Puts up service temperature and reduces creep. Converts product into rigid three-dimensional molecular structure.	The actual incorporation of relevant agent during compounding.	Controlling factor is the degree of cross-linking, i.e. cross-linking density. The cross-linking takes place with actual moulding or extruding process. The composition thus enters the machine as thermoplastic, the product leaves as a rigid (three-dimensional) thermoset unit. Degree of cure depends on time and temperature applied, and the curing agent used.
Orientation	Uniaxial Biaxial Mechanical orientation of properties may be effected by the use of reinforcing agents (see Table 3.1.6). Stereo regular polymers isotactic, syndiotactic, atactic.	To improve properties, increase strength axially. Ditto and reduce environmental stress cracking and improve the overall physical properties. These are strategically placed to cause uniformity of flow of the plastic within the mould. To impart specific regulation to chemical groups within the polymer molecule thereby controlling structure and influencing physical properties.	The alignment of the constituent polymer chains effected during extrusion is uniaxial giving increased longitudinal strength. Simultaneous (biaxial) orientation increases strength at right angles. One or both may be applied depending on the physical properties of the end-product required. Use of special polymerisation catalysts.	Moulding and extrusion of all types must impart a degree of orientation. Moulds must be designed to give the greatest possible freedom of molecular flow, so that maximum alignment is attained. 'Faulting' in a mould where one stream of material flows contra to another must be avoided at all costs. Design of mould 'gates' influences the flow of the plastic into and in the mould.
Expansion	Open structure (cell interconnection)	Gives absorbent material, natural sponge is a typical example. Reduces weight, gives high strength/weight ratio.	Blowing agent (gas generator) incorporated in the polymer mass remains stable under normal conditions of storage.	Process of conversion (temp. and pressure) so adjusted to release the gas chemically, which at the temperature of the operation will cause the thermoplastic mass to expand in continuous form throughout the composition.
	Closed structure (no cell interconnection)	Necessary for heat insulation by preventing movement of air within the plastics mass.	Incorporation as above.	As above, but the conditions are such that the blowing is localised so that there is no interconnection between the individual air 'cells'.
	Beads	To produce beads (air pockets) within the polymer mass, thereby reducing density.	Beads (e.g. styrene) produced by suspension. Blowing agent (e.g. 6% of low-boiling petroleum ether) incorporated before polymerisation or to impregnate the beads. Volatilises during injection releasing gas which inflates the plastic locally and regularly.	Beads are heated in the final converting operation so that the volatile solvent they contain expands and 'blows' individual beads which then aggregate and form light-weight mass of the desired shape and density.

Table 3.1.4

TABLE 3.1.4 Characteristics of resin modification and combination—*continued*

Material modification	*Types*	*Function*	*Aspects of incorporation*	*Aspects of converting (moulding extrusion, etc.)*
Expansion *(continued)*	Structural foams	To produce rigid foamed polymers which are structured *per se* as a homogeneous mass. Alternatively to produce product with a central foamed mass surrounded by a continuous skin of the polymer to give smooth surface.		Recently patented ICI process which by injection gives a foamed centre with continuous load-bearing surface. The expanded material is 'encapsulated' with outer layer of solid plastic.
Coextrusion	Extrude two polymers simultaneously.	All processes for manufacturing sheet or sections incorporating surface material with properties differing from the substrate (or substrates). This may economise in expensive material, provide strength *and* surface properties, ensure impermeability, facilitate joining or improve performance.	Coextrusion requires complete compatibility at interface, also equality of shrinkage to avoid buckling. Thickness control is essential for properties and cost. All processes may require priming and use of adhesives.	Coextrusions can be treated as normal thermoplastic sheet. Removal of volatiles is essential to prevent shrinkage and buckling. Term lamination usually applied to reinforced sheet (see Table 3.1.6).
Extrusion coating	Extrude substrate between sheets held by rolls.			
Melt roll coating	Feed coating material over substrate between rolls.			
Laminating	Combine two films by rolling together using heat and/or adhesive.			
Coating	Wide variety of thermosetting and thermoplastic materials available.	Surface protection and/or decoration, usually of metals. Plastics coatings may modify the surface properties (e.g. friction) of metals. Improvement of barrier properties of some polymeric films.	May be applied as conventional solution of plastics (lacquer) Powdered thermo-plastics may be applied by fluidised-bed-dip coating or flame spraying.	Apart from conventional solution application, the difficulty in applying solid plastics is the need to obtain this in uniformly finely divided form. Adhesion between layers has to be maintained.

Table 3.1.4—*continued*

TABLE 3.1.5 General summary of the effect of fillers on the performance, properties and processing characteristics of thermoplastics

Advantages	*Limitations*
Improved dimensional stability, e.g. better creep resistance, higher elastic modulus.	Reduced tensile and impact strength with some fillers.
Lower shrinkage and post mould distortion.	Surface finish generally inferior in terms of smoothness and gloss, to unfilled polymer.
Lower coefficient of thermal expansion.	
Faster production cycle times (components can be ejected from the mould at higher temperatures due to improved hot strength and after shorter times due to improved thermal conductivity).	Long, thin complex shapes can present moulding problems.
Improved bearing properties and abrasion resistance with certain fillers.	Higher wear rates of machinery caused by some fillers.
Improved fire resistance due to decreased heat of combustion.	Increase in density with most solid fillers compared with unfilled polymer.
Lower total costs of components.	

TABLE 3.1.6 Polymer additives and reinforcement

Material modification	*Types*		*Function*	*Aspects of incorporation*	*Aspects of converting (moulding extrusion, etc.)*
Additives Aim to correct deficiencies inherent in individual polymers. Essential to all polymers for practical use but vary widely with different polymers.	1. Stabilisers	Heat (2.0% addition)	Wide variation—some materials cannot be processed without stabiliser addition—others need none.	Toxicity— some stabilisers are toxic, thus need care in selection according to end use.	Care to be taken that stabilisers do not suffer decomposition during final conversions, e.g. extrusion or moulding. Heat stabilisers counter this, but they must not interact with any others present.
		UV (0.1% addition).	In the case of light stabilisers only small addition necessary		
	2. Processing aids		Additions to assist flow, prevent adhesion to the mould surface and/or improve surface quality of final product.	Added with other ingredients during compounding.	No difficulty as they are specifically added to facilitate processing and give uniform flow in and through the mould.
	3. Plasticisers (flexibilisers)		Internal lubricants, the need and quantity varies with the polymer. They are essential with cellulosics which otherwise cannot be moulded, also much needed with PVC for flexible grades.	Toxicity— again select with care according to the end use of product. Plasticisers must not exude in final product. Plasticisers may induce proneness to biodegradation in otherwise resistant polymers.	Correctly compounded plasticiser facilitates all forms of conversion.
	4. Flame resistant additives		Almost all polymers need some flammability reducing agent. Non-toxic products should be used if possible and in the minimum effective quantity.	Must be compatible with other ingredients of composition. In the case of sheets and foils must be non-toxic.	Converting temperature should be such that additive suffers no chemical change.
	5. Colourants—Dyes Pigments		Dyes and pigments (transparent and opaque) usually needed. Colour matching in the polymer under processing conditions is essential.	Complete uniformity throughout the polymer mass is essential to avoid striation or mottling.	Conversion temperature must be within the colour stability of the additive.
	6. Impact improvers		To adjust the glass transition temperature to meet the needs of end use. Below this temperature amorphous polymers are subject to brittle fracture but toughen as the temperature is raised to approach the glass transition temperature (Tg.)	Compounding important, rubbery materials added to glass-like polymers enhance toughness.	Before converting, the tough–brittle transition of commercial polymer composition should be determined.
	7. Friction reducers slip agents, anti-tack, anti-block agents		To reduce surface/surface friction of the final products, also to render surface of final product tack-free, and in the case of films and foils easy separation of sheet, i.e. non-blocking.	Must be compatible with other ingredients of composition. In the case of sheets and foils must be non-toxic.	Must not decompose at conversion temperatures, must not exude to surface during these processes.
	8. Anti-static agents		Essential with almost all plastics, in particular films, foils and extruded products. Static-films show blocking, static in end-product inhibits easy handling. Attracts surface dust.	Additives must be compatible with all other ingredients of composition. Non-toxic essential for films and foils and similar products.	Added effect by operating conversion (especially film and sheet formation) in atmosphere of maximum humidity.

Table 3.1.6

TABLE 3.1.6 Polymer additives and reinforcement—*continued*

Material modification	*Types*	*Function*	*Aspects of incorporation*	*Aspects of converting (moulding extrusion, etc.)*
Additives *(continued)*	9. Conducting	Polymers are good electrical insulators so that to convert them to conductors is difficult. Some degree of conductivity by additives such as carbon black, lead powder and powdered conducting metals.	Difference of density between polymer and some conducting additives makes production of homogeneous composition more difficult.	
	10. Extenders (cost reducers) and other fillers	Main function to give bulk to low-cost products. May be incorporated scrap.	Uniformity of incorporation essential to avid stress or weak spots.	
Reinforcement	General	Increase stiffness; increases strength; reduces creep; increases service temperature and control over directional properties. Reinforcing material is arranged to give maximum strength where this is needed. Orientation of the fibre gives added unidirectional strength.	Essential to have complete union (wettability) between polymer and reinforcing agent. This is not so difficult where the fabric, fibre, etc., is impregnated with polymer solution. More difficult where the polymer is paste or dough. Essential that the polymer is completely incorporated in the reinforcing medium e.g. by 'stippling' glass fibre with polyester in GRP.	The modified flow properties must be compensated by the processing properties; e.g. injection of reinforced and non-reinforced polymer is quite different.
	Continuous Cellulosic, glass, carbon and other fibres	Used where very high strength and/or modulus needed.		
	Metallic components (wire and woven)	Only used where strength, toughness and considerable flexibility required.	Complete wetting of the reinforcing metal with the polymer is essential.	Specialist materials, e.g. conveyor belts, treated according to individual needs.
	Solid wood	Strong, rigid insulating. Used when high specific strength material required.	Impregnate solid (or laminated) wood.	
	Fibre bundles Rovings Fabrics (type of weave varies with product)	Reinforced sheet (high and low pressure laminates constituted virtually a separate industry for the production of sheets, rods and tubes 'laminate').		
	Porous metal	Self-lubricated bearings	Porous metal impregnated with fluoroplastic.	
	Discontinuous Chopped strand Flakes Mats. Normally penetrated to give uniform sheets	Dough moulding compound (DMC)	Can be oriented during moulding process.	
	Comminuted wood (wood chips, natural fibres, ground nut shells etc.)	Comminuted wood bonded with polymer gives sheet material (chipboard).	No longer a means of using scrap sawdust. The shape of the wood particle determines the physical property of the sheet.	Control of particle shape and size also: moisture content temperature pressure time of cure are vital to the properties of the board.

Table 3.1.6—*continued*

Table 3.1.7

TABLE 3.1.7 Common fillers for plastics

Filler	Form	Characteristics	Effects on processing
1. Glass	(a) Spheres	(i) Economical and inert (ii) Increase elongation and impact strength (iii) Improve abrasion resistance of ABS (iv) Added to nylon to produce uniform moulded parts giving higher flexural modulus, compressive and tensile strength and melt index (v) Also used with PVC, styrene copolymers, nylons, epoxies and polyesters (thermoset)	(i) Control of shrinkage and warpage (ii) Can cause wear to moulds and machines (iii) Reduced cycle times because of higher heat distortion temperature.
	(b) Spheres + mica flakes, fibrous glass, irregular mineral particles	Optimal packing geometry for improved processability and performance as (a)	As above
	(c) Microballoons	Reduce density (see also syntactic foams; Section 3.6.3)	
	(d) Flakes	Improve resistance of thermosets to moisture and acids	
2. Carbon	(a) Carbon black	(i) Improved weather resistance (ii) Promotes electrical conductivity (iii) Improves lubricity (iv) Increases modulus of elastomers (v) Decreases tensile strength in thermoplastics (vi) Raises embrittlement temperature in thermoplastics (vii) Improves heat resistance Crystalline polymers (e.g. HDPE) are more sensitive to effects	Improved mouldability Products are black (see 5 below)
	(b) Hollow carbon spheres ('carbospheres')	(i) More resistant to hydrolytic environments than hollow glass (ii) Require resistance from oxidation at elevated temperatures in matrix	
	(c) Graphite	(i) Electrical conductivity (ii) Lubricity	
	(d) Ground petroleum coke	Added to PTFE to form inexpensive anti-friction composite	
	(e) Powdered coal	Chemical resistance and moisture resistance	
3. Cellulosics	(a) Woodflour	(i) Used in thermosets (melamines, phenolics, ureas) to improve electrical properties and impact resistance (ii) Up to 60% woodflour can be added to polypropylene to produce extrudable composite for applications not involving high compressive forces	(i) Low cost filler reducing mould shrinkage (ii) Care must be taken during thermal processing not to degrade this organic filler
	(b) Cork	(i) Used as filler in linoleum, gaskets, mastics, paving compositions and insulating composites using asphalt, PVC or SBR rubber as continuous phases for tiles or floor covering, or alkyds or polyesters for electrical insulation	
4. Starch	Starch	(i) Up to 10% can be used as filler using standard processing techniques (ii) Starch composites degrade in soil	Little effect on polymer viscosity
5. Chalk	Calcium carbonate	(i) Very common filler to reduce costs (ii) Highly-filled PVC used for floor tiles (iii) Improves physical properties of PVC, PE, PP, epoxy, polyester, phenolic and polyurethane foams (iv) Stearate coated $CaCO_3$ can increase impact strength (not so if uncoated) (v) High specific gravity: high density composites	No known problems Ultra-fine precipitated while calcium carbonate available to reinforce synthetic elastomers as alternative to carbon black
6. Metallic oxides	(a) Zinc oxide	(i) Used in polypropylene, elastomers and polyesters (ii) Polypropylene/zinc oxide composites have good resistance to mildew and weathering	

TABLE 3.1.7 Common fillers for plastics—*continued*

6. Metallic oxides *(continued)*	(b) Hydrated aluminium oxide	Increases flame resistance of polyesters	
	(c) Magnesium and titanium oxides	(i) Increased stiffness, hardness and creep resistance in composites (ii) MgO used to increase viscosity of polyester resin mixes	
	(d) Beryllium oxide (microspheres)	(i) Increased thermal conductivity with 70% by weight filled epoxies	
	(e) Zirconia and γ-iron oxide	(i) Increased specific gravity of polypropylene resins (ii) Hard composites with high moduli	
7. Metallic powders or fibres	General	Impart dimensional stability Promote electrical and thermal conductivity Greater strength Sound damping Radiation resistance Magnetic properties Bearing properties	
	(a) Bronze powder	(i) Added to acetals or nylons to produce thermally and electrically conducting mouldings which can be electroplated	
	(b) Aluminium powders	(i) As 7(a) (ii) Used to improve physical properties of polypropylene (iii) Reinforcing filler for polyethylene and epoxy resins, the latter for cast forming tools, metal cements and coatings	Improves machinability
	(c) Silver	As 7(a)	
	(d) Zinc	Used in polymers employed as protective coatings	
	(e) Copper	Used to stabilise and reinforce nylon 6/6	
8. Powdered polymers	(a) PTFE and fluoropolymers	Added to nylons, acetals, polyamide-imide, polycarbonate, polyamide, epoxies and phenolics for lubricity and enhancement of impact strength	
	(b) Polystyrene and phenolics	Added to styrene-butadiene to increase storage modulus at elevated temperatures	
9. Silica products	(a) Sand and gravel	Used with epoxies for gap filling, but heavy	All silica products can cause wear to machines and mould
	(b) Natural silica	(i) Reduces cost (ii) Dimensional stability (iii) Thermal conductivity (iv) Electrical insulation (v) Moisture resistance (vi) Added to PMMA to improve dielectric properties and resistance to heat (vii) Also increases glass transition temperature, modulus, compressive strength (vii) Reduces swelling in solvents	
	(c) Quartz	(i) Used with phenolics and epoxies for ablative insulation of space capsules (ii) Quartz-epoxy (60% filled) has similar expansion coefficient to aluminium or brass (iii) Added to polyethylene to reduce crack formation, shrinkage, electrical properties and moisture resistance (15–45% filled) (iv) Dielectric, physical and processing characteristics of acrylics are improved by the addition of 30–40% quartz	See (iv)

Table 3.1.7—*continued*

TABLE 3.1.7 Common fillers for plastics—*continued*

Filler	*Form*	*Characteristics*	*Effects on processing*
10. Silicates	(a) Mineral fillers	(i) Greater dimensional stability (ii) Elevated heat resistance and fire retardance (iii) Improved wear resistance (iv) Used with nylons, fluoroplastics, polyamide-imide and most thermosets	
	(b) Asbestos	(i) Very common usage as heat-resistant filler in phenolics, polypropylene, polystyrene, alkyds, DAP, melamines, epoxy, polyarylsulphone, PPS (ii) Higher strength and rigidity (iii) Improves chemical resistance (iv) Improves electrical insulation	Improved mouldability
	(c) Kaolin clay	(i) Widely used because of good electrical properties and melt flow (ii) High surface gloss and smoothness to PVC (iii) Increases decomposition temperature of PVC and the insulation resistance of PVC wire coatings (iv) Used as filler in SMC and DMC	Difficult to get good bonding of filler and polymer, especially when coupling agents are not used
	(d) Mica	(i) Electrical insulation (ii) Moisture resistance (iii) Lubricity (iv) Dimensional stability (v) Less hazardous than asbestos (vi) Compatible with most resins	
	(e) Talc	(i) Stiffness increases appreciably even at small percentage loadings, and high temperature rigidity imparted to PP (ii) Dimensional stability and moisture resistance (iii) Dielectric properties (iv) Cost reduction (v) Some reduction in tensile and impact properties (vi) Increases wear resistance of PVC tiles (vii) Used with polypropylene (extensively), ABS and polyethylene	(i) Fast moulding cycles possible; one of the easiest mineral fillers to process (ii) Talc-filled polypropylene can be vacuum-formed owing to improved toughness of pure polymer (iii) Reduced shrinkage
11. Inorganic compounds	(a) Barium sulphate	Increases specific gravity, friction resistance and X-ray opaqueness of composites	
	(b) Barium ferrite	Imparts magnetic property to PVC	
	(c) Molybdenum disulphide	(i) Lower coefficient of friction (improved lubricity) (ii) Raises strength (iii) Used with fluoroplastics, nylons, polyimide and phenolics	
	(d) Lithium aluminium silicate	Increase thermal expansion coefficient in epoxy resins	
12. Ceramic powders		(i) Improved dimensional stability (ii) Improved compressive qualities (iii) Wear resistance and bearing properties improved (iv) Employed on fluoroplastics, alkyds, DAP, epoxies	
13. Chopped cloth		Raises impact strength	Reduces mould shrinkage
14. Magnesium carbonate	Dolomite magnesium	(i) Reductions in tensile and impact properties (ii) Increases stiffness on polyethylene, PPP, PVC, ABS (iii) Auxiliary fire retardant with antimony trioxide	Reduced shrinkage No difficulties

TABLE 3.1.8 Fibre reinforcement of plastics

	Forms and characteristics	***Polymers***
Asbestos fibres	Increased hardness, impact resistance and heat resistance. Mainly chrysotile (white) asbestos. Health hazards associated with asbestos fibres.	PVC, nylon, polypropylene, phenolic, urea, melamine, unsaturated polyesters, DAP prepolymers, epoxies, silicones.
Carbon fibres	Very high specific strength, high modulus and low thermal expansion coefficient. Based on graphite filaments and whiskers, mainly for high-performance applications.	Nylon 6/6, polycarbonate, PPS, polysulphone, PBT, polyimides, acetals, polyamide-imide, epoxies, polybutadienes.
Cellulose fibres	Cotton flock: fair impact resistance, high bulk factor and low moisture resistance. Cotton cloth and cord composites have higher impact values. Jute and yarn: low cost, moderate impact resistance.	Polyester, phenolics.
Glass fibres	Predominant reinforcement offering increased stiffness, strength (can be double unreinforced value), creep resistance, high temperature performance and heat resistance. Forms include short, continuous, rovings, mats, chopped strands. Most widely used type of glass is E and C glass (lime–alumina–borosilicate) for electrical and chemical performance, respectively. S glass for heat resistance and low-cost soda lime where high durability not required.	Almost all common thermoplastics and thermosets except acrylics, cellulosics, ureas, PVC/acrylics, PVC/ABS.
Inorganic fibres	Brass-plated wire steel sheet: significant strength improvement. Can also contain glass. Boron. Continuous ceramic fibres. Silicon carbide, silicon nitride.	Polyester. Thermoplastic (TP) and thermosetting (TS).
Potassium titanate	6 x 0.1 μm. Platable, stiff glossy parts with good resistance to warping and heat. Acts as pigment.	Nylon, ABS, polypropylenes.
Synthetic organic fibres	Nylon—wear resistance. Aramides (aromatic nylons)—very high specific strengths. Polyester: high specific strength. PETP: improved abrasion resistance and impact resistance. PVA (polyvinyl alcohol): improved impact and weathering properties.	Electrical grade laminates. Phenolics, polyesters, (concrete). Polyesters. Polyesters, polyurethane.

3.1.2 Mechanical properties of plastics

3.1.2.1 Short-term properties

This section provides sufficient information on the mechanical properties of plastics to enable a preliminary assessment of the suitability of a specific material for preliminary design. Information in diagrammatic or graphical form is given on: tensile strength, flexural strength, modulus and impact properties of reinforced and non reinforced plastics, at room and elevated temperatures.

The following short-term properties are illustrated in:

Fig 3.1.1 *Room temperature tensile strengths of plastics*
Fig 3.1.2 *Variation of tensile strength of plastics with temperature*
Fig 3.1.3 *Room temperature flexural strengths of plastics*
Fig 3.1.4 *Room temperature flexural moduli of plastics*
Fig 3.1.5 *Variation of flexural moduli of plastics with temperature*
Fig 3.1.6 *Room temperature impact strengths of plastics*
Fig 3.1.7 *Variation of impact strengths of thermoplastics with temperature*
Fig 3.1.8 *Typical notch sensitivities of some thermoplastics at 23°C*

3.1.2.2 Creep

The long-term deformation properties are summarised in five comparative figures in this section and additional data can be obtained by consulting individual sections.

The long-term deformation of plastics will depend on the stress, its duration and the temperature and nature of the environment under which the stress is applied. Many curves, illustrating the effects of these variables on creep are drawn in terms of (apparent) creep modulus (M_t) in tension or compression at time t

$$M_t = \frac{\text{initial applied stress}}{\text{total strain at time } t}$$

against time for a range of values of initial stress.

The total strain at a given time for an applied stress may, of course, be derived from these curves. An alternative presentation—the stress to produce a given strain for a range of times and temperature—may also be employed. All curves are intended as a general guide and individual suppliers should be consulted for safe levels of stress, time, temperature and moisture content for a given maximum strain, for individual grades or formulations. The comparative creep data figures include:

Fig 3.1.9 *Comparative creep data of unreinforced thermoplastics (initial applied stress 6.9 MN/m^2; 23°C)*
Fig 3.1.10 *Comparative creep data of GR thermoplastics (initial applied stress 14 MN/m^2; 23°C)*
Fig 3.1.11 *Comparative creep data of reinforced and unreinforced plastics (initial applied stress 3.5 MN/m^2; 83°C)*
Fig 3.1.12 *Comparative creep data of reinforced and unreinforced plastics (initial applied stress 3.5 MN/m^2; 116°C)*

3.1.2.3 Fatigue

For unreinforced polymers the tensile compressive stress fatigue limits for 10^6 cycles range between about 25 and 50% of the UTS although there are materials such as for example, the oriented polypropylene used for integral hinges that have exceptional resistance to bending fatigue (see Section 3.2.10).

Some available fatigue data are listed in sections dealing with specific plastics, notably nylon and polyacetal (Section 3.2.1 and Section 3.2.2) and PPO (Section 3.2.14).

Fatigue of reinforced plastic materials (particularly high rigidity/high strength reinforcements such as carbon or aramid fibre) has more engineering significance. Materials with a substantial degree of fibre reinforcement in the direction longitudinal to the applied stress do not fail by crack propagation in the same way as do metals. When subjected to flexural fatigue the matrix degrades progressively, starting at the points of maximum strain, and the longitudinal fibres remain intact but become unbonded. The flexural modulus of the component progressively decreases, while the tensile strength is not seriously impaired.

'Fatigue failure' is considered to have occurred when the flexural modulus has decreased by a specific amount, say 5 or 10%. If fatigue failure of this kind occurs in service the results are not catastrophic, because the resultant softening of a leaf spring or the increase in deflection of an aerogenerator blade is already detectable before the effects are serious. Recent work has shown a tendency to report the relationship between cyclic strain amplitude and cycles to failure, as in:

Fig 3.1.13 *Applied strain versus cycles to failure for epoxy resin compared with glass and carbon fibre reinforced epoxy*

Glass fibre reinforced composites tend to the same strain limit as the epoxy matrix at high N values. (Carbon fibre composites fail at lower strain values than the matrix, but at much higher stresses.)

With correctly designed plastic components reinforced with low modulus glass or aramid fibres, design is almost invariably strain or buckling limited and not fatigue limited.

Only in a very exceptional case will a component that has acceptably low distortion, and does not collapse when stressed, fail by fatigue. As a general rule a design limit of 0.2% strain will usually ensure freedom from fatigue failure, particularly if reinforcement is continuous and is oriented to have a significant longitudinal component.

Materials reinforced with short lengths, or with fibres normal to the principle stress are more fatigue sensitive. However, even in these cases the fatigue limit is usually in the region of 35% of the U T S.

The design of carbon fibre reinforced plastics may very well be fatigue limited. The high moduli of these materials make it relatively easy to design against buckling, and repeated application of high design stresses may lead to fatigue.

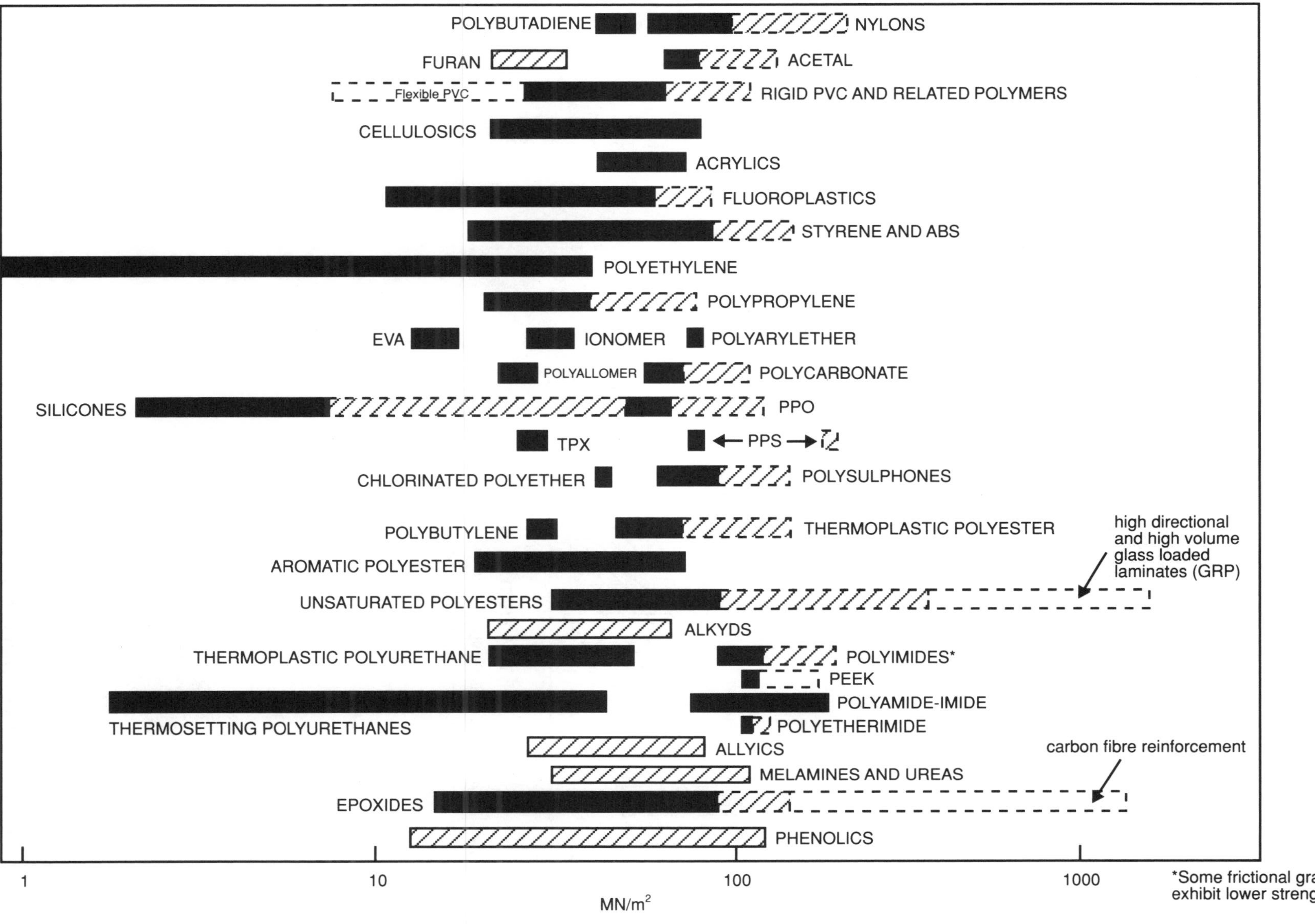

FIG 3.1.1 Room temperature tensile strengths of plastics. Partially shaded areas represent reinforced plastics and the additional strength thereby obtained

Fig 3.1.1

Thermoplastics: Nylons, Polyacetals, PVC

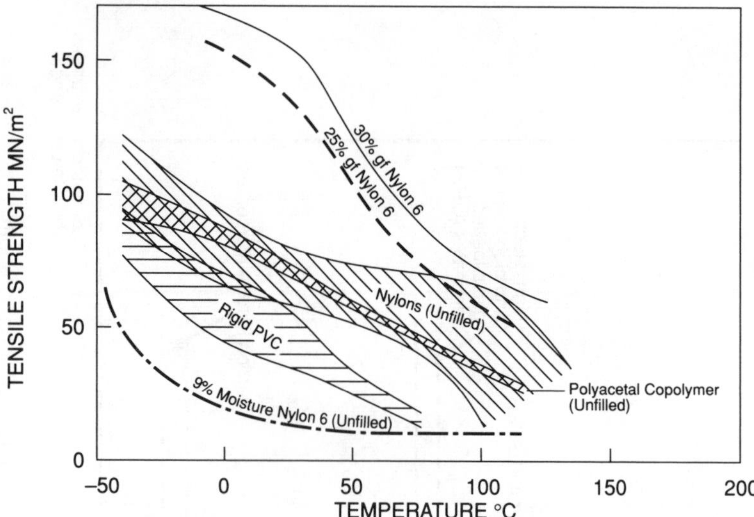

Thermoplastics: Styrenes, Fluoroplastics

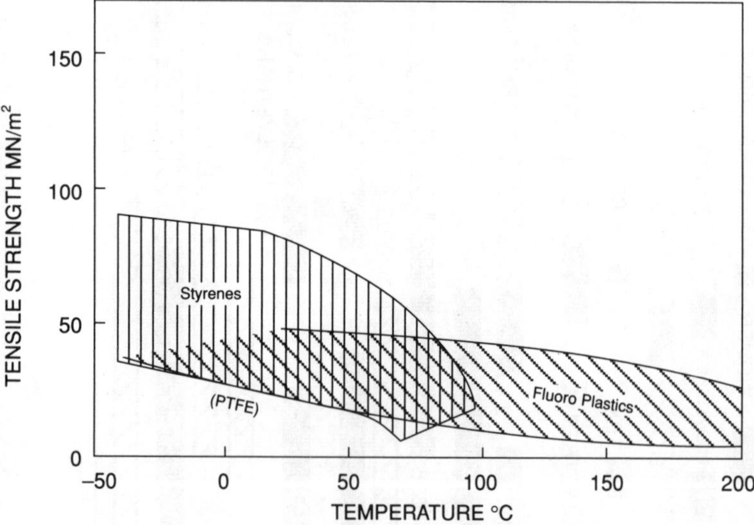

Thermoplastics: Polyethylene, Polypropylene, EVA, Polycarbonate

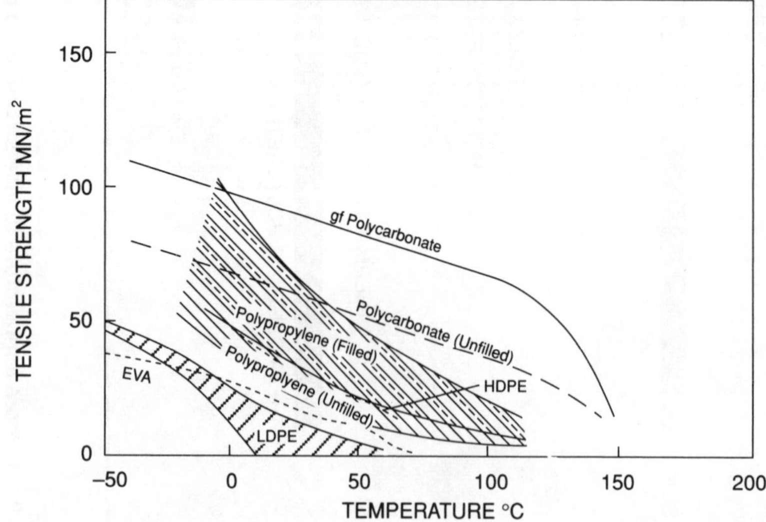

FIG 3.1.2 Variation of tensile strength of plastics with temperature

Fig 3.1.2

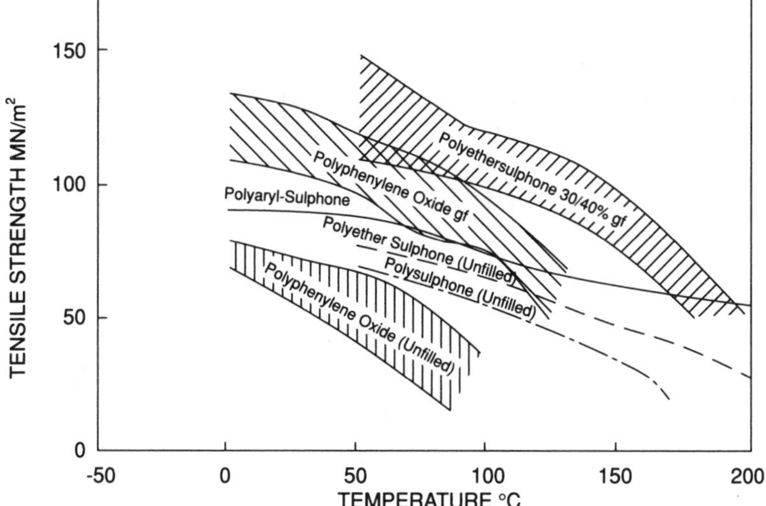

Thermoplastics: TPX, PPS, PBT, Polyamide-Imide

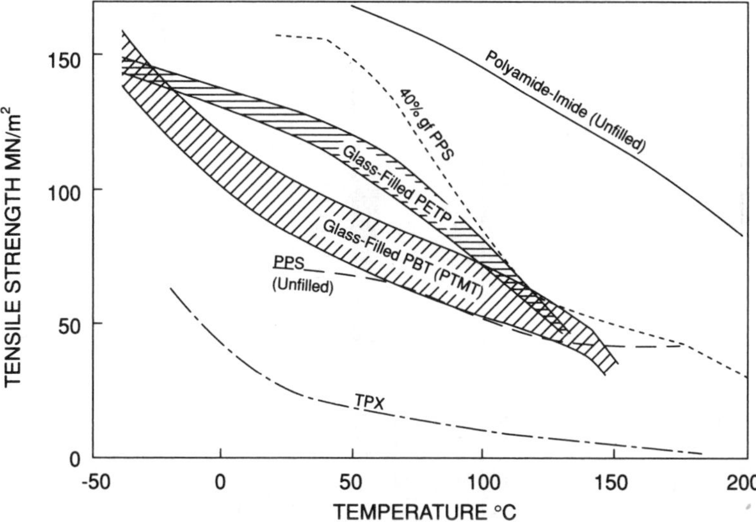

Thermoplastics: Polyester, Polyimide, Polyurethane

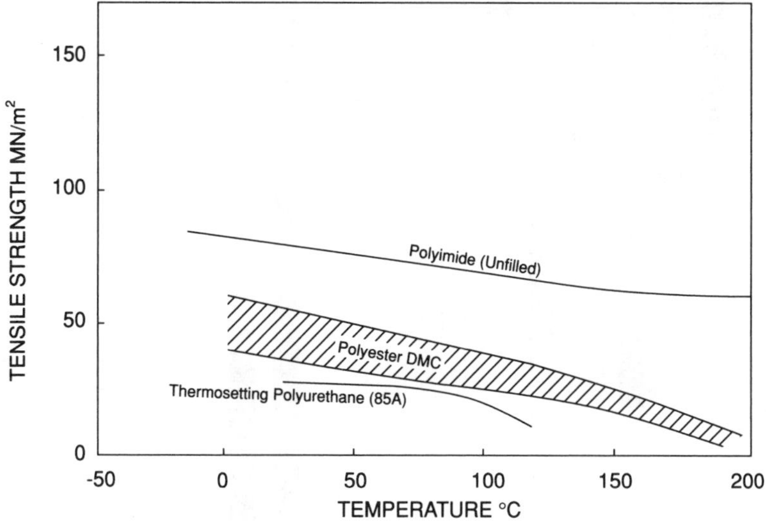

FIG 3.1.2 Variation of tensile strength of plastics with temperature—*continued*

Fig 3.1.2—*continued*

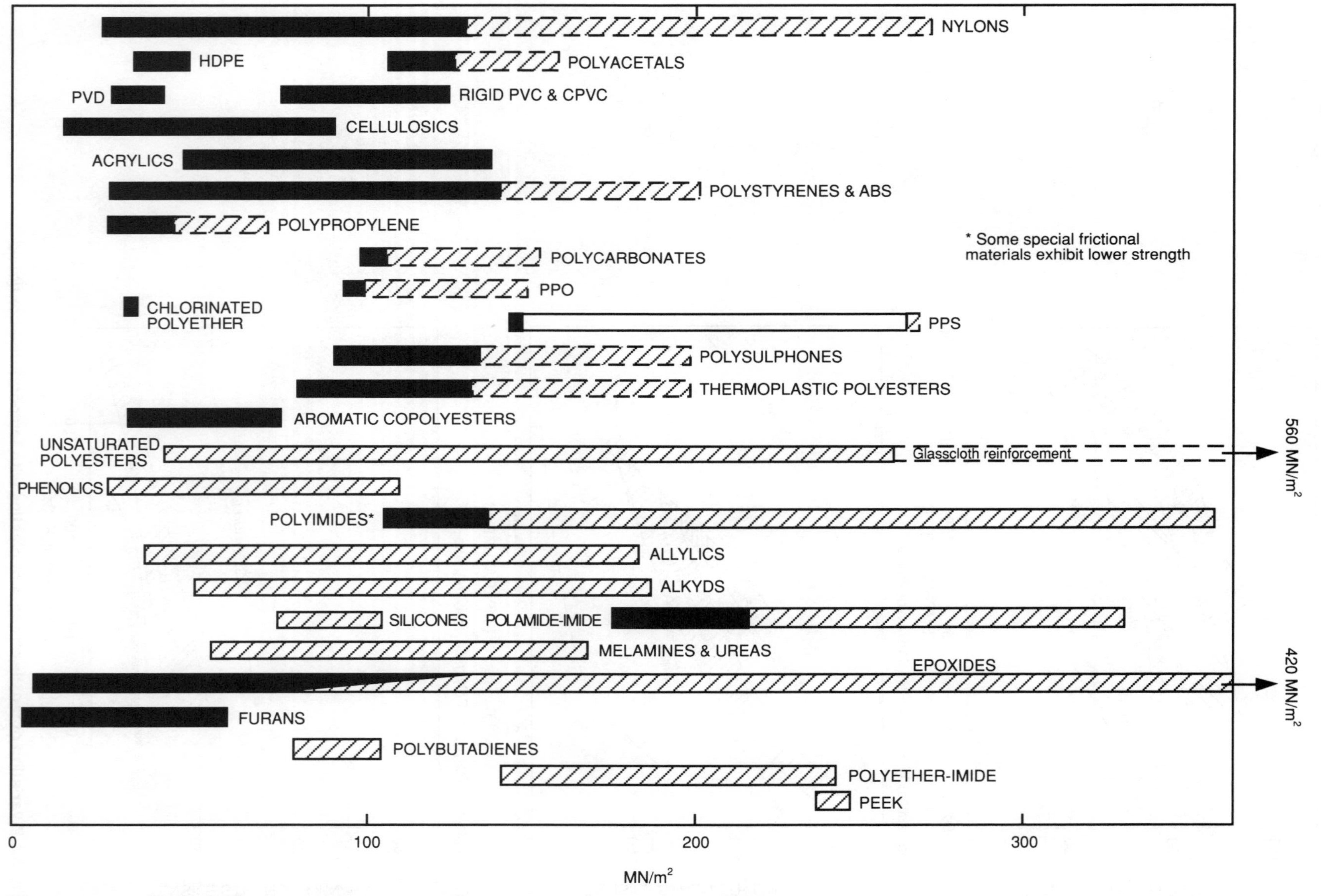

FIG 3.1.3 Room temperature flexural strengths of plastics. Hatched areas denote reinforcement

Fig 3.1.3

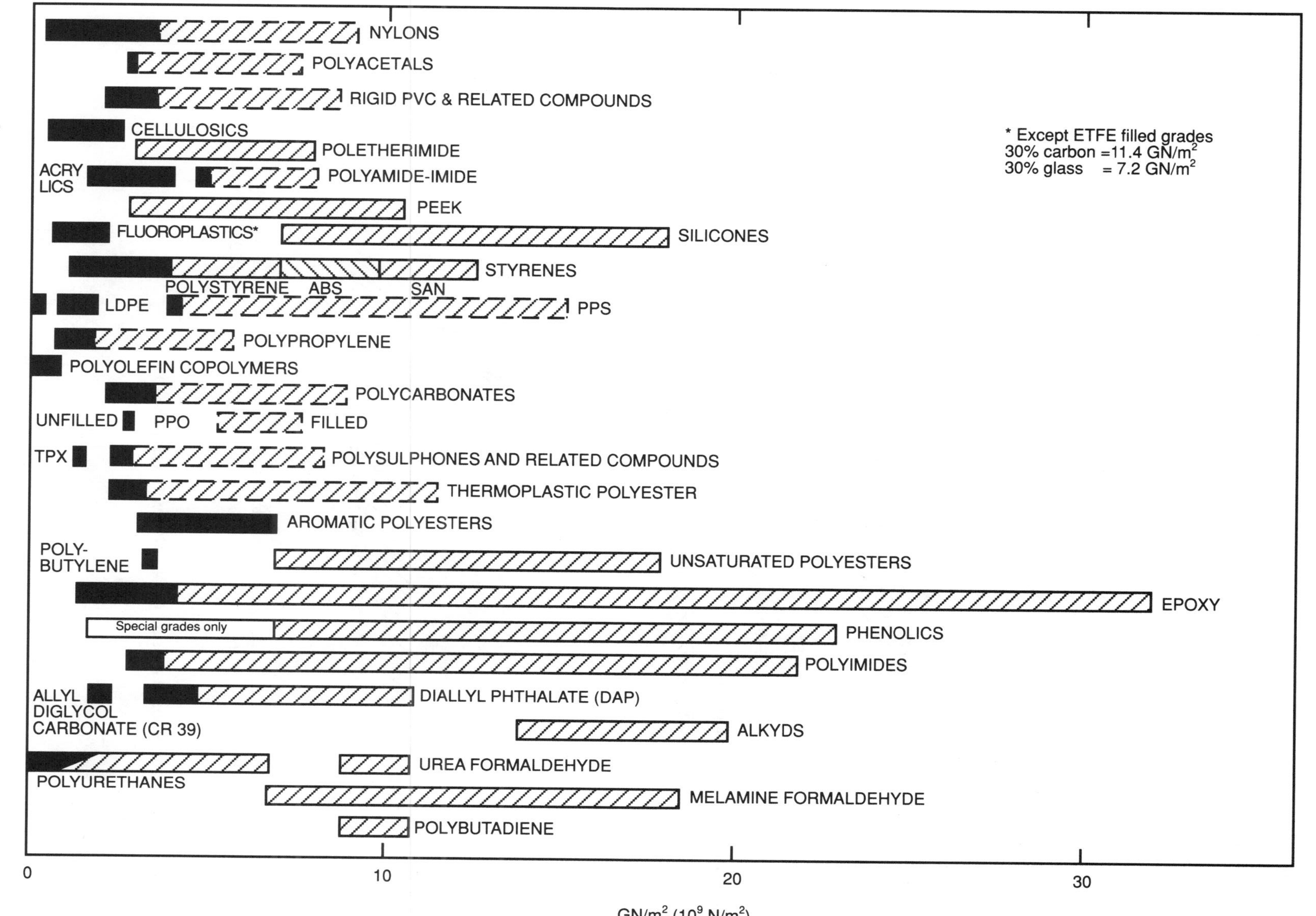

FIG 3.1.4 Room temperature flexural moduli of plastics. Hatched areas represent reinforcement

Fig 3.1.4

Nylons, Polyacetals, ABS

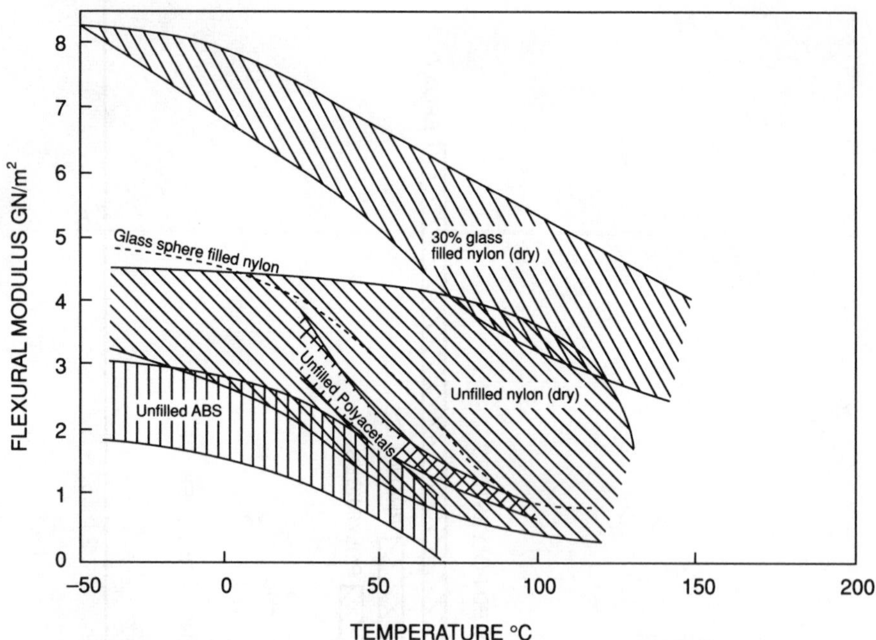

Polycarbonates, PPO, Polysulphones

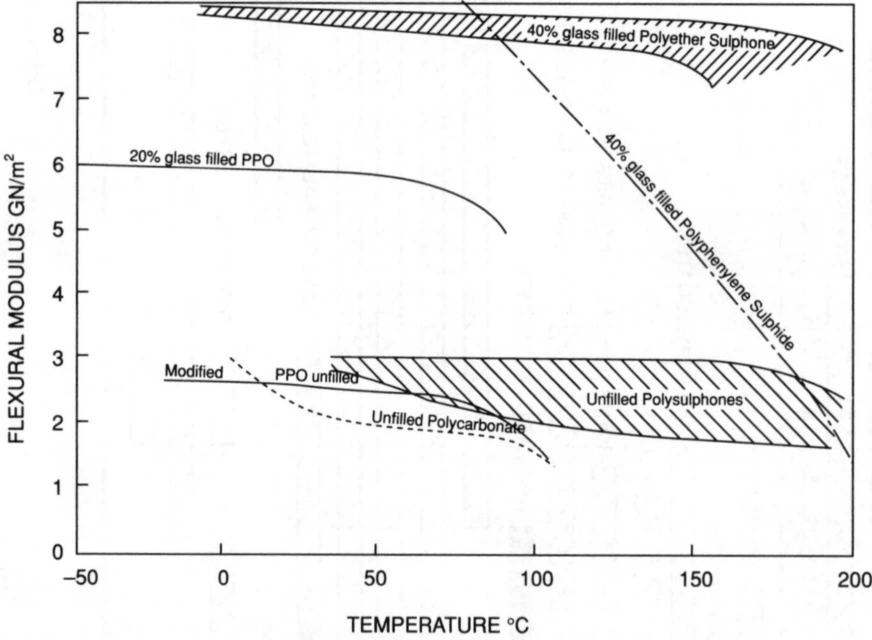

FIG 3.1.5 Variation of flexural moduli of plastics with temperature

PEEK, PBT, Polyamide-Imide

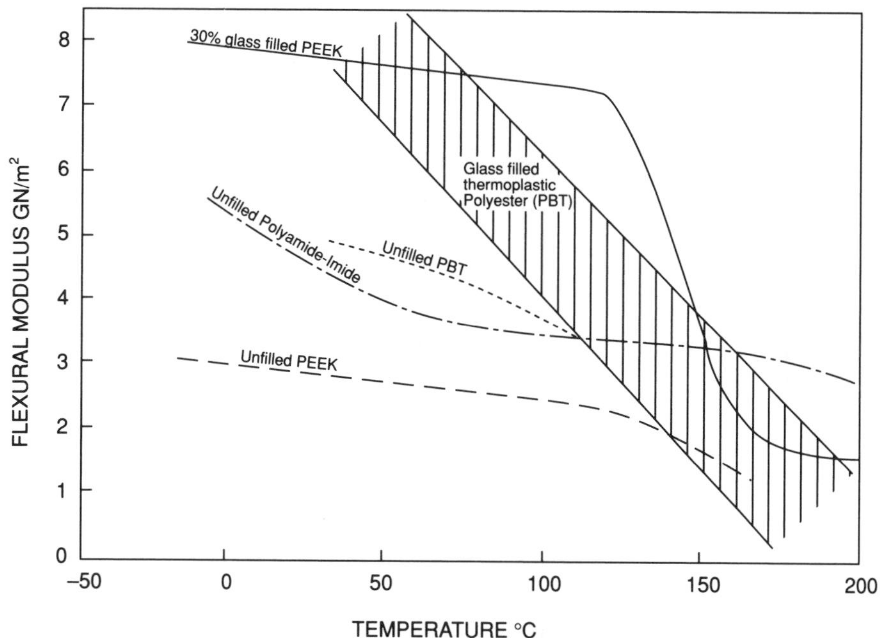

Thermosets

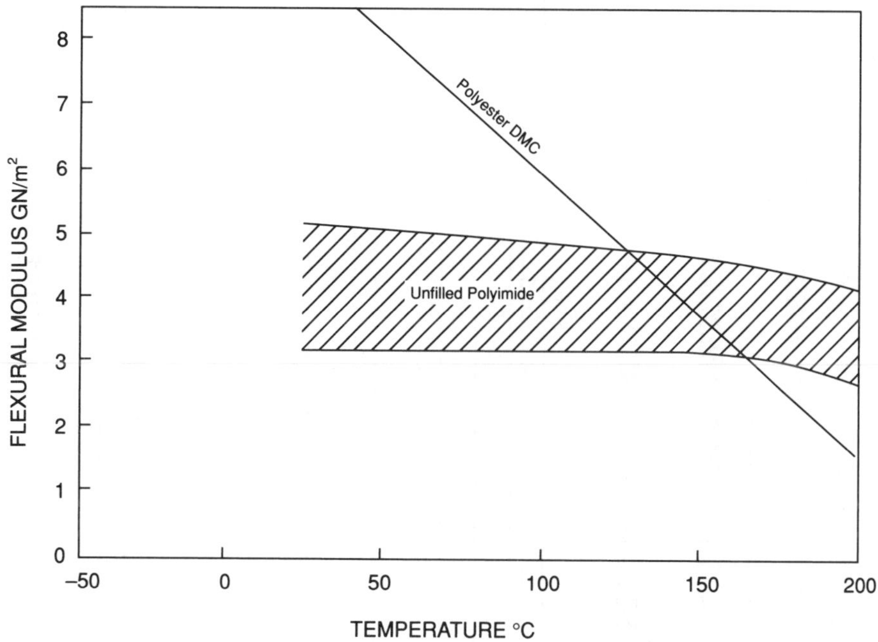

FIG 3.1.5 Variation of flexural moduli of plastics with temperature—*continued*

Fig 3.1.5—*continued*

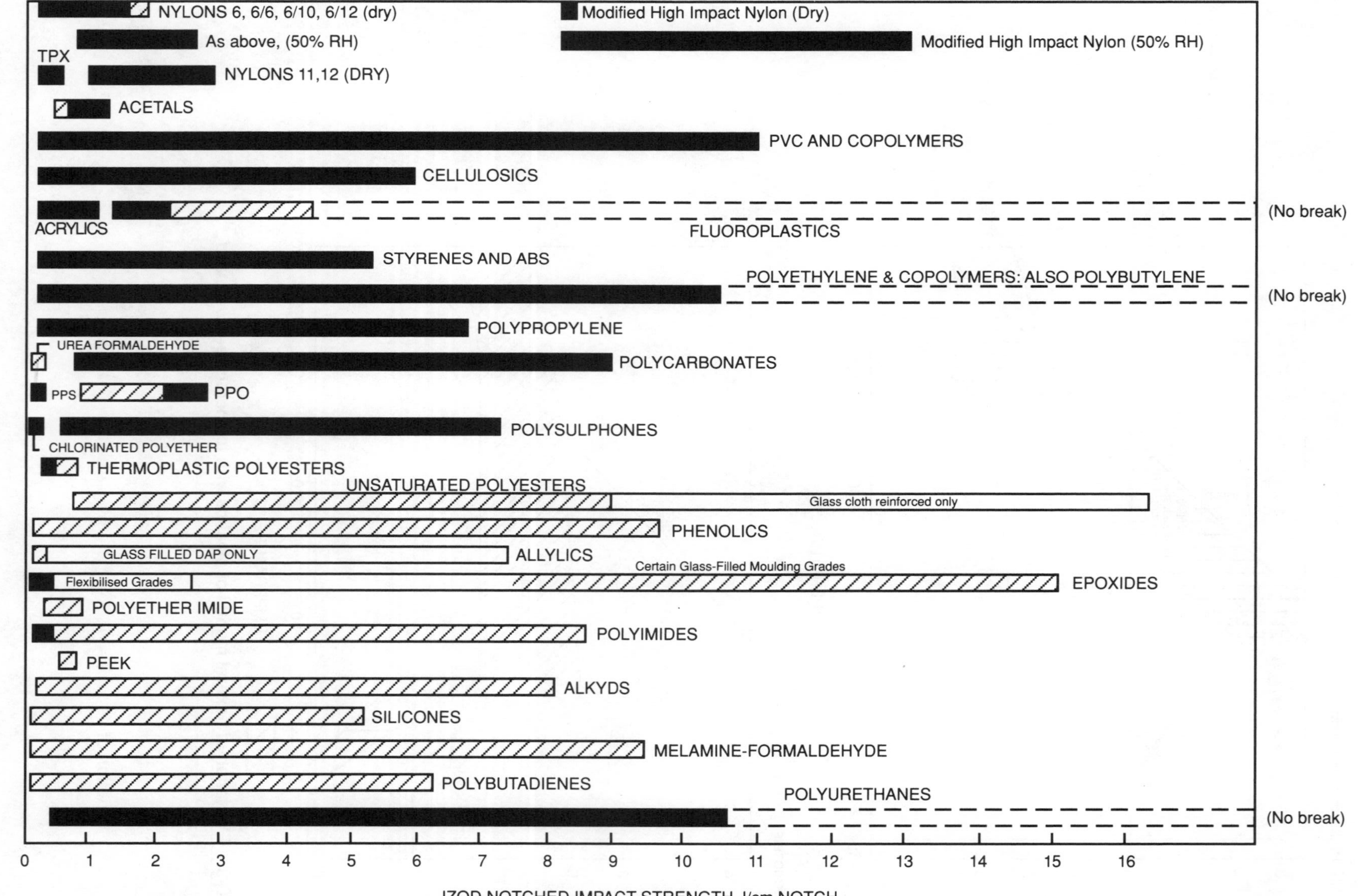

FIG 3.1.6 Room temperature impact strengths of plastics. Hatched areas represent increased range afforded by reinforcement

Fig 3.1.6

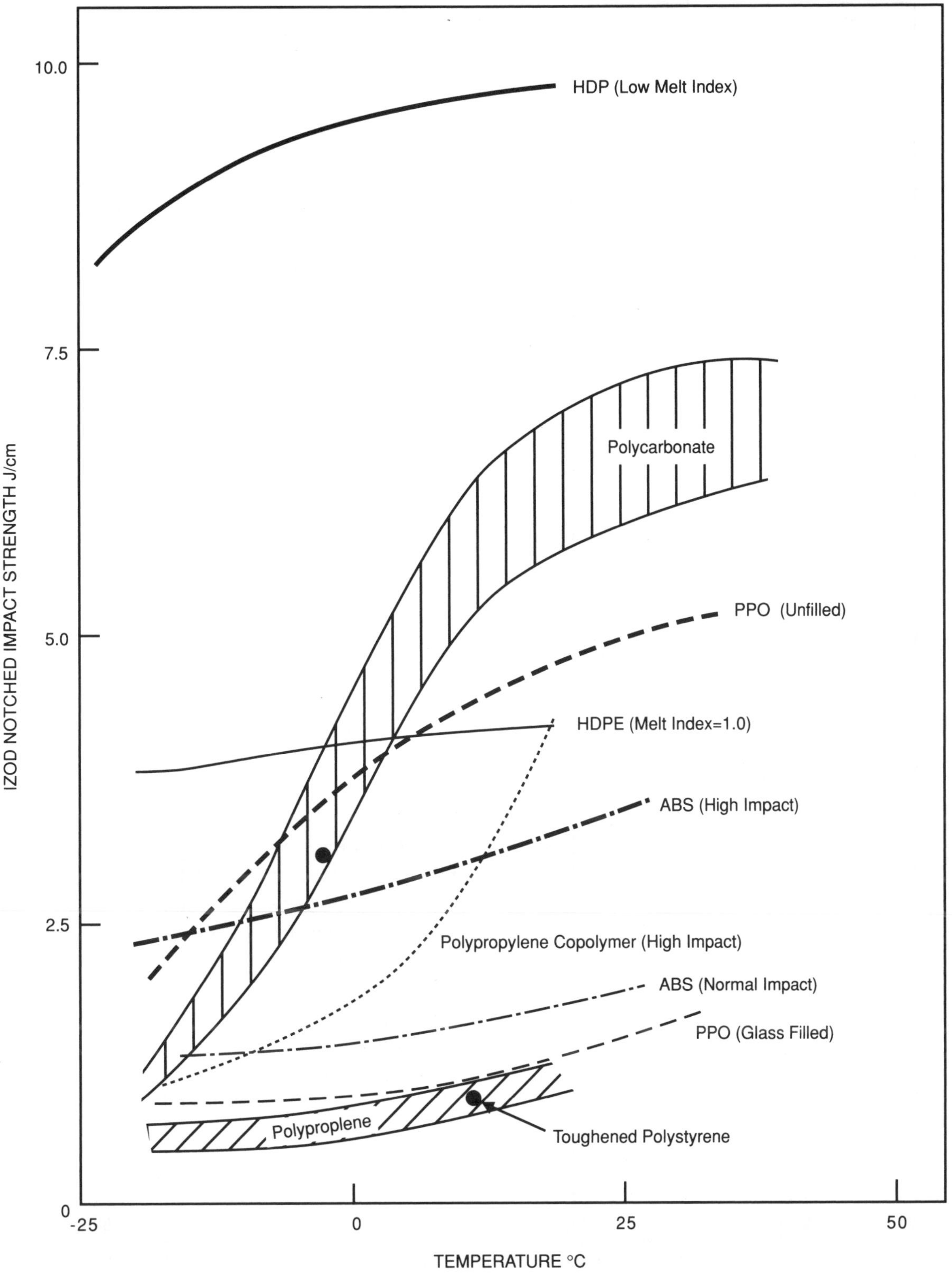

FIG 3.1.7 Variation of impact strength of thermoplastics with temperature

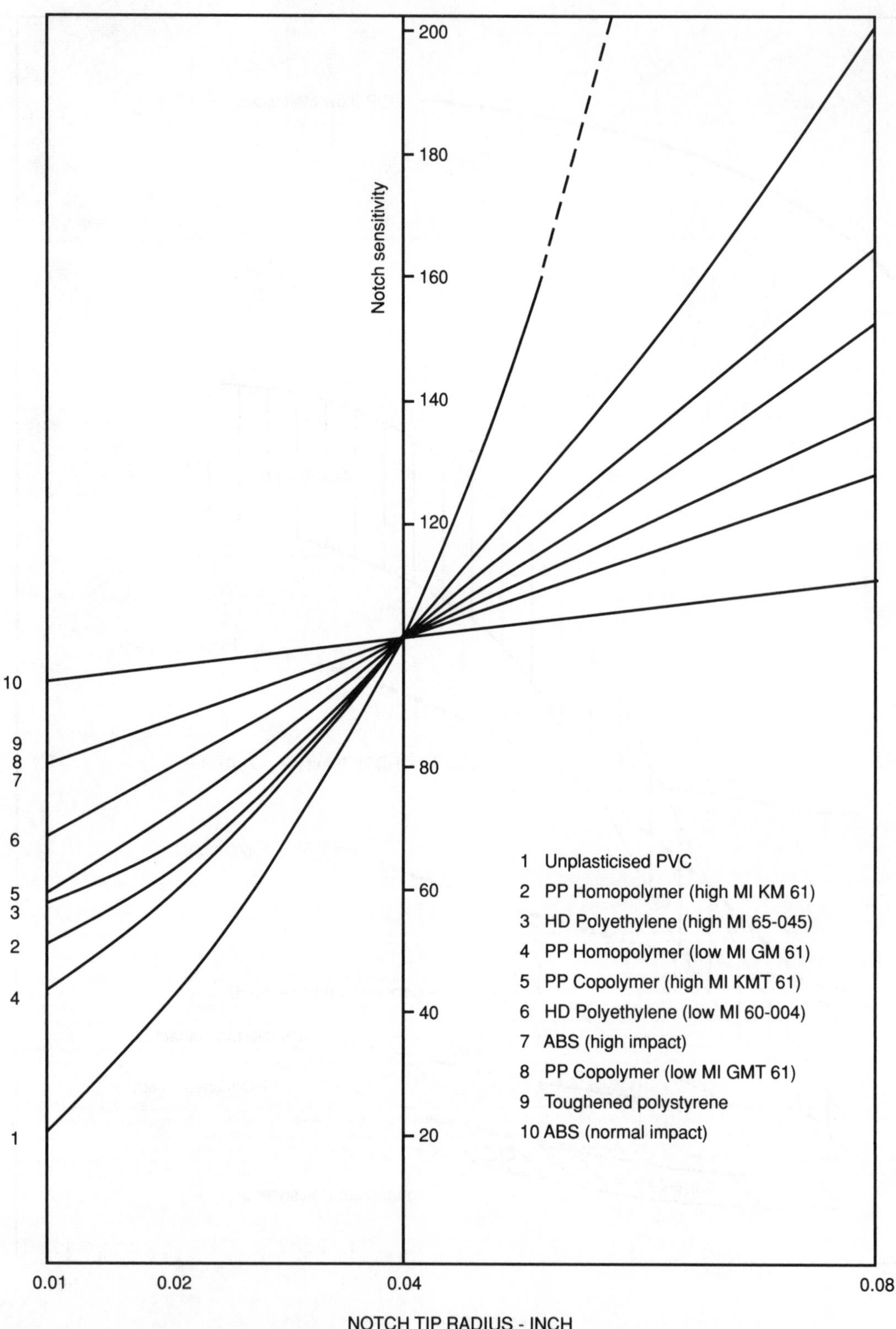

FIG 3.1.8 Typical notch sensitivities of some thermoplastics at 23°C

Fig 3.1.8

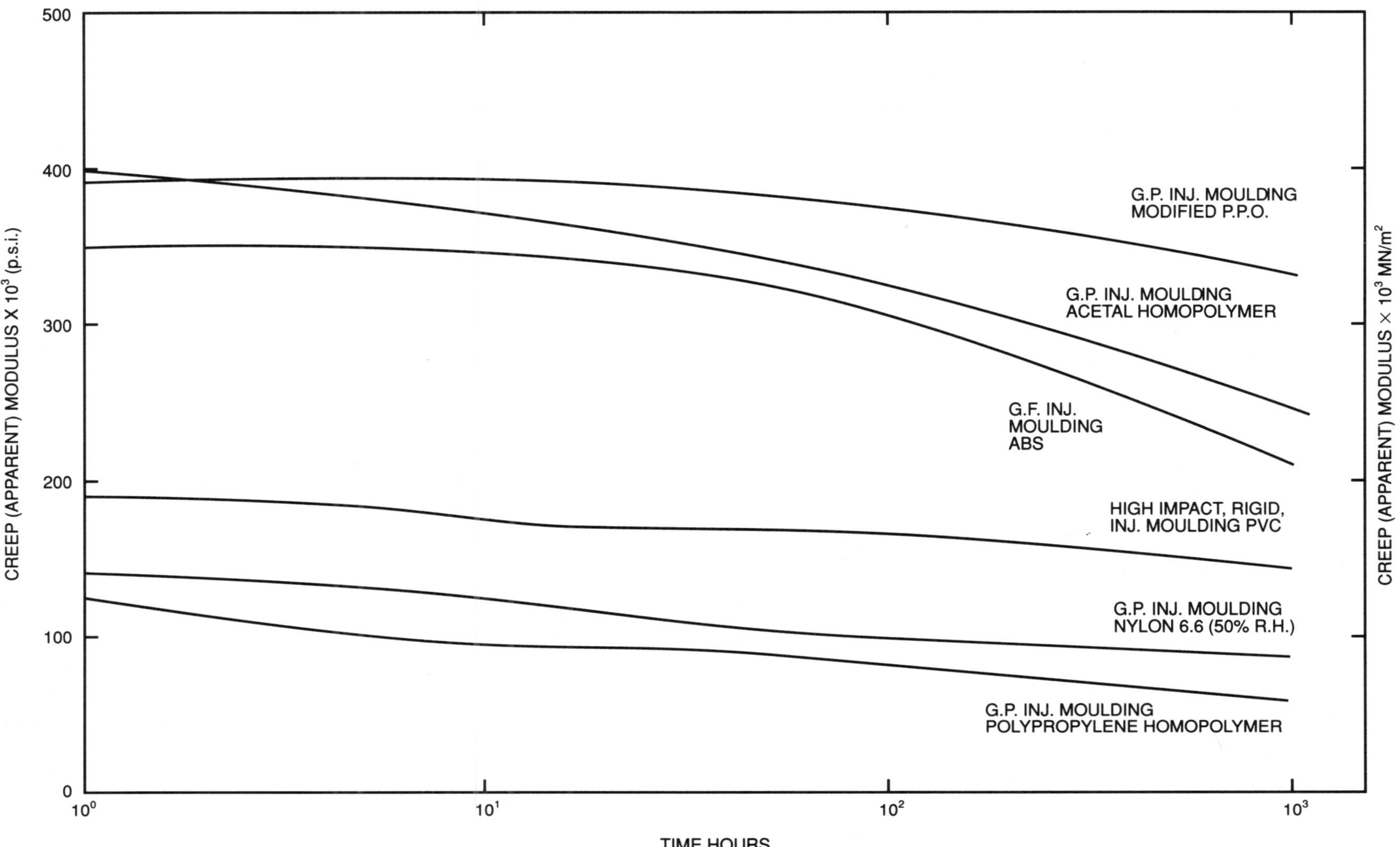

FIG 3.1.9 Comparative creep data of unreinforced thermoplastics. (Initial applied stress 6.9 MN/m^2; 23°C)

Fig 3.1.9

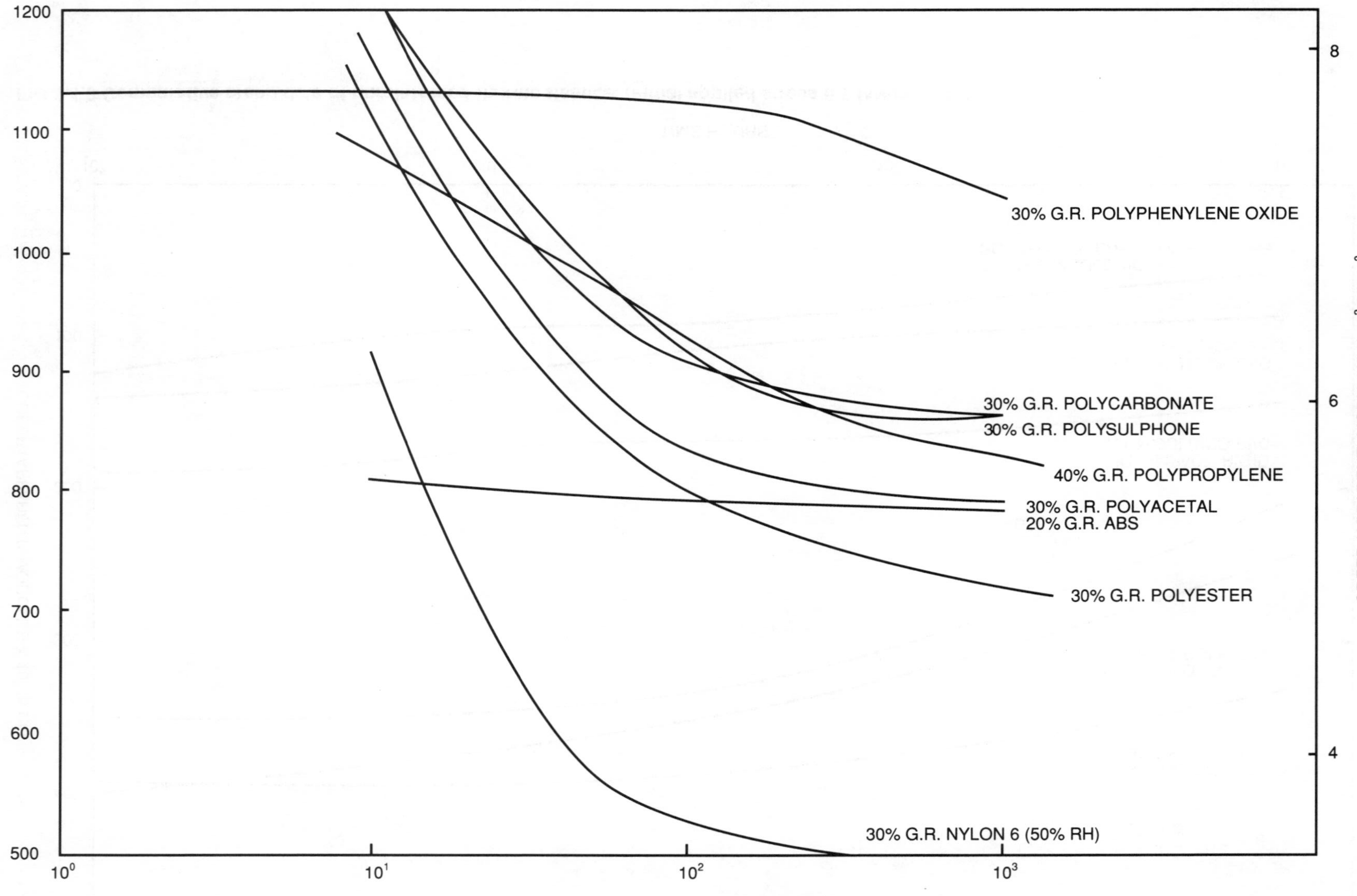

FIG 3.1.10 Comparative creep data of glass reinforced thermoplastics. (Initial applied stress 14 MN/m^2; 23°C)

Fig 3.1.10

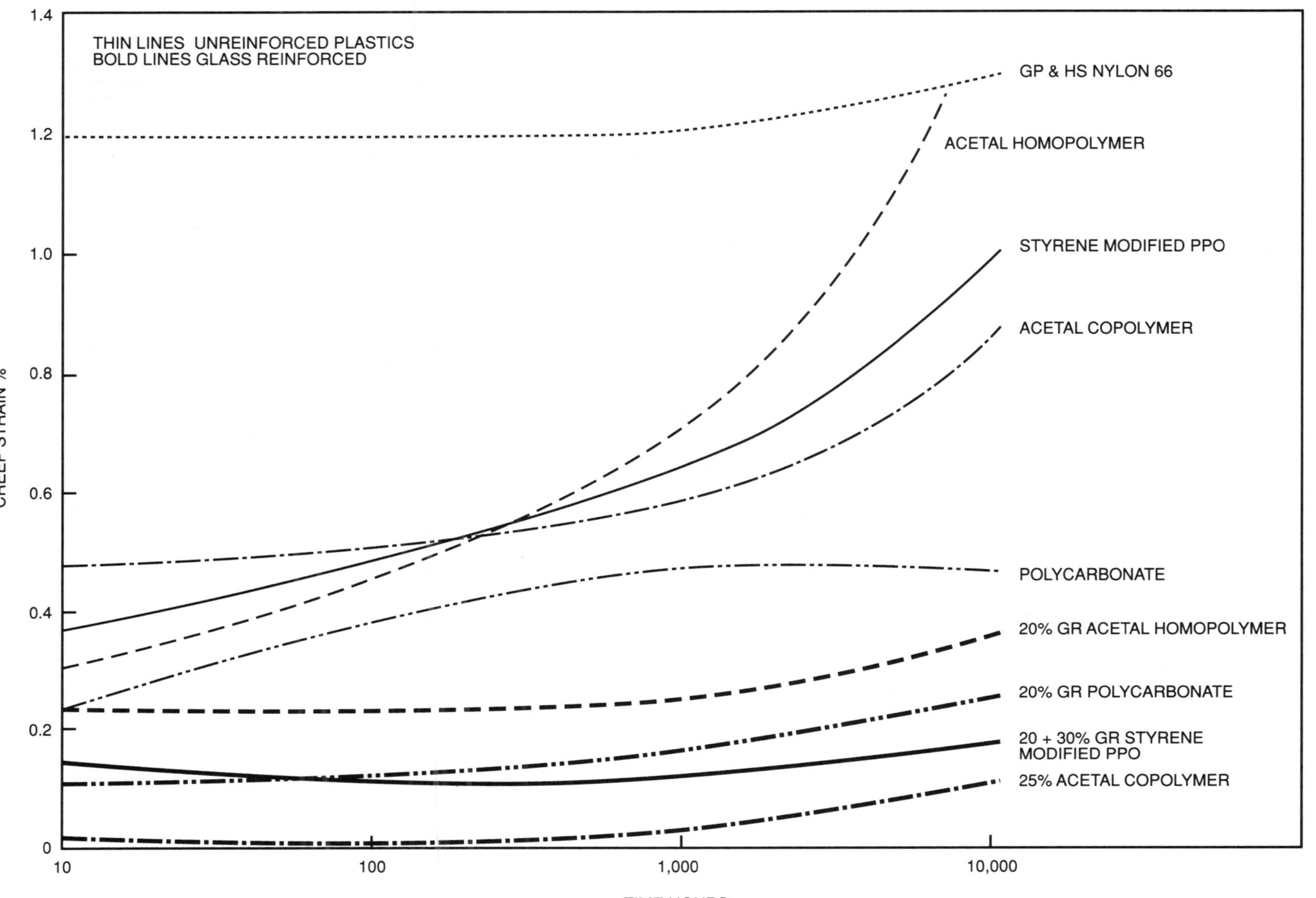

FIG 3.1.11 Comparative creep data of reinforced and unreinforced plastics. (Initial applied stress 3.5 MN/m^2; 83°C)

Fig 3.1.11

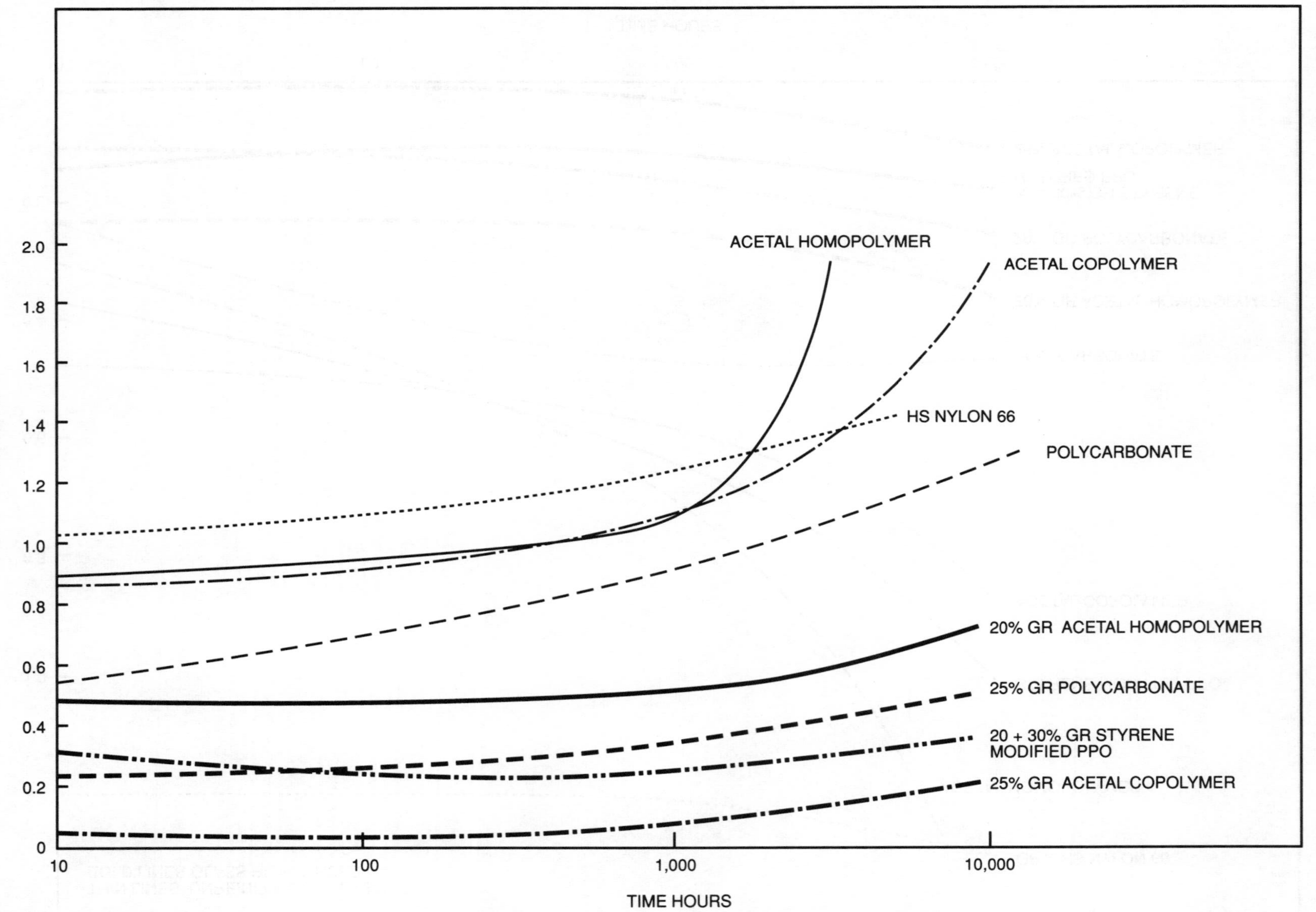

FIG 3.1.12 Comparative creep data of reinforced and unreinforced plastics. (Initial applied stress 3.5 MN/m^2; 116°C)

Fig 3.1.12

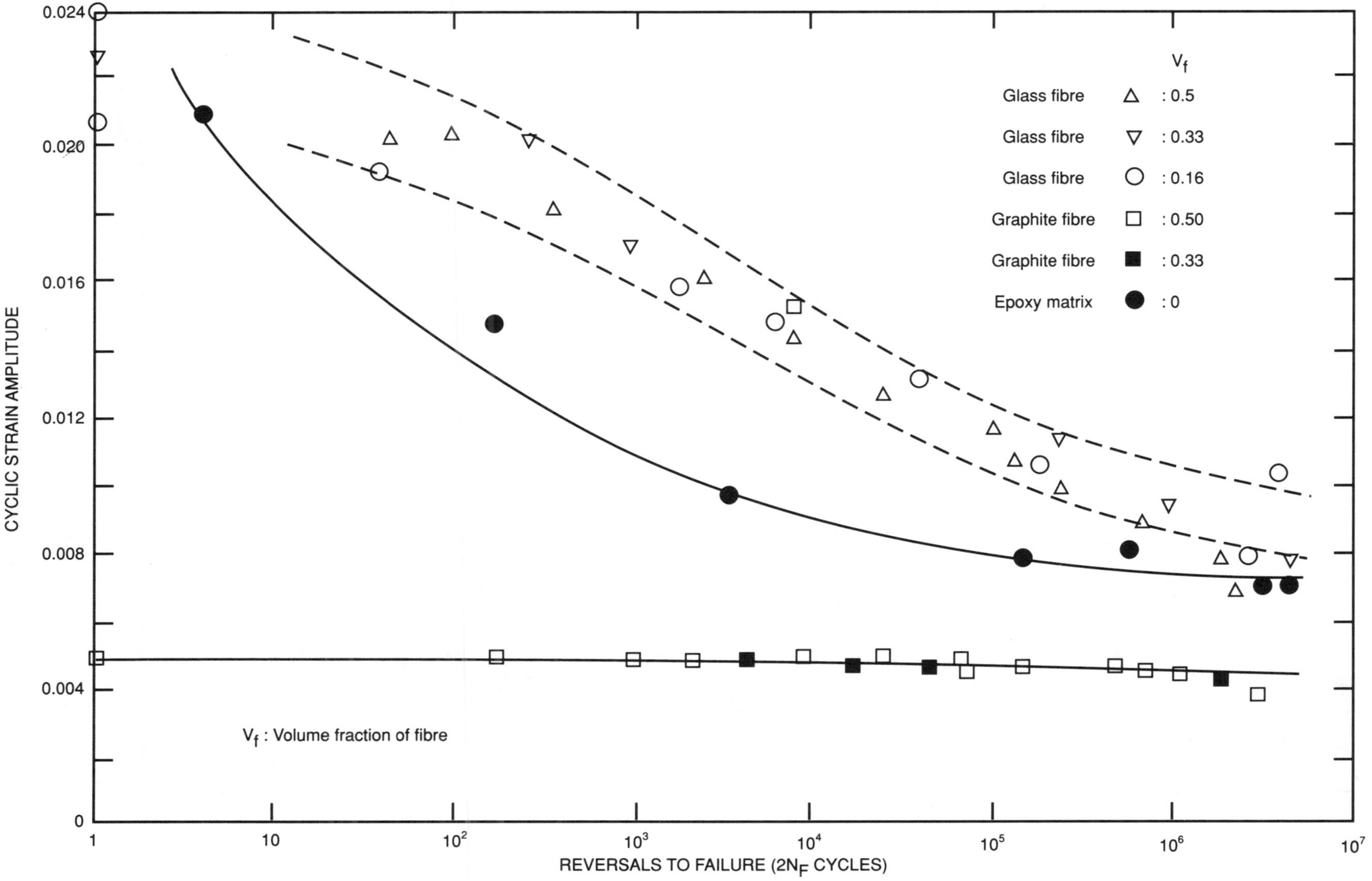

FIG 3.1.13 Applied strain versus cycles to failure for epoxy resin compared with glass and carbon fibre reinforced epoxy

Fig 3.1.13

3.1.3 Physical properties of plastics

3.1.3.1 Specific gravity

The specific gravity of polymers group around 1.25 (although TPX is only 0.83). These values can be lowered, by for example filling with microballoons, or raised to above 4, by filling for example with metal. Specific gravity is illustrated diagrammatically in:
Fig 3.1.14 *Specific gravity of plastics*

3.1.3.2 Thermal properties

The thermal expansion coefficients of polymers vary over a wide range but are normally higher than those of metals. Reinforcement with low thermal expansion fibres very significantly reduces the thermal expansion of plastics. Thermal expansion is illustrated diagrammatically in:
Fig 3.1.15 *Linear thermal expansion coefficients of plastics*

The thermal conductivities of polymers average about 0.4 W/m per K, much lower than those of metals but may be increased to above 4 W/m per K by the addition of appropriate fillers (metals or graphite), as is illustrated diagrammatically in:
Fig 3.1.16 *Thermal conductivity of plastics*

Thermal conductivity is temperature dependent and its variation is illustrated in:
Fig 3.1.17 *Thermal conductivity of a number of thermoplastic materials as a function of temperature*

Raising temperature from ambient to 150°C reduces the thermal conductivity of the more highly conductive plastics to about 0.25 W/m per K. That of the less conductive plastics is unchanged.

3.1.3.3 Electrical properties

Polymers are usually non-conductive of electricity and their use in electrical engineering makes their insulating and dielectrical properties important. The more important electrical properties are: dielectric strength, dry arc resistance, dielectric constant and dissipation factor.

The dielectric strength, measured in accordance with ASTM D149, is an indication of the electrical strength of a material as an insulator. The dielectric strength of an insulating material is the voltage gradient at which electric failure or breakdown occurs as a continuous arc (the electrical property analogous to tensile strength in mechanical properties). The dielectric strength of materials varies greatly with several conditions, such as humidity and geometry and it is not possible to apply the standard best values directly to field use unless all conditions, including specimen dimension, are the same. Because of this, the dielectric strength test results are of relative rather than absolute value as a specification guide.

Dielectric strength varies inversely with the thickness of the specimen and is illustrated diagrammatically in:
Fig 3.1.18 *Dielectric strength of plastics*

These values will drop sharply if holes, bubbles or contaminants are present in the material. The dielectric strength of polythene is approximately 2×10^4 kV/m.

The high voltage low current dry arc resistance of solid electrical insulation is tested in accordance with ASTM D495 which subjects the material to high AC voltage across the surface. Types of electrically induced failure for plastics and elastomers include ignition, tracking and carbonisation.

Test results for a number of plastics are included in Table 3.1.13, Section 3.1.4, and also in the sections dealing with specific plastics.

Dielectric constant is the ratio of the capacity of a condenser made with a particular dielectric to the capacity of the same condenser with air as the dielectric. For a material used to support and insulate components of an electrical network from each other and the ground, it is generally desirable to have a low level of dielectric constant. For a material to function as the dielectric of a capacitor it is desirable to have a high value of dielectric constant, so the capacitor may be physically as small as possible. The dielectric constants of insulating materials are compared diagrammatically in:

Fig 3.1.19 *Dielectric constants of non-metallic materials*

The dissipation factor (or loss tangent) is a measure of total losses in the dielectric material and of the conversion of the reactive power to real power, showing as heat. This parameter is shown in:

Fig 3.1.20 *Variation of loss tangent of typical polymers with temperature*

The insulating properties of plastics are sometimes modified to eliminate static electricity.

3.1.3.4 Optical Properties

Plastics materials find extensive application in engineering and related fields for their optical properties. The refractive index (n) of a material for a given wavelength, in vacuum, or electromagnetic radiation is defined by the ratio of the velocity of light in a vacuum and the velocity in the material. The refractive index is a function of wavelength and temperature.

Birefringence is a measure of optical anisotropy. Some materials are isotropic unless stressed elastically; permanent birefringence may be introduced by processing, or by dispersing one isotropic material within another.

Relative dispersion is the ratio of angular dispersion of two colours at the extremes of the visible spectrum (red and blue) to their mean dispersion. The reciprocal dispersion v is defined as:-

$$v = \frac{n \text{ yellow} - 1}{n \text{ red} - n \text{ blue}} \qquad \text{where } n \text{ is refractive index}$$

Transparency, which is a general quality of a plastics material depending largely upon bulk and superficial homogeneity, is measured by loss of clarity and loss of contrast. These two effects are closely interconnected and constitute the two different aspects of one phenomenon.

Haze characterises the loss of contrast which results when objects are seen through a scattering medium. Deterioration of contrast is mainly due to the light scattered forward at high angles to the undeviated transmitted beam. The 'milkiness' of translucent samples when viewed from the side on which light is incident is largely due to the backward scattering.

Clarity is a measure of the capacity of the sample for allowing details in the object to be resolved in the image which it forms. It is strongly dependent upon angular distribution of scattering intensity between object and sample.

Gloss is a property of the surface of a plastic specimen. Its magnitude depends on the refractive index of the plastic, the smoothness of the surface and the occurrence of the subsurface optical features. The relative importance of these factors depends upon the angle of incidence of light falling on the surface.

The total transmission factor (transmittance) is defined by the ratio of the total transmitted flux to the incident flux. The total reflection factor (reflectance) is defined (for normal incidence) as the ratio of the backward scattered flux to the incident flux. The more important optical properties of transparent and translucent plastics are listed in:

Table 3.1.9 *Properties of optical and other transparent plastics,* and

Table 3.1.10 *Optical properties of plastics compared with those of inorganics*

Table 3.1.9

TABLE 3.1.9 Properties of optical and other transparent plastics

Plastic	Density	Upper limit of optical stability	Refractive Index ($\eta D/20$)	Reciprocal dispersion	ASTM haze	D1003 Total luminous transmittance on 0.125" sheet	Typical applications of transparent grades
	(g/cm^3)	(°C)			(%)	(%)	
ABS	1.07	—	1.536	—	10	72	Similar to standard ABS where transparency required; also blending resins for PVC bottle compounds (added impact).
Acrylics	1.18–1.19	70–100	1.49–1.56	57.8 (at n_o = 1.492)	0.8–8	88–92	Internally illuminated outdoor signs, structural applications, glazing, skylights and sunscreens, furniture, lamps, vandal resistant lighting lenses, display cabinets, aircraft windows (covers).
Allyl diglycol carbonate (CR-39)	1.32	60–70	1.498	60	—	—	Eyeglasses, face shields, transparent panels.
Butadiene rubber	0.9	—	—	—	—	—	Pharmaceutical products
Cellulose acetate Cellulose acetate butyrate Cellulose propionate	1.28 1.19 1.22	50–60	1.46–1.50 1.46–1.50 —	45–50	<1	88	Extruded tape, toys, tool handles. Pen and pencil barrels, signal lenses. Toothbrush handles, blister packs.
Epoxy	1.2	—	1.54–1.60	—	1–1.2	94–96	General purpose impregnant and potting resin.
Ethyl cellulose	1.14	—	1.47	—	—	—	Coatings
Flurocarbons CTFE PTFE PVDF	2.1 2.1 1.76	— — —	1.42 1.35 1.42	— — —		94–96	Packaging and cable jacketing Chemical and electrical equipment Chemical and electrical equipment
Ionomer	0.94	—	—	—	2.1	75–85	Composite packaging structures
Nylons (Polyamides) Amorphous Nylon 6 Nylon 12	1.12 1.14 1.015	— — —	1.52 1.53 —	— — —	0.7	85	Sight glasses, cups, covers. Electronic. medical, industrial, meter parts. Mainly packaging.
Phenolic (wood-flour filled)	1.42	—	1.59	—	Opaque	—	General purpose
Polyacrylonitrile	1.17	—	1.51	—	—	—	Barrier resin for bottles and film
Polyallomer	0.9	—	1.49	—	—	—	Packaging and general purpose
Polyarylate	1.2	—	1.62	—	1–7	75–88	Lenses, lighting, tinted glazing, solar energy collecters.

TABLE 3.1.9 Properties of optical and other transparent plastics—*continued*

Polycarbonate	1.2	120–135	1.54–1.59 (optical grade 1.586)	29.9	1–3	85–88	Automotive lighting and lenses, safety glass, sunglasses, ski goggles, face protection, helmets, bottles and packaging.
			>1.6				Camera lenses (developmental)
Polyester (styrene-modified)	1.12	50–120	154–1.57	43	0.5	90	Display embedments, glazing, skylighting, structural panels, machine covers and guards.
Polyethylene/vinyl acetate	0.95	—	1.48–1.49	—	19	—	Film, sheet, packaging.
Polyethylene	0.92	—	1.50–1.51	—	—	—	General purpose, film and packaging, electrical parts.
Polyethylene terephthalate (PETP)	1.37	—	1.62	—	—	—	—
Polypropylene	0.9	—	1.50	—	—	—	Automotive, packaging, closures, battery caps, hair curlers.
Polymethylpentene (TPX)	0.83–0.91	—	1.463	—	—	—	Baking cartons
Polysulphone Polyethersulphone Polyarylsulphone	1.24 1.37 1.36	— — —	1.57–1.59 1.65 1.67	— — —	3.6–8.8	80	Medical instrumentation, blow moulded bottles. Reflectors, bulb housings, spotlight components, lenses and frames (primarily operating at high temperature) and as polysulphone.
Polystyrene	1.08	70	1.59 (1.57 copol)	30.8	—	—	Automotive, appliance housings, packaging, furniture, cassettes, shoe heels, radio and television components, hot vending containers, bath panels, lampshades, trays, boxes, cosmetic packaging, vacuum flasks.
Polyurethane	1.05–1.50	—	1.5–1.6	—	—	—	Seals, tubing, cable jackets.
Polyvinylacetate (PVA)	—	—	1.47–1.49	—	—	—	—
PVC	1.35	—	1.54–1.56	—	—	—	High quality items for outdoor use, medical and food containers, oxygen therapy equipment, film and sheet, food packaging.
PVC/vinyl acetate	—	—	1.525–1.529	—	—	—	—
Poly-p-xylene	1.11	—	1.661–1.669	—	—	—	Electrical applications.
Syrene-acrylonitrile (SAN)	1.08	90	1.569	35.7	—	—	Caps, tumblers, trays, radio knobs and scales, cosmetic and other packaging items, instrument lenses.
Styrene butadiene rubber	—	—	1.52	—	—	—	—
Styrene-methacrylate	1.14	95	1.533–1.57	42.4	—	—	Dishwasher—safe transparent housewares, lenses, meter faces.

TABLE 3.1.10 Optical properties of plastics compared with those of inorganics

Advantages	*Limitations*
1. Cheapness of manufacturing processes. Moulding as opposed to grinding and polishing.	1. Low selection of refraction dispersion relations available.
2. Higher relative dispersions for a given refractive index than inorganic materials, i.e. lower values of $\nu = (n_o - 1)/(N_f - N_c)$ are advantageous for prisms. Disadvantageous for lenses.	2. Higher refractive index than inorganics, but lack of high refractive index materials.
	3. High coefficient of thermal expansion.
	4. Inferior resistance to surface abrasion.
3. Relatively light weight.	5. Grinding and polishing more difficult than with inorganics.
4. Tough and non-brittle.	
5. Good ultra-violet transmittance.	

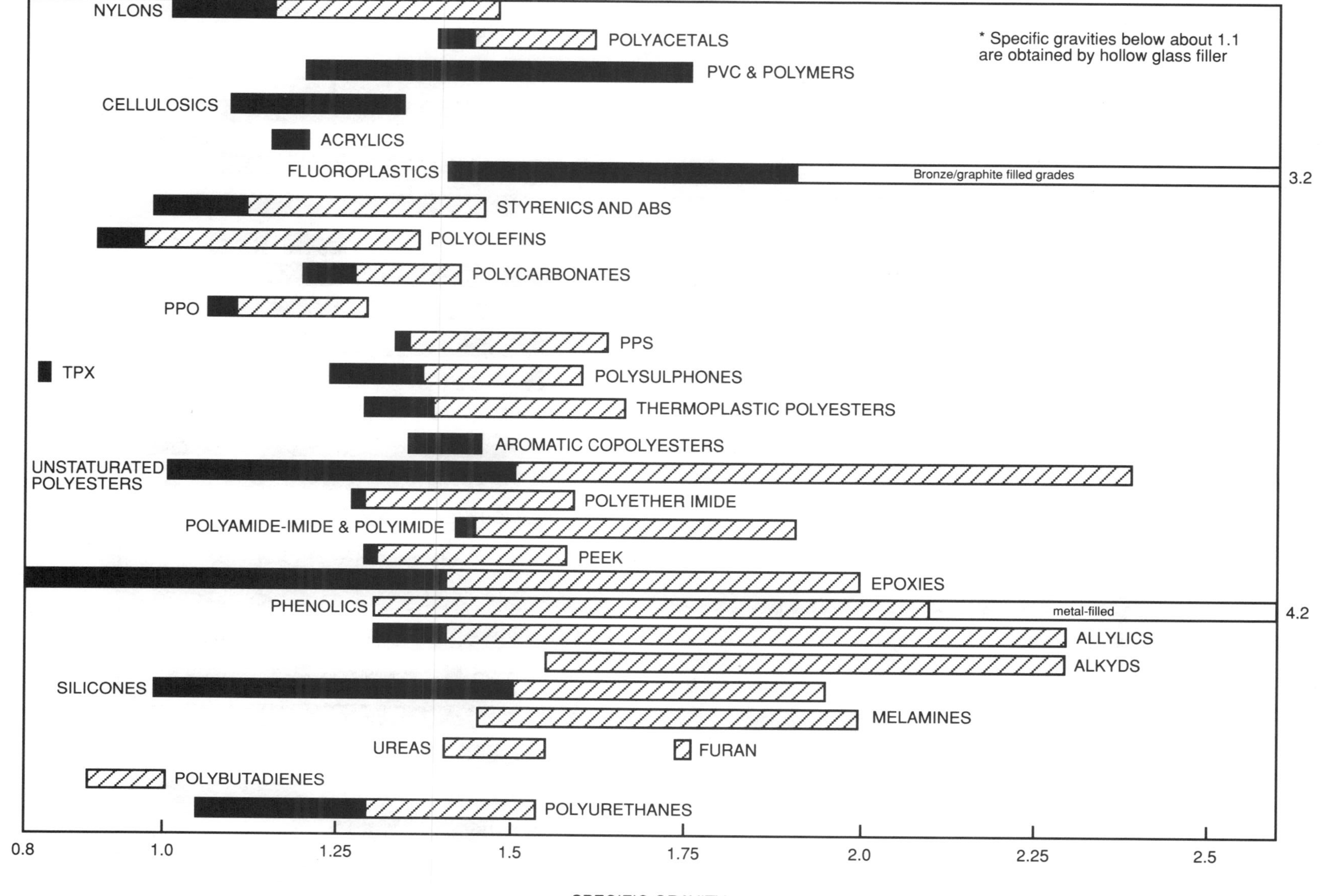

FIG 3.1.14 Specific gravity of plastics. Hatched areas represent higher values obtained by reinforcement

Fig 3.1.14

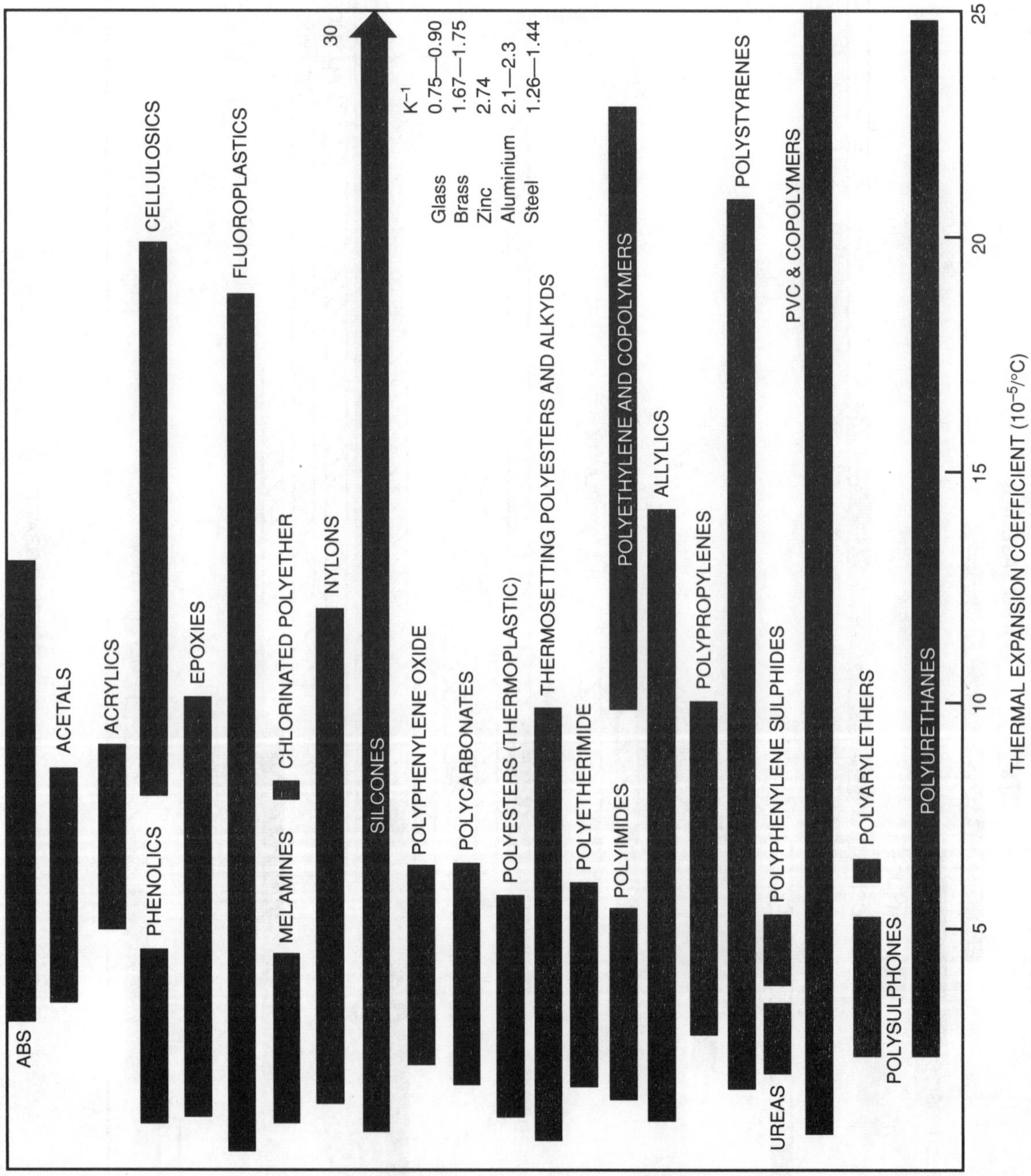

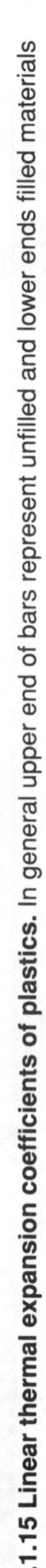

	K^{-1}
Glass	0.75—0.90
Brass	1.67—1.75
Zinc	2.74
Aluminium	2.1—2.3
Steel	1.26—1.44

FIG 3.1.15 Linear thermal expansion coefficients of plastics. In general upper end of bars represent unfilled and lower ends filled materials

Fig 3.1.15

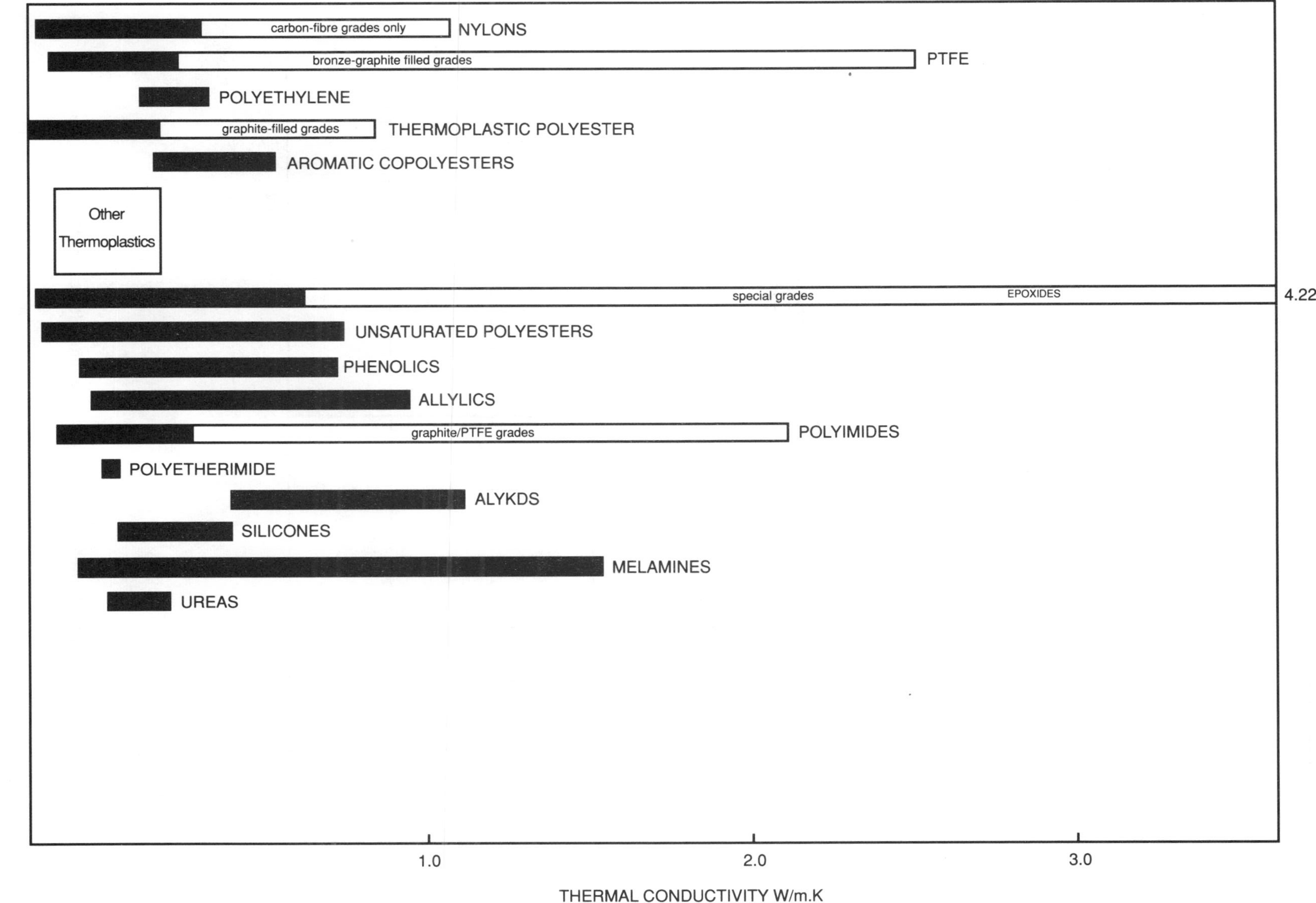

FIG 3.1.16 Thermal conductivity of plastics

Fig 3.1.16

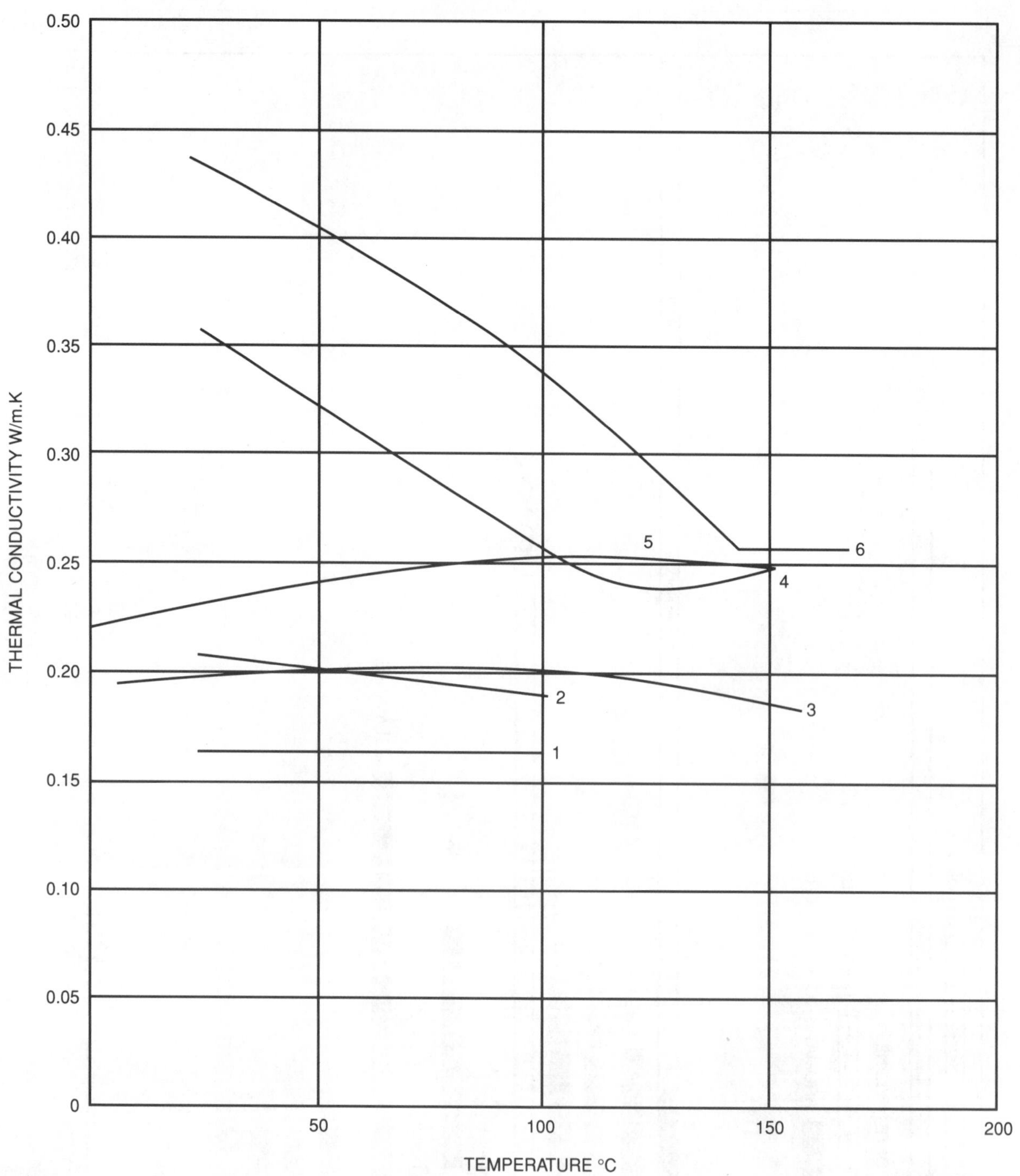

1 PVC and Polystyrene
2 Polypropylene
3 Acrylic
4 HDPE
5 Polycarbonate
6 LDPE

FIG 3.1.17 Thermal conductivity of a number of thermoplastic materials as a function of temperature

Fig 3.1.17

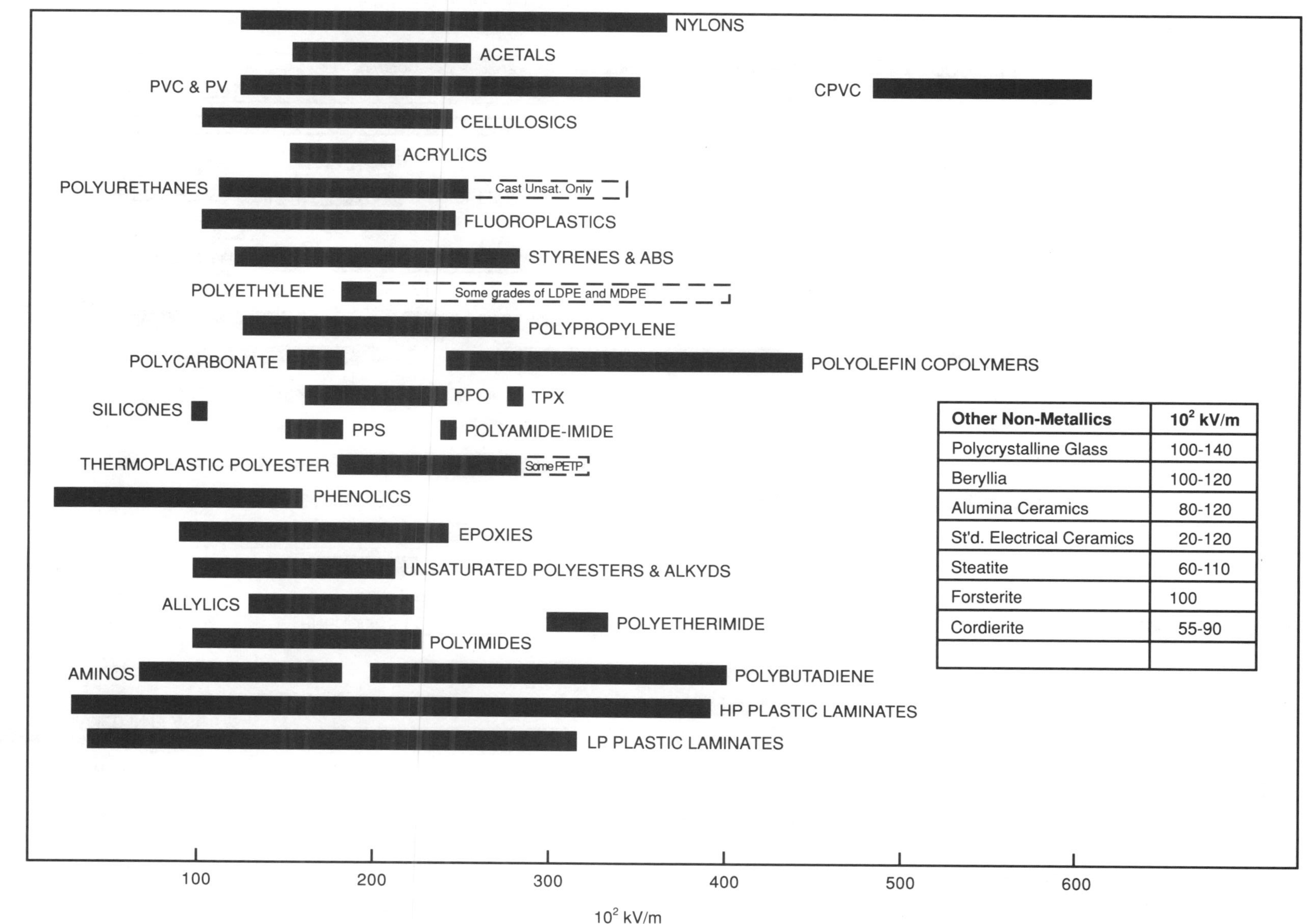

Other Non-Metallics	10^2 kV/m
Polycrystalline Glass	100-140
Beryllia	100-120
Alumina Ceramics	80-120
St'd. Electrical Ceramics	20-120
Steatite	60-110
Forsterite	100
Cordierite	55-90

FIG 3.1.18 Dielectric stength of plastics (ASTM 0149, 3mm thickness, short time)

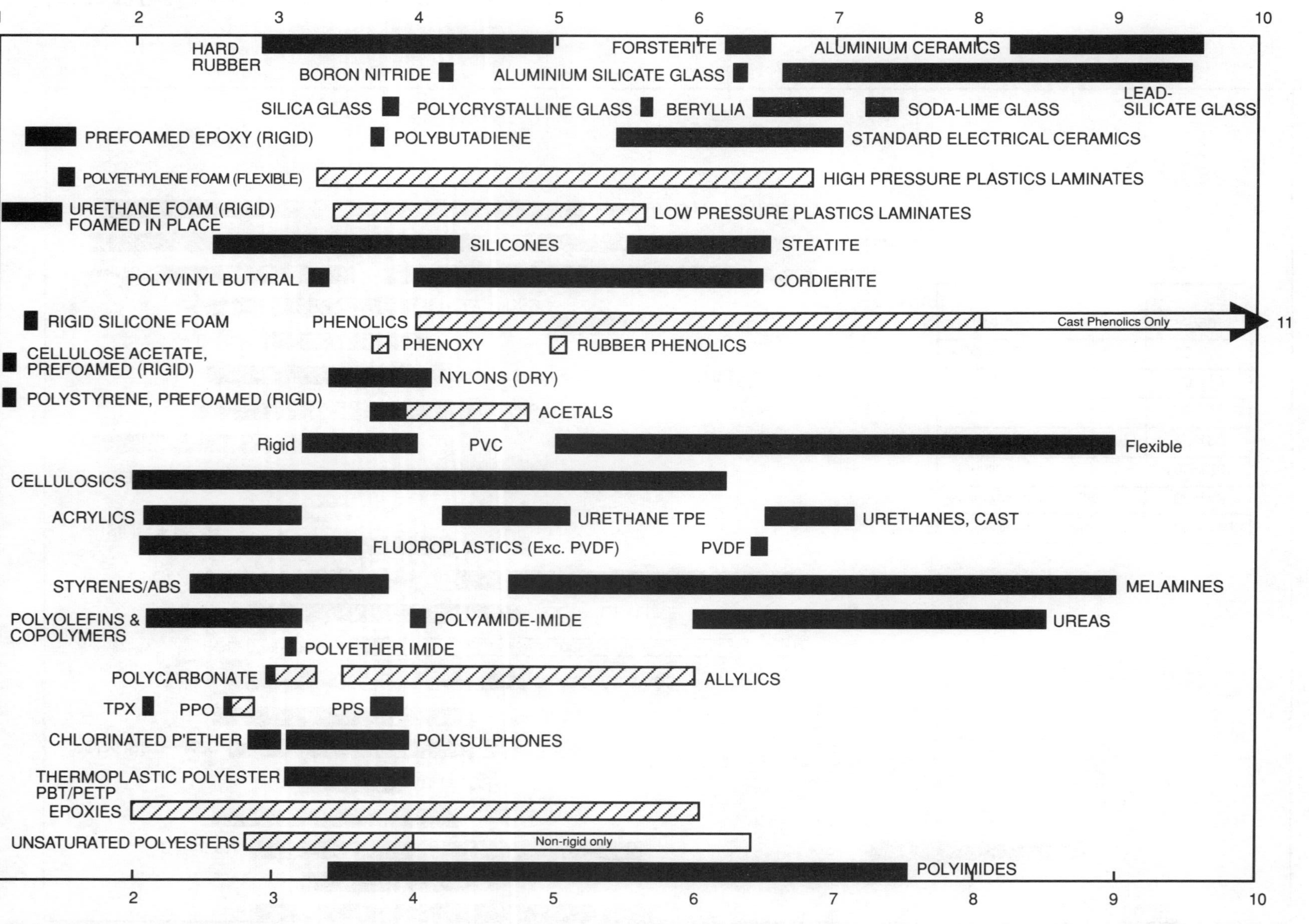

FIG 3.1.19 Dielectric constants of non-metallic materials. Hatched areas denote filled grades

Fig 3.1.19

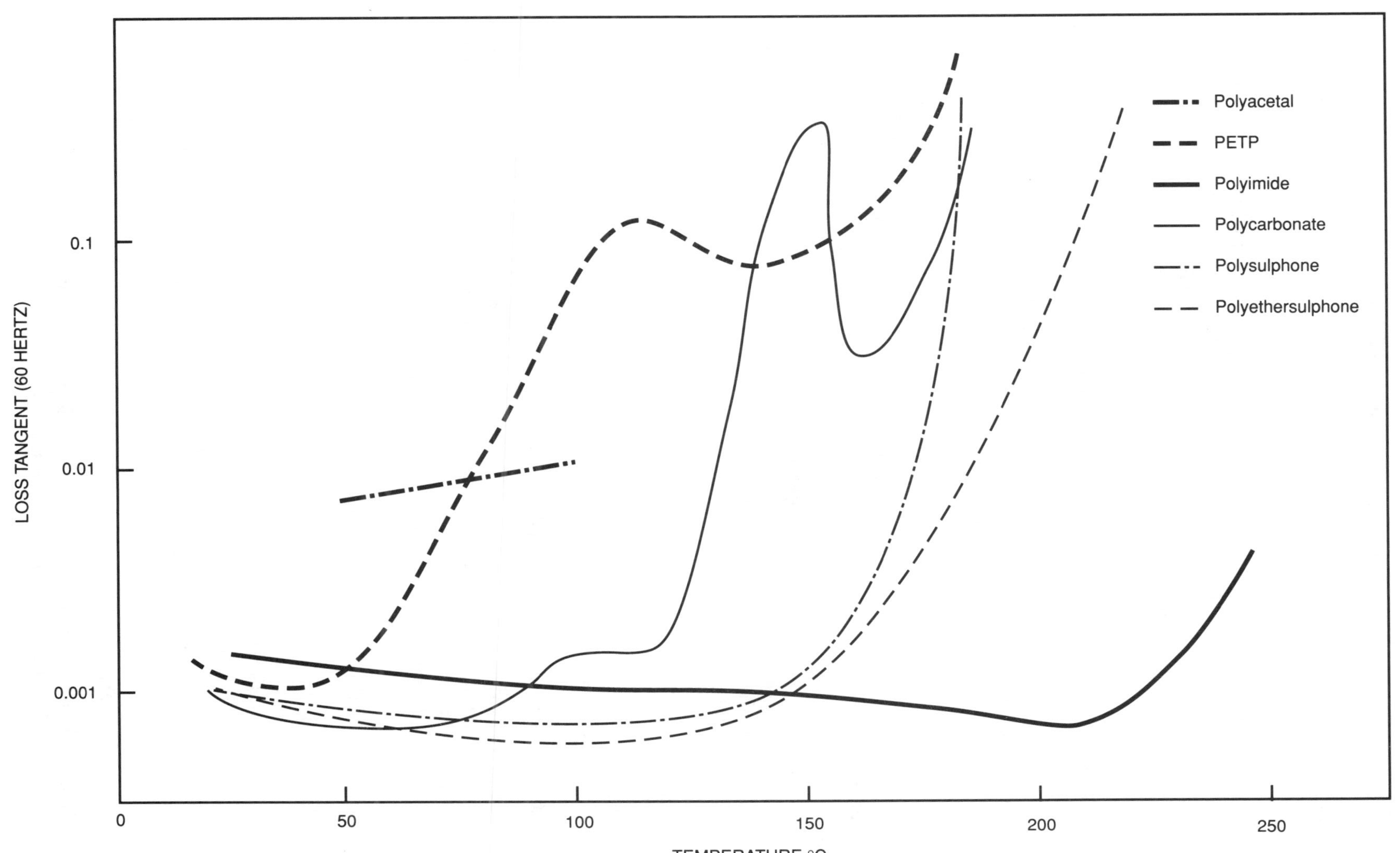

FIG 3.1.20 Variation of loss tangent of typical polymers with temperature

Fig 3.1.20

3.1.4 Environmental performance of plastics

3.1.4.1 Temperature limits

UPPER TEMPERATURE LIMIT

The main factors limiting the temperature that a plastic will withstand are softening and burning at high temperature and embrittlement at low temperatures.

Most commonly used thermoplastics have softening points defined as the temperature at which some physical property is reduced (usually) to some arbitrarily defined value, which is low by most standards. Typical 'Softening Point' values are listed in:
Table 3.1.11 *Softening points of polymers by various standard methods*

Few thermoplastics have a sharp melting point. They soften gradually, and the maximum safe design temperature will be well below the softening point. Maximum service temperatures for thermoset materials are usually higher, but the limitation on service temperature still exists.

Maximum service temperatures depend on the service conditions and particularly on whether the high temperature is continuous or intermittent and the nature of service loads. Maximum service temperatures, assessed by various methods, are given in:
Table 3.1.12 *Comparison of heat resistance values for common plastics*

Maximum service temperatures are illustrated diagrammatically in:
Fig 3.1.21 *Recommended maximum service temperatures (no load)*

Properties of polymers which are commonly used at high temperatures are given in:
Table 3.1.13 *Properties of high-temperature polymers*

Plastics oxidise, usually at much lower temperatures than metals. Degradation may also occur: the long-chain molecules of many plastics materials gradually 'unzip' at quite moderate temperatures, although this phenomenon can be delayed or stopped by the incorporation in their compounding of various protective chemicals, antioxidants, stabilisers, etc., which react with the chain end to impede the degradative process.

Plastics subjected to excessive heating in air are liable to burn, and it is essential to assess the tendency for this to occur, and to evaluate the consequences. Flammability is discussed in Section 3.1.5.

LOWER TEMPERATURE LIMIT

Plastics materials also exhibit a loss of flexibility, or embrittlement, as their temperatures are reduced, and a variety of tests are used to indicate their characteristics in this respect. Recommended minimum operating temperatures are given in:
Fig 3.1.22 *Recommended minimum operating temperatures*

3.1.4.2 Chemical resistance

The resistance of plastics to solvents, aqueous solutions and other materials and phenomena is given in:
Table 3.1.14 *General environmental performance of plastics*

More detailed information is given in:
Fig 3.1.23 *The resistance of plastics to a range of solvents*

Information on the resistance of plastics film is given in chapter 3.7

3.1.4.3 Abrasion resistance

Preliminary information on wear resistance is provided in:
Table 3.1.15 *Abrasion resistance of plastics*

TABLE 3.1.11 Softening points of polymers determined by various standard methods

Material	*Cantilever BS 2782: M.102C (°C)*	*Deflection temperature under load (DTL) (ASTM D648)*		*Vicat BS 2782: M.102D (°C)*
		1.8 MN/m² fibre stress (°C)	*0.45 MN/m² fibre stress (°C)*	
Polystyrene	95	90	97	98
Toughened polystyrene	84	72	85	86
ABS	94	84	96	95
Polymethyl methacrylate	95	80	97	90
Cellulose acetate	76	64	77	72
Rigid PVC	78	70	82	82
Polyethylene (Low density)	(too flexible)	(too flexible)	45	85
(Medium density)	90	35	69	105
(High density)	115	45	75	125
Polypropylene	145	60	140	150
Nylon 6.6	180	75	183	185
Acetal	170	120	165	175

TABLE 3.1.12 Comparison of heat resistance values for common plastics

Material	*Heat deflection temperature*		*Maximum service temperature*		*Underwriters' laboratory[a] Thermal index*			
	0.45 MN/m²	*1.8 MN/m²*	*Intermittent*	*Continuous*	*Electrical*	*Mechanical with impact*	*Mechanical without impact*	*Generic*
ABS unfilled	90–120	90–110	—	75–100	60	60	60	
Acetal unfilled	153–170	110–124	110–126	90–120	105	90	90	
Acetal glass-filled	165–175	163	—	—	110	95	105	
Acrylic	80–100	75–100	—	54–95	50	50	50	
Cellulosics	52–105	48–86	—	55–110	—	—	—	
PTFE, CTFE, FEP and TFE	120	—	260	230–260	180	180	180	
Ionomer	44–52	—	—	70	—	—	—	150
Nylon 6/6, unfilled	216–240	71–105	—	80–110	65–130	65–105	65–130	65
Nylon 6/6, glass filled	255–265	240–257	—	80–120	65–130	65–105	65–130	65
PPO (Noryl), unfilled	110–140	100–133	—	—	110	105	110	
PPO (Noryl), glass-filled	140–150	135–150	—	—	110	105	110	
Polyallomer	—	49–51	—	110	—	—	—	
Polyamide-Imide	—	272	—	—	220	50–200	50–200	
Polycarbonate	135–154	127–149	120–130	104–130	125	115	125	65
Polyester PBT, unfilled	160–170	50–70	—	120–137	130	130	140	
Polyester PBT, 30% glass	215–220	205	—	180	—	—	—	
Polyester PETP, unfilled	72–115	66–82	—	—	65	65	65	
Polyester Polyarylate	—	174	149	129	—	—	—	
Polyethersulphone	—	203–215	—	175–200	—	—	—	180
Polyethylene LD	40–52	—	—	—	—	—	—	
Polyethylene MD	60	—	—	—	—	—	—	
Polyethylene HD	77–132	—	—	—	50	50	50	
PPS (Ryton)	—	>260	—	232	170–200	170–200	170–200	
Polyphenylsulphone	—	204	204	179	—	—	—	
Polypropylene, unfilled	90–135	48–65	—	90–120	115	115	115	
Polypropylene, 30% glass	146–163	130–150	—	90–120	—	—	—	
Polypropylene, 25% talc	115–135	90	—	—	—	—	—	
Polystyrene	—	65–115	—	80–115	50	50	50	
Polysulphone	187	174	—	150	150–180	140–170	150–180	
PVC	56–82	54–80	—	66–80	80 (high impact)	—	—	
SAN	—	87–103	—	90–105	50	50	50	
Alkyd	—	180–260	—	150–235	130	130	130	130
Diallylphthalate	—	150–260	—	150–230	130	130	130	130
Epoxy (moulding)	—	95–260	—	150–260	130	130	130	130
Melamine	—	125–180	—	100–120	130–150	130–150	130–150	130
Melamine-Phenolic	—	140–155	—	135–165	130–150	130–150	130–150	130
Phenolic, various fillers	—	130–315	—	100–290	150–155	150–155	150–155	150
Polyester DMC	—	190–260	—	150–180	130	180	180	130
Polyester SMC	—	190–230	—	150–205	130	130	130	130
Polyimide	—	250–350	—	260	240	210	210	
Silicone	—	>500	—	260–>300	105–260	105–260	105–240	105
Urea	—	130–145	—	75	100	100	100	100

[a] Absolute values are a function of thickness and grade.

Table 3.1.12

TABLE 3.1.13 Properties of high-temperature polymers

Property	*Units*	*Polyester PBT*	*Polyethersulphone*		*Polyamide-Imide*	*Polysulphones*		*Polyaryl-sulphone*	*Polyphenylene sulphide*			*Polyimide*	
			Unfilled	*20–30%g.f.*		*Unfilled*	*30% glass*		*Unfilled*	*40% glass*	*40% glass*[a]	*Unfilled*	*50% glass*
Tensile strength 20°C	MN/m^2	96.5	84	124–140	185	60–74	117	90	75	150	83	86	190
Tensile strength 150°C	MN/m^2		55	78–100	105		15	60		48			
Tensile strength 260°C	MN/m^2	21	0	0	52			22	14	20		42	154
Elongation at 20°C	%	8	40–80	3	12	50–100		13	3	3	1	5	
Flexural strength at 20°C	MN/m^2	117	129	172–190	182	106	172	119	140	260		131	147
Flexural strength at 260°C	MN/m^2	18			76			62				58	
Flexural modulus at 20°C	GN/m^2	4.8	2.6	5.9–8.4	4.6	2.7	8.3	2.72	4.1	15.2	15.2	3.1	13.65
Flexural modulus at 150°C	GN/m^2		2.5	5.8–8.2									
Flexural modulus at 260°C	GN/m^2				3.0			1.8		1.4		1.8	9.8
Impact strength IZOD	J/cm notch	0.5	0.8	0.8	0.6	0.7	0.96	0.8	0.2	0.8		0.8	3.0
Heat distortion temperature 1.8 MN/m^2	°C	293	203	210–216	274	174	185	260	137	260		360	349
Coefficient of linear expansion	$\times 10^{-5} K^{-1}$	2.9	5.5	2.3–2.6	3.6	5.6		4.70	4.9	4		5.6	1.3
Thermal conductivity	W/mK	0.98	0.17	0.33	0.24	0.26		0.22	0.29	0.32–0.45		0.35	0.36
Max. cont. service temperature	°C	188	200		210	150	177	260		260		260	>260
Dielectric constant 10^3 Hz	—	3.2	3.5	3.8	3.5	3.1	3.6	3.8	3.1	3.8		3.6	4.5
Power factor 10 Hz	—	0.01	0.004		0.001	0.001	0.0015	0.003	0.0005	0.001		0.0018	0.017
Dielectric strength	10^2 kV/m	138	160	160	236			190	150	177		224	177
Specific gravity	—	1.4	1.37	1.51–1.60	1.4	1.24	1.41	1.36	1.3	1.64		1.43	1.7
Water absorption %	—	0.01	0.43	0.28–0.35	0.28	0.3	0.22	1.1	<0.02	0.01		0.24	0.2–0.3
Remarks		Ekkcel 1200 Carborundum Co.) Linear accurate heat stabilised grade. Most grades of PBT have max. service temperatures around 120°C (unfilled) or 180°C (30% glass).	ICI 300P	ICI 420/430P	Amoco Torlon 4203			Astrel 360 (3M Co)	Ryton R4 (Phillips Petroleum)	Ryton R4 (Phillips Petroleum)	[a]Aged at 232°C for 4 months	Vespel SP-1 (Du Pont)	Kinel 5514 (Rhodia)

Table 3.1.13

TABLE 3.1.14 General environmental performance of plastics

Material		Aliphatic hydrocarbons	Aromatic hydrocarbons	Oils fats waxes	Fully halogenated hydrocarbons	Partially halogenated hydrocarbons
Acrylics		Fair	Poor	Good	Poor	Poor
Cellulosics		Fair	Variable	Good	Fair to Poor	Fair to Poor
Chlorinated polyether		Good	Good	Good	Fair	Fair
Fluoroplastics	PTFE	Excellent	Excellent	Excellent	Excellent	Good
	FEP	Excellent	Excellent	Excellent	Excellent	Good
	PFA	Excellent	Excellent	Excellent	Excellent	Good
	PVDF	Excellent	Good	Excellent	Good	Fair/Good
	PTFCE	Excellent	Fair	Good	Good	Good
Polyacetals[a]		Good	Variable	Good	Fair/Good	Good
Nylons (polyamides)		Good	Fair	Good	Fair	Poor
Polyamide-imide		Good	Good	Good	Good	Good
Polycarbonate		Fair	Poor	Poor/Fair	Poor	Poor
Thermoplastic polyesters		Good	Poor/good at RT	Good	Poor/Good	Poor
Polyethylene		Fair	Fair/poor	Excellent	Poor	Poor
Polyolefin copolymers		Good/poor	Poor/fair	Good	Poor	Poor
Polyphenylene oxide (mod.)			Poor	Variable	Poor	Poor
Polyphenylene sulphide		Excellent	Good	Good	Good	Good
Polypropylene		Fair	Generally Good	Good	Good at RT	Good at RT
Polysulphones	Polysulphone	Poor	Poor	Good	Poor	Poor
	Polyethersulphone	Good	Variable	Good	Poor	Poor
	Polyarylsulphone	Fair	Poor	Good	Poor[c]	Poor[c]
	Polyphenylsulphone	Good	Good		Poor	Poor
Polyurethanes		Fair	Fair to Poor	Good	Poor	Poor
Polyvinyl chloride		Good	Poor	Good	Poor	Poor
Vinyl esters		Good at RT	Good at RT		Good at RT	Good at RT
Styrene-based polymers	ABS	Fair	Poor	Good	Poor	Poor
	PS	Fair/poor	Poor	Mainly poor	Poor	Good
	SAN	Good	Poor	Good	Poor	Poor
TPX			Poor	Good	Poor	Poor
Polybutylene		Good at RT	Fair/poor		Poor	Poor
Polyesters and Alkyds		Good	Fair to poor	Good	Poor	Poor
Aminos	Melamine formaldehyde	Good	Good/exc.	Fair to good	Excellent	Excellent
	Urea formaldehyde	Good	Good	Good	Excellent	Excellent
Diallyl phthalate			Good	Good	Poor	Poor
Epoxy resins		Excellent	Good to exc.	Excellent	Excellent	Good
Furan		Good	Good	Good	Excellent	Excellent
Phenolics		Excellent	Excellent	Excellent	Excellent	Good
Polybutadiene		Good	Good	Good	Good	Good
Polyimides		Excellent	Excellent	Excellent	Excellent	Excellent
Silicones		Fair	Fair to poor	Excellent	Poor	Poor

[a] Wide variations exist since copolymers are more inert than homopolymers.
[b] Stress-cracks.
[c] Freons have little effect.
RT = room temperature.

Table 3.1.14

TABLE 3.1.14 General environmental performance of plastics—*continued*

Material		*Alcohols monohydric*	*Alcohols polyhydric*	*Phenols*	*Ketones*	*Esters*	*Ethers*
Acrylics		Poor to fair	Good	Poor	Poor	Poor	Poor
Cellulosics		Fair to poor	Excellent	Poor	Poor	Poor	Fair
Chlorinated polyether		Good	Excellent	Good	Fair/Poor	Fair	Good
Fluoroplastics	PTFE	Excellent	Excellent	Excellent	Excellent	Excellent	Good
	FEP	Excellent	Excellent	Excellent	Excellent	Excellent	Good
	PFA	Excellent	Excellent	Excellent	Excellent	Excellent	Good
	PVDF	Excellent	Excellent	Fair	Fair	Fair	Good
	PTFCE	Excellent	Excellent	Good	Fair	Fair	Poor
Polyacetals[a]		Good	Good	Poor	Good	Good	Good
Nylons (polyamides)		Good	Good	Poor	Good	Fair	Good
Polyamide-imide		Good	Good	Poor?	Good	Good	Good
Polycarbonate		Mainly good	Good	Poor	Poor	Poor	Poor
Thermoplastic polyesters		Good/exc.	Excellent	Poor	Good	Good	Good
Polyethylene		Good to fair	Good	Excellent	Good	Good	Good
Polyolefin copolymers		Good	Good		Poor	Good at RT	Poor
Polyphenylene oxide (mod.)		Excellent	Excellent		Poor		
Polyphenylene sulphide		Excellent	Excellent	Good	Good	Excellent	Good
Polypropylene		Good	Excellent	Good	Good	Good	Fair
Polysulphones	Polysulphone	Good to fair	Good	Poor	Poor	Poor	Poor
	Polyethersulphone	Good			Poor	Poor	Poor
	Polyarylsulphone				Poor[b]	Poor	
	Polyphenylsulphone				Poor	Poor	
Polyurethanes		Fair to poor	Fair to good	Poor	Fair to poor	Poor	Poor
Polyvinyl chloride		Good	Good	Poor	Poor	Poor	Poor
Vinyl esters					Poor	Poor	
Styrene-based polymers	ABS	Fair	Good	Poor	Poor	Poor	Poor
	PS	Good	Good	Poor	Poor	Poor	Poor
	SAN	Fair	Good	Poor	Poor	Poor	Poor
TPX					Fair to poor		
Polybutylene					Good at RT	Good at RT	
Polyesters and Alkyds		Fair	Fair	Poor	Poor	Poor	Good
Aminos	Melamine formaldehyde	Excellent	Excellent	Good	Good	Good	Excellent
	Urea formaldehyde	Good to exc.	Excellent	Good	Good	Fair to exc.	Excellent
Diallyl phthalate		Fair			Poor		
Epoxy resins		Excellent	Excellent	Fair	Fair	Good	Fair
Furan		Excellent	Excellent	Good	Good	Good	Excellent
Phenolics		Good	Excellent	Excellent	Good	Good	Excellent
Polybutadiene		Excellent	Excellent	Good	Good	Good	Good
Polyimides		Excellent	Excellent	Good	Excellent	Excellent	Excellent
Silicones		Good	Excellent	Poor	Good	Fair	Fair

[a] Wide variations exist since copolymers are more inert than homopolymers.
[b] Stress-cracks.
[c] Freons have little effect.
RT = room temperature.

Table 3.1.14—*continued*

TABLE 3.1.14 General environmental performance of plastics—*continued*

Material		*Inorganic acids*		*Bases*		*Salts*		
		Conc.	*Dilute*	*Conc.*	*Dilute*	*Acid*	*Neutral*	*Basic*
Acrylics		Poor to good	Good	Good	Good	Good	Excellent	Good
Cellulosics		Poor	Fair	Poor	Fair	Good	Good	Fair
Chlorinated polyether		Fair	Good	Fair	Good	Good	Good	Good
Fluoroplastics	PTFE	Excellent	Excellent	Excellent	Excellent	Excellent	Excellent	Excellent
	FEP	Excellent	Excellent	Excellent	Excellent	Excellent	Excellent	Excellent
	PFA	Excellent	Excellent	Excellent	Excellent	Excellent	Excellent	Excellent
	PVDF	Excellent	Excellent	Excellent	Excellent	Excellent	Excellent	Excellent
	PTFCE	Excellent	Excellent	Excellent	Excellent	Excellent	Excellent	Excellent
Polyacetals[a]		Poor	Poor	Fair to poor	Fair to poor	Fair	Good	Good
Nylons (polyamides)		Poor	Good	Good	Excellent	Poor	Good	Fair
Polyamide-imide		Good	Good	Poor	Poor	Good	Good	Poor
Polycarbonate		Fair	Good	Poor	Poor	Good	Excellent	Fair
Thermoplastic polyesters		Good/fair	Good	Poor	Poor/fair	Good	Good	Poor/fair
Polyethylene		Fair to good	Excellent	Good	Good	Excellent	Excellent	Excellent
Polyolefin copolymers		Fair to poor	Good at RT Variable at elevated temperature					
Polyphenylene oxide (mod.)		Fair	Good	Excellent	Excellent	Good	Good	Good
Polyphenylene sulphide		Good	Good	Good	Excellent	Good	Good	Good
Polypropylene		Excellent	Excellent	Excellent	Excellent	Excellent	Excellent	Excellent
Polysulphones	Polysulphone	Good	Good	Good	Excellent	Good	Excellent	Good
	Polyethersulphone	Variable	Good	Good	Excellent	Good	Good	Fair
	Polyarylsulphone		Good		Excellent			
	Polyphenylsulphone		Excellent		Excellent			
Polyurethanes		Poor	Fair	Poor	Fair	Good	Excellent	Excellent
Polyvinyl chloride		Good	Excellent	Excellent	Excellent	Excellent	Excellent	Excellent
Vinyl esters		Good at RT	Good at RT	Good at RT	Good at RT	Good at RT	Good at RT	Good at RT
Styrene-based polymers	ABS	Poor	Good	Good	Good	Good	Excellent	Good
	PS	Poor to fair	Good	Fair	Excellent	Good	Excellent	Good
	SAN	Good	Good	Good	Excellent	Good	Excellent	Good
TPX		Good	Good	Good	Good	Good	Good	Good
Polybutylene			Good at RT		Good at RT	Good at RT	Good at RT	Good at RT
Polyesters and Alkyds		Good	Poor	Fair	Good	Excellent	Excellent	Good
Aminos	Melamine formaldehyde	Poor	Good to fair	Poor	Good	Excellent	Excellent	Good
	Urea formaldehyde	Poor	Fair	Poor	Fair	Good	Good	Good
Diallyl phthalate		Fair	Good	Fair	Good			Poor
Epoxy resins		Fair	Excellent	Excellent	Excellent	Excellent	Excellent	Excellent
Furan		Good	Good	Good	Good	Excellent	Excellent	Good
Phenolics		Fair to poor	Fair to good	Poor	Poor	Excellent	Excellent	Fair
Polybutadiene		Fair	Excellent	Good	Good	Excellent	Excellent	Good
Polyimides		Poor	Fair	Poor	Fair	Good	Good	Poor
Silicones		Depends on acid	Good	Good	Good	Excellent	Excellent	Excellent

[a] Wide variations exist since copolymers are more inert than homopolymers.
[b] Withstands prolonged immersion in boiling water.
[c] CPVC commonly used for hot-water pipes.
[d] Highly swollen by hot acetic acid.
RT = room temperature.

Table 3.1.14—*continued*

TABLE 3.1.14 General environmental performance of plastics—*continued*

Material		Organic acids		Oxidising acids		Sunlight and weathering	Fungus and bacteria
		Conc.	Dilute	Conc.	Dilute		
Acrylics		Poor	Fair to poor	Poor	Good	Excellent	
Cellulosics		Poor	Poor	Poor	Poor	Some grades OK	
Chlorinated polyether		Fair	Good	Poor	Fair	Fair	
Fluoroplastics	PTFE	Good/exc.	Excellent	Excellent	Excellent	Excellent	Excellent
	FEP	Good/exc.	Excellent	Excellent	Excellent	Excellent	Excellent
	PFA	Excellent	Excellent	Excellent	Excellent	Excellent	Excellent
	PVDF	Fair	Excellent	Good	Excellent	Good	Excellent
	PTFCE	Excellent	Excellent	Excellent	Excellent	Excellent	Excellent
Polyacetals[a]		Poor	Fair	Poor	Poor	Generally good	
Nylons (polyamides)		Poor	Fair	Poor	Poor	Fair to good	Very resistant
Polyamide-imide		Generally good	Generally good	Good	Good	Good	
Polycarbonate		Fair	Fair	Poor	Good	Good	
Thermoplastic polyesters		Poor/fair	Good	Poor	Fair/poor	Some good	
Polyethylene		Excellent	Excellent	Poor	Good	Poor	Resistant
Polyolefin copolymers		Generally good	Generally good	Good	Good	Good (esp. EVA)	
Polyphenylene oxide (mod.)						Poor	
Polyphenylene sulphide		Good	Good	Poor	Fair	Good	
Polypropylene		Good	Excellent	Poor	Good	Poor	
Polysulphones	Polysulphone	Fair	Good	Poor	Good	Good to exc.	
	Polyethersulphone	Fair	Fair	Good	Good	Not rec. without pigment	
	Polyarylsulphone		Exc.[d]				
	Polyphenylsulphone						
Polyurethanes		Poor	Fair	Poor	Poor	Good	
Polyvinyl chloride		Poor	Good	Good	Excellent	Good	Poor
Vinyl esters							
Styrene-based polymers	ABS	Poor	Fair to poor	Poor	Good	Fair to good	
	PS	Poor	Fair	Poor	Fair	Poor to fair	
	SAN	Fair	Good	Poor	Good	Fair	
TPX				Poor		Fair	
Polybutylene				Good at RT	Good at RT		
Polyesters and Alkyds		Good	Good	Poor	Fair	Fair to good	
Aminos	Melamine formaldehyde	Good	Good	Poor	Fair	Good	
	Urea formaldehyde	Good	Good	Poor	Fair	Good	
Diallyl phthalate		Poor to fair				Good	
Epoxy resins		Fair	Fair to good	Poor	Fair	Good	
Furan		Good	Excellent	Poor	Fair	Good	
Phenolics		Good	Fair	Poor	Poor	Good	
Polybutadiene		Fair/good	Excellent	Fair	Good	Very good	
Polyimides		Good	Excellent	Poor	Good	Excellent	Very resistant
Silicones		Fair	Good	Poor	Excellent	Excellent	

[a] Wide variations exist since copolymers are more inert than homopolymers.
[b] Withstands prolonged immersion in boiling water.
[c] CPVC commonly used for hot-water pipes.
[d] Highly swollen by hot acetic acid.
RT = room temperature.

Table 3.1.14—*continued*

TABLE 3.1.14 General environmental performance of plastics—*continued*

Material		*Compatibility with foodstuffs*	*Hot water*	*Sterilisability (medical applications)*	*Detergents*
Acrylics			Depends on grade		Excellent
Cellulosics			No	No	Fair to excellent
Chlorinated polyether			Good		Excellent
Fluoroplastics	PTFE		Excellent		Excellent
	FEP		Excellent		Excellent
	PFA		Excellent		Excellent
	PVDF	Excellent	Excellent		Excellent
	PTFCE		Excellent		Excellent
Polyacetals[a]		Not recommended	Good		Fair to excellent
Nylons (polyamides)		Resistant but stains	Some grades fair	By steam or irradiation	Fair to excellent
Polyamide-imide					
Polycarbonate		Approved grades	Degrades >65°C	No	Fair
Thermoplastic polyesters			Degrades >70°C	No	Good
Polyethylene		Non-toxic grades	HDPE has been used at 90°C	No	Fair
Polyolefin copolymers				By low level radiation or chemicals	Good
Polyphenylene oxide (mod.)		O.K.	Excellent	Excellent	Excellent
Polyphenylene sulphide		Good	Excellent		
Polypropylene		Non-toxic grades	Excellent	Steam sterilised	Excellent
Polysulphones	Polysulphone	Approved grades	Good[b]	Repeatedly steam sterilisable	Good
	Polyethersulphone				Good
	Polyarylsulphone				
	Polyphenylsulphone		Excellent[b]		
Polyurethanes		Restricted applications	Not recommended		Fair to poor
Polyvinyl chloride		Special grades. Some concern exists.	Not recommended[c]		Moderate to excellent
Vinyl esters					
Styrene-based polymers	ABSS	Special grades only		No	Excellent
	PS	O.K.	No	No	Great variation
	SAN	O.K.	No	No	Good to excellent
TPX		Yes	Yes	Yes	Good
Polybutylene					
Polyesters and Alkyds			Good		Good
Aminos	Melamine formaldehyde	O.K.	Some effect on properties		Good
	Urea formaldehyde	O.K.	Some effect on properties		Very good
Diallyl phthalate					Good
Epoxy resins		Fair	Fair		Good
Furan					
Phenolics					Fair to good
Polybutadiene					
Polyimides			Excellent		Good
Silicones					

[a] Wide variations exist since copolymers are more inert than homopolymers.
[b] Withstands prolonged immersion in boiling water.
[c] CPVC commonly used for hot-water pipes.
[d] Highly swollen by hot acetic acid.

Table 3.1.14—*continued*

TABLE 3.1.15 Abrasion resistance of plastics

ASTM D 1044 (Taber CS-17 test)	*mg/1000 cycles*
Thermoplastic polyurethane	0.08–0.6
Ionomer	2.4
Nylon 6/10	3.2
Nylon 11	5
Nylon 6/12	5–7
Nylon 6/6	2–12
Nylon 6/6; 30% glass	7–36
Polyethersulphone	6
Polyethersulphone (glass filled)	8
Polycarbonate	9–10
Polyacetal copolymer	14
Polyacetal homopolymer	14–20
Polyacetal terpolymer	14
Polyacetal homopolymer; 22% PTFE	9
Polyacetal homopolymer; 20% glass	33
Polyacetal copolymer; 20% glass	40
Polypropylene	13–16
Polypropylene; talc/asbestos reinforced	20–23
PVC, high impact	18
PBT	6–21
PBT (glass filled)	40
Acrylic	12–40
PPO	20
PPO (glass filled)	35
PVDF	24
Natural rubber	29
Phenolic	24–96
SBR	36
ABS	60–75
Cellulose acetate	65
Cellulose acetate butyrate	55–120
Polystyrene	60–200

N.B. Resistance to abrasion is an important factor in many applications of plastics, including transparent plastics. The principal limitation of this test is the poor reproducibility. Lab to lab variation is significant although intra-lab data has been fairly good.

Further information on the selection and performance of polymeric materials for abrasive wear resistance is given in Vol. 1, Chapter 1.5

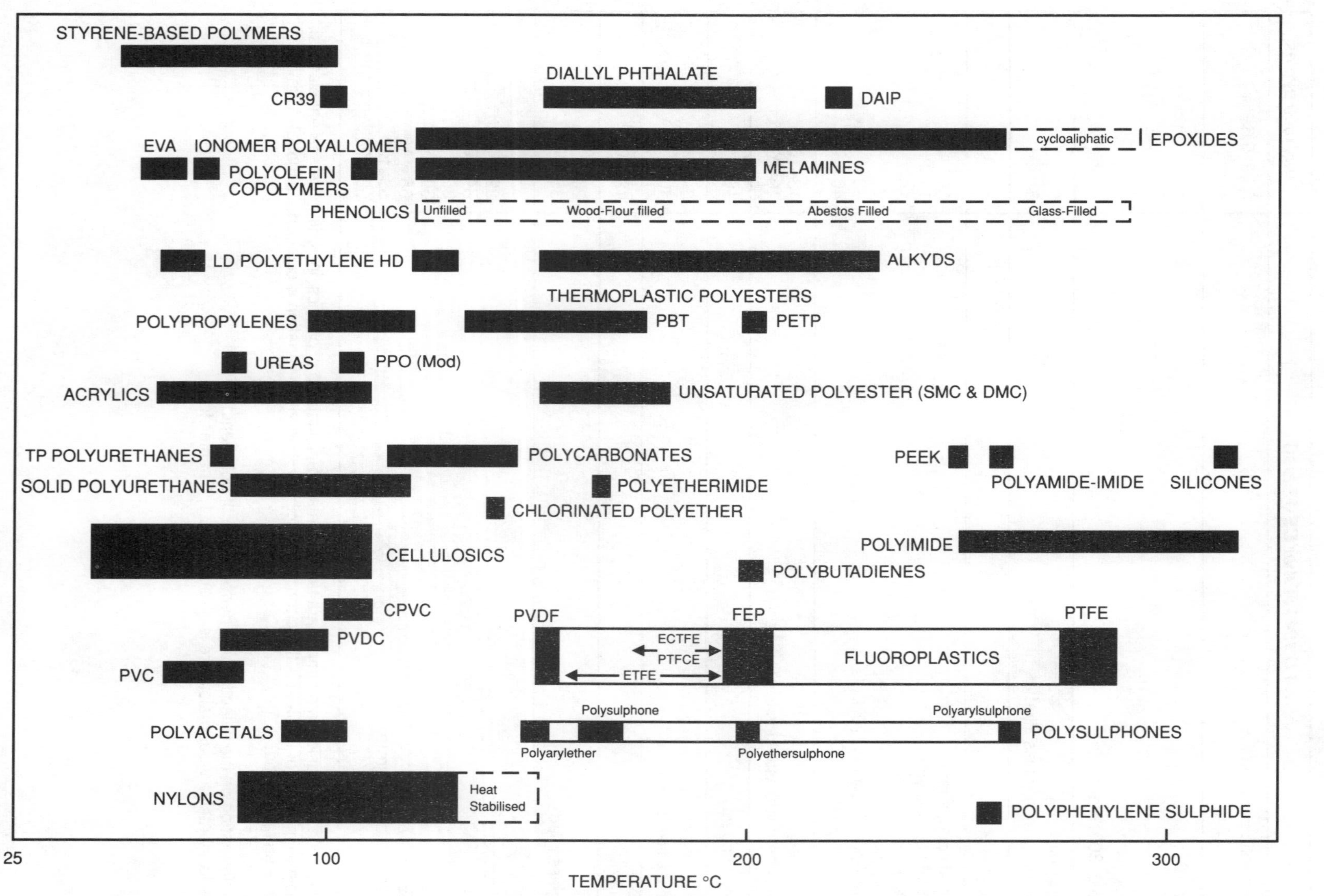

1. In many cases short-term exposure to higher temperatures may be tolerated (consult specific materials or manufacturers). Figures relate to long-term continuous exposure.
2. Figures relate to continuous exposure in AIR; for other media see also FIG 3.1.23

FIG 3.1.21 Recommended maximum service temperatures (no load)

Fig 3.1.21

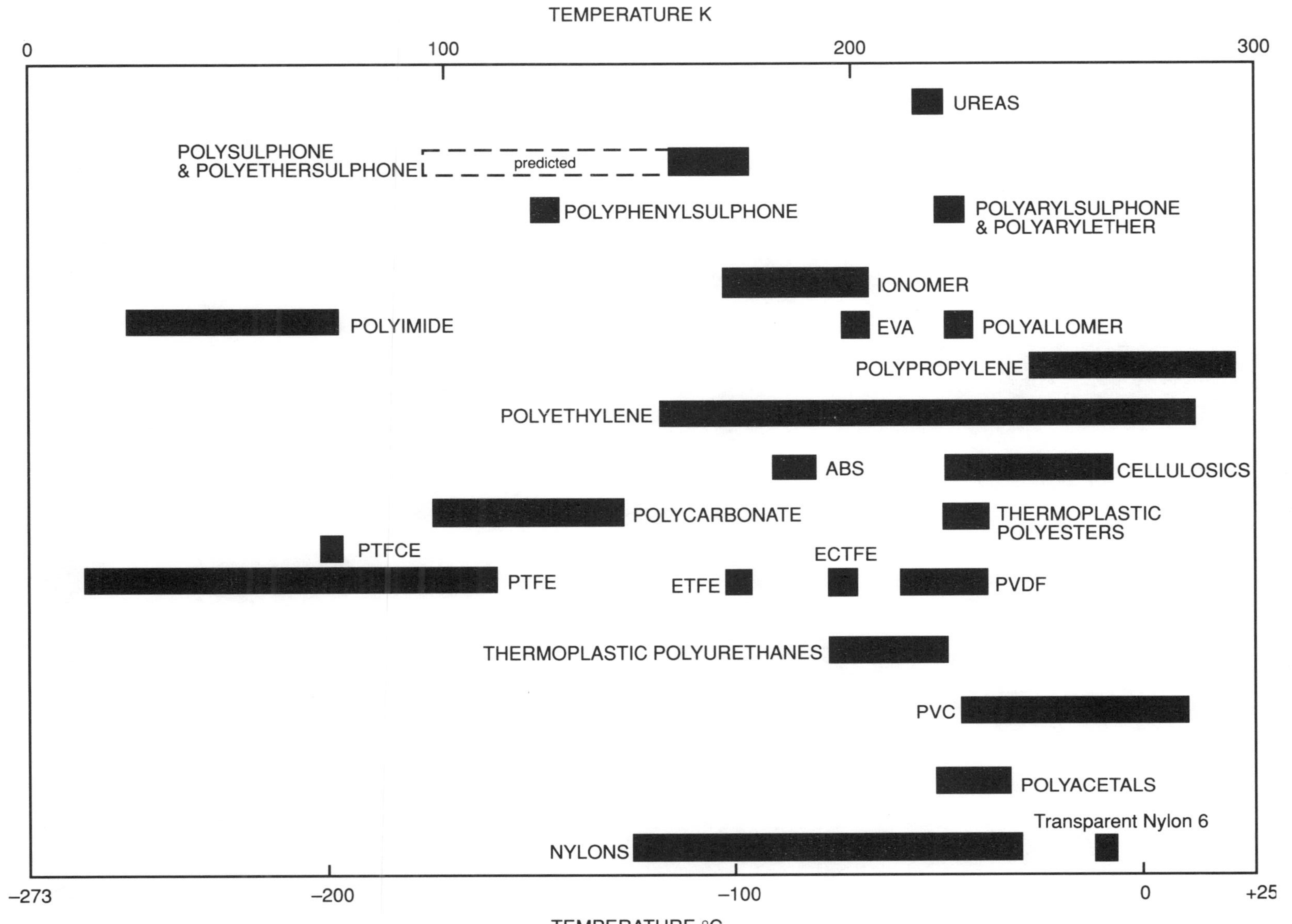

FIG 3.1.22 Recommended minimum operating temperatures

Fig 3.1.22

FIG 3.1.23 The resistance of plastics to a range of solvents

Key to Fig 3.1.23

Resistance of materials to solvents

Resistant		Weight increase	< 3%	on long-term exposure
	or	Weight loss	< 0.5%	
	and/or	Decrease in UTS	< 15%	
Moderate resistance		3% < Weight increase	< 8%	on long-term exposure
	or	0.5% < Weight loss	< 3%	
	and/or	15% < Decrease in UTS	< 30%	
Not resistant		Weight increase	> 8%	
	or	Weight loss	> 3%	
	and/or	Decrease in UTS	> 30%	

Fully shaded entries imply that the material performs as described at all temperatures between room temperature and its normal maximum service temperature.

Susceptible to stress corrosion cracking but may be suitable for non-load bearing applications depending upon shading of border.

60

Figures relate to temperatures in degrees Celsius unless otherwise indicated. In this example the material is suitable at room temperature; unsuitable above 60°C.

60 Resistant

60 Moderate resistance

60 Not resistant

This does not mean that the material performance is guaranteed at intermediate temperatures.

20

A numerical entry with no border indicates that data for higher temperatures in not available.

Blank entries indicate absence of data.

Manufacturers advice should be obtained prior to the selection of a material coded

Fig 3.1.23a lists materials according to number in Fig 3.1.23.

Fig 3.1.23b lists solvents according to number in Fig 3.1.23.

Fig 3.1.23c lists solvents alphabetically with corresponding number in Fig 3.1.23.

Fig 3.1.23d List of abbreviations.

Fig 3.1.23e Special notes.

Fig 3.1.23

THERMOPLASTICS

SOLUTES \ SOLVENTS		1	2	3	4	5	6	7	8	9	10	11	12	13	14	15	16	17	18	19	20
1. ACRYLICS					60							60			60						
2. CELLULOSICS	2.1																			TX	
	2.2																				
	2.3								ME												
	2.4																				
3. CHLORINATED POLYETHER		100															20		20	20	
4. FLUOROPLASTICS	4.1																				
	4.2																				
	4.3				•																
	4.4		100														70	20	20		
	4.5	60	100		100	50				60						100	60				
	4.6	100															•		•		
	4.7		100	100																	
5. POLYACETALS	5.1					60														X^{B}_{T}	
	5.2								A								60				
	5.3	AS 5.1 EXCEPT WHERE OTHERWISE INDICATED																			
6. POLYAMIDES	6.1								M											$T^{B}X$	
	6.2											60			60					$T^{B}X$	
	6.3																			$T^{B}X$	
	6.4								BE	A BE					60					$T^{B}X$	
	6.5									A BE										$T^{B}X$	
	6.7									A BE										^{B}X	

FIG 3.1.23 The resistance of plastics to a range of solvents—*continued*

Fig 3.1.23 (1)—***continued***

THERMOPLASTICS

SOLUTES \ SOLVENTS		1	2	3	4	5	6	7	8	9	10	11	12	13	14	15	16	17	18	19	20
7. POLYAMIDE-IMIDE																					
8. POLYCARBONATES				60			20				20	20		20		20		20			20
9. POLYESTERS, THERMOPLASTIC	9.1			80					80	50					60					B	
	9.2			60					P E M	E										B X T	
10. POLYETHYLENE HIGH & LOW DENSITY		60	60			L H •	H	H	L H •	L H										L H	
11. POLYOLEFIN COPOLYMERS	11.1																70				
	11.2																				
	11.3																				
12. POLYPHEN. OXIDE MOD.																					
13. POLYPHEN. SULPHIDE										A B E											
14. POLYPROPYLENE			60		60				60	20							60	60	60	T X	
15. POLYSULPHONES	15.1					5% H_2O			B E M												
	15.2									A E										B X T	
	15.3																				
	15.4																				
16. POLYURETHANES, THERMOPLASTIC	16.1								SEE SECTION 46.1												
	16.2								SEE SECTION 46.2												
17. POLYVINYL CHLORIDE	17.1					AS RIGID PVC EXCEPT WHERE OTHERWISE INDICATED															
	17.2		20						A E B									60	60		60
	17.3								60										60		
	17.4		60	60	50	60			50	E A B									50		

FIG 3.1.23 The resistance of plastics to a range of solvents—*continued*

Fig 3.1.23 (2)—*continued*

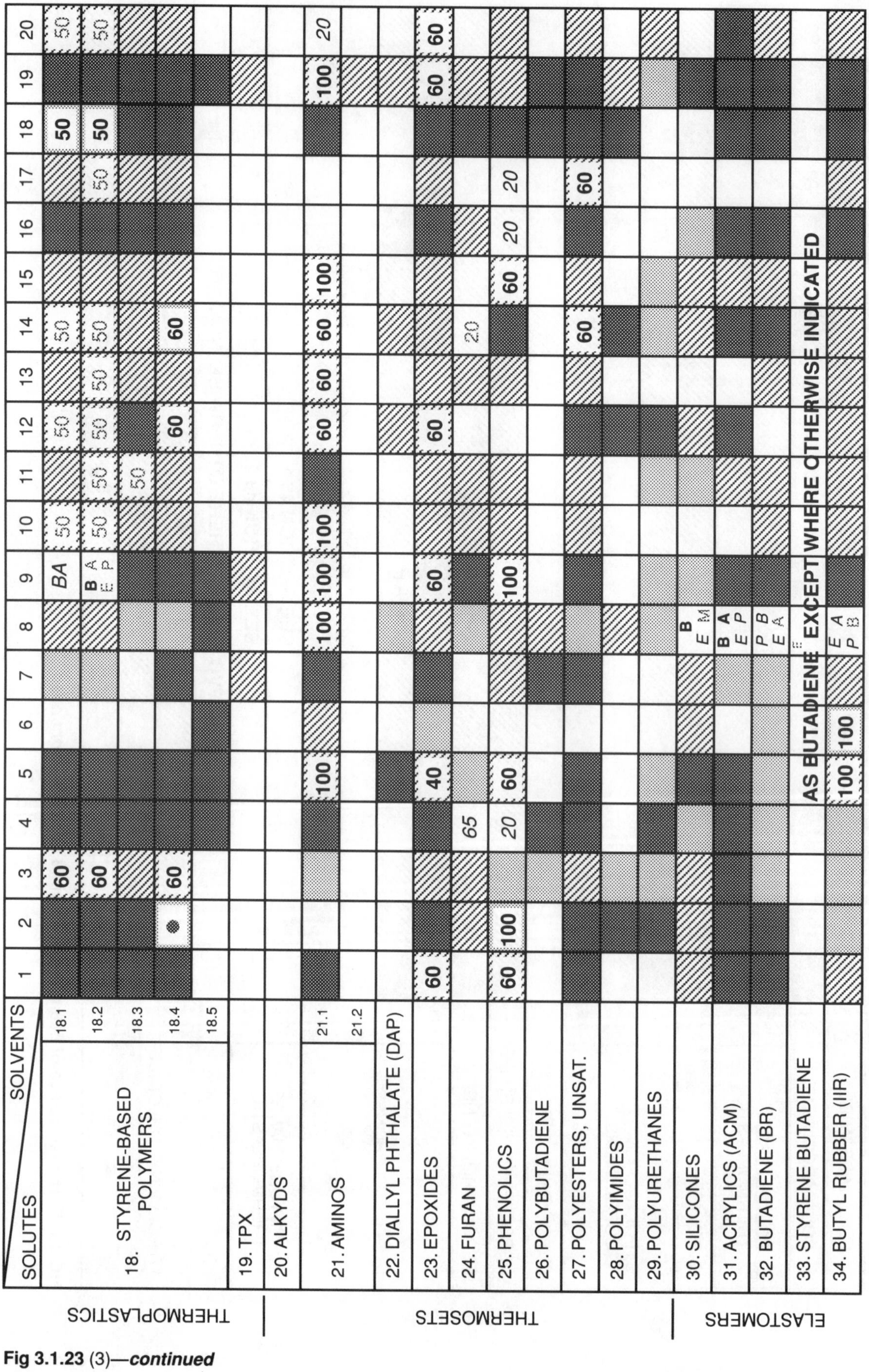

	SOLUTES \ SOLVENTS		1	2	3	4	5	6	7	8	9	10	11	12	13	14	15	16	17	18	19	20
THERMOPLASTICS	18. STYRENE-BASED POLYMERS	18.1			60						BA	50		50		50				50		50
		18.2			60						B A E P	50	50	50	50	50			50	50		50
		18.3											50									
		18.4		●	60									60		60						
		18.5																				
	19. TPX																					
THERMOSETS	20. ALKYDS																					
	21. AMINOS	21.1					100			100	100	100		60	60	60	100				100	20
		21.2																				
	22. DIALLYL PHTHALATE (DAP)																					
	23. EPOXIDES		60				40				60			60							60	60
	24. FURAN					65										20						
	25. PHENOLICS		60	100		20	60				100						60	20	20			
	26. POLYBUTADIENE																					
	27. POLYESTERS, UNSAT.															60			60			
	28. POLYIMIDES																					
	29. POLYURETHANES																					
ELASTOMERS	30. SILICONES									B E M												
	31. ACRYLICS (ACM)									B A E P												
	32. BUTADIENE (BR)									P B E A												
	33. STYRENE BUTADIENE						AS BUTADIENE EXCEPT WHERE OTHERWISE INDICATED			E												
	34. BUTYL RUBBER (IIR)						100	100		E A P B												

FIG 3.1.23 The resistance of plastics to a range of solvents—*continued*

Fig 3.1.23 (3)—***continued***

	SOLUTES \ SOLVENTS		1	2	3	4	5	6	7	8	9	10	11	12	13	14	15	16	17	18	19	20
ELASTOMERS	35. CHLOROSULPHONATED POLYETHYLENE			60			20			E				60				50%		60		
ELASTOMERS	36. EBONITE					100																
ELASTOMERS	37. EPICHLOROHYDRIN (CO, ECO)									E P A B												
ELASTOMERS	38. ETHYLENE PROPYLENE										A B E											
ELASTOMERS	39. FLUOROELASTOMERS (FKM)									B E P A												
ELASTOMERS	40. FLUOROSILICONE																				B T X	
ELASTOMERS	41. SILICONE									A B E P	E A B		100									
ELASTOMERS	42. ISOPRENE (NATURAL R.)									B E P A												100
ELASTOMERS	43. NITRILE RUBBER (NBR)			20	100			20	60					100		100					60	
ELASTOMERS	44. POLYSULPHIDE									A B P E	A B E										T B X	
ELASTOMERS	45. CHLOROPRENE (NEOPRENE)			20	100																	
ELASTOMERS	46. URETHANES	46.1								E A B P												
ELASTOMERS	46. URETHANES	46.2		AS 46.1 BUT WITH BETTER HYDROLYTIC STABILITY FOR SOLVENT RESISTANCE 46.1 (POLYETHER) USUALLY PREFERRED																			
THERMOPLASTIC ELASTOMERS	47. THERMOPLASTIC ELASTOMERS	47.1																				
THERMOPLASTIC ELASTOMERS	47. THERMOPLASTIC ELASTOMERS	47.2								ME	A E											
THERMOPLASTIC ELASTOMERS	47. THERMOPLASTIC ELASTOMERS	47.3																				
THERMOPLASTIC ELASTOMERS	47. THERMOPLASTIC ELASTOMERS	47.4																				

FIG 3.1.23 The resistance of plastics to a range of solvents—*continued*

Fig 3.1.23 (4)—***continued***

THERMOPLASTICS

SOLUTES \ SOLVENTS		21	22	23	24	25	26	27	28	29	30	31	32	33	34	35	36	37	38	39	40
1. ACRYLICS			*20*											**60**	*20*						
2. CELLULOSICS	2.1																				
	2.2																				
	2.3																				
	2.4																				
3. CHLORINATED POLYETHER											*20*							*20*	*20*		
4. FLUOROPLASTICS	4.1																*most grades*				
	4.2																				
	4.3										●						*most grades*				
	4.4						*70*		*20*												
	4.5	*60*							*60*		**60**							**60**	**60**	*60*	
	4.6											● at **100**									
	4.7																**150**	**150**	**150**	**150**	
5. POLYACETALS	5.1		**60**								**60**							60			
	5.2				60						60										
	5.3	**AS 5.1 EXCEPT WHERE OTHERWISE INDICATED**																			
6. POLYAMIDES	6.1																				
	6.2											**K***N*				*K* N **B**					
	6.3																				
	6.4																				
	6.5																				
	6.7																				

FIG 3.1.23 The resistance of plastics to a range of solvents—*continued*

Fig 3.1.23 (5)—***continued***

THERMOPLASTICS

SOLUTES \ SOLVENTS		21	22	23	24	25	26	27	28	29	30	31	32	33	34	35	36	37	38	39	40
7. POLYAMIDE-IMIDE																					
8. POLYCARBONATES				20	80	60				20					60	20					
9. POLYESTERS, THERMOPLASTIC	9.1				100																
	9.2																				
10. POLYETHYLENE HIGH & LOW DENSITY			80H		●						20H	30%	20H	20H			20H		20H		L/60 H
11. POLYOLEFIN COPOLYMERS	11.1																				
	11.2																				
	11.3																				
12. POLYPHEN. OXIDE MOD.																					
13. POLYPHEN. SULPHIDE																			100		
14. POLYPROPYLENE									20								60	20	20		60
15. POLYSULPHONES	15.1																				
	15.2																				
	15.3																				
	15.4																				
16. POLYURETHANES, THERMOPLASTIC	16.1								SEE SECTION 46.1												
	16.2								SEE SECTION 46.2												
17. POLYVINYL CHLORIDE	17.1					AS RIGID PVC EXCEPT WHERE OTHERWISE INDICATED															
	17.2	60											20	dilute			60			20	20%
	17.3																				
	17.4										50	NK		50							

FIG 3.1.23 The resistance of plastics to a range of solvents—*continued*

Fig 3.1.23 (6)—***continued***

	SOLUTES \ SOLVENTS		21	22	23	24	25	26	27	28	29	30	31	32	33	34	35	36	37	38	39	40
THERMOPLASTICS	18. STYRENE-BASED POLYMERS	18.1	60	60	60																	
		18.2		60									50			50	50					
		18.3										50									50	
		18.4		60																		60
		18.5																				
	19. TPX																					
THERMOSETS	20. ALKYDS																					
	21. AMINOS	21.1	60	60			20		20	60		100	60				100		100	100		
		21.2																				
	22. DIALLYL PHTHALATE (DAP)																					
	23. EPOXIDES						60				80	60	60									
	24. FURAN																	20				
	25. PHENOLICS									100		60				20						
	26. POLYBUTADIENE																					
	27. POLYESTERS UNSAT.			40							80		60					60				
	28. POLYIMIDES																					
	29. POLYURETHANES																					
ELASTOMERS	30. SILICONES																					
	31. ACRYLICS (ACM)												KN									
	32. BUTADIENE (BR)												KN									
	33. STYRENE BUTADIENE		AS BUTADIENE EXCEPT WHERE OTHERWISE INDICATED																			
	34. BUTYL RUBBER (IIR)																	60				

FIG 3.1.23 The resistance of plastics to a range of solvents—*continued*

Fig 3.1.23 (7)—***continued***

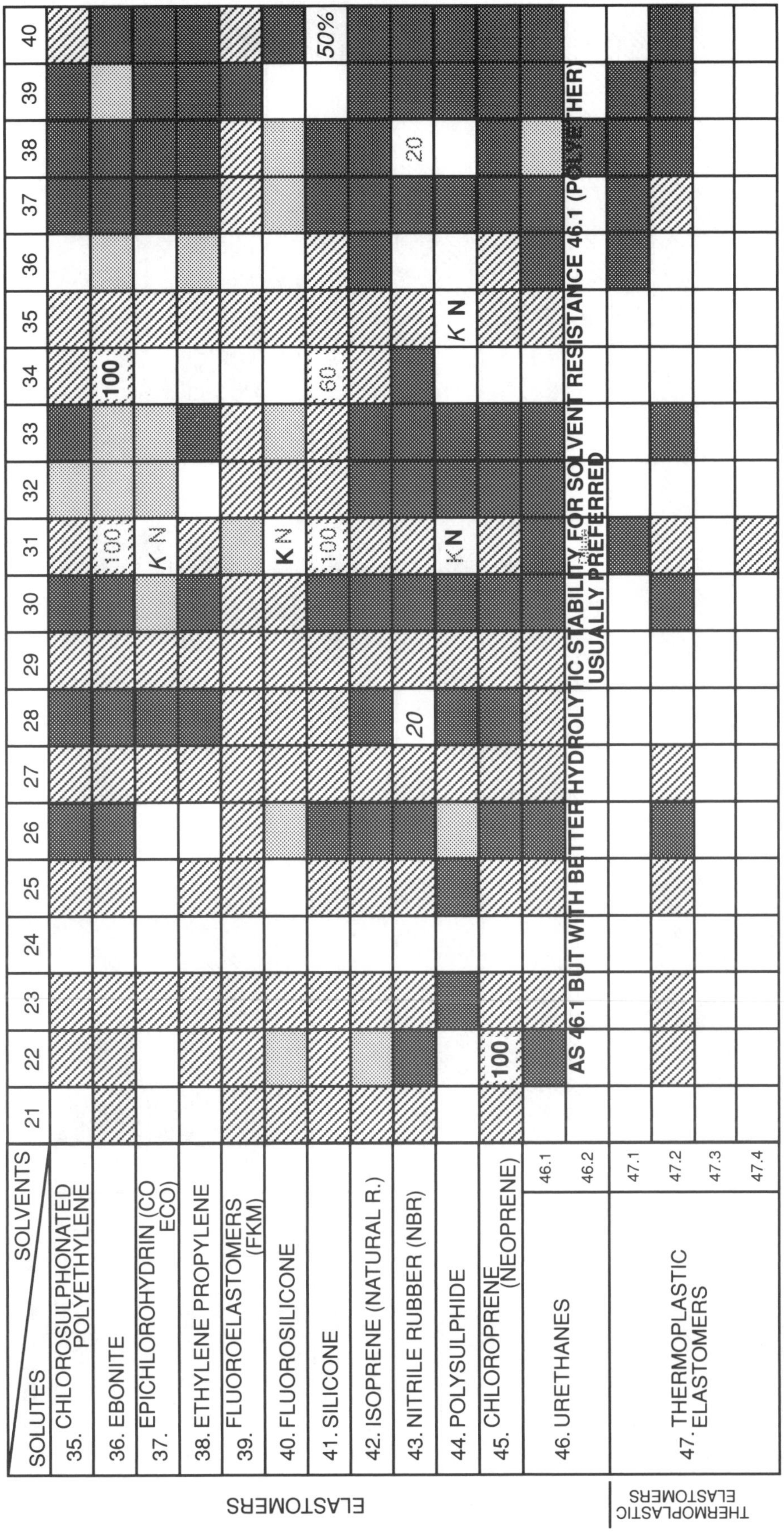

	SOLUTES \ SOLVENTS		21	22	23	24	25	26	27	28	29	30	31	32	33	34	35	36	37	38	39	40
ELASTOMERS	35. CHLOROSULPHONATED POLYETHYLENE																					
	36. EBONITE												100			100						
	37. EPICHLOROHYDRIN (CO ECO)												K N									
	38. ETHYLENE PROPYLENE																					
	39. FLUOROELASTOMERS (FKM)																					
	40. FLUOROSILICONE												K N									
	41. SILICONE												100			60						50%
	42. ISOPRENE (NATURAL R.)																					
	43. NITRILE RUBBER (NBR)									20										20		
	44. POLYSULPHIDE												K N				K N					
	45. CHLOROPRENE (NEOPRENE)			100																		
	46. URETHANES	46.1																				
		46.2		AS 46.1 BUT WITH BETTER HYDROLYTIC STABILITY FOR SOLVENT RESISTANCE 46.1 (POLYETHER) USUALLY PREFERRED																		
THERMOPLASTIC ELASTOMERS	47. THERMOPLASTIC ELASTOMERS	47.1																				
		47.2																				
		47.3																				
		47.4																				

FIG 3.1.23 The resistance of plastics to a range of solvents—*continued*

Fig 3.1.23 (8)—***continued***

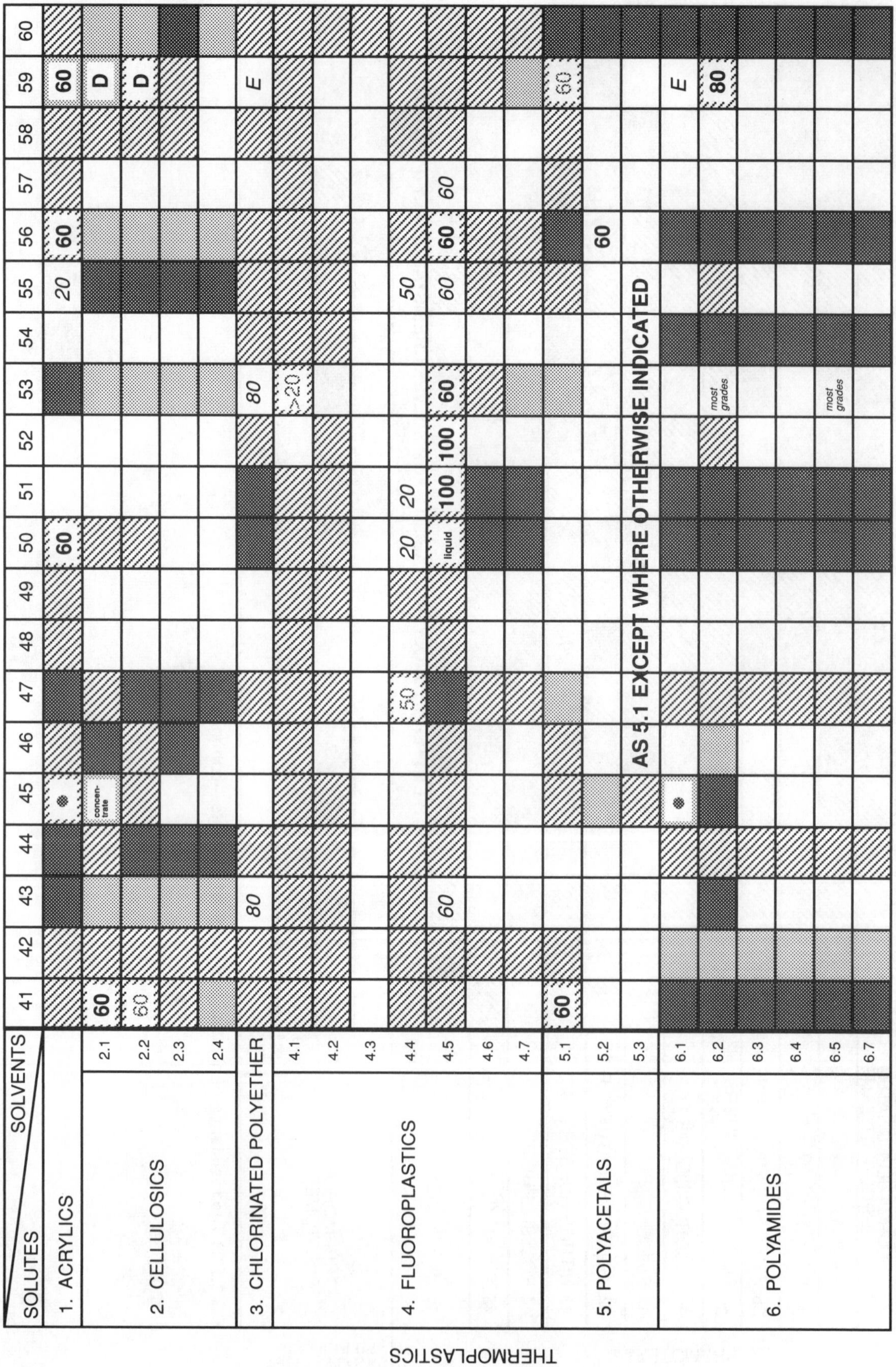

THERMOPLASTICS

SOLUTES \ SOLVENTS		41	42	43	44	45	46	47	48	49	50	51	52	53	54	55	56	57	58	59	60
1. ACRYLICS						●					**60**					*20*	**60**			**60**	
2. CELLULOSICS	2.1	**60**				concen-trate														**D**	
	2.2	60																		**D**	
	2.3																				
	2.4																				
3. CHLORINATED POLYETHER				*80*										*80*						*E*	
4. FLUOROPLASTICS	4.1													>20							
	4.2																				
	4.3																				
	4.4							50			*20*	*20*				*50*					
	4.5			*60*							liquid	**100**	**100**	**60**		*60*	**60**	*60*			
	4.6																				
	4.7																				
5. POLYACETALS	5.1	**60**																		60	
	5.2																**60**				
	5.3						**AS 5.1 EXCEPT WHERE OTHERWISE INDICATED**														
6. POLYAMIDES	6.1					●														*E*	
	6.2													*most grades*						**80**	
	6.3																				
	6.4																				
	6.5													*most grades*							
	6.7																				

FIG 3.1.23 The resistance of plastics to a range of solvents—*continued*

Fig 3.1.23 (9)—***continued***

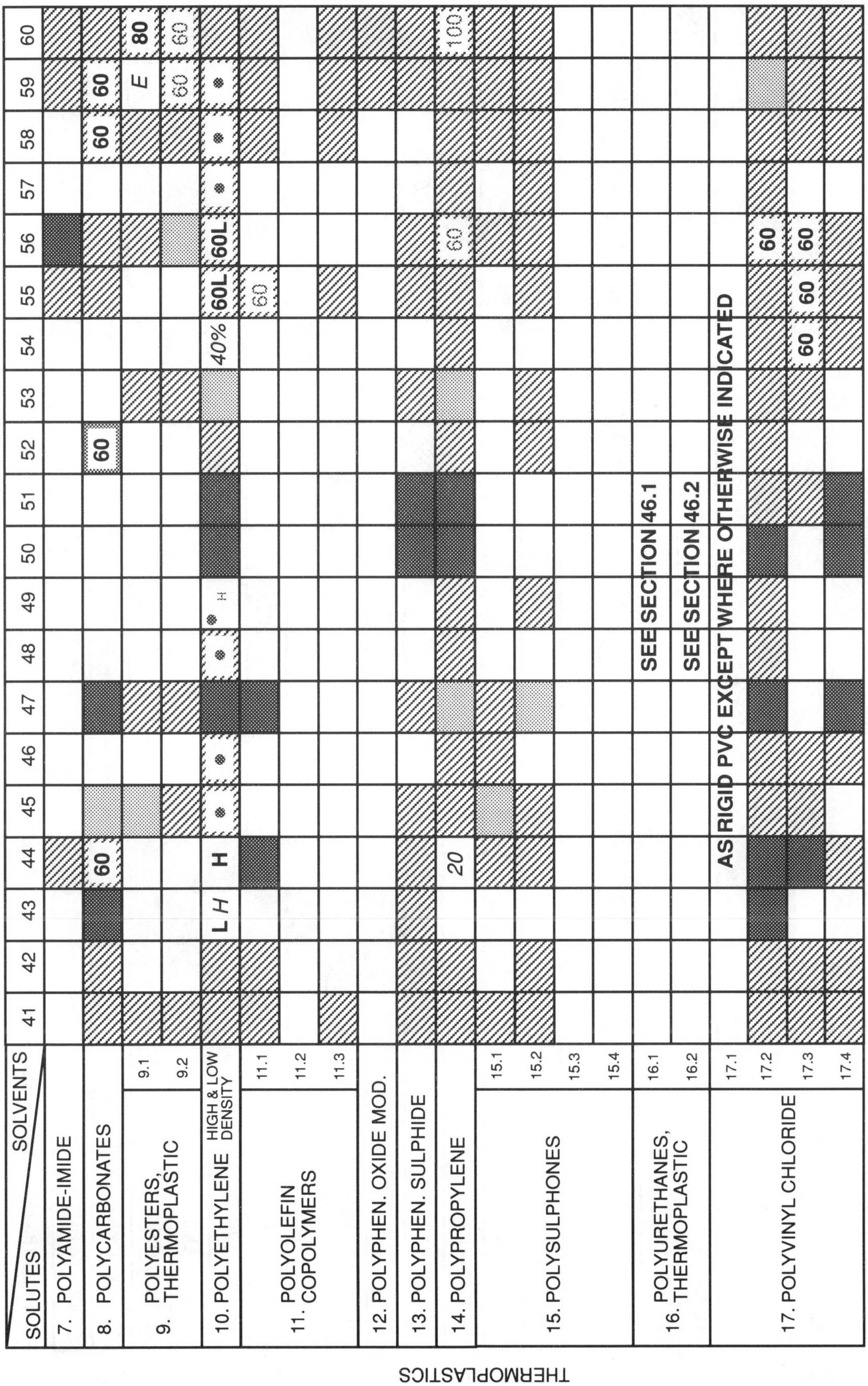

THERMOPLASTICS

SOLUTES \ SOLVENTS		41	42	43	44	45	46	47	48	49	50	51	52	53	54	55	56	57	58	59	60
7. POLYAMIDE-IMIDE																					
8. POLYCARBONATES					60								60						60	60	
9. POLYESTERS, THERMOPLASTIC	9.1																			*E*	80
	9.2																			60	60
10. POLYETHYLENE	HIGH & LOW DENSITY			L *H*	H	•	•		•	• H					*40%*	60L	60L	•	•	•	
11. POLYOLEFIN COPOLYMERS	11.1															60					
	11.2																				
	11.3																				
12. POLYPHEN. OXIDE MOD.																					
13. POLYPHEN. SULPHIDE																					
14. POLYPROPYLENE					*20*												60				100
15. POLYSULPHONES	15.1																				
	15.2																				
	15.3																				
	15.4																				
16. POLYURETHANES, THERMOPLASTIC	16.1	SEE SECTION 46.1																			
	16.2	SEE SECTION 46.2																			
17. POLYVINYL CHLORIDE	17.1	AS RIGID PVC EXCEPT WHERE OTHERWISE INDICATED																			
	17.2																60				
	17.3														60	60	60				
	17.4																				

FIG 3.1.23 The resistance of plastics to a range of solvents—*continued*

Fig 3.1.23 (10)—***continued***

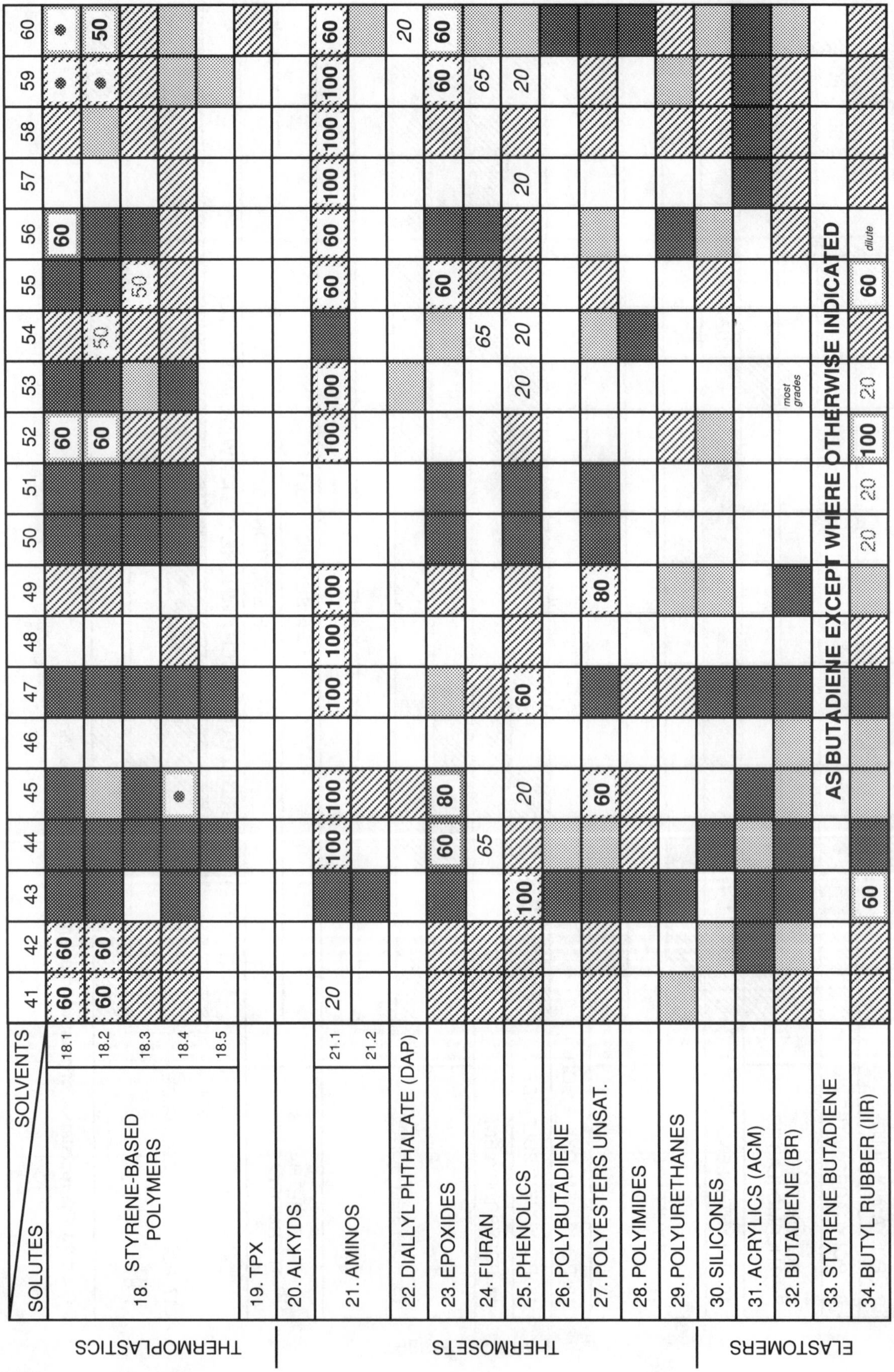

	SOLUTES \ SOLVENTS		41	42	43	44	45	46	47	48	49	50	51	52	53	54	55	56	57	58	59	60
THERMOPLASTICS	18. STYRENE-BASED POLYMERS	18.1	60	60										60				60			●	●
		18.2	60	60										60		50					●	50
		18.3															50					
		18.4					●															
		18.5																				
	19. TPX																					
THERMOSETS	20. ALKYDS																					
	21. AMINOS	21.1	20			100	100		100	100	100			100	100		60	60	100	100	100	60
		21.2																				
	22. DIALLYL PHTHALATE (DAP)																					20
	23. EPOXIDES					60	80										60				60	60
	24. FURAN					65										65					65	
	25. PHENOLICS				100		20		60						20	20			20		20	
	26. POLYBUTADIENE																					
	27. POLYESTERS UNSAT.						60				80											
	28. POLYIMIDES																					
	29. POLYURETHANES																					
ELASTOMERS	30. SILICONES																					
	31. ACRYLICS (ACM)																					
	32. BUTADIENE (BR)														most grades							
	33. STYRENE BUTADIENE					AS BUTADIENE EXCEPT WHERE OTHERWISE INDICATED																
	34. BUTYL RUBBER (IIR)				60							20	20	100	20		60	dilute				

FIG 3.1.23 The resistance of plastics to a range of solvents—*continued*

Fig 3.1.23 (11)—***continued***

	SOLUTES \ SOLVENTS		41	42	43	44	45	46	47	48	49	50	51	52	53	54	55	56	57	58	59	60
ELASTOMERS	35. CHLOROSULPHONATED POLYETHYLENE				60										20		60					100
	36. EBONITE									20	100			60					60		60	
	37. EPICHLOROHYDRIN (CO ECO)																					
	38. ETHYLENE PROPYLENE																					60
	39. FLUOROELASTOMERS (FKM)																					
	40. FLUOROSILICONE														most grades							
	41. SILICONE																	20				
	42. ISOPRENE (NATURAL R.)		100												most grades		20	20				100
	43. NITRILE RUBBER (NBR)										60						60	60			100	100
	44. POLYSULPHIDE														most grades							
	45. CHLOROPRENE (NEOPRENE)																					
	46. URETHANES	46.1																				
		46.2		AS 46.1 BUT WITH BETTER HYDROLYTIC STABILITY FOR SOLVENT RESISTANCE 46.1 (POLYETHER) USUALLY PREFERRED																		
THERMOPLASTIC ELASTOMERS	47. THERMOPLASTIC ELASTOMERS	47.1																				
		47.2																				
		47.3																				
		47.4																				

FIG 3.1.23 The resistance of plastics to a range of solvents—*continued*

Fig 3.1.23 (12)—***continued***

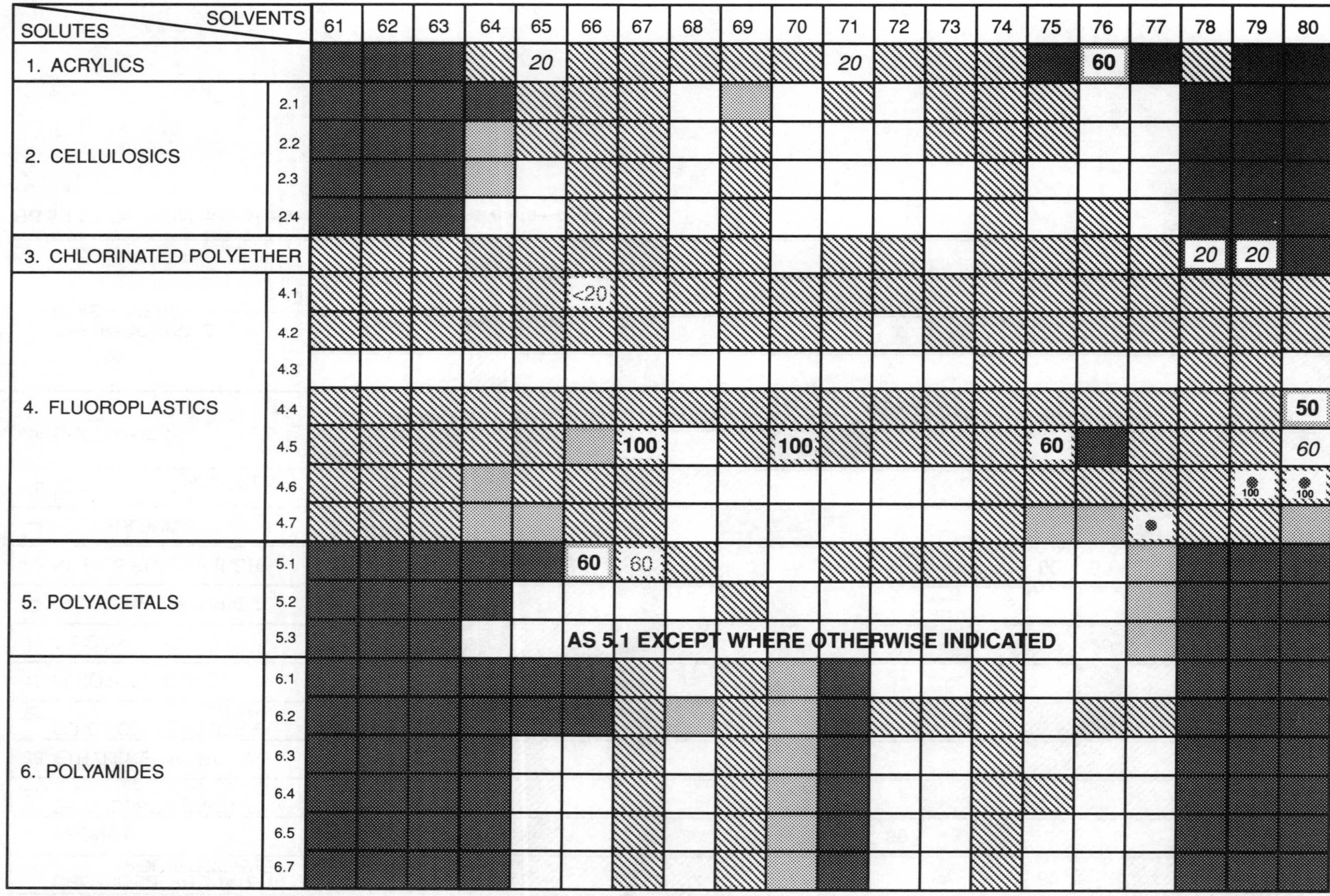

FIG 3.1.23 The resistance of plastics to a range of solvents—*continued*

Fig 3.1.23 (13)—***continued***

SOLUTES \ SOLVENTS		61	62	63	64	65	66	67	68	69	70	71	72	73	74	75	76	77	78	79	80
THERMOPLASTICS																					
7. POLYAMIDE-IMIDE																					
8. POLYCARBONATES			60										60	60	60				60		
9. POLYESTERS, THERMOPLASTIC	9.1																				
	9.2		60																60		
10. POLYETHYLENE	HIGH & LOW DENSITY	LH	LH	LH				● H			H				H	LH	LH	H		H60/L	
11. POLYOLEFIN COPOLYMERS	11.1				60											60					
	11.2																				
	11.3																				
12. POLYPHEN. OXIDE MOD.																					
13. POLYPHEN. SULPHIDE																					
14. POLYPROPYLENE		100	60		60 20							40%				60	20				
15. POLYSULPHONES	15.1																				
	15.2																				
	15.3																				
	15.4																				
16. POLYURETHANES, THERMOPLASTIC	16.1								SEE SECTION 46.1												
	16.2								SEE SECTION 46.2												
17. POLYVINYL CHLORIDE	17.1					AS RIGID PVC EXCEPT WHERE OTHERWISE INDICATED															
	17.2		60																	20	
	17.3																				
	17.4																				

FIG 3.1.23 The resistance of plastics to a range of solvents—*continued*

Fig 3.1.23 (14)—***continued***

	SOLUTES \ SOLVENTS		61	62	63	64	65	66	67	68	69	70	71	72	73	74	75	76	77	78	79	80
THERMOPLASTICS	18. STYRENE-BASED POLYMERS	18.1		50		60					60				60							
		18.2		50		60		50			60		50		60							
		18.3		50													50	50		50	50	
		18.4	60	60		60							60		60		60			50		
		18.5																				
	19. TPX																					
THERMOSETS	20. ALKYDS																					
	21. AMINOS	21.1				60	100			20	60	20	100	100	60	100						
		21.2																				
	22. DIALLYL PHTHALATE (DAP)																					
	23. EPOXIDES														60					60		
	24. FURAN																65			5%		
	25. PHENOLICS		20	20			20		60	20			100									
	26. POLYBUTADIENE																					
	27. POLYESTERS UNSAT.										60									60		
	28. POLYIMIDES																					
	29. POLYURETHANES																					
ELASTOMERS	30. SILICONES																					
	31. ACRYLICS (ACM)																					
	32. BUTADIENE (BR)																					
	33. STYRENE BUTADIENE		AS BUTADIENE EXCEPT WHERE OTHERWISE INDICATED																			
	34. BUTYL RUBBER (IIR)																			5%		

FIG 3.1.23 The resistance of plastics to a range of solvents—*continued*

Fig 3.1.23 (15)—***continued***

	SOLUTES \ SOLVENTS		61	62	63	64	65	66	67	68	69	70	71	72	73	74	75	76	77	78	79	80
ELASTOMERS	35. CHLOROSULPHONATED POLYETHYLENE		60	60	60		20								100		60			100	100	
	36. EBONITE		60	100																		
	37. EPICHLOROHYDRIN (CO ECO)																					
	38. ETHYLENE PROPYLENE																					
	39. FLUOROELASTOMERS (FKM)				60																	
	40. FLUOROSILICONE																					
	41. SILICONE		20							60												
	42. ISOPRENE (NATURAL R.)		100	100					20													
	43. NITRILE RUBBER (NBR)			20	20	60	60		100				100				100			60		
	44. POLYSULPHIDE																					
	45. CHLOROPRENE (NEOPRENE)		60	20	60													60		100		
	46. URETHANES	46.1																				
		46.2		AS 46.1 BUT WITH BETTER HYDROLYTIC STABILITY FOR SOLVENT RESISTANCE 46.1 (POLYETHER) USUALLY PREFERRED																		
THERMOPLASTIC ELASTOMERS	47. THERMOPLASTIC ELASTOMERS	47.1																				
		47.2																				
		47.3																				
		47.4																				

FIG 3.1.23 The resistance of plastics to a range of solvents—*continued*

Fig 3.1.23 (16)—***continued***

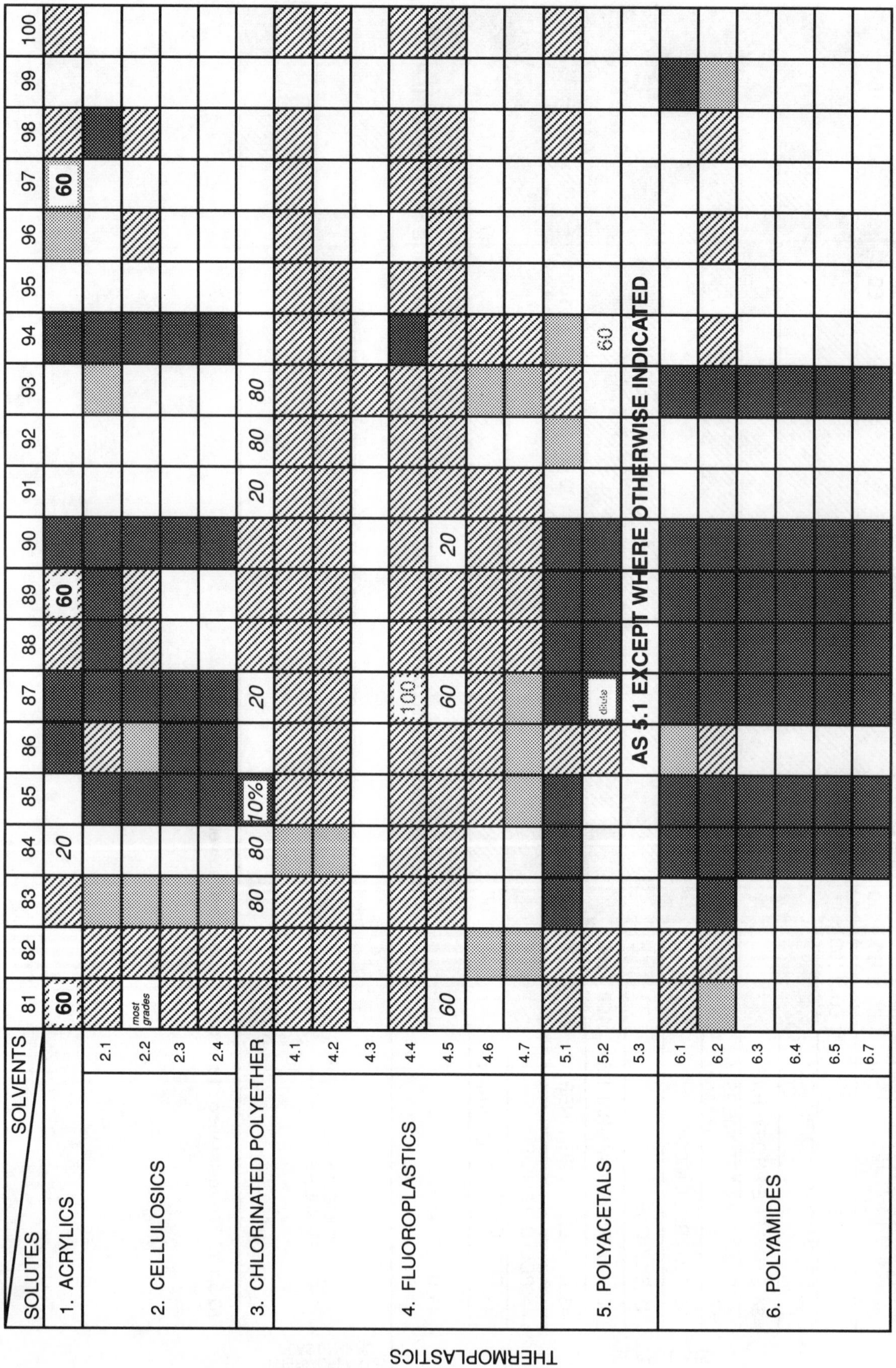

SOLUTES \ SOLVENTS		81	82	83	84	85	86	87	88	89	90	91	92	93	94	95	96	97	98	99	100
1. ACRYLICS		**60**			20					**60**								**60**			
2. CELLULOSICS	2.1																				
	2.2	most grades																			
	2.3																				
	2.4																				
3. CHLORINATED POLYETHER				80	80	10%		20				20	80	80							
4. FLUOROPLASTICS	4.1																				
	4.2																				
	4.3																				
	4.4							100													
	4.5	60						60			20										
	4.6																				
	4.7																				
5. POLYACETALS	5.1																				
	5.2							dilute							60						
	5.3							AS 5.1 EXCEPT WHERE OTHERWISE INDICATED													
6. POLYAMIDES	6.1																				
	6.2																				
	6.3																				
	6.4																				
	6.5																				
	6.7																				

THERMOPLASTICS

FIG 3.1.23 The resistance of plastics to a range of solvents—*continued*

Fig 3.1.23 (17)—***continued***

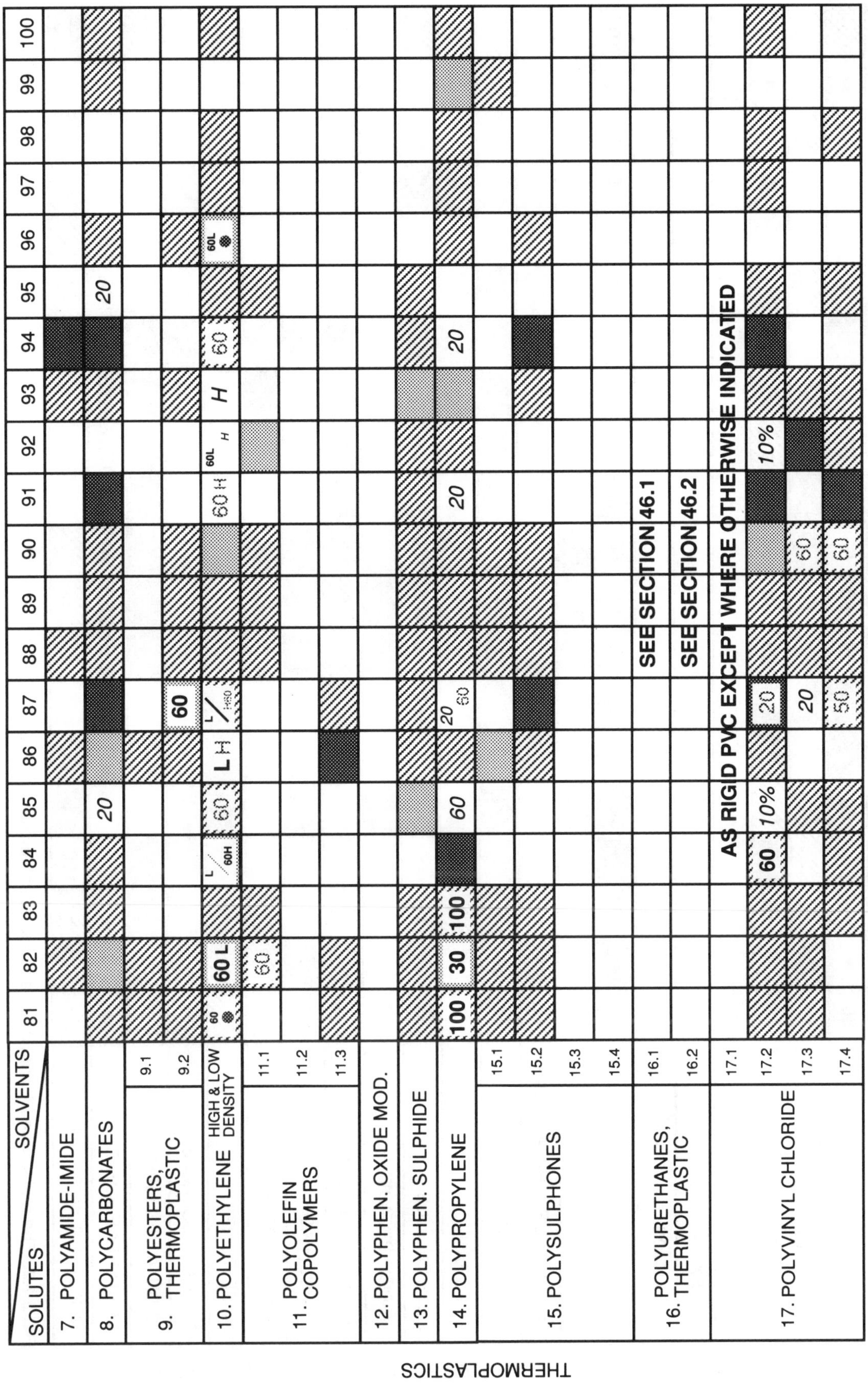

THERMOPLASTICS

SOLUTES \ SOLVENTS		81	82	83	84	85	86	87	88	89	90	91	92	93	94	95	96	97	98	99	100
7. POLYAMIDE-IMIDE																					
8. POLYCARBONATES						20										20					
9. POLYESTERS, THERMOPLASTIC	9.1																				
	9.2							60													
10. POLYETHYLENE	HIGH & LOW DENSITY	60 ●	60 L		L / 60H	60	L H	L / H60				60 H	60L H	H	60		60L ●				
11. POLYOLEFIN COPOLYMERS	11.1		60																		
	11.2																				
	11.3																				
12. POLYPHEN. OXIDE MOD.																					
13. POLYPHEN. SULPHIDE																					
14. POLYPROPYLENE		100	30	100		60		20 60				20			20						
15. POLYSULPHONES	15.1																				
	15.2																				
	15.3																				
	15.4																				
16. POLYURETHANES, THERMOPLASTIC	16.1								SEE SECTION 46.1												
	16.2								SEE SECTION 46.2												
17. POLYVINYL CHLORIDE	17.1				AS RIGID PVC EXCEPT WHERE OTHERWISE INDICATED																
	17.2				60	10%		20					10%								
	17.3							20			60										
	17.4							50			60										

FIG 3.1.23 The resistance of plastics to a range of solvents—*continued*

Fig 3.1.23 (18)—*continued*

Fig 3.1.23 (19)—***continued***

	SOLUTES \ SOLVENTS		81	82	83	84	85	86	87	88	89	90	91	92	93	94	95	96	97	98	99	100
THERMOPLASTICS	18. STYRENE-BASED POLYMERS	18.1	60	•			50							50							•	
		18.2	60	•			50			50	50	50		50						50	•	
		18.3									50				50							
		18.4			60					60	60	60							20	60	•	
		18.5																				
	19. TPX																					
THERMOSETS	20. ALKYDS																					
	21. AMINOS	21.1	60		100					60	60	60		20		20	100		100	60		100
		21.2																				
	22. DIALLYL PHTHALATE (DAP)																					
	23. EPOXIDES			60				60														
	24. FURAN												65									
	25. PHENOLICS				20				100	100	100	100	20	20						60		
	26. POLYBUTADIENE																					
	27. POLYESTERS UNSAT.														60							
	28. POLYIMIDES																					
	29. POLYURETHANES																					
ELASTOMERS	30. SILICONES																					
	31. ACRYLICS (ACM)							100														
	32. BUTADIENE (BR)																					
	33. STYRENE BUTADIENE						AS BUTADIENE EXCEPT WHERE OTHERWISE INDICATED															
	34. BUTYL RUBBER (IIR)					100																

FIG 3.1.23 The resistance of plastics to a range of solvents—*continued*

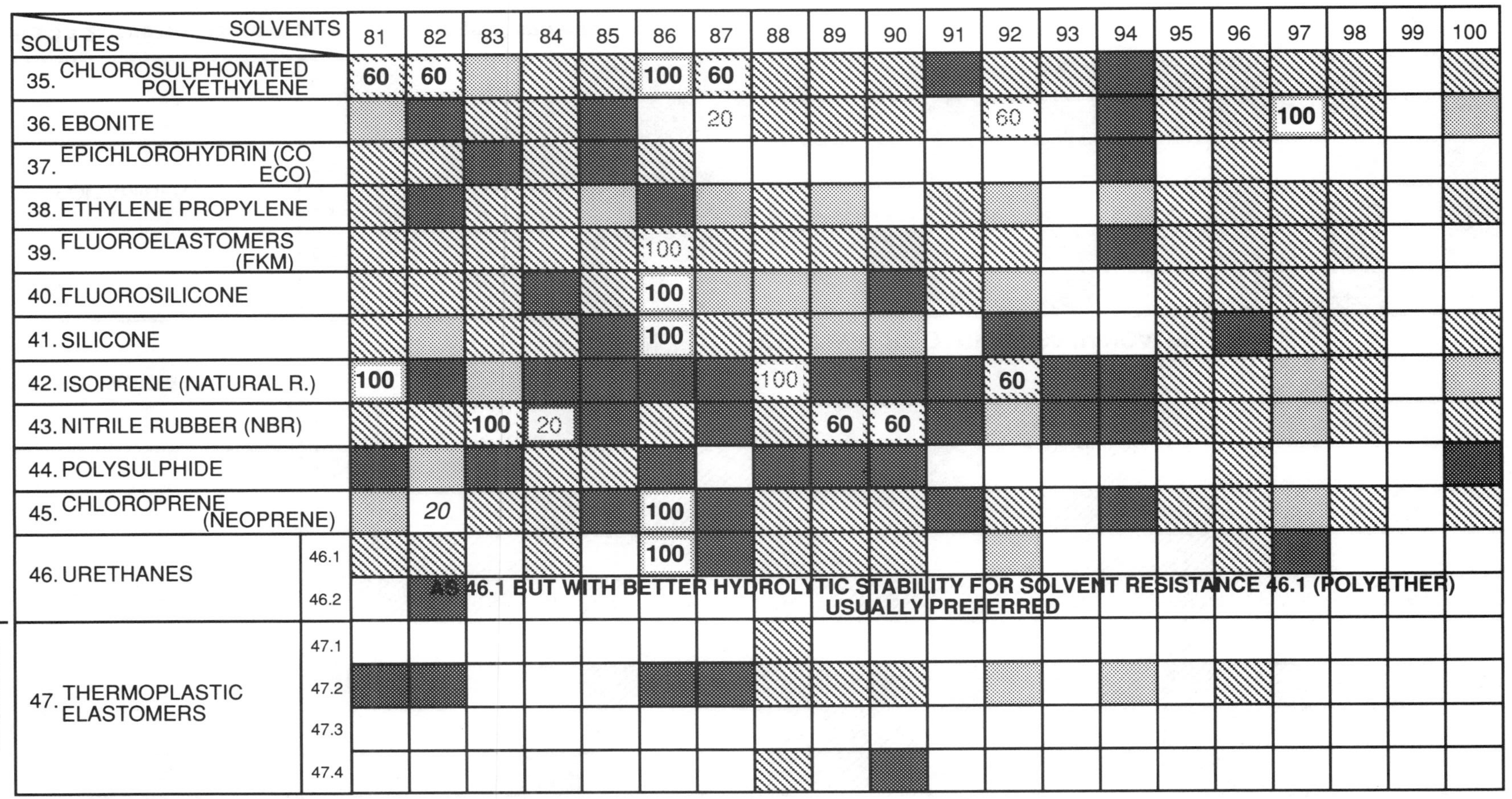

FIG 3.1.23 The resistance of plastics to a range of solvents—*continued*

Fig 3.1.23 (20)—***continued***

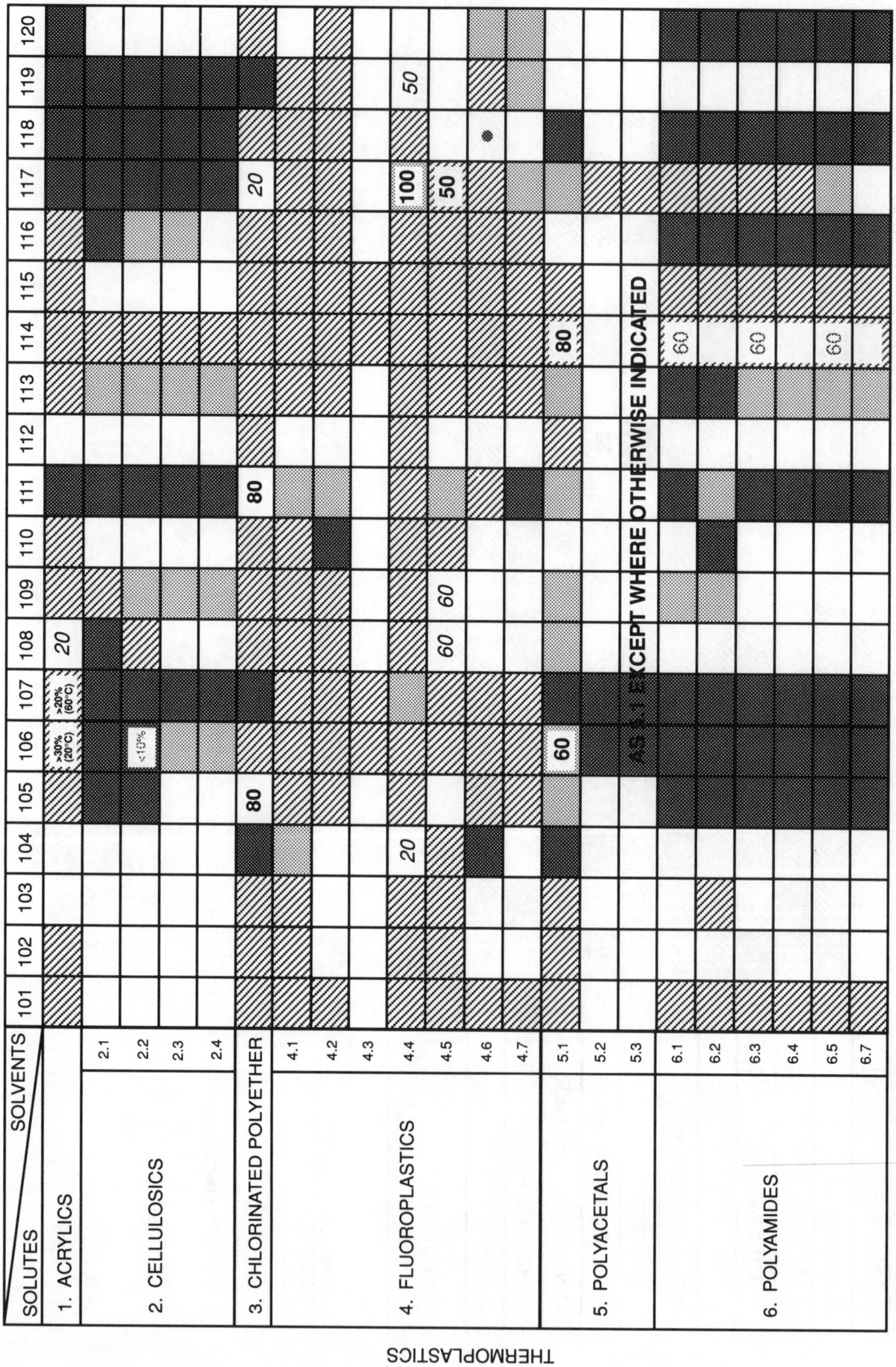

SOLUTES \ SOLVENTS		101	102	103	104	105	106	107	108	109	110	111	112	113	114	115	116	117	118	119	120
1. ACRYLICS							>30% (20°C)	>20% (60°C)	20												
2. CELLULOSICS	2.1																				
	2.2						<10%														
	2.3																				
	2.4																				
3. CHLORINATED POLYETHER						80						80						20			
4. FLUOROPLASTICS	4.1																				
	4.2																				
	4.3																				
	4.4				20													100		50	
	4.5								60	60								50			
	4.6																		●		
	4.7																				
5. POLYACETALS	5.1						60								80						
	5.2																				
	5.3						AS 5.1 EXCEPT WHERE OTHERWISE INDICATED														
6. POLYAMIDES	6.1														60						
	6.2																				
	6.3														60						
	6.4																				
	6.5														60						
	6.7																				

THERMOPLASTICS

FIG 3.1.23 The resistance of plastics to a range of solvents—*continued*

Fig 3.1.23 (21)—***continued***

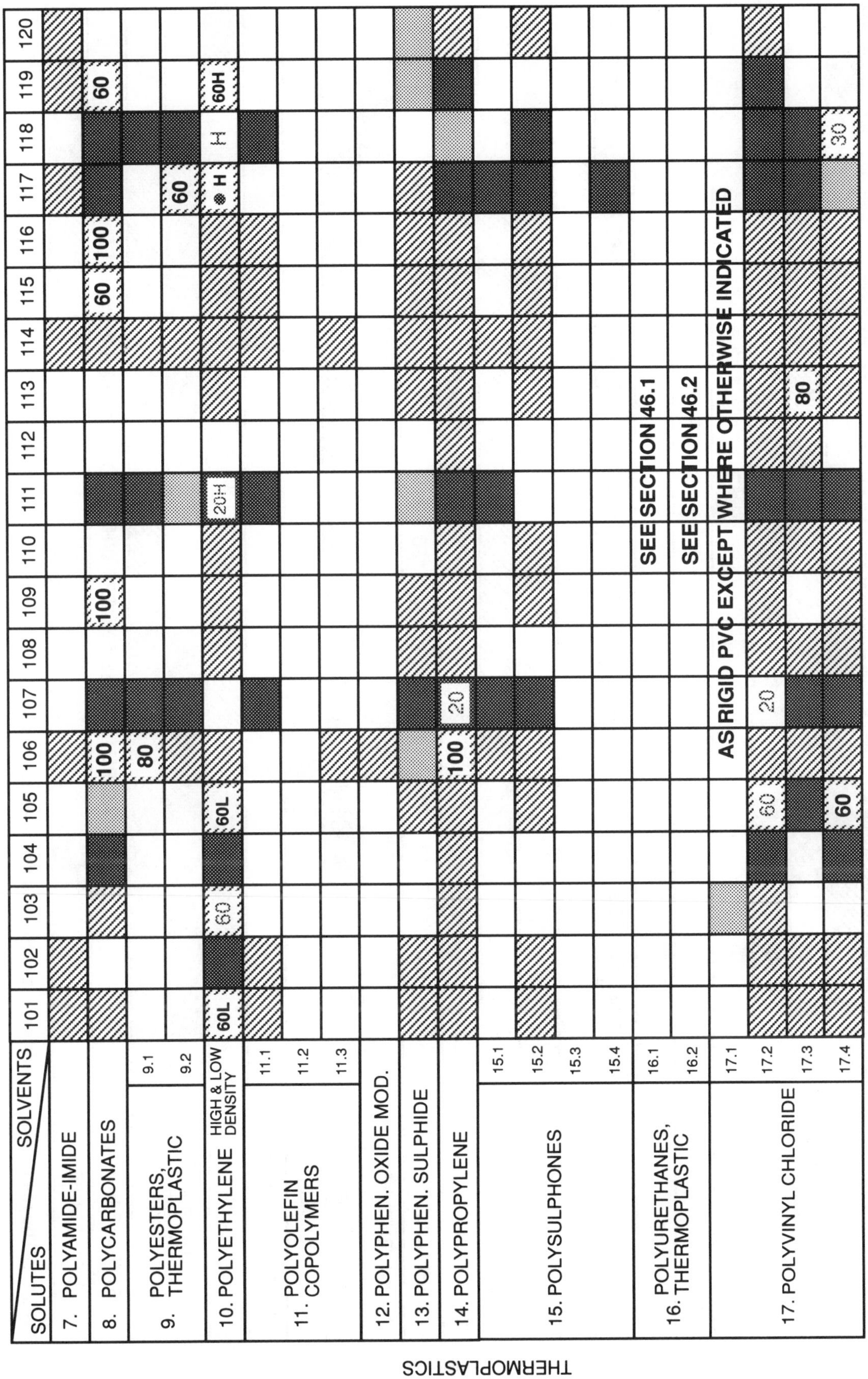

THERMOPLASTICS

SOLUTES \ SOLVENTS		101	102	103	104	105	106	107	108	109	110	111	112	113	114	115	116	117	118	119	120
7. POLYAMIDE-IMIDE																					
8. POLYCARBONATES							100			100						60	100			60	
9. POLYESTERS, THERMOPLASTIC	9.1						80														
	9.2																	60			
10. POLYETHYLENE HIGH & LOW DENSITY		60L		60		60L						20H						● H	H	60H	
11. POLYOLEFIN COPOLYMERS	11.1																				
	11.2																				
	11.3																				
12. POLYPHEN. OXIDE MOD.																					
13. POLYPHEN. SULPHIDE																					
14. POLYPROPYLENE							100	20													
15. POLYSULPHONES	15.1																				
	15.2																				
	15.3																				
	15.4																				
16. POLYURETHANES, THERMOPLASTIC	16.1										SEE SECTION 46.1										
	16.2										SEE SECTION 46.2										
17. POLYVINYL CHLORIDE	17.1						AS RIGID PVC EXCEPT WHERE OTHERWISE INDICATED														
	17.2					60		20													
	17.3													80							
	17.4					60													30		

FIG 3.1.23 The resistance of plastics to a range of solvents—*continued*

Fig 3.1.23 (22)—*continued*

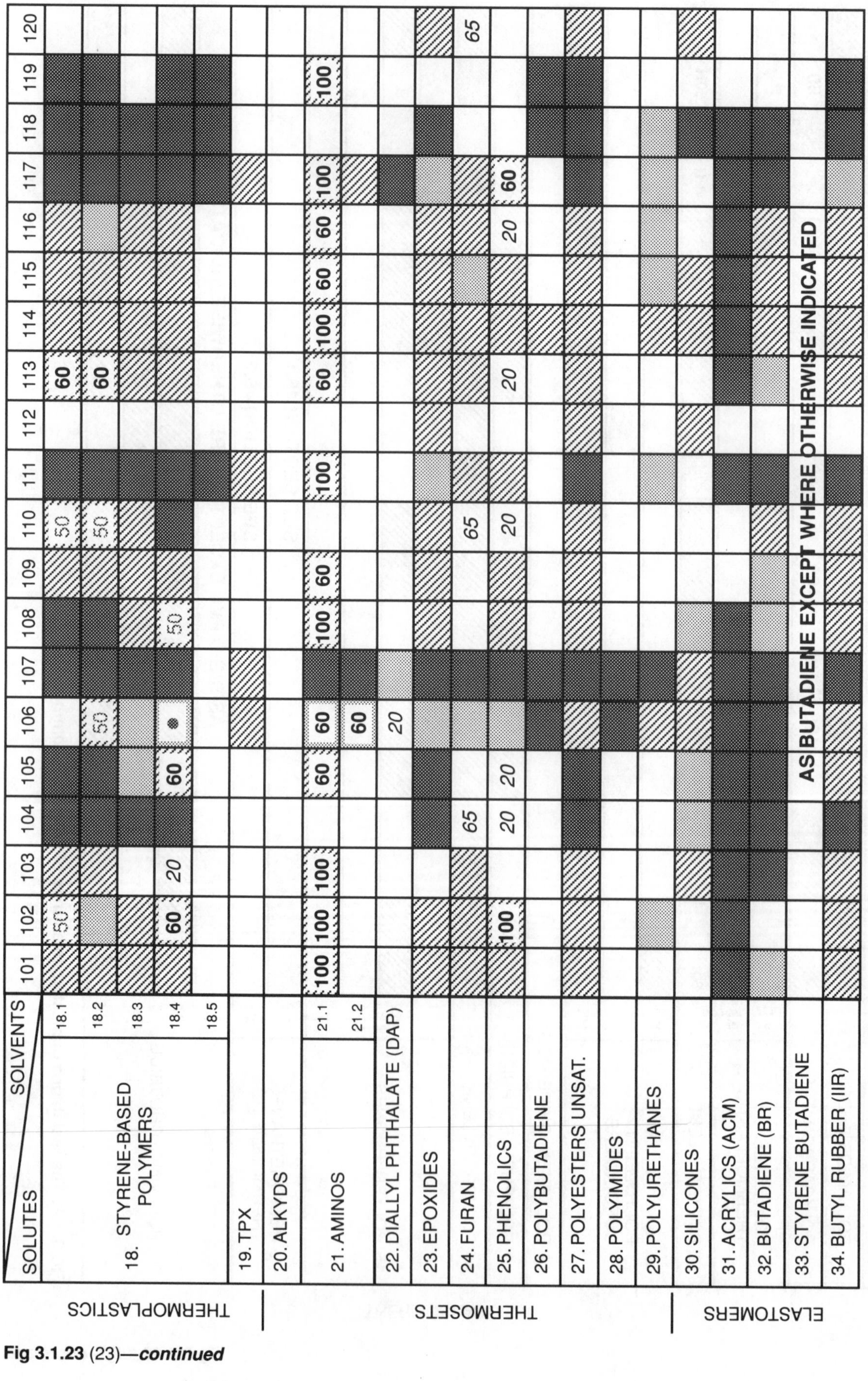

	SOLUTES \ SOLVENTS		101	102	103	104	105	106	107	108	109	110	111	112	113	114	115	116	117	118	119	120
THERMOPLASTICS	18. STYRENE-BASED POLYMERS	18.1		50								50			60							
		18.2						50				50			60							
		18.3																				
		18.4		60	20		60	●		50												
		18.5																				
	19. TPX																					
THERMOSETS	20. ALKYDS																					
	21. AMINOS	21.1	100	100	100		60	60		100	60		100		60	100	60	60	100		100	
		21.2						60														
	22. DIALLYL PHTHALATE (DAP)							20														
	23. EPOXIDES																					
	24. FURAN					65						65										65
	25. PHENOLICS			100		20	20					20			20			20	60			
	26. POLYBUTADIENE																					
	27. POLYESTERS UNSAT.																					
	28. POLYIMIDES																					
	29. POLYURETHANES																					
ELASTOMERS	30. SILICONES																					
	31. ACRYLICS (ACM)																					
	32. BUTADIENE (BR)																					
	33. STYRENE BUTADIENE		AS BUTADIENE EXCEPT WHERE OTHERWISE INDICATED																			
	34. BUTYL RUBBER (IIR)																					

FIG 3.1.23 The resistance of plastics to a range of solvents—*continued*

Fig 3.1.23 (23)—***continued***

	SOLUTES / SOLVENTS		101	102	103	104	105	106	107	108	109	110	111	112	113	114	115	116	117	118	119	120
ELASTOMERS	35. CHLOROSULPHONATED POLYETHYLENE																					
	36. EBONITE						100					60										
	37. EPICHLOROHYDRIN (CO ECO)																					
	38. ETHYLENE PROPYLENE																					
	39. FLUOROELASTOMERS (FKM)																					
	40. FLUOROSILICONE																					
	41. SILICONE																					
	42. ISOPRENE (NATURAL R.)			100	60			100							100							
	43. NITRILE RUBBER (NBR)						100	60		100	100											
	44. POLYSULPHIDE																					
	45. CHLOROPRENE (NEOPRENE)								20													
	46. URETHANES	46.1																				
		46.2	AS 46.1 BUT WITH BETTER HYDROLYTIC STABILITY FOR SOLVENT RESISTANCE 46.1 (POLYETHER) USUALLY PREFERRED																			
THERMOPLASTIC ELASTOMERS	47. THERMOPLASTIC ELASTOMERS	47.1																				
		47.2																				
		47.3																				
		47.4																				

FIG 3.1.23 The resistance of plastics to a range of solvents—*continued*

Fig 3.1.23 (24)—***continued***

THERMOPLASTICS

SOLUTES / SOLVENTS		121	122	123	124	125	126	127	128	129	130	131	132	133	134	135	136
1. ACRYLICS														●			
2. CELLULOSICS	2.1																
	2.2																
	2.3																
	2.4																
3. CHLORINATED POLYETHER															20		
4. FLUOROPLASTICS	4.1																
	4.2																
	4.3																
	4.4			50	60	60							50				
	4.5																
	4.6																
	4.7																
5. POLYACETALS	5.1																
	5.2					60											
	5.3	AS 5.1 EXCEPT WHERE OTHERWISE INDICATED															
6. POLYAMIDES	6.1																
	6.2																
	6.3																
	6.4																
	6.5																
	6.7																

FIG 3.1.23 The resistance of plastics to a range of solvents—*continued*

Fig 3.1.23 (25)—***continued***

THERMOPLASTICS

SOLUTES / SOLVENTS		121	122	123	124	125	126	127	128	129	130	131	132	133	134	135	136				
7. POLYAMIDE-IMIDE																					
8. POLYCARBONATES																					
9. POLYESTERS, THERMOPLASTIC	9.1													T							
	9.2														60						
10. POLYETHYLENE	HIGH & LOW DENSITY				H									HL	50L	50	H				
11. POLYOLEFIN COPOLYMERS	11.1												60								
	11.2																				
	11.3																				
12. POLYPHEN. OXIDE MOD.																					
13. POLYPHEN. SULPHIDE																					
14. POLYPROPYLENE					50									TR							
15. POLYSULPHONES	15.1																				
	15.2																				
	15.3																				
	15.4																				
16. POLYURETHANES, THERMOPLASTIC	16.1										SEE SECTION 46.1										
	16.2										SEE SECTION 46.2										
17. POLYVINYL CHLORIDE	17.1							AS RIGID PVC EXCEPT WHERE OTHERWISE INDICATED													
	17.2								60	60											
	17.3																				
	17.4					50		50	50	50						50					

FIG 3.1.23 The resistance of plastics to a range of solvents—*continued*

Fig 3.1.23 (26)—*continued*

	SOLUTES \ SOLVENTS		121	122	123	124	125	126	127	128	129	130	131	132	133	134	135	136
THERMOPLASTICS	18. STYRENE-BASED POLYMERS	18.1																
		18.2							50									
		18.3																
		18.4																
		18.5																
	19. TPX																	
THERMOSETS	20. ALKYDS																	
	21. AMINOS	21.1		20														
		21.2																
	22. DIALLYL PHTHALATE (DAP)																	
	23. EPOXIDES				40					60				80	60			
	24. FURAN			65		20		45			45							
	25. PHENOLICS																	
	26. POLYBUTADIENE																	
	27. POLYESTERS UNSAT.													80	60			
	28. POLYIMIDES																	
	29. POLYURETHANES																	
ELASTOMERS	30. SILICONES																	
	31. ACRYLICS (ACM)																	
	32. BUTADIENE (BR)																	
	33. STYRENE BUTADIENE		AS BUTADIENE EXCEPT WHERE OTHERWISE INDICATED															
	34. BUTYL RUBBER (IIR)																	

FIG 3.1.23 The resistance of plastics to a range of solvents—*continued*

Fig 3.1.23 (27)—***continued***

	SOLUTES \ SOLVENTS		121	122	123	124	125	126	127	128	129	130	131	132	133	134	135	136
ELASTOMERS	35. CHLOROSULPHONATED POLYETHYLENE																	
	36. EBONITE																	
	37. EPICHLOROHYDRIN (CO ECO)													P S				
	38. ETHYLENE PROPYLENE																	
	39. FLUOROELASTOMERS (FKM)																	
	40. FLUOROSILICONE													P S				
	41. SILICONE																	
	42. ISOPRENE (NATURAL R.)																	
	43. NITRILE RUBBER (NBR)																	
	44. POLYSULPHIDE													P S				
	45. CHLOROPRENE (NEOPRENE)																	
	46. URETHANES	46.1																
		46.2		AS 46.1 BUT WITH BETTER HYDROLYTIC STABILITY FOR SOLVENT RESISTANCE 46.1 (POLYETHER) USUALLY PREFERRED														
THERMOPLASTIC ELASTOMERS	47. THERMOPLASTIC ELASTOMERS	47.1																
		47.2																
		47.3																
		47.4																

FIG 3.1.23 The resistance of plastics to a range of solvents—*continued*

Fig 3.1.23 (28)—***continued***

Fig 3.1.23 (a)

FIG 3.1.23a List of polymeric materials

		Material number
THERMOPLASTICS		
Acetals	see Polyacetals	
1. **Acrylics**	Polymethyl methacrylate (PMMA)	1
2. **Cellulosics**	Cellulose Acetate	2.1
	Cellulose Acetate Butyrate (CAB)	2.2
	Cellulose Propionate	2.3
	Ethyl Cellulose	2.4
3. **Chlorinated Polyether**		3
4. **Fluoroplastics**	PTFE	4.1
	FEP	4.2
	PFA	4.3
	PVDF	4.4
	PTFCE	4.5
	ETFE	4.6
	ECTFE	4.7
Noryl	see Polyphenylene Oxide (PPO)	
Nylons	see Polyamides	
5. **Polyacetals**	Acetal Copolymer	5.1
	Acetal Homopolymer	5.2
	Acetal Terpolymer	5.3
6. **Polyamides**	Nylon 6	6.1
	Nylon 6.6	6.2
	Nylon 6.10	6.3
	Nylon 6.12	6.4
	Nylon 11	6.5
	Nylon12	6.7
7. **Polyamide-Imide**		7

		Material number
THERMOSETS		
20. **Alkyds**		20
21. **Aminos**	Melamine formaldehyde resins (MF)	21.1
	Urea formaldehyde resins (UF)	21.2
22. **Diallyl Phthalate (DAP)**		22
23. **Epoxides**	Epoxy resins or polyepoxides	23
24. **Furan**		24
Melamines	see Aminos	
25. **Phenolics**	Phenol formaldehyde resins (PF)	25
26. **Polybutadiene**		26
27. **Polyesters, Unsaturated**	Base resin	27
28. **Polyimides**		28
29. **Polyurethanes**		29
Ureas	see Aminos	
ELASTOMERS		
30. **Silicones**		30
31. **Acrylic (ACM)**	(Polyacrylate)	31
32. **Butadiene**	(BR)	32

Material	Type	Fig.
8. Polycarbonates		8
9. Polyesters, Thermoplastic	PBT (PTMT)	9.1
	PET (PETP)	9.2
10. Polyethylene		10
Polyimides	see under Thermosets	
Poly-4-Methyl Pentene-1	see TPX	
11. Polyolefin Copolymers	Ethylene Vinyl Acetate (EVA)	11.1
	Ionomer	11.2
	Polyallomer	11.3
12. Polyphenylene Oxide (modified)	(PPO)	12
13. Polyphenylene Sulphide	(PPS)	13
14. Polypropylene		14
Polystyrene	see Styrene-based Polymers	
15. Polysulphones	Polysulphone	15.1
	Polyethersulphone (PES)	15.2
	Polyarylsulphone (PAS)	15.3
	Polyarylether	15.4
16. Polyurethanes, Thermoplastic	Polyester-based	16.1/46.1
	Polyether-based	16.2/46.2
17. Polyvinyl Chloride	PVC Flexible	17.1
	PVC Rigid	17.2
	CPVC	17.3
	PVDC	17.4
PTFE	see Fluoroplastics	
18. Styrene-based Polymers	Polystyrene (PS)	18.1
	HIPS	18.2
	SAN	18.3
	ABS	18.4
	BDS	18.5
19. TPX		19

Material	Type	Fig.
33. Styrene Butadiene	(SBR)	33
34. Butyl Rubber	(IIR)	34
35. Chlorosulphonated Polyethylene (CSM, Hypalon)		35
36. Ebonite		36
37. Epichlorohydrin	(CO, ECO)	37
38. Ethylene Propylene	(EPM, EPDM)	38
39. Fluoroelastomers	(FKM, Fluorocarbon polymers)	39
40. Fluorosilicone		40
41. Silicone	(MQ, PMQ, UMQ, PVMQ)	41
42. Natural Rubber	(Isoprene)	42
43. Nitrile Rubber	(NBR)	43
44. Polysulphide		44
45. Chloroprene	(CR)	45
46. Urethanes	Polyether (AU)	46.1
	Polyester (EU)	46.2
47. THERMOPLASTIC ELASTOMERS		
Copolymers		47.1
Olefin Based		47.2
Polyurethanes		47.3/46.1 and 46.2
Styrene-based		47.4

Fig 3.1.23 (a)—***continued***

Fig 3.1.23 (b)

FIG 3.1.23b Numerical list of solvents

No.	Solvent	No.	Solvent
1	Acetaldehyde	41	Citric acid (aqueous solution)
2	Glacial acetic acid	42	Copper (II) sulphate (Cupric Sulphate) (aqueous solution)
3	10% acetic acid	43	Cresylic acid (aqueous solution)
4	Acetic anhydride	44	Cyclohexane
5	Acetone (propanone)	45	Detergent solutions: anionic, cationic, non-ionic
6	Acetylene	46	Developing fluids
7	Acid fumes	47	Diethyl ether
8	Alcohols: methanol, ethanol, butan-l-ol, propan-l-ol	48	Emulsifiers: anionic, cationic, non-ionic
9	Aliphatic esters: amyl acetate, butyl acetate, ethyl acetate	49	Fatty acids
10	Alum (aqueous solution)	50	Fluorine (dry)
11	Aluminium chloride (aqueous solution)	51	Fluorine (wet)
12	Anhydrous ammonia	52	Fruit juices
13	Ammonium chloride (aqueous solution)	53	Fluorinated refrigerants, fluorocarbons, chlorofluorocarbons
14	Ammonium hydroxide	54	Fluosilicic acid
15	Nitrates of K, Na, Li, Ba, Mg, Ca (aqueous solutions)	55	Formaldehyde
16	Aniline (aminobenzene)	56	Formic acid (aqueous solution)
17	Antimony trichloride (aqueous solution)	57	Gelatin
18	Aqua Regia	58	Glycerin
19	Aromatic solvents: xylene, benzene, toluene	59	Ethylene glycol, diethylene glycol, propylene glycol
20	Beer	60	Hydrochloric acid (10%)
21	Benzoic acid (aqueous solution)	61	Hydrochloric acid (conc)
22	Bleach (sodium hypochlorite)	62	Hydrofluoric acid (40%)
23	Boric acid	63	Hydrofluoric acid (75%)
24	Brake fluid	64	Hydrogen peroxide
25	Brine	65	Hydrogen sulphide
26	Anhydrous bromine	66	Ferric (iron III) chloride
27	Calcium chloride (aqueous solution)	67	Lactic acid (aqueous solution)
28	Carbon disulphide (bisulphide)	68	Lead acetate (aqueous solution)
29	Carbonic acid, carbon dioxide water	69	Calcium oxide (hydroxide) and limewater
30	Carbon tetrachloride (tetrachloromethane)	70	Meat juices
31	Caustic soda and caustic potash (sodium and potassium hydroxide)	71	Mercuric chloride (aqueous solution)
32	Dry chlorine	72	Mercury
33	Wet chlorine	73	Milk
34	Chlorates of Na, K, Ba, Mg, Li, Ca (aqueous solutions)	74	Moist air
35	Chlorites of Na, K, Ba, Mg, Li, Ca (aqueous solutions)	75	Naphtha
36	Chloroacetic acids	76	Naphthalene
37	Chlorobenzene	77	Ethyl chloride (chloroethane)
38	Chloroform (trichloromethane)	78	Nitric acid (25%)
39	Chlorosulphonic acid	79	Nitric acid (50%)
40	Chromic acid (80%)	80	Nitric acid, fuming

81 Animal oil and vegetable oil
82 Mineral oil
83 Oxalic acid (aqueous solution)
84 Ozone
85 Perchloric acid (aqueous solution)
86 Petroleum (gasoline)
87 Phenol
88 Phosphoric acid 25%
89 Phosphoric acid 50%
90 Phosphoric acid 95%
91 Phosphorus trichloride
92 Picric acid
93 Potassium permanganate (aqueous solution)
94 Pyridine
95 Silicic acid
96 Silicone fluids
97 Sodium peroxide (aqueous solution)
98 Sodium silicate
99 Solder flux
100 Sucrose solution

101 Sulphates of K, Na, Mg, Li (aqueous solution)
102 Sulphites of K, Na, Mg, Li (aqueous solution)
103 Sulphur
104 Sulphur chlorides
105 Sulphur dioxide
106 Sulphuric acid (dilute)
107 Sulphuric acid (conc.)
108 Tannic acid (aqueous solution)
109 Tartaric acid (aqueous solution)
110 Stannous and stannic (tin (II) and tin (IV)) chlorides
111 Trichloroethylene
112 Urine
113 Vinegar
114 Distilled water
115 Salt (sea) water
116 Zinc chloride
117 Methyl ethyl (ethyl methyl) ketone
118 Methylene dichloride (dichloromethane)
119 Acetyl chloride
120 Chromates of Na, K (aqueous solutions)

121 Dichromates (aqueous solutions)
122 Ethylene dichloride
123 Methylamine
124 Dimethylamine
125 Trimethylamine
126 Ethanolamine
127 Diethanolamine
128 Triethanolamine
129 Ethylene glycol monomethyl and monoethyl ethers
130 Toluene diisocyanate (TDI)
131 Diphenylmethane diisocyanate (MDI)
132 Phthalate esters — dioctyl phthalate and dioctyl sebacate
133 Terpene derivatives (turpentine, diterpene, rosin)
134 Dioxan
135 Propylene oxide
136 Epichlorohydrin

Fig 3.1.23 (b)—***continued***

FIG 3.1.23c Alphabetical list of solvents

	Solvent number
Acetaldehyde	1
Acetic acid glacial	2
Acetic acid 10%	3
Acetic anhydride	4
Acetone	5
Acetyl chloride	119
Acetylene	6
Acid fumes	7
Air, moist	74
Alcohols M=methanol; E=ethanol; P=propanol; B=butanol	8
Aliphatic esters A=amyl acetate; B=butyl acetate; E=ethyl acetate	9
Alum (aq. soln.)	10
Aluminium, chloride (aq. soln.)	11
Aminobenzene	16
Ammonia, anhydrous	12
Ammonium chloride (aq. soln.)	13
Ammonium hydroxide (aq. soln.)	14
Amyl acetate	9
Aniline	16
Animal oil	81
Antimony trichloride (aq. soln.)	17
Aqua regia	18
Aromatic solvents; X=xylene; B=benzene; T=toluene	19
Beer	20
Benzene	19
Benzoic acid (aq. soln.)	21
Bleach (sodium hypochlorite)	22
Boric acid	23
Brake fluid	24
Brine	25
Bromine, anhydrous	26
Butan-l-ol	8
Butyl acetate	9
Calcium chloride (aq. soln.)	27
Carbon disulphide (bisulphide)	28
Calcium oxide (hydroxide) (aq. soln.)	69
Carbonic acid, carbon dioxide water	29
Carbon tetrachloride	30
Caustic soda, potash (P=KOH; N=NaOH)	31
Cellosolve	129
Chloracetic acid	36
Chlorates of Na, K, Ba, Mg, Li, Ca (aq. solns.)	34
Chlorides of Na, K, Ba, Mg, Li, Ca (aq. solns.)	35
Chlorine (dry)	32
Gasoline	86
Gelatin	57
Glycerin	58
Glycol ethers	129
Glycols D=diethylene; E=ethylene; P=propylene	59
Hydrochloric acid (10%)	60
Hydrochloric acid (conc)	61
Hydrofluoric acid (40%)	62
Hydrofluoric acid (75%)	63
Hydrogen peroxide	64
Hydrogen sulphide	65
Iron (III) chloride (aq. soln.)	66
Lactic acid (aq. soln.)	67
Lead acetate (aq. soln.)	68
Limewater	69
MDI	131
Meat juices	70
Mercuric chloride (aq. soln.)	71
Mercury	72
Methanol (methyl alcohol)	8
Methylamine	123
Methyl cellosolve	129
Methyl ethyl ketone	117
Methylene dichloride	118
Methyl oxitol	129
Milk	73
Mineral oil	82
Moist air	74
Monoethanolamine	126
Naphtha	75
Naphthalene	76
Nitrates of K, Na, Ba, Li, Mg, Ca (aq. soln.)	15
Nitric acid (25%)	78
Nitric acid (50%)	79
Nitric acid (fuming)	80
Oils, animal and vegetable	81
Oils, mineral	82
Oxalic acid (aq. soln.)	83
Oxitol	129
Ozone	84
Perchloric acid (aq. soln.)	85
Petroleum	86

Fig 3.1.23 (c)

Chemical	No.
Chlorine (wet)	33
Chlorobenzene	37
Chloroethane	77
Chlorofluorocarbons	53
Chloroform	38
Chlorosulphonic acid	39
Chromates of Na, K (aq. solns.)	120
Chromic acid (80%)	40
Citric acid	41
Copper (II) sulphate (aq. soln.)	42
Cresylic acid (aq. soln.)	43
Cupric sulphate (aq. soln.)	42
Cyclohexane	44
Detergent solutions	45
Developing fluids	46
Dichloromethane	118
Dichromates (aq. soln.)	121
Diethyl ether	47
Diethanolamine	127
Dimethylamine	124
Dioctyl phthalate	132
Dioctyl sebacate	132
Dioxane	134
Diphenyl methane diisocyanate (MDI)	131
Emulsifiers	48
Epichlorohydrin	136
Esters (Aliphatic)	9
Ethanol	8
Ethanolamine	126
Ether	47
Ethyl acetate	9
Ethyl alcohol	8
Ethyl chloride	77
Ethyl methyl ketone	117
Ethylene dichloride	122
Ethylene glycol	59
Ethylene glycol monoethyl ether	129
Ethylene glycol monomethyl ether	129
Fatty acids	49
Ferric chloride	66
Fruit juices	50
Fluorine (dry)	51
Fluorine (wet)	52
Fluorinated refrigerants	53
Fluorocarbons	53
Fluosilicic acid	54
Formaldehyde	55
Formic acid (aq. soln.)	56
Phenol	87
Phosphoric acid 25%	88
Phosphoric acid 50%	89
Phosphoric acid 95%	90
Phosphorus trichloride	91
Picric acid	92
Potassium aluminium sulphate	10
Potassium hydroxide (aq. soln.)	31
Potassium permanganate (aq. soln.)	93
Propan-l-ol	8
Propanone	5
Propylene oxide	135
Pyridine	94
Salt (sea) water	115
Silicic acid	95
Silicone fluids	96
Sodium hydroxide (aq. soln.)	31
Sodium hypochlorite	22
Sodium peroxide (aq. soln.)	97
Sodium silicate	98
Solder flux	99
Stannic chlorides	110
Sucrose solution	100
Sulphates of Na, K, Mg, Li (aq. soln.)	101
Sulphites of Na, K, Mg, Li (aq. soln.)	102
Sulphur	103
Sulphur chlorides	104
Sulphur dioxide	105
Sulphuric acid (dilute)	106
Sulphuric acid (conc)	107
Tannic acid (aq. soln.)	108
Tartaric acid (aq. soln.)	109
TDI	130
Tetrachloromethane	30
Tin (II) tin (III) chloride	110
Toluene	19
Toluene diisocyanate (TDI)	130
Trichloroethylene	111
Trichloromethane	38
Trethanolamine	128
Trimethylamine	125
Turpentine	133
Urine	112
Vegetable oil	81
Vinegar	113
Water distilled	114
Water, salt (sea)	115
Xylene	19
Zinc chloride	116

Fig 3.1.23 (c)—***continued***

FIG 3.1.23d Lettered abbreviations

Solvent 8 (Alcohols)	M — Methanol E — Ethanol B — Butan-l-ol A — Amyl Alcohol P — Propan-l-ol
Solvent 9 (Esters)	A — Amyl Acetate B — Butyl Acetate E — Ethyl Acetate
Solvent 19 (Acromatic Solvents)	B — Benzene T — Toluene X — Xylene
Solvent 31 (Strong Alkalis)	N — Sodium Hydroxide K — Potassium Hydroxide
Solvent 35 (Chlorides)	K — Potassium Chloride N — Sodium Chloride B — Barium Chloride
Solvent 59 (Glycols)	E — Ethylene Glycol D — Diethylene Glycol P — Propylene Glycol
Solvent 132 (Phthalate Esters)	P — Dioctyl Phthalate S — Dioctyl Sebacate
Material 10 (Polyethylene)	L — LDPE H — HDPE

FIG 3.1.23e Special notes

Column 53 (Fluorinated Refrigerants)

Great variations exist in the effect of different refrigerants on some materials. It is recommended to seek manufacturer's advice at an early stage.

Row 29 (Polyurethane, Thermosetting)

Great variations exist in the suitability and corrosion resistance of materials in this class. A ▒ coding should be interpreted 'seek manufacturer's advice'.

Fig 3.1.23 (d) **and Fig 3.1.23** (e)

3.1.5 Flammability

3.1.5.1 Fire testing

The results of fire tests are acutely dependent on the method used. As a result no clear-cut statements can be made concerning the flammability, combustibility, etc., of most materials, and the best that can be achieved is a statement of the way in which a particular material or material combination (which must be closely defined) will behave in a particular fire test. The full significance of the data related to its particular test environment must therefore be appreciated if the risk of fire is to be minimised.

In assessing fire test results for plastics and similar materials it should be remembered that they are all quite specific to the test situation and that quite small changes in formulation, thickness, method of application, etc., may significantly alter performance.

Fire tests generally fall into four categories: small-scale laboratory tests, small-scale tests, large-scale tests and full-scale tests.

SMALL-SCALE LABORATORY TESTS

Many tests are included in British Standard, Underwriters Laboratory, ASTM and other authoritative specifications. As previously described they are useful as quality control techniques and can provide an indication to the knowledgeable user of those materials that may ignite very easily, for example, from a match flame. They give no indication of the behaviour of materials subjected to larger ignition sources, and extrapolation from such small tests is not valid. In a recent amendment to #BS 2782 this limitation is recognised by the BSI, and reports now have to state:

'The following test results relate only to the behaviour of the test specimens under the particular conditions of test; they are not intended as a means of assessing the potential fire hazard of the material in use.'

In addition the reports are now limited to factual statements describing the performance of the material under test. However, the designer should beware of the many items of sales literature produced from earlier tests, where data derived from the test cited in the first paragraph of this section could justify such sweeping conclusions as 'that the material is self-extinguishing' (BS 2782, 508A) or 'the material is of very low flammability' (508D).

Two small-scale tests which are widely used to characterise polymers are the Oxygen Index (LOI) and the Underwriters Laboratory tests.

The oxygen index ASTM 02863-70 test was developed to provide a graduated, numerical measure of flammability and thus overcome the ambiguity of categorical ratings. The oxygen index (LOI) is defined as the percentage of oxygen in an oxygen–nitrogen mixture that will just sustain combustion of a vertically mounted specimen bar ignited by application of an external flame to its upper end. The oxygen content of ordinary air is approximately 20%.

The Underwriter's Laboratory UL94 Burning Rating Code for specimens conditioned and tested in accordance with UL standard 94 is as follows:

Requirement	*UL94V-0*	*94V-1*	*94V-2*	*94-5V*	*HB*
Extinction of flaming combustion in a vertical bar test within: (average 10 ignitions on 5 specimens)	5s	25s	25s	60s	—

Requirement	*UL94V-0*	*94V-1*	*94V-2*	*94-5V*	*HB*
For each individual ignition extinction within:	10s	30s	30s	60s	—
No glowing combustion after: each specimen	30s	60s	60s	60s	—
Additional demands	No flaming dripping that ignites cotton	As V-0	None	Figures relate to after fifth ignition on each specimen No dripping. No significant destruction of specimen.	Max. burning rates specified dependent upon thickness. Refer to UL test Methods.

The Underwriters Laboratory tests also incorporate tests to estimate the susceptibility of plastics to ignition from electrical sources (hot wire ignition and high current arc ignition tests). Arc resistance and high voltage arc tracking rate are also measured.

Thickness and volume of test specimens usually have a very significant effect on flammability and arc/track data. For proper interpretation flammability and arc-track test values and ratings need to be accompanied by specimen dimensions. This factor, in addition to the great variation between different commercial grades and sources of the same plastic, result in the general categorisations listed above.

SMALL-SCALE TESTS

The more scientific of the small-scale tests can be exemplified by BS 476: Part 6 (*Fire propagation test for materials*), in which a 228mm square specimen of the material or composite of materials is exposed to radiation from electrical heating elements and small gas flames whilst enclosed in a small asbestos box. The hot gases produced by the heat sources and by any combustion of the specimen rise through a chimney where their temperature is measured at intervals and recorded. Prior to this stage of the test a similar run is made using a standard, non-combustible specimen, and the temperatures throughout the range are compared. It is obvious that a comparison of this type can yield positive evidence of the amount of heat the specimen will emit under this sort of fire situation.

LARGE-SCALE TESTS

For most purposes the results obtained from the relatively large-scale tests are the most relevant.

The most commonly quoted result in this category is for the surface spread of flame test for materials (BS 476: Part 7). This test procedure is usually applied to lining materials for ceilings or walls, and measures the rate at which a flame front will progress along an exposed surface in standard conditions. In essence, a 900 × 230mm panel of the material is mounted, with the long axis horizontal and the short axis vertical, at right angles to one edge of a 900mm square furnace, which is operated under standard conditions. The tendency of the material to support the spread of flame across its surface is assessed from the rate and distance of flame spread along the test-piece, according to the following four classes:

Classification	Flame spread at 90 s		Final flame spread	
	Limit (mm)	Tolerance for one specimen in sample (mm)	Limit (mm)	Tolerance for one specimen in sample (mm)
Class 1	165	25	165	25
Class 2	215	25	455	45
Class 3	265	25	710	75
Class 4	(exceeding Class 3 limits)			

FULL-SCALE TESTS

Full-scale tests fulfil the need to simulate as nearly as possible real fire conditions, which have in recent years included several major experimental fires using houses due for demolition or occasionally purpose-constructed buildings. For obvious reasons, relatively few data from this source are available, and for commercial and logistical reasons it is unlikely that individual products will generally be evaluated by such techniques. The technique is valuable because it will gradually enable predictions based on the laboratory tests to be confirmed (or otherwise) in practical situations.

3.1.5.2 Fire performance of plastics

The flammability of plastics under the Underwriters Laboratory classification system is given in:
Table 3.1.16 *Flammability of plastics*

The tendency of different polymers to generate smoke is represented diagrammatically in:
Fig 3.1.24 *Smoke emission of some plastics on burning*

Most plastics and similar materials will burn if the combustion conditions are severe enough, yet the greater proportion can be modified by the inclusion of additives or by clever design so that they will satisfy the fairly stringent conditions of the large-scale British Standard tests. The inclusion of additives to reduce fire risk is usually accompanied by a deterioration in other properties.

The effect of additives is shown in:
Table 3.1.17 *Typical properties of some flame resistant thermoplastics*
and for thermoplastic sheet in:
Table 3.1.18 *Properties, application data and relative costs of fire-rated thermoplastic sheet.*

TABLE 3.1.16 Flammability of plastics

Material	*Limiting oxygen index*	*Underwriter's laboratory ratings*			*Flash ignition temperature ASTM D1929*	*Self ignition temperature ASTM D1929*	*Arc resistance*
		Best	*Most*	*Worst*			
Units	%	*UL94*	*UL94*	*UL94*	°C	°C	s
ABS	18–20	V–O	HB	HB	—	—	20–130
Acetal	15–16	HB	HB	HB	—	—	113–240
Acrylic	17–19	HB	HB	HB	280–300	450–462	—
Alkyd	31–63	V–O	V–O	HB	—	—	149–241
Cellulose acetate	19	HB	HB	HB	305	475	—
Cellulose butyrate	—	HB	HB	HB	—	—	—
Diallyl phthalate	—	V–O	Range	HB	—	—	115–186
Epoxies	20–30	V–O	Range	HB	—	—	120–220
FEP	>95	V–O	V–O	—	—	—	—
PTFE	95	V–O	V–O	V–O	—	—	—
PFA	>95	—	—	—	—	—	—
ECTFE	60	—	—	—	—	—	—
Melamine–formaldehyde	40	V–O	V–O	V–O	—	—	130–>180
Nylon 6	22–28	V–O	V–2/HB	HB	—	—	33–190
Nylon 6/6	21–32	V–O	Range	HB	—	—	4–500
Nylon 6/10	25–28	V–O	HB	HB	—	—	120–140
Nylon 6/12	—	—	V–2	—	—	—	121
Nylon 11	—	V–O	V–2	HB	—	—	122–123
PETP	20–31	V–O	HB	HB	—	—	31–81
PBT	20–37	V–O	V–O	HB	—	—	10–150
Phenolic	25–36	V–O	Range	HB	—	—	14–185
PPO	28–29	V–O	Range	HB	—	—	3–96
Polycarbonate	25–32.5	V–O	Range	HB	—	—	5–130
Polyester	20–80	V–O	V–O	HB	—	—	28–242
Polyethersulphone	34–40	V–O	—	—	—	—	70–80
Polyethylene	17–18	V–2	HB	HB	340	440–445	140–150
Polyamide	44	—	SE–O	—	>540	>540	125–180
PPS	38–47	—	VE–O	—	—	—	34–123
Polypropylene	17–28	—	V–O	HB	—	430–440	69–183
Polystyrene	18	V–O	HB	HB	345–360	488–496	48–120
Polysulphone	30–39	V–O	V–O (VE–O)	V–2	—	—	39–122
Polyurethane	29 max.	V–O	HB	HB	—	—	90
Polyurethane (rigid foam)	—	—	—	—	310	416	—
PVC	45–49	V–O	V–O	V–O	390	454	5
PVDC	60	—	—	—	>530	>530	—
SAN	28	V–O	HB	HB	366	454	60–70
Urea formaldehyde	30	V–O	V–O	HB	—	—	—
GRP (glass-reinf. polyester)	—	—	—	—	345–400	480–490	—
ETFE	30	—	—	—	—	—	—
Polyamide–imide	43	—	—	—	—	—	—
Aromatic copolyester	37	—	—	—	—	—	—
			Other materials				
Wool	24–25.6	—	—	—	200	590	—
Cotton	17.5–18.0	—	—	—	230–266	254	—
Paraffin wax	16	—	—	—	—	—	—

Table 3.1.16

TABLE 3.1.17 Typical properties of some flame resistant thermoplastics

Property \ Material	*Polypropylene*			*Nylon*		*Poly-carbonate UL V–O*	*PPO UL V–O*	*PBT UL V–O*
	Unmodified	*UL V–2*	*UL V–O*	*UL V–2*	*UL V–O*			
Specific gravity	0.905	0.94	1.25–1.39	1.07–1.20	1.20–1.27	1.21–1.25	1.06–1.10	1.40–1.44
Tensile strength, MN/m^2	28–37	30	19–22	62–90	61–70	63–67	48–76	55–61
Elongation to break, %	>200	190	3.5–100	10–300	30–65	15–110	35–60	15–100
Flexural Modulus, GN/m^2	1.24–1.52	1.48	1.52–2.48	1.21–2.86	2.83–2.97	2.23–3.45	2.23–2.48	2.55–2.76
UL temperature index, °C	95–115	115	105	105	90–105	115–125	80–110	120–140
Heat deflection temperature (455 kN/m^2)	92–122	122	109–118	166–243	204–221	138–179	138–157	149–179
Hot wire ignition (0.25 in), s	6–24	33	22–243	7–20	15–300	16–134	9–300+	20–23
High-current arc ignition, number of arcs to ignite	200+	200+	200+	200+	200+	17–200+	29–105	200+
Dielectric strength 10^2 kV/m	236	463	207	236	217	157	157	138–228
Volume resistivity Ω cm	1×10^{17}	5×10^{15}	5×10^{15}	1×10^{14}	1×10^{14}	1×10^{14}	1×10^{16}	1×10^{15}

Table 3.1.18

TABLE 3.1.18 Properties, application data, and relative costs of fire-rated thermoplastic sheet

Property	*Resin*						
	Polysulphones	*Polycarbonates*	*PVC (modified)*	*ABS/PVC alloys*	*ABS (unmodified)*	*Polystyrenes (unmodified)*	*Polyolefins (unmodified) High density polyethylene and polypropylene*
Fire-rating test[a] On 0.050-in-thick sample FAR, Par. A. UL 94, V–O FAR, Par. B MVSS	Pass Pass Pass Pass	Pass Pass Pass Pass	Pass Pass Pass Pass	Fail Fail Fail Pass	Fail Fail Fail Pass	Fail Fail Fail Pass	Fail Fail Fail Pass
On 0.125-in-thick sample FAR, Par. A UL 94, V–O FAR, Par. B MVSS	Pass Pass Pass Pass	Pass Pass Pass Pass	Pass Pass Pass Pass	Fail Pass Pass Pass	Fail Fail Fail Pass	Fail Fail Fail Pass	Fail Fail Fail Pass
Smoke generation using NBS tests	Low	Low	Moderate	High	High	High	Low
Heat distortion temperature (1.82 kN/m^2) °C	162–176	129–140	71–77	77–88	77–107	49–104	HDPE: 43–54
Flexural Modulus (RT), GN/m^2	2.1–2.7	2.2–2.4	1.7–2.1	1.5–1.9	0.9–2.9	2.8–3.2	HDPE: 0.7–1.8
Impact resistance	Low (unless modified)	High	High	Moderate	Low to high	Low	Moderate to high
Specific gravity	1.20–1.24	1.2	1.3–1.4	1.17–1.25	1.04–1.09	1.04–1.09	HDPE: 0.94–0.96 PP: 1.7–2.5
Formability	Very difficult	Moderately difficult	Moderately difficult	Easy to difficult	Easy	Easy	Difficult
Reprocessability	Yes	Yes	Generally not reprocessable	Yes	Yes	Yes	Yes
Cost: relative price per unit mass of sheet	11–16	6–8	3.5–5	3–4	2–2.5	1–1.5	1–1.5
Typical applications	Potential in aircraft and rapid transit, chemical industry	Aircraft, rapid transit, impact resistant glazing, exterior signs	Rapid transit, aircraft, wall panels	Electronic instrument housings, rapid transit, aircraft	Components, concrete forms, recreational products, trays	Food packaging, trays, point-of-purchase displays	Pallets, boxes, dustbins, liners

[a] Fire-rating tests.

Notes:

(a) US Federal Aviation Regulation 25,853 Paragraph A.
Specimen vertically mounted in frame, flame applies for 60 s. Material must self-extinguish within 15 s or less.

(b) Underwriters Laboratory UL 94 V–O.
Specimen suspended over flame for two separate 10-s ignition intervals. After both, material must self-extinguish within 15 s or less.

(c) US Federal Aviation Regulation 25,853 Paragraph B. As (a) above but flame applies for 12 s only.

(d) US Motor Vehicle Safety Standard 302. Specimen horizontal, flame applied to one end. Material must burn less than 4 in in 1 min.

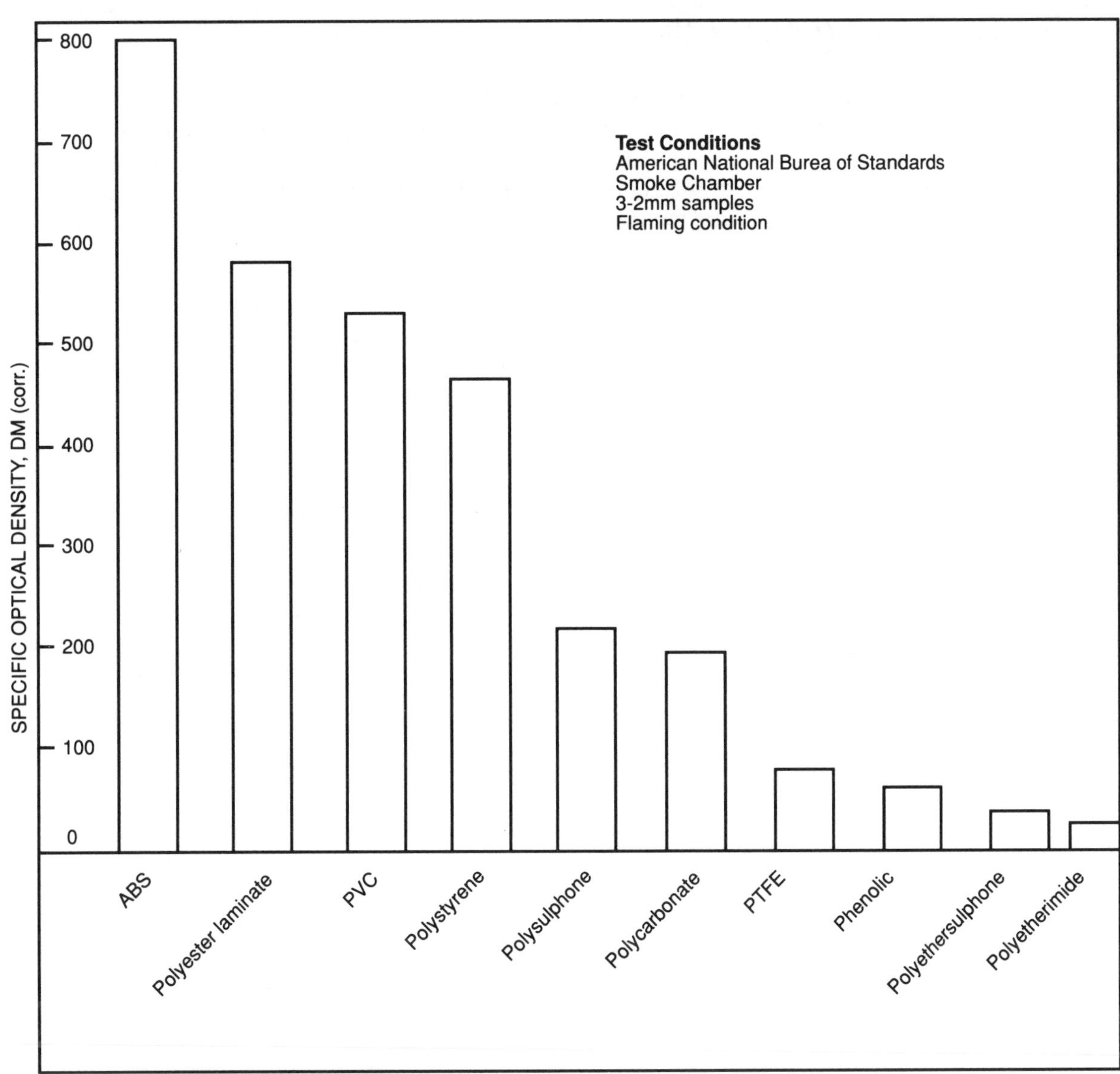

FIG 3.1.24 Smoke emission of some plastics on burning
Reproduced by permission of ICI Ltd. Plastics Division

3.1.6 Testing methods

Wherever possible, the values quoted in the tables for individual plastics in the following chapters have been evaluated by ASTM standard test methods. In those few cases where this has not been possible values evaluated by BSS or DIN standard methods have been listed instead. Standard methods by the three institutions are listed in:
Table 3.1.19 *Standard methods of testing plastics*

TABLE 3.1.19 Standard methods of testing plastics

	Property	Castings and mouldings: ASTM	Castings and mouldings: BS2782			Castings and mouldings: DIN		Film & sheet: ASTM
		Method	Part	Method	Date	Method	Date	Method
Mechanical	Tensile strength at break	D638	3	301		53455		D882
	Elongation at break	D638	3	301		53455	1968	
	Tensile yield strength	D638				53455		
	Compressive strength (rupture or yield)	D695	3	303		53454	1972	
	Flexural strength (rupture or yield)	D790	3	304		53452	1977	
	Tensile modulus	D638	3	302		53457		
	Compressive modulus	D695			1957	53457	1968	
	Flexural modulus	D790				53457		
	Izod impact	D256A	3	306		53453	1975	
	Hardness Rockwell Shore/Barcol	D785 D2240/ D2583						
	Shear strength		3	305				
	Bursting strength							D774
	Tearing strength initial							D1922
	Folding endurance							D2196
Physical	Specific gravity	D792				53479		
	Coef. of linear thermal expansion	D696				(VDE) 0304		
	Thermal conductivity	C177						
Electrical	Volume resistivity	D257	2	202		43482	1979	
	Surface resistivity		2	203				
	Dielectric strength	D149			1957			
	Dielectric constant	D150				53483	1955 and 1970	
	Dissipation factor		2	206				
	Arc resistance	D495						
Thermal and environmental	Flow temperature		1	105	1956	53460	1976	
	Heat deflection temperature	D648				53458	1968	
	Max. recommended service temperature	D542						
	Flammability	D635				53461	1969	
	Water absorption	D570				53495	1973	
	Rate of water vapour transmission							E96E
	Gas permeability							D1434
	Tabor abrasion resistance	D1044						
Processing	Mould shrinkage/post-mould shrinkage	D995	1	106	1956	53464	1962	
	Heat sealing temperature range							D759

3–2

Thermoplastics

Contents

List of tables

List of figures

3.2.1 Nylons (polyamides)

3.2.1.1 General characteristics and properties

The advantages and limitations of nylons compared to other materials are given in:
Table 3.2.1 *Characteristics of nylons*

AVAILABLE TYPES AND THEIR APPLICATIONS AND PROPERTIES

The commercially available nylons form a group of crystalline thermoplastics with the one exception of the transparent grade which is amorphous. The degree of crystallinity of nylons, which can be controlled by processing methods, affects the performance properties in that increased crystallinity imparts increased tensile strength, stiffness, heat distortion temperatures and resistance to chemicals, coupled with lower impact strength, toughness and elongation, and decreased thermal expansion and lower heat distortion temperature. The crystallite size also affects the performance properties and by nucleation of large numbers of small crystallites higher strength and toughness can be achieved.

The system of identification used in relation to the nylon family is based on reference to the number of carbon atoms in the materials, from which the polymer is produced. Nylon 6, for example, is produced from capro-lactam, which contains six carbon atoms; and nylon 6.6 is formed from hexamethylene diamine, with six carbon atoms and adipic acid, also having six carbon atoms. They form a most versatile family of engineering thermoplastics. The grades in general use include nylon 6, 6.6, 6.10, 6.12, 11 and 12, 4.6 and 12.12 and also a transparent grade. They are compared in:
Table 3.2.2 *Comparative characteristics of different nylon types*, and
Table 3.2.3 *Typical comparative properties of polyamides (unfilled)*
Fig 3.2.1 *Impact strength vs tensile strength for nylons*
Fig 3.2.2 *Flexural strength vs impact strength for dry nylons*
Fig 3.2.3 *Elongation vs tensile strength for nylons*
Fig 3.2.4 *Flexural modulus vs impact strength for dry nylons*

Nylons with a much higher impact strength (toughness) than normal nylons are produced by a number of companies. Characteristics of one grade are given in:
Table 3.2.4 *Properties and applications of a toughened nylon*

Nylons are also available as film. Nylon films exhibit high tensile strength and elongation, good impact strength, gas impermeability, high melting point and good low temperature (–70°C) flexibility and strength, abrasion resistance, good flex-crack resistance, good clarity. Increasing moisture content reduces barrier properties but increases elongation and flexibility.

Film can be sealed by all conventional methods, usually hot bar and thermal impulse. Major application is vacuum packaging of processed foods, frequently combined with ionomer or polyethylene; it is sometimes PVDC coated to enhance barrier properties.

Additives and fillers

Plasticisers, additives and fillers are used to produce a wide range of grades within each material type, tailored to specific needs of performance or processing. (One supplier has almost 90 grades of nylon 11 alone.)
Additives are used with all types of nylons to:

impart resistance to degradation by heat and light;
impart resistance to attack by fungi, etc;
act as flame retardants;
improve processing characteristics.

A large number of fillers is available to improve the properties of each type of nylon or to produce specific application characteristics. A summary of the advantages and limitations of the more commonly used fillers is given in:

Table 3.2.5 *Effect of fillers commonly incorporated in nylons*

Typical properties of filled (mostly reinforced) nylons are given in:

Table 3.2.6 *Typical comparative properties of polyamides (filled)*

3.2.1.2 Mechanical properties of nylons

The stress which a nylon will withstand depends on the long-term creep at operating temperatures and for very special applications on the fatigue properties. It may be strongly influenced by the environment, particularly by the absorption of water.

The presence and orientation of fibre reinforcement strongly influences the mechanical properties of polymers. Advantage can be taken of this phenomenon in those cases when the reinforcement is built into the polymer by a 'laying up' process. Reinforced nylon is, however, usually produced by moulding a compound which already contains the reinforcement.

Glass fibre-reinforced nylons are anisotropic materials, which means that their properties are dependent on the direction in which the glass fibres are lying. When glass fibre-reinforced nylon is moulded, the fibres tend to align themselves in the direction of flow of the melt into the cavity. The extent to which this happens depends on a number of processing variables, the most important of which are the position of the gate, the melt temperature and the speed of injection. In most mouldings, however, because of complex flow paths in the cavity, there is no pronounced orientation and properties are substantially uniform. Where there is pronounced orientation, as in a simple rod moulded with the gate at one end, it is found that the material has a higher strength when the measurement is made in the direction of fibre orientation.

A serious problem associated with the anisotropy of glass fibre-reinforced nylon is warpage. Layers of different orientation exist in a moulding and warping can be reduced by increasing the randomly orientated core at the expense of the orientated layers just beneath the surfaces. The same effect can be achieved by increasing section thickness and the degree of orientation in surface layers can be reduced by introducing ribs, or more than one gate. Ribs do not substantially reduce orientation but they do increase rigidity which reduces warpage.

Alternatively, modification of the filler (i.e glass beads) is employed.

The effect of orientation on tensile properties is shown in:

Fig 3.2.5 *Effect of glass fibre orientation on tensile strength of nylon 6.6 (dry)*, and

Fig 3.2.6 *Effect of glass fibre orientation on tensile modulus of nylon 6.6 (dry)*

The variation of flexural modulus, apparent creep modulus, and notched impact strength of reinforced and unreinforced nylons with temperature and humidity are shown in:

Fig 3.2.7 *Variations of flexural moduli of nylons with temperature and humidity*,

Fig 3.2.8 *Effect of glass fibre reinforcement on creep of different types of nylons (room temperature)*, and

Fig 3.2.9 *Variation of impact strength of nylons with temperature (also showing acetal copolymer)*

3.2.1.3 Chemical and environmental resistance

The most significant environmental influence on nylons is the effect of moisture.

Nylons absorb moisture until equilibrium is achieved with the surroundings. The moisture causes swelling, impairs some properties (creep, stiffness, tensile and flexural strength and electrical characteristics) but improves others (ductility and impact strength). The effect is most marked in Types 6 and 6.6 and decreases progressively in Types 6.10, 6.12, 11 and 12, respectively. At saturation the strengths of Types 6.10 and 6.12 exceed those of Types 6 and 6.6. Fillers are particularly effective in reducing the deleterious influence of moisture but filled grades are still affected by moisture to a degree.

As moisture absorption is a diffusion controlled process, the swelling is most serious in thin-sectioned components, but in others the effect is often less important as increased flexibility and resilience allow the component to deform slightly and compensate the change in dimensions.

In addition, the absorption of water is a very slow process and day to day variations in humidity have an insignificant effect on performance except in very thin sections and where flexure is important. The rate of water absorption depends on the relative humidity, type of nylon, degree and type of reinforcement, the dimensions of the component and the temperature. Water absorption is illustrated in:

Table 3.2.7 *The influence of section thickness on water absorption of nylon 6.6*, and

Fig 3.2.10 *Equilibrium water content vs relative humidity of environment*

The effects of moisture on linear dimensions and flexural modulus of nylons is shown in:

Table 3.2.8 *Effect of moisture on properties of nylons*, and

Fig 3.2.11 *Equilibrium change in linear dimensions vs relative humidity of environment*

For the glass-filled material, the change of dimensions depends on the orientation of the fibres—maximum changes occurring perpendicular to fibre orientation and minimum parallel to the fibres. Absorbed moisture also impairs the efficiency of nylon as an electrical insulator, as shown in:

Fig 3.2.12 *Steady state volume resistivity vs relative humidity of environment*

The effect of solvents and other chemicals on nylons is listed in:

Table 3.2.9 *Chemical and environmental resistance of nylons*

The information in the table is a general guide and only applicable up to 60°C. While more detailed information can be obtained from suppliers, guarantees of the suitability of material will often be dependent on the results of testing in the intended environment.

3.2.1.4 Friction and wear properties (see also Vol. 1, Chapter 1.5)

The low coefficient of friction in dry rubbing situations and the ability to absorb abrasive foreign particles are responsible for widespread use of nylons in gears and bearings. Initial lubrication of nylon in gears and bearings is beneficial in extending the life of components but will not extend load bearing–velocity parameters. Water does not act as a lubricant. Continuous lubrication will produce the best results and no degradation occurs in lubricants, which are not normally required for slower speeds.

Reinforcement with glass fibre imparts higher load-bearing properties particularly at lower speeds. PTFE does not change its viscosity with temperature (cf. oil base lubricant) and load–velocity limits are increased at all speeds. PTFE is much more effective in lowering wear rates than glass fibres. However, PTFE lowers strength and it is common to add both PTFE and glass fibre to achieve optimum performance in extreme wear regimes.

The reduction in wear and improvement in PV limits by adding PTFE and glass fibre are summarised in:

Table 3.2.10 *Wear rate and friction of filled and unfilled nylon rubbing against various materials*, and

Fig 3.2.13 *Limiting P–V curves for different types of nylons showing effect of PTFE and glass fillers.*

Note the high wear of unfilled nylon rubbing on unfilled nylon and, as with metals, like rubbing on like should be avoided. However, nylon and acetal, in combination,

produce very low friction and wear rates. Common mating materials are steel and cast iron. Softer materials such as brass are not recommended, as nylon will tend to wear them rapidly away.
NB: P–V limited data in Fig 3.2.13 should not be used as design data as the test conditions under which this data is collected may not simulate service conditions and material suppliers should be consulted for recommended design limits. However, the data give a useful relative guide.

3.2.1.5 Processing of nylons

a. The nylons can be processed by most methods used in the plastics industry, although processes such as casting, coating and rotational moulding are restricted to specific types.
b. Each nylon is capable of variation in terms of melt viscosity to suit a particular conversion process. A blow moulding grade is designed to be of greater viscosity than, for example, a grade used to produce a fine detail injection moulding.
c. Nylons, being semi-crystalline polymers, have sharp melting points and melt to low viscosity, which necessitates care in processing, but also make possible thin section work and reproduction of fine detail. A nozzle with a shut-off valve is necessary to prevent drooling.
d. Wide variations exist in cycle times between grades. Fastest injection moulding times are with nucleated or controlled crystallisation grades and further improvements can be obtained by improving mould release. There is some evidence that the nucleated grades, produced by the addition of microscopic inert particles to stimulate nucleation, may be more brittle in the dry as-moulded condition.
e. With mineral-filled grades the surface finish is very good on small parts, but can vary on larger ones. In thick sections flow resembles that of glass-reinforced grades; equipment wear may be reduced by up to 30%.
f. The glass sphere-filled grades are well suited for producing thin-section components of a highly intricate nature. Sections of 0.005 in thickness have been successfully moulded by injection.
g. Glass-filled nylon requires high (60–100°C) mould temperatures and high injection pressures for optimum surface properties.

Typical processing temperatures and mould (linear) shrinkage for different nylon types are summarised in:
Table 3.2.11 *Processing characteristics of different nylon types*

TOLERANCES

Dimensions of 6.6 mouldings can, with best practice, be kept within ± 0.02%. Closer tolerances are possible if control of the moulding process is given special consideration.

SHRINKAGE

a. Post-moulding shrinkage and water absorption may cause variations in dimensions.
b. A warm mould will reduce post-moulding shrinkage to a very low level.
c. An annealing treatment in, for example, a refined oil will further minimise post-moulding shrinkage though change occurs during annealing.
d. Shrinkage will be greater in the case of thick mouldings due to the centre remaining molten for a longer time allowing a higher degree of crystallisation to take place.
e. The shrinkage in mineral-filled materials is uniform in all directions.

f. The post-moulding shrinkage of the transparent grade is only 0.02–0.03% (because it is non-crystalline).

MACHINABILITY

Most nylons can be readily machined using techniques akin to those used for light alloys. A machined component will be different in characteristics from a moulded component of similar configuration.

FABRICATION

Nylons can be joined with adhesives, induction bonding and ultrasonic welding.

FINISHING

Electroplating grades have been developed. It is possible to print on all types of nylon. Types 11, 12 can be vacuum metallised, as can mineral-reinforced 6.6.

DRYING OF MOULDING POWDER

The granular material is hygroscopic and, if left exposed for long periods of time, will absorb moisture to cause defects on mouldings. If a quantity of material is allowed to absorb moisture it should be redried before use. There is a danger of degradation and discolouration at temperatures above 80°C if material is dried in the presence of air.

CONDITIONING

When dimensional stability is important, the process of conditioning (absorption of moisture into the mouldings) is an essential part of the production of a satisfactory nylon 6 or 6.6 moulding; at equilibrium a moulding will absorb 2½–3½% of water. Whilst this can be allowed to take place naturally for very thin mouldings, the average part will condition too slowly in this way. It is therefore common practice to accelerate conditioning by immersion in water, generally at elevated temperature (60–80°C), until the necessary uptake of water has been achieved.

Mouldings are often conditioned for much shorter periods of time to achieve a minimum degree of toughening, the remaining conditioning being left to absorption of moisture from the air.

3.2.1.6 'Nylons' produced by reaction injection moulding (nylon polyoxypropylene copolymer)

The nylon RIM system is a copolymer, consisting of nylon and polyoxypropylene blocks. The nylon is said to give the system good high temperature modulus behaviour; the polyoxypropylene block gives good low temperature impact strength and toughness. Nylon RIM is said to offer better heat sag resistance than the more common polyurethane RIM systems and holds up well in a paint bake oven up to 160°C, compared with the urethane limit of about 120°C under the same conditions. Processing benefits include no need for post-curing and, as a thermoplastic material, the ability to re-use the scrap in conventional injection moulding operations.

Typical applications are expected to be front and rear automobile ends, valve covers and under-bonnet applications. Properties are shown in:

Table 3.2.12 *Typical properties of nylon RIM*

3.2.1.7 Trade names and suppliers

Suppliers and trade names of nylons are listed in:

Table 3.2.13 *Major suppliers of nylon grades*

TABLE 3.2.1 Characteristics of nylons

	Advantages	***Limitations***
Mechanical properties	Good combination of mechanical properties—fatigue and creep strength, stiffness, toughness and resilience—only slightly inferior to polyacetals.	All nylons absorb or give up moisture (the magnitude varies with different types) to achieve equilibrium with ambient conditions—moisture acts as a plasticiser and decreases tensile and creep strength and stiffness and increases impact strength and the dimensions of the component. The effect is most serious in thin-sectioned components. Because nylons depend upon moisture for impact performance, embrittlement can occur in desiccated air.
Wear	Good abrasion resistance (ability to absorb foreign particles) and self-lubricating properties are responsible for widespread use in gears and bearings.	
Thermal	Suitable for prolonged service temperatures of 80–100°C and this can be increased to 140°C with heat stabilised grades.	Thermal expansion varies with temperature and moisture content.
Electrical properties	Good commercial insulator but...	Electrical properties are greatly influenced by moisture content and/or temperature increase.
Environmental	All nylons are resistant to fuels, oils, fats and most technical solvents such as aliphatic and aromatic hydrocarbons, chlorinated hydrocarbons, esters, ketones and alcohols. All have good alkali resistance.	All nylons are attacked by strong mineral acids, acetic acid and dissolved by phenols. Some types are dissolved by formic acid. UV attacks unstabilised nylons causing embrittlement in a comparatively short time.
Food and medicine	Can be used in contact with most foodstuffs at room temperature and sterilised by steam or infra-red radiation.	
Fillers	Wide range of fillers and additives to improve specific properties and reduce limitations of unmodified materials, e.g. glass fibre filler greatly reduces effects of moisture on dimensions and properties compared with unfilled materials.	
Processing	Most material types are available in grades suitable for injection, blow and rotational moulding and extrusion, with additional possibilities of fluidised bed coatings, sintering and casting for special grades. The latter (casting from monomer) is particularly useful for producing large stress-free sections in small economical batches.	Nylons have a sharply defined melting point and high shrinkage values occur on moulding thick sections. Nylons (with one exception) are crystalline; this results in longer cycle times in moulding. Conditioning of mouldings is frequently necessary. Those fast cycling grades reliant on a nucleation mechanism may be especially brittle in the dry as-moulded condition.

TABLE 3.2.2 Comparative characteristics of different nylon types

Nylon type	*Advantages*	*Limitations*	*Typical uses*
6.6	Most widely used nylons—strongest widest range of temperature and moisture of any unfilled nylon. Good resistance to abrasion. With Type 6 the cheapest of all nylon types. Lowest permeability rate for petrol, mineral oil and fluorocarbon refrigerants.	High moisture absorption. Low impact strength and ductility when dry. Most sensitive of all nylon types to UV and oxidative degradation. Lower resistance to weak acids than types 6.10, 6.12, 11 and 12.	Bearings, bushes, cams, gears, hydraulic fluid lines, hoses, petrol piping, compressed air lines, propellers, impellers, rollers. Growth area in electrical components, stimulated by improved fire retardancy. Replacement for die-cast hand tool bodies.
6	Better impact strength and flex-fatigue life than Type 6.6 under moist conditions. Other mechanical properties only slightly inferior to 6.6. Good resistance to abrasion. Same raw material cost as 6.6. Can be processed at lower temperature and is less crystalline than Type 6.6 which means mould shrinkage is low and closer tolerance possible. Transparent grades available.	Highest moisture absorption of any type with attendant dimensional instability and changes in mechanical and electrical properties. Effects can be substantially reduced by alloying with LDPE. Low impact and ductility when dry. Lower resistance to weak acids than 6.10, 6.12, 11 and 12.	Castors, curtain runners, door catches, fasteners, bearings, gears—components requiring higher impact strength than type 6.6 but not requiring higher yield strength. Most widely employed grade in West Germany.
6.10	Lower moisture absorption than 6 or 6.6 (½ Type 6 and ½ Type 6.6) and therefore smaller property and dimensional changes and better electrical properties. Very low embrittlement temperature.	Lower strength than 6 and 6.6. Higher cost than 6 and 6.6.	Electrical insulation and similar components to 6 and 6.6 where lower moisture absorption justifies extra cost.
6.12	Expected to replace 6.10 - (better heat resistance and cheaper). Properties generally between 6 and 6.6 with better moisture resistance. Slightly superior creep properties to 6.6 under moist conditions.	Higher cost than 6 and 6.6. Lower strength than 6.6.	
11 and 12	Lower moisture absorption than other types (Type 12 lowest of all unfilled nylons) and consequently better electrical properties and dimensional stability, even when plasticised. Greater flexibility and impact strength in dry condition than 6, 6.6, 6.10 or 6.12. Flexible grade of 11 available. Type 12 second only to flexible copolymers. Type 12 has excellent stress-crack resistance, and lowest density of all nylons. Both types resistant to formic acid.	Lower strength than 6.10 or 6.12. Type 12 has lowest strength and impact strength. Higher cost than 6.10 or 6.12. About twice the price of nylon 6.6. Lower maximum service temperature than 6, 6.6, 6.10 or 6.12 (same as flexible copolymers).	Electrical insulation, corrosion resistant coatings and moulded components similar to 6 and 6.6 where lower strength can be tolerated and need for lowest moisture absorption (i.e. dimensional stability and constant properties) justify higher cost—e.g. automotive nylon 11 is used for air brake hose on French lorries, where, at low temperatures, nylon 12 would fail in a brittle mode. Flexible grade expected to be used in bearing runners, sealing gaskets, bicycle seats and

Table 3.2.2

TABLE 3.2.2 Comparative characteristics of different nylon types—*continued*

Nylon type	*Advantages*	*Limitations*	*Typical uses*
11 and 12 *(continued)*	Lower melting points and processing temperatures. Both types used as coatings Weathering resistance better than all nylons except super tough grades of nylon 6.6.		pneumatic tubing. Small amounts of transparent 11 and 12 used in packaging.
4.6	High impact strength. Lower creep at higher temperatures than 6.6. Better fatigue behaviour than 6.6. Much higher stiffness at high temperatures than 6.6.	Lower processing window. Impact properties influenced by processing temperature. Higher moisture absorption than 6.6. Cost penalty of 1.5–2.5 times 6.6. Mould shrinkage of 1.5–2.4% (unfilled).	Bearing cages, gearwheels and related applications demanding wear resistance. Clutch pin for Volvo 300 series.
12.12	Better properties than 6 or 6.6. Less expensive than 11 or 12. Higher m.p. and better retention of part shape than nylon 12. Similar low moisture absorption to nylon 12. Better resistance to hydrolysis and common chemicals than nylon 6.6 and polyesters.	More expensive than 6 and 6.6.	Replacement of metals in valves fittings and similar fluid applications, and for flexible extrusions to replace metal in automotive fuel lines.
Cast nylon 6	Complex parts of large section size can be processed in economical small batches. Properties similar to Type 6.6, but with better resistance to wear and fatigue. Better dimensional stability and lower moisture absorption. Improved impact strength in larger sizes.	Lowest ductility of unfilled types. Dimensional tolerances inferior to injection moulded parts.	Bearings, wear plates, bushings, rollers, stock shapes—components requiring thicker sections than can be obtained by moulding or small batch production.
Type 6 and 6.6 Flexible copolymers	Higher elongation and impact strength than general purpose Type 6 and 6.6. Cheaper than 6.10, 6.12, 11 and 12.	Less resistant to weak acids than 6 and 6.6. High moisture absorption—comparable with 6 and 6.6. Not self extinguishing. Lower maximum service temperature than 6, 6.6, 6.10 and 6.12 (same as Type 12).	Parts requiring high impact strength and flexibility. Adhesive applications. Very small percentage of nylon market.
Transparent amorphous type	Easy to process with lower shrinkage values than other nylons. Scratch and abrasion resistant. Low moisture absorption—comparable with 6.10 with consequent electrical and dimensional characteristics. High resistance to stress cracking and crazing.	Lower deflection temperatures than other nylons.	Lenses, containers, gauges, instrument windows.

Table 3.2.2—*continued*

TABLE 3.2.2 Comparative characteristics of different nylon types—*continued*

Nylon type		Advantages	Limitations	Typical uses
Sintered grades		High crystallinity which improves wear and deformation resistance and lowers hygroscopic and thermal expansion. Incorporation of molybdenum disulphide further increases dimensional stability and minimizes internal strains. Porosity offers oil impregnation and retention possibilities.	Lower tensile and impact strength than moulded components.	Gears, bearings, bushings, etc., requiring greater degree of lubrication.
Filled and reinforced grades	General	Much improved heat resistance, stiffness, yield and creep strength, reduction of deleterious effects of moisture, possibility of improving friction and wear characteristics.	Loss of self extinguishing properties. Higher cost.	As above but components requiring better strength, stiffness or friction and wear properties.
	Glass fibre	Glass-filled grades available with highest impact resistance of any thermoplastic (Zytel ST –Du Pont Nylon 6.6).	Impact strength (measured by energy to break) is reduced, but notch sensitivity also reduced, giving improved performance in some cases. Warpage and anisotropic properties.	Glass-filled grades used in metal-replacement applications in the automotive, agricultural industries and in consumer durables.
	Glass bead	Employed where low warpage/anisotropy is desirable at the expense of mechanical properties.		
	Mineral	Improved performance over unreinforced nylons and better than most glass-reinforced grades on toughness. Cheaper than glass and mould rapidly. Good dimensional stability, fire and wear resistance. Can withstand storing temperatures.	Inferior to glass-reinforced grades except on toughness. Toughest grades have lowest strength.	Licence plate holders. Power tool housings. Textile bollin flanges. Outboard motor propellers. Snap and press fits. Carburettor parts in lawn mowers. Replacements for automotive die-cast components in zinc and aluminium, and for phenolic resins.

Table 3.2.2—*continued*

TABLE 3.2.3 Typical comparative properties of polyamides (unfilled)

Property		Mechanical properties									Thermal properties		
		Tensile yield strength	*Ultimate elongation*	*Modulus of elasticity in tension*	*Flexural strength*	*Modulus of elasticity in flexure*	*Impact strength (IZOD notched)*	*Compressive strength (1% offset)*	*Hardness*	*Abrasion resistance (Taber CS-17)*	*Thermal conductivity*	*Thermal expansion coefficient*	*Maximum recommended service temperature*
Units		(MN/m^2)	(%)	(GN/m^2)	(MN/m^2)	(GN/m^2)	(J/m)	(MN/m^2)	(Rockwell)	(mg/1000 cyc.)	(W/mK)	(10^{-5} K^{-1})	(°C)
ASTM test method		D638	D638	D638	D790	D790	D256	D695	D785	D1044	C117	D696	D542
Material	Condition	1	2	3	4	5	6	7	8	9	10	11	12
Type 6.6 General purpose (GP)	Dry	51–81	60	3.2	98–137	2.8–3.3	50 (35–37[b, d])	34	R118	3–5	0.2–0.3	7.7–11.0	80–110[a]
	50% RH	58 (40[d])	300–540	2.6	40	1.2 (0.93[d])	100 (90[d])	—	R108	6–8	—	—	
Modified high impact Type 6.6	Dry	52–66	40–80	—	68	1.57–2.0 (1.45[d])	187–900	—	R112	5–6	—	—	—
	50% RH	35–52	210–270		—	0.6–1.1	800–1330	—	R89	7–8			
Type 6 General purpose	Dry	65–86	20	2.6–3.0	—	2.7–3.2	53	66	R119	5	0.2–0.3	4.8–8.6	80–110[a]
	50% RH	24–53	180–300	1.4	35	0.75–1.1	160–490[c]	—	—	—		—	
Type 6.10 General purpose	Dry	50–70	85	1.9	100	1.9	32	21	R111	—	0.2	9.0	108
	50% RH	49	300	1.1	—	1.1–1.5	85	—	—	—	—	—	
Type 6.12 General purpose	Dry	60	150	2.0	87–90	2.0–2.6	53	16	R114	—	0.2	9.0	—
	50% RH	51	340	1.2	—	1.2	75	—	—	—	—	—	
Type 11	Dry	55	100–300	1.3	—	0.3–1.3	70–110	50	R108	—	0.3	10.0	100–121[a]
	50% RH	—	380	—	—	—	—	—	—	—	—	—	
Type 12	Dry	49–65	120–350	1.1–1.4	—	1.2–1.55	49–274	—	R106	—	0.25–0.35	10.0	80–110[a]
Plasticised 12	Dry					0.3–0.42	No break						
Type 6 Cast	Dry	75–96	10–100	2.4–3.1	113	3.5	50	96	R112–120	2.7	—	10.0	95–100
Type 6 Flexible copolymer	Dry	51–68	200–320	1.6–2.8 (Dry) 0.3–0.7 (Moist)	23–112	2.6	80–No break	—	R72 R119	4.5	—	2	80–93[a]

[a] Heat stabilised grade.
[b] Typical nucleated fast cycling grade.
[c] High impact grade.
[d] Fire retardant grade.
[e] High impact types tend to be lower.

TABLE 3.2.3 Typical comparative properties of polyamides (unfilled)—*continued*

Thermal properties						*Electrical properties*									
Deflection temperature			*Water absorption*					*Dielectric constant*		*Dissipation factor*				
0.45 MN/m²	*1.8 MN/m²*	*Flammability*	*24 h*	*50% RH*	*Saturation*	*Volume resistivity*	*Dielectric strength (short time)*	*60 cycles*	*10⁶ cycles*	*60 cycles*	*10⁶ cycles*	*Arc resistance*	*Specific gravity*	*Possible processing methods*
(°C)	*(°C)*		*(%)*	*(%)*	*(%)*	*(ohm–m)*	*(10² kV/m)*	—	—	—	—	*(s)*	—	—
D648		D635	D570	D570	D570	—	—	D150	D150	D150	D150	D495	D792	—
13	14	15	16	17	18	19	20	21	22	23	24	25	26	27
[e]216–240	[e]71–105	UL94 V–2 (UL94 V–0[d])	1.5	2.5	8.9	10^{13}	160–250	4.0	3.4–3.6	0.014	0.04	120	1.13–1.15	Injection moulding Blow moulding Extrusion
						10^{11}	—	6.0	4.7	0.04	—	—	—	
216	71–75	UL94 HB (UL94 V–0[d])	1.2	2.3	6.7	10^{12}–10^{13}	120	3.2	2.9	0.01	0.02		1.09	Injection moulding
						10^{11}	70	5.5	3.2	0.18	0.05			
145–205	60–95	Self Ex.	1.6–2.0	2.7–3.0	9.5	10^{12}	140–250	3.8–5.3	3.4	0.01–0.06	0.03	—	1.12–1.14	Injection moulding Blow moulding Rotational moulding Extrusion
						—	—	—	—	—	—	—	—	
155	57	Self Ext.	0.4	1.0–1.5	3.1–3.5	10^{11}–10^{12}	300	3.9	3.5	0.04	0.03	120	1.07–1.09	Injection moulding Blow moulding Extrusion
						—	—	—	—	0.04	—	—	—	
160	66	Self Ext. UL94 V–2	0.4	1.0–1.5	3.0	10^{12}–10^{13}	160–200	3.8–4.0	3.5	0.02–0.04	0.02	—	1.06–1.08	Injection moulding
						10^{11}	—	6.0	4.0	0.02	0.03	—	—	
150	45–60	Self Ext.	0.3	—	1.9	10^{9}–10^{12}	250	3.7–3.8	—	0.03	0.02	—	1.04	Injection, blow and rotational moulding Extrusion
						—	—	—	—	—	—	—	—	
141	55–70	Self Ext. (Moist) to slow burning	0.25	0.9	1.4	10^{12}	180	4.2	3.1	0.04	0.03	110	1.01	Injection, blow and rotational moulding Extrusion
	48–52												1.02–1.03	
216	210	Self Ext. (Moist)	0.8–1.4	—	6–7	—	120–160	4.0	3.5	0.015	0.05	—	1.15	Casting
127–177	45–55	—	1.5–2.4	—	—	—	175–180	3.2–4.0	3.0–3.6	0.01	0.010–0.015	—	1.12–1.14	Injection moulding Extrusion

Table 3.2.3—*continued*

TABLE 3.2.3 Typical comparative properties of polyamides (unfilled)—*continued*

Property		Mechanical properties									Thermal properties		
		Tensile yield strength	Ultimate elongation	Modulus of elasticity in tension	Flexural strength	Modulus of elasticity in flexure	Impact strength (IZOD notched)	Compressive strength (1% offset)	Hardness	Abrasion resistance (Taber CS–17)	Thermal conductivity	Thermal expansion coefficient	Maximum recommended service temperature
Units		(MN/m²)	(%)	(GN/m²)	(MN/m²)	(GN/m²)	(J/m)	(MN/m²)	(Rockwell)	(mg/1000 cyc.)	(W/mK)	(10^{-5} K^{-1})	(°C)
ASTM test method		D638	D638	D638	D790	D790	D256	D695	D785	D1044	C117	D696	D542
Material	Condition	1	2	3	4	5	6	7	8	9	10	11	12
Transparent amorphous	Dry	67–75	72–130	2.8	91	2.1–2.6	60	23	M93	21	—	5–7	—
Polyamide 6 (high impact)	Dry	45	40			0.8	1060+		R95			10	65
Type 4.6	Dry	100	30			3.5	100		D85			7.4	100
Type 6.9	Dry	50	15			1.4	60		D78			9	80
Polyamide/ABS alloy	Dry	47	270			2.14	850		R99			17	70
Polyamide Type 6 elastomer copolymer	Dry	37	165			0.75	590		D76			7	80
Polyamide Type 6 6.6	Dry	80	200			1.2	230		M87			12	80

[a] Heat stabilised grade.
[b] Typical nucleated fast cycling grade.
[c] High impact grade.
[d] Fire retardant grade.
[e] High impact types tend to be lower.

Table 3.2.3—*continued*

TABLE 3.2.3 Typical comparative properties of polyamides (unfilled)—*continued*

Thermal properties						Electrical properties								
Deflection temperature		Flammability	Water absorption			Volume resistivity	Dielectric strength (short time)	Dielectric constant		Dissipation factor		Arc resistance	Specific gravity	Possible processing methods
0.45 MN/m²	1.8 MN/m²		24 h	50% RH	Saturation			60 cycles	10^6 cycles	60 cycles	10^6 cycles			
(°C)	(°C)		(%)	(%)	(%)	(ohm–m)	(10^2 kV/m)	—	—	—	—	(s)	—	—
D648		D635	D570	D570	D570	—	—	D150	D150	D150	D150	D495	D792	—
13	14	15	16	17	18	19	20	21	22	23	24	25	26	27
140–145	124–135	Self Ext.	0.41	—	—	>5 × 10^{13}	270	3.77	3.3	0.031	0.01	—	1.12	Injection, blow and compression moulding Extrusion
216	71	HB	1.2			10^{11}	140						1.09	
220	160	V2	2.3			10^{13}	200						1.18	
183	61	V2	0.48			10^{12}	190						1.08	
94	85	HB	1.1			10^{12}	170						1.06	
160	65	HB	1.6			10^{13}	200						1.11	
210	70	HB	1.6			10^{12}	240						1.11	

Table 3.2.3—*continued*

TABLE 3.2.4 Properties and applications of a toughened nylon (ZYTEL ST)

Grade		*Tensile strength (MN/m²)*	*Elongation (%)*	*Impact strength (J/m)*	*Applications/remarks*
ST 801	Dry 50% RH	55 48	40 200	740 740–1325	First (standard) grade—sprockets for motocross bikes, helmets, housings, fasteners (replacement for opaque polycarbonate).
ST 811	Dry 50% RH	48 41	415 575	1174 —	Flexible grade—ski boots, ice hockey skates, flexible couplings, cable jacketing, seals and gaskets.
80G—33 (33% glass filled)		144	3.7	221	Higher degree of impact strength and toughness than provided by standard reinforced grades—wheels, chain-saw housings, conveyor belts, heavy-duty engine fans, under-the-bonnet automotive components. 14 and 43% glass-filled grades are also available.
		(The impact strength is about twice, and the tensile strength about 75%, those of a conventional 33% glass-filled nylon.)			
12T (36% mineral-filled)	Dry 50% RH	79 61	20 45	130 190	Claimed to be toughest available mineral-filled nylon—car licence-plate holders, appliance housings, motor propellers, computer tape reels. Not available in Europe.

Data courtesy of Du Pont.

Table 3.2.4

TABLE 3.2.5 Effect of fillers commonly incorporated in nylons

	Advantages	*Limitations*
Glass fibres	Good interfacial bond between glass and nylon causes improvement in compressive, tensile, flexural and fatigue strength, stiffness, toughness and high temperature load bearing characteristics (dimensional stability greater over wide temperature range). Effect of moisture on mechanical and electrical properties reduced compared with unfilled nylon. Reduction of linear coefficient of expansion down to 1/5th. 33% glass reinforcement optimum for upgrading properties without greatly impairing processing characteristics.	Large reduction in ductility. Lower in impact strength than unfilled nylons at higher moisture contents. Surface finish can vary on large mouldings and a good finish can be difficult to obtain. Addition of fillers can cause self-extinguishing nylons to become slow burning. Special flame retardant additives can reduce strength. Anisotropic properties unless design modifications or glass sphere grade used. Increased cost.
Glass microspheres	Improvement in compressive strength. Easy to process. Well suited for producing thin section components of intricate nature. Mouldings possess low creep values, good abrasion resistance, improved stiffness, uniform or isotropic shrinkage and low water absorption. Typical quantities 25–50%.	As for glass fibres. Lesser improvement in tensile strength, stiffness, and toughness than for glass fibres. Increased cost.
Asbestos fibres	Lower cost than glass fibres. Improves rigidity, strength, dimensional stability and abrasion resistance of nylons. Typical quantities 25%.	As for glass fibres. Less effective than glass fibre in improving strength and creep resistance.
Talc	Improves rigidity and dimensional stability. Easy to process.	
Titanate whiskers	Increase in rigidity. Improvement in heat distortion temperature. Reduction in thermal expansion (1/3 unfilled nylon). Good surface finish on mouldings (comparable with unfilled grades).	Increased cost.
Aluminium powder	Improves heat distortion resistance and minimises electrical charge build-up.	Increased cost.
Lubricants, e.g. graphite, molybdenum disulphide fluoropolymer powder	Improves self lubrication qualities and produces lower friction and wear. Imparts greater bearing strength. Graphite and MoS_2 used for low–medium wear resistance. Fluoropolymer powder for higher wear resistance.	Increased cost.
Carbon black	Stabilises nylon against UV degradation. Increases tensile strength and hardness. Promotes electrical conductivity. Improves self-lubricating properties.	Decreases ductility and toughness. (Black only.)
Mineral fibres	Greater impact strength than unmodified and glass reinforced grades, and cheaper. Stiffer than unreinforced nylon. Lower warpage than glass-reinforced nylon. Dimensional stability, heat resistance, and wear resistance improved.	Grades with highest toughness have lower strength and stiffness. Glass-filled grades have superior mechanical properties except impact.

TABLE 3.2.6 Typical comparative properties of polyamides (filled)

Property		*Mechanical properties*									*Thermal properties*		
		Tensile yield strength	*Ultimate elongation*	*Modulus of elasticity in tension*	*Flexural strength*	*Modulus of elasticity in flexure*	*Impact strength (IZOD notched)*	*Compressive strength (1% offset)*	*Hardness*	*Abrasion resistance (Taber CS-17)*	*Thermal conductivity*	*Thermal expansion coefficient*	*Maximum recommended service temperature*
Units		*(MN/m²)*	*(%)*	*(GN/m²)*	*(MN/m²)*	*(GN/m²)*	*(J/m)*	*(MN/m²)*	*(Rockwell)*	*(mg/1000 cyc.)*	*(W/mK)*	*(10^{-5} K^{-1})*	*(°C)*
ASTM test method		D638	D638	D638	D790	D790	D256	D695	D785	D1044	C117	D696	D542
Material	Condition	1	2	3	4	5	6	7	8	9	10	11	12
30% Glass fibre Type 6	Dry	144–157	2–4	6.8–8.2	178–232	6.8–8.2	123–160	130–137	R121	14	0.2	2.0–3.0	80–100 120[b]
30% Glass fibre Type 6.6	Dry	117–200	3.0	9.6	260	9.1	45–110	245	E60	—	0.2	1.7–3.6	93–150[b]
	50% RH	160–126	3.6–5.0	5.6	—	4.7–6.3	130			13–25			
30% Glass fibre Type 6.10	Dry	116–144	2	5.5–6.5	157	5.8	181	123	E40–50	18	0.2	9.0	93–150[b]
30–35% Glass fibre Type 6.12	Dry	138–164	4.5	8.2	260	6.8–8.4	—	—	E40–50	—	—	—	93–150[b]
	50% RH	137	5.0	6.2	—	6.3	149	—	—	—	—	—	
30% Glass fibre Type 11	Dry	96	3–5	—	—	6.0	90–120	—	—	—	0.2	3.0	93–150[b]
30% Glass fibre Type 12	Dry	119	3–5	3–6	—	6.8	160–220	—	R113	—	0.16	7.5	93–150[b]
30% Glass fibre + MoS_2 Type 6.6	Dry	130–150	1.9	—	178–192	7.5–8.5	—	—	M95–100	—	—	8.5	93–150[b]
40% Mineral fibre Type 6.6	Dry	79–96	3–14	—	82–109	4.6–5.9	32–130[a]	—	R119–121	12–30	—	1.4–2.5	—
	50% RH	57–64	16–45			1.75–3.0	37–189[a]						
25% Mineral fibre, 15%glass fibre reinforced Type 6	Dry	136	2–3	—	186	7.6	37	159	R121	—	0.43	3.1	—
40% Glass sphere-filled Type 6.6	Dry	94	—	4.7–6.8	137	7.8	45	—	—	—	—	4.9	80
	50% RH	60		2.45	64								
20% Carbon fibre Type 6.6	Dry	196	3–4	—	290	16.8	50	—	—	—	0.2	2.5	—
UV Stabilised Type 6	Dry	38	50			1	150		D75			9	80
Fire retardant Type 6	Dry	38	40			1	150		D75			11	80
Glass fibre and bead-reinforced Type 6	Dry	85	3			3.5	100		D75			4	80

[a] Zytel 12T (36% mineral reinforced). [b] Heat stabilised grade. [c] Flame retardant grades are available with slightly improved physical properties.

Table 3.2.6

TABLE 3.2.6 Typical comparative properties of polyamides (filled)—*continued*

Thermal properties						*Electrical properties*								
Deflection temperature			*Water absorption*					*Dielectric constant*		*Dissipation factor*				
0.45 MN/m²	*1.8 MN/m²*	*Flammability^c*	*24 h*	*50% RH*	*Saturation*	*Volume resistivity*	*Dielectric strength (short time)*	*60 cycles*	*10⁶ cycles*	*60 cycles*	*10⁶ cycles*	*Arc resistance*	*Specific gravity*	*Possible processing methods*
(°C)	*(°C)*		*(%)*	*(%)*	*(%)*	*(ohm–m)*	*(10² kV/m)*	—	—	—	—	*(s)*	—	—
D648		D635	D570	D570	D570			D150		D150		D495	D792	
13	14	15	16	17	18	19	20	21	22	23	24	25	26	27
205–218	190–216	Slow burning	1.3	—	6.5–7	5×10^9	160	4.2	3.6–3.9	0.018	0.017	81–135	1.31–1.36	Injection moulding
255–265	240–257	Slow burning	0.9	1.8	2.8–5.8	5×10^{13}	210	4.5	3.7	0.006	0.019	80–148	1.34–1.42	Injection moulding
						2×10^7	—	25 (100% RH)	10.7 (100% RH)	—	0.11	—		
220	215	Slow burning	0.2	—	1.85–2.1	10^{10}	200	4.2	3.5	0.026	0.022	98–125	1.31–1.38	Injection moulding
—	210–218	Slow burning	0.2	—	2.0	10^{13}	210	3.7	3.4	0.024	0.016	—	1.31–1.38	Injection moulding
						3×10^{10}	173	7.8	4.0	0.136	0.100	—		
—	172–180	Slow burning	0.12	0.54	1.30	10^{12}	200	—	—	—	—	—	1.24	Injection moulding
95–175	160–170	Slow burning	0.07	—	1.0	10^{11}	180	—	3.0	—	0.06	—	1.23	Injection moulding
—	—	—	0.5–0.7	—	—	—	—	—	—	—	—	135	1.37–1.41	Injection moulding
224–250	154–215	Self Ext.	0.5–0.8	—	—	10^{18}	190	4	—	0.01	—	115	1.47	Injection moulding
						10^{13}	150	15		0.4				
215	204	UL94 94HB	0.9	—	—	4.9×10^{13}	200	—	—	—	—	—	1.46–1.48	Injection moulding
232	138–>200	—	0.73	1.2	—	5×10^{14}	100	4.6	3.8	0.02	0.02	—	1.45	Injection moulding
						10^{11}	100	—	4.0		0.08			
—	255	—	0.6	—	—	—	—	—	—	—	—	—	1.23	
180	66	HB	1.8			10^{11}	200						1.14	
195	90	VO >1.5	1.5			10^{13}	200						1.34	
205	200	HB	1.0			10^{13}	250						1.48	

Table 3.2.6—*continued*

TABLE 3.2.6 Typical comparative properties of polyamides (filled)—*continued*

Property		Mechanical properties									Thermal properties		
		Tensile yield strength	Ultimate elongation	Modulus of elasticity in tension	Flexural strength	Modulus of elasticity in flexure	Impact strength (IZOD notched)	Compressive strength (1% offset)	Hardness	Abrasion resistance (Taber CS-17)	Thermal conductivity	Thermal expansion coefficient	Maximum recommended service temperature
Units		(MN/m^2)	(%)	(GN/m^2)	(MN/m^2)	(GN/m^2)	(J/m)	(MN/m^2)	(Rockwell)	(mg/1000 cyc.)	(W/mK)	(10^{-5} K^{-1})	(°C)
ASTM test method		D638	D638	D638	D790	D790	D256	D695	D785	D1044	C117	D696	D542
Material	Condition	1	2	3	4	5	6	7	8	9	10	11	12
30% Carbon fibre Type 6	Dry	220	3.3			18	80		D75			2.5	80
MoS_2 lubricated Type 6	Dry	38				1	150		D75			9	80
2% Silicone Type 6	Dry	38	50			1	200		D75			10	80
Glass fibre-reinforced sheet Type 6	Dry	190	2			10	400		E65			2	80
60% Glass fibre-reinforced Type 6	Dry	220	1			19	190		M101			1.5	80
50% Glass bead-filled Type 6	Dry	75	2			5	60		D80			5	80
UV Stabilised Type 6.6	Dry	56	50			1.2	100		R99			8	80
Fire retardant Type 6.6	Dry	40	40			1.0	90		R117			8	80
Glass fibre and bead-reinforced Type 6.6	Dry	100	3			4.5	60		M95			2.5	80
Super tough fire retardant Type 6.6	Dry	40	30			0.9	900		R95			10	65
Super tough 33% Glass fibre-reinforced Type 6.6	Dry	110	4			4.5	240		R98			6	65
Fire retardant Type 6.10	Dry	50	200			1.1	60		R100			14	70
30% Carbon fibre-reinforced Type 6.10	Dry	145	7.5			7.5	110		M97			3	70
Fire retardant Type 6.12	Dry	47	200			1.4	30		R105			9	70
30% Carbon fibre-reinforced Type 6.12	Dry	200	3			15	100		R120			1.7	70

[a] Zytel 12T (36% mineral reinforced). [b] Heat stabilised grade. [c] Flame retardant grades are available with slightly improved physical properties.

Table 3.2.6—*continued*

TABLE 3.2.6 Typical comparative properties of polyamides (filled)—*continued*

Thermal properties						*Electrical properties*									
Deflection temperature			*Water absorption*					*Dielectric constant*		*Dissipation factor*				
0.45 MN/m²	*1.8 MN/m²*	*Flammability^c*	*24 h*	*50% RH*	*Saturation*	*Volume resistivity*	*Dielectric strength (short time)*	*60 cycles*	*10⁶ cycles*	*60 cycles*	*10⁶ cycles*	*Arc resistance*	*Specific gravity*	*Possible processing methods*
(°C)	*(°C)*		*(%)*	*(%)*	*(%)*	*(ohm–m)*	*(10² kV/m)*	—	—	—	—	*(s)*	—	—
D648		D635	D570	D570	D570			D150		D150		D495	D792	
13	14	15	16	17	18	19	20	21	22	23	24	25	26	27
213	205	HB	0.7			10	—						1.28	
150	75	HB	2.0			10^{11}	150						1.17	
200	80	HB	1.5			10^{12}	250						1.14	
220	215	HB	0.6			10^{13}	250						1.5	
218	215	HB	0.7			10^{13}	220						1.7	
210	130	HB	0.7			10^{13}	200						1.5	
200	100	HB	1.2			10^{12}	200						1.15	
180	70	VO >1.5	1.0			10^{11}	150						1.3	
250	245	HB	1.0			10^{12}	150						1.4	
216	71	VO >1.5	1.2			10^{11}	130						1.25	
210	200	HB	1			10^{12}	140						1.34	
157	66	VO >1.5	0.35			10^{10}	170						1.30	
220	210	HB	0.12			10^{13}	170						1.30	
160	80	VO >1.5	0.3			10^{10}	180						1.20	
225	216	V2 >3	0.15			1	—						1.22	

Table 3.2.6—*continued*

TABLE 3.2.6 Typical comparative properties of polyamides (filled)—*continued*

Property		*Mechanical properties*									*Thermal properties*		
		Tensile yield strength	*Ultimate elongation*	*Modulus of elasticity in tension*	*Flexural strength*	*Modulus of elasticity in flexure*	*Impact strength (IZOD notched)*	*Compressive strength (1% offset)*	*Hardness*	*Abrasion resistance (Taber CS-17)*	*Thermal conductivity*	*Thermal expansion coefficient*	*Maximum recommended service temperature*
Units		*(MN/m²)*	*(%)*	*(GN/m²)*	*(MN/m²)*	*(GN/m²)*	*(J/m)*	*(MN/m²)*	*(Rockwell)*	*(mg/1000 cyc.)*	*(W/mK)*	*(10⁻⁵ K⁻¹)*	*(°C)*
ASTM test method		D638	D638	D638	D790	D790	D256	D695	D785	D1044	C117	D696	D542
Material	Condition	1	2	3	4	5	6	7	8	9	10	11	12
20% PTFE lubricated Type 6.12	Dry	45	200			1.4	40		R100			10	70
UV Stabilised Type 11	Dry	46	250			0.9	50		R105			9	70
Fire retardant Type 11	Dry	46	50			1.1	40		R110			9	70
30% Carbon fibre-reinforced Type 11	Dry	150	5			13	70		R116			2.5	70
UV Stabilised Type 12	Dry	50	200			1.4	60		R105			11	70
Fire retardant Type 12	Dry	45	100			1.4	50		R105			10	70
50% Glass bead-reinforced Type 12	Dry	55	25			3	40		R113			8	70
60% Glass fibre Type 6.6	Dry	220	2			15	120		M104			2	80
2% Silicone lubricated Type 6.10	Dry	55	280			1.1	30		R95			11	70
2% Silicone lubricated Type 6.12	Dry	49	250			1.5	40		R100			9	70
30% Glass fibre Amorphous nylon	Dry	145	3.5			7.8	60		M93			3.2	90
30% Long glass fibre-reinforced Type 6.6	Dry	195				15.8	270						145
50% Long glass fibre-reinforced Type 6.6	Dry	230				15.8	270						145

[a] Zytel 12T (36% mineral reinforced). [b] Heat stabilised grade. [c] Flame retardant grades are available with slightly improved physical properties.

Table 3.2.6—*continued*

TABLE 3.2.6 Typical comparative properties of polyamides (filled)—*continued*

Thermal properties						Electrical properties									
Deflection temperature		Flammability[c]	Water absorption			Volume resistivity	Dielectric strength (short time)	Dielectric constant		Dissipation factor		Arc resistance	Specific gravity	Possible processing methods
0.45 MN/m²	1.8 MN/m²		24 h	50% RH	Saturation			60 cycles	10^6 cycles	60 cycles	10^6 cycles			
(°C)	(°C)		(%)	(%)	(%)	(ohm–m)	(10^2 kV/m)	—	—	—	—	(s)	—	—
D648		D635	D570	D570	D570			D150		D150		D495	D792	
13	14	15	16	17	18	19	20	21	22	23	24	25	26	27
160	80	V2 >3	0.3			10^{12}	33						1.19	
150	55	V2 >3	0.4			10^{11}	250						1.05	
150	55	VO >3	0.3			10^{10}	150						1.16	
185	175	V2 >3	0.2			10	—						1.2	
150	55	V2 >3	0.25			10^{11}	400						1.04	
150	55	VO >3	0.25			10^{11}	450						1.1	
155	110	V2 >3	0.1			10^{11}	450						1.44	
260	260	HB	0.4			10^{12}	160						1.7	
150	51	HB	0.35			10^{11}	200						1.07	
150	54	V2 >3	0.28			10^{12}	200						1.05	
210	140	V2 >1.5	0.19			10^{13}	200						1.35	
	261	HB											1.37	
	261	HB											1.57	

Table 3.2.6—*continued*

TABLE 3.2.7 The influence of section thickness on water absorption of nylon 6.6

Thickness of nylon 6.6		*Time to attain equilibrium*	
(cm)	*(in)*	*In water 20°C*	*In water 100°C*
0.625	1/4	900 days	100h
0.313	1/8	150 days	30h
0.156	1/16	45 days	6h
0.078	1/32	—	2h

TABLE 3.2.8 Effect of moisture on properties of nylons

Nylon type	*Water absorption (%)*		*Dimensional change (%)*		*Flexural modulus (GN/m²)*		
	50% RH	*Saturation*	*50% RH*	*Saturation*	*Dry*	*50% RH*	*Saturation*
6.6	2.5	9.0	0.6	2.8	2.83	1.21	0.59
6	2.7	9.5	0.7	3.0	2.72	0.96	0.48
6.10	1.5	3.5	0.2	0.7	1.93	1.10	0.69
6.12	1.3	3.0	0.2	0.6	2.00	1.24	0.83
11	0.8	1.9	0.1	0.4	1.32	0.98	0.95
12	0.7	1.4	0.1	0.3	1.41	1.21	1.02

TABLE 3.2.9 Chemical and environmental resistance of nylons

Attacking agent	*Nylon 6*	*Nylon 6.6*	*Type 6 and 6.6 copolymers*	*Nylon 6.10*	*Nylon 6.12*	*Nylon 11*	*Nylon 12*
Weathering	Nylons are embrittled by prolonged exposure to sunlight but stabilised grades are available. Finely dispersed carbon black is the most effective stabiliser.						
Weak acids[a]	Dilute solutions of weak acids may be satisfactory	Less resistant than 6.	Less resistant than 6 & 6.6.	More resistant than type 6.			
Strong acids[a]	All nylons are attacked by strong and oxidising acids, at any concentration.						
Weak alkalis[a]	All nylons are resistant to weak alkalis.						
Strong alkalis[a]	Attacked by conc. NaOH and KOH.	Resistant to conc. NaOH. Attacked by conc. KOH.	No data	Attacked by conc. NaOH and KOH		Generally resistant.	
Oils and greases	Generally resistant although chemical resistance to transformer oil depends on composition of oil.						
Foodstuffs	Generally resistant although moisture absorption tendency can cause staining.						
Organic solvents and others	All nylons generally resistant to most aromatic and aliphatic hydrocarbons, common esters, ketones, ethers and amides.						
	Resistant to carbon tetrachloride.					Nylon 11 and 12 attacked by carbon tetrachloride.	
	Attacked by trichlorethylene, xylenes, phenols, chloroform, glycols, hydrogen peroxide, chlorine, bromine, fluorine, iodine, ozone, methylene chloride, zinc chloride, mercuric chloride, cupric chloride, calcium chloride, formic, citric, and acetic acid, sodium hypochlorite, potassium permanganate and nitrobenzene.						
Hot water and steam	Generally resistant. Oxidative attack and embrittlement occur more severely at high temperatures. Internal stress, aeration and chlorine have deleterious effects. Impact and fatigue do not suffer.						

[a] The term 'weak' implies an acid or base which does not ionize in solution to any considerable extent. The term 'strong' implies a fully ionised solution. These terms are not the same as 'dilute' and 'concentrated'.

Table 3.2.10

TABLE 3.2.10 Wear rate and friction of filled and unfilled nylon rubbing against various materials

Component Material				*Mating material*	*Wear factor K 10^{-10}in^3 min/ft lb h*	*Coefficient of friction*		*Limiting PV (ft lb/in^2 min)*			*Thermal expansion coefficient (10^{-5} K^{-1})*	*Thermal conductivity (W/mK)*
Base Resin	*Glass (%)*	*Carbon fibre (%)*	*Other (%)*			*Static*	*Dynamic*	*10 fpm*	*100 fpm*	*1000 fpm*		
Nylon 6	—	—	—	Steel	200	0.22	0.26	—	—	—	4.8–8.6	0.2–0.3
Nylon 6	—	—	15 PTFE	Steel	15	0.09	0.19	—	—	—	—	—
Nylon 6	30	—	15 PTFE	Steel	17	0.20	0.25	—	—	—	2.0–3.0	0.2
Nylon 6	—	—	MoS_2	Steel	160	0.28	0.30	12,500	12,500	3,000	6.7	0.36
Nylon 6.6	—	—	—	Steel	200	0.20	0.28	3,000	2,500	2,500	7.7–11.0	0.2–0.3
Nylon 6.6	—	—	5 PTFE	Steel	60	0.13	0.20	—	—	—	—	—
Nylon 6.6	—	—	20 PTFE	Steel	12	0.10	0.18	14,000	17,500	8,000	8.3	—
Nylon 6.6	—	—	20 PTFE + silicone	Steel	6	0.06	0.08	14,000	30,000	12,000	8.3	—
Nylon 6.6	—	—	MoS_2	Steel	165	0.28	0.30	12,500	12,500	3,000	6.3	0.43
Nylon 6.6	—	—	Graphite	Steel	55	0.15	0.20	—	—	—	7.6	—
Nylon 6.6	10	—	—	Steel	80	0.21	0.26	—	—	—	4.8	0.39
Nylon 6.6	30	—	—	Steel	25	0.25	0.31	—	—	—	1.7–3.6	0.2
Nylon 6.6	40	—	—	Steel	70	0.26	0.33	—	—	—	2.3	0.52
Nylon 6.6	60	—	—	Steel	44	0.27	0.35	—	—	—	1.6	0.58
Nylon 6.6	20		15 PTFE	Steel	16	0.19	0.26	—	—	—	—	—
Nylon 6.6	—	20	—	Steel	40	0.16	0.20	—	—	—	2.5	0.8
Nylon 6.6	—	40	—	Steel	14	0.16	0.18	—	—	—	1.45	1.2
Nylon 6.10	30	—	15 PTFE	Steel	15	0.23	0.31	—	—	—	—	—
Nylon 6.10	—	—	MoS_2	Steel	180	0.30	0.33	7,000	5,000	2,500	5.6	0.45
Nylon 6.12	—	—	MoS_2	Steel	190	0.31	0.33	7,500	6,000	3,000	5.6	—
Nylon 6.6	—	—	—	Nylon 6.6	1,150	0.17	0.11	—	—	—	7.7–11.0	0.2–0.3
Nylon 6.6	—	—	—	Nylon 6.6 + 20% PTFE	30	0.08	0.05	—	—	—	7.7–11.0	0.2–0.3
Nylon 6.6	—	—	—	Acetal	69	0.04	0.05	—	—	—	7.7–11.0	0.2–0.3
Nylon 6.6	—	—	20 PTFE	Acetal + 20% PTFE	12	0.03	0.04	—	—	—	8.3	—
								10 ft/min ≡ 0.05 m/s 100 ft/min ≡ 0.05 m/s 1000 ft/min ≡ 5.1 m/s			10^{-10} in^3/min ft lb hr ≡ 2.0 10^{-17} m^3/J 1 ft lb/in^2 ≡ 35 J/m^2 s	

TABLE 3.2.11 Processing characteristics of different nylon types

Nylon type		*Mould (linear)[a] shrinkage (%)*	*Processing temperature (°C)*	*Mould temperature (°C)*
6	Unfilled	0.5–2.0	225–300	20–120[c]
	25% glass fibre	0.4–1.2	225–250	20–120
	50% glass sphere	1.0–1.5		
6.6	Unfilled	1.2–1.8	260–300	70–120
	25–30% glass fibre	0.5	210–300	70–120
	40% mineral	0.6–1.3[b]	285–305	95–120
6.10	Unfilled	1.2–1.8	230–260	60–90
	Glass filled	0.4	230–260	60–90
6.12	Unfilled	0.7–1.0	235–245	38–72
	Glass filled	0.3	235–245	38–72
11	Unfilled	1.2–1.3	186–250	40
	Glass filled	0.3–0.4	200–270	60–90
12	Unfilled	1.0–1.7	180–250	20–100
	30% glass fibre	0.4–0.8	200–260	20–100
	Plasticised	1.5	—	—
	50% glass sphere	0.1	—	—
Transparent		0.5	250–320	—

[a] Shrinkage is usually greatest in direction of flow.
[b] High toughness grade.
[c] Non-nucleated grades, when moulded at low mould temperatures (20–40°C) produce amorphous translucent mouldings; mould temperatures of 7–120°C give opaque, crystalline stronger mouldings. Nucleated grades can employ temperatures of 25–95°C to achieve crystallinity.

TABLE 3.2.12 Typical properties of nylon RIM

Property		*Units*	*20% polyol unreinforced*	*20% polyol reinforced*	*25% polyol unreinforced*	*25% polyol reinforced*	*30% polyol unreinforced*	*30% polyol reinforced*
Glass content[a]		%	0	25	25	30	30	30
Hardness		Shore d	70	—	—	—	—	—
Flexural modulus	−30°C	GN/m^2	2.17	—	—	—	—	—
	20°C		0.79	1.89	1.58	2.04	1.69	1.14
	70°C		0.47	—	—	—	—	—
Tensile strength		MN/m^2	41	52	28	48	43	32
elongation		%	250	13	65	24	47	30
Heat sag (1 h, 6 in cantilever)	120°C	mm/h	2.25–3.75	1.5	—	—	—	—
	160°C		5.75–7.25	2.5	—	—	—	—
Coefficient of thermal expansion		10^{-6} K^{-1}	126	—	—	39.6	—	—
Impact strength Izod notch		J/m	648	211	281	205	227	259

[a] 1/16 inch milled glass Data supplied by Monsanto.

Table 3.2.13

TABLE 3.2.13 Major suppliers of nylon grades

Company	Trade names	Nylon 6	Nylon 6/6	Nylon 6/9	Nylon 6/10	Nylon 6/12	Nylon 11	Nylon 12	Amorphous Nylon	Nylon 4/6	Notes:
Adell	Adell	U, G, GM	U, G, W, GW								
Akzo	Akulon	U, G, M, V	U, G, M, V								
Allied Chem.	Capron	U, P. H, C, G, M									Stamped nylon 6 designated STX.
ATO Chimie	Rilsan (11, 12) Orgamide (6)	U, G, V, N, S					U, CT, W, Y, V GY, N, B	U, V, N, G, S	U		
BASF	Ultramid Ultralon	U, G, V, N, C Cast	U, G, K, V, N, W	U	U, N				U		
Bayer	Durethan 8	U, UT, G, V, N, C	U, V, G						U		
BIP	Beetle	U, G, X, V, N	U, V, G								
Calanese/Amcel	Celanese	—	U., G								
Chemische Werke (Huls UK)	Vestamid							U, UT, S, G, W, Y, V, N			
Chemplast	Ertalon, Vekton (66)	U, W, Y	U, G				U				
Customer Resins	CRI	U									
Danco/Dixon	Interpact		U								Modified high impact
Du Pont	Zytel, Minlon, Elvamine	(U)	U, G, M, V, N			U, G, N					Zytel ST is a modified high impact 6/6.
Dynamit Nobel	Trogamid T								UT, G		
Edison	Renyl	U									
Emser /Werke/ Grilon (UK)	Grilon (6) Grilamid (12)	U, UT, G, S, V, N, M, Cast, C	U, N, G, S					U, G, S, V, N	UT, G		
Fabelta/Grilon (UK)	Fabelnyl	U, G, M	V, M, W			U					
Fiberfil Dart (UK)	Nylafil	G	G			G					
Fiberite (Vigilant Plastics)	RTP	G, S, F	G, B, N, M, S, W, F		G, F, J	G, F, J	G				
Foster Grant	Fostanylon	U									
ICI	Maranyl	U, G, V, N, S	U, G, W, D, S, V, N								
LNP	Thermocomp Ny-Kon	G, W, F, Y, J, S Q	W, B, G, F, Y, J S		G, B, W, R, F,J Q, J	G, W, F, J B	G, Z, B	G, F, S	G, F		

KEY: U = unfilled; G = glass fibre-filled; M = mineral-filled; W = molybdenum disulphide-filled; P = plasticised; H = homopolymer; C = copolymer; T = transparent; S = glass sphere-filled; V = UV stabilised; N = Fire retardant; Q = C fibre.

TABLE 3.2.13 Major suppliers of nylon grades — *continued*

Company	*Trade names*	*Nylon 6*	*Nylon 6/6*	*Nylon 6/9*	*Nylon 6/10*	*Nylon 6/12*	*Nylon 11*	*Nylon 12*	*Amorphous Nylon*	*Nylon 4/6*	*Notes*
Mitsui	Novamid	U, G, M									
Monsanto	Vydyne, Triax	—	U, C with 6, V, N, M	U	U						Triax – polyamide ABS alloy
Montecatini Edison	Renyl	—	U								
Nylacast	Nylacast	Cast	—			Cast					Also modified (Orlan grade)
Nypel	Nypel	U, G, W	U, G, W, C with 6								
Polypenco	Monocast, Nylatron Nylasint	Cast, V, G	G, W, F								Sintered
Rhone Poulenc	Technyl, Nailon	U, V, G	U, G, S, V, N		U, N						
Snia Viscosa	Sniamid, Sniavitrid	U, UT, W, G, A, S, M, V, N	U, W, Y, G, S						U		
TBA	Arpylene	G	S, G, A								
Thermofil	Thermofil		G, J, F, W								
Union Carbide	Amidel								UT		
Wellman	Well-Fibre, -Korr, -Sphere, Wellamid		U, G, M, S, C with 6								
Freeman Chemicals	Latamid	U, V, N, GS, Q, W, J, F	U, V, N, M, GS, B, W, F					U, V, G			
Jonylon	Jonylon	U, V, N, GS	U, V, N, MG, S, W								
Ferro (GB) Ltd.	Staramide Star	M, G, Q, F	M, G, B, F								
Wilson Fibrefil International SA	Nylode, Nylafil Plaslube	M, G W, J	N, M, G, B, F			G					
Allied Corp. International	STX	Stampable Sheet with G									
DSM UK Ltd	Nyrim, Stanyl	RIM								U	
Courtaulds	Grafil		B								
Borg Warner	Elemid										Polyamide/ ABS alloy

KEY: U = unfilled; G = glass fibre-filled; M = mineral-filled; W = molybdenum disulphide-filled; P = plasticised; H = homopolymer; C = copolymer; T = transparent; S = glass sphere-filled; V = UV stabilised; N = Fire retardant; Q = C fibre.

Table 3.2.13—*continued*

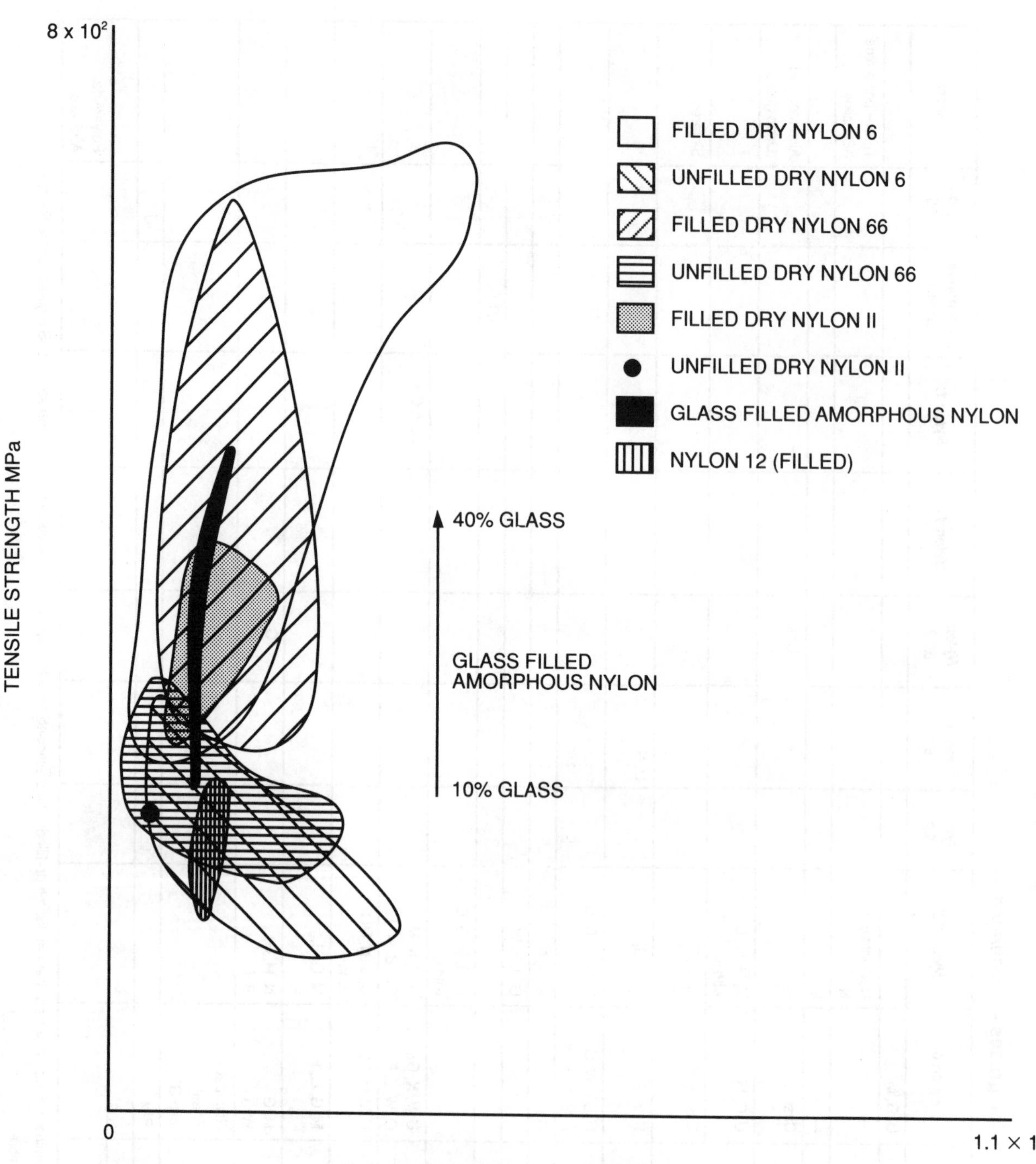

FIG 3.2.1 Impact strength vs tensile strength for nylons

Fig 3.2.1

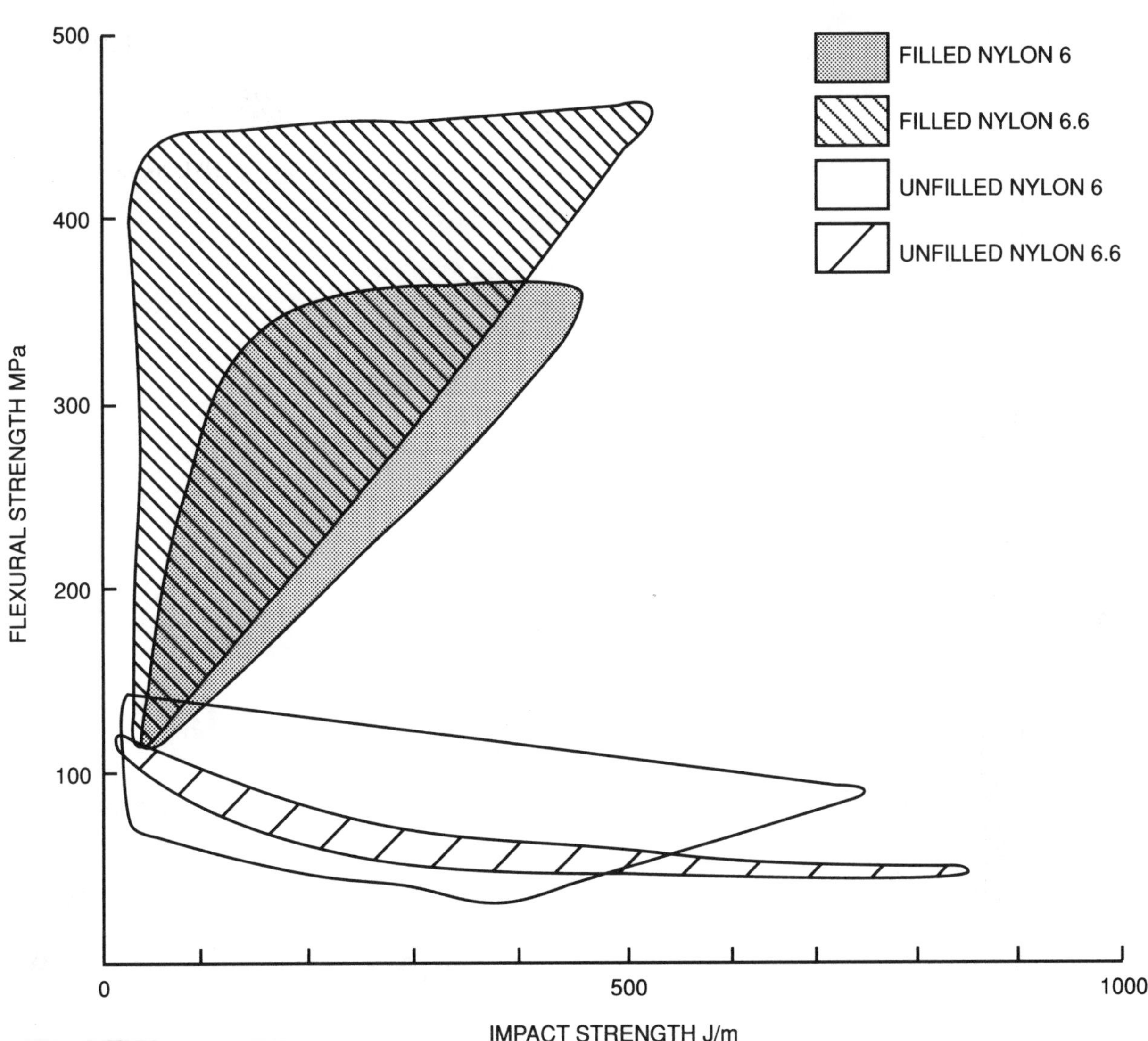

FIG 3.2.2 Flexural strength vs impact strength for dry nylons

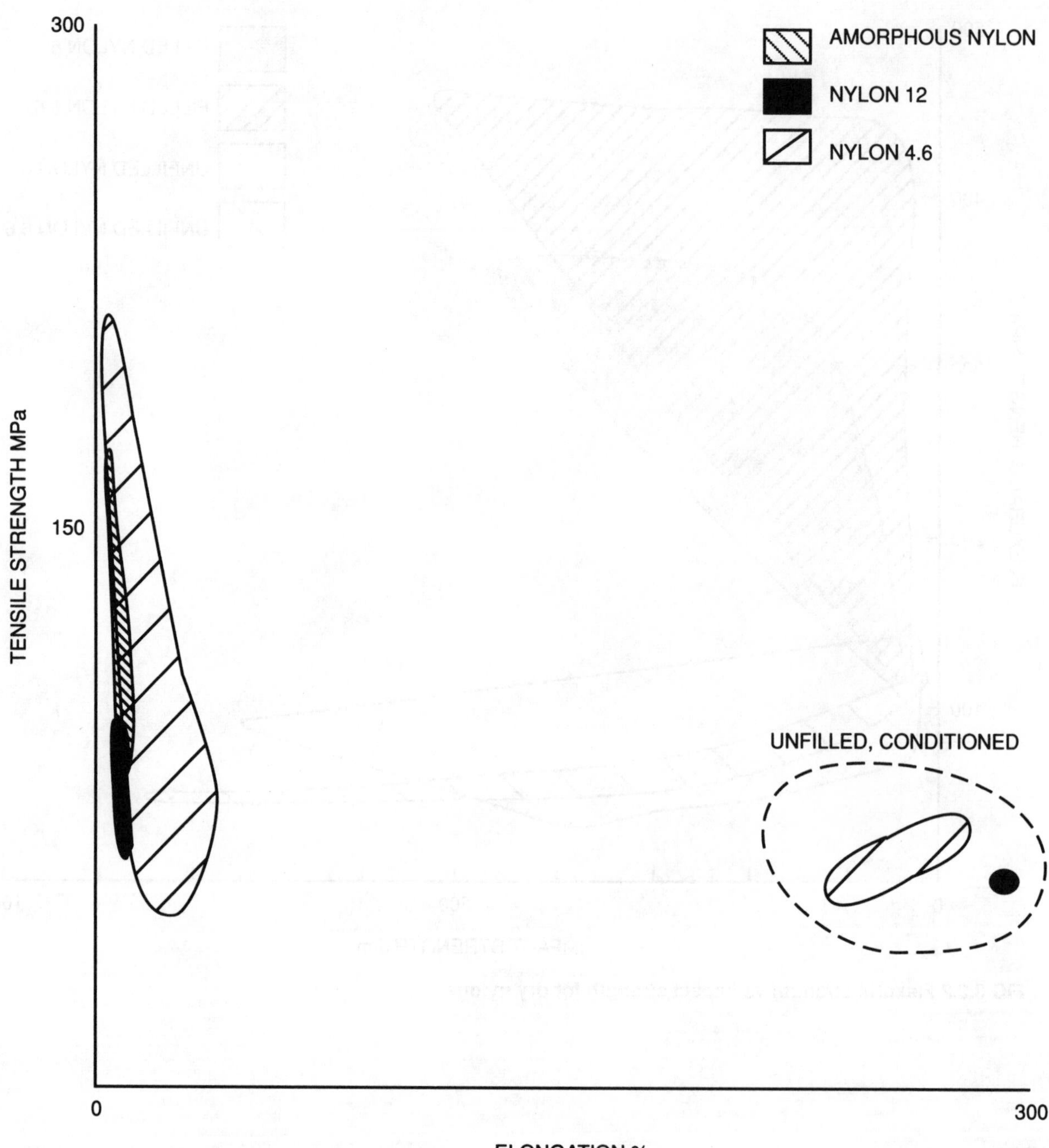

FIG 3.2.3 Elongation vs tensile strength for nylons

Fig 3.2.3

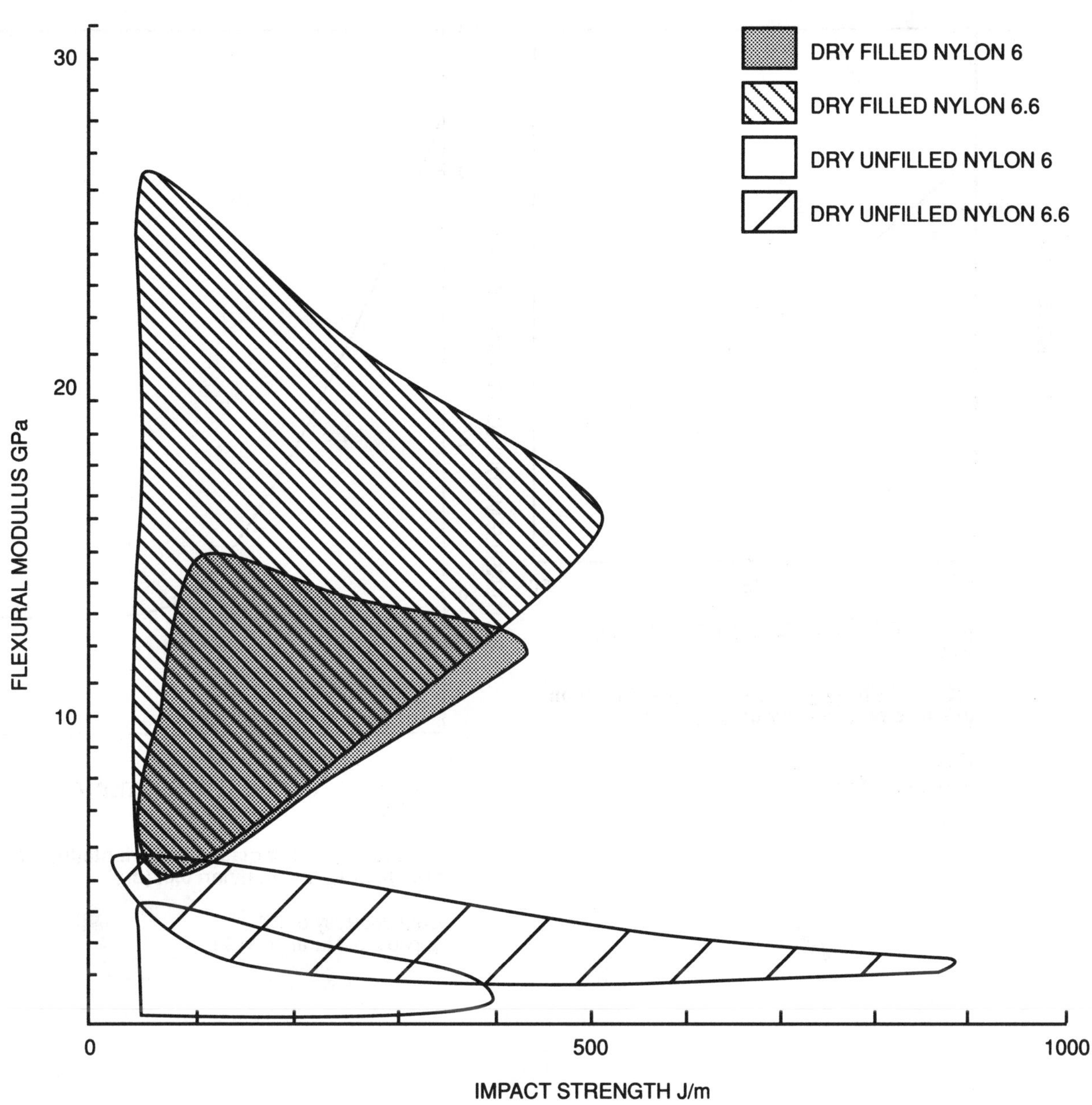

FIG 3.2.4 Flexural modulus vs impact strength for dry nylons

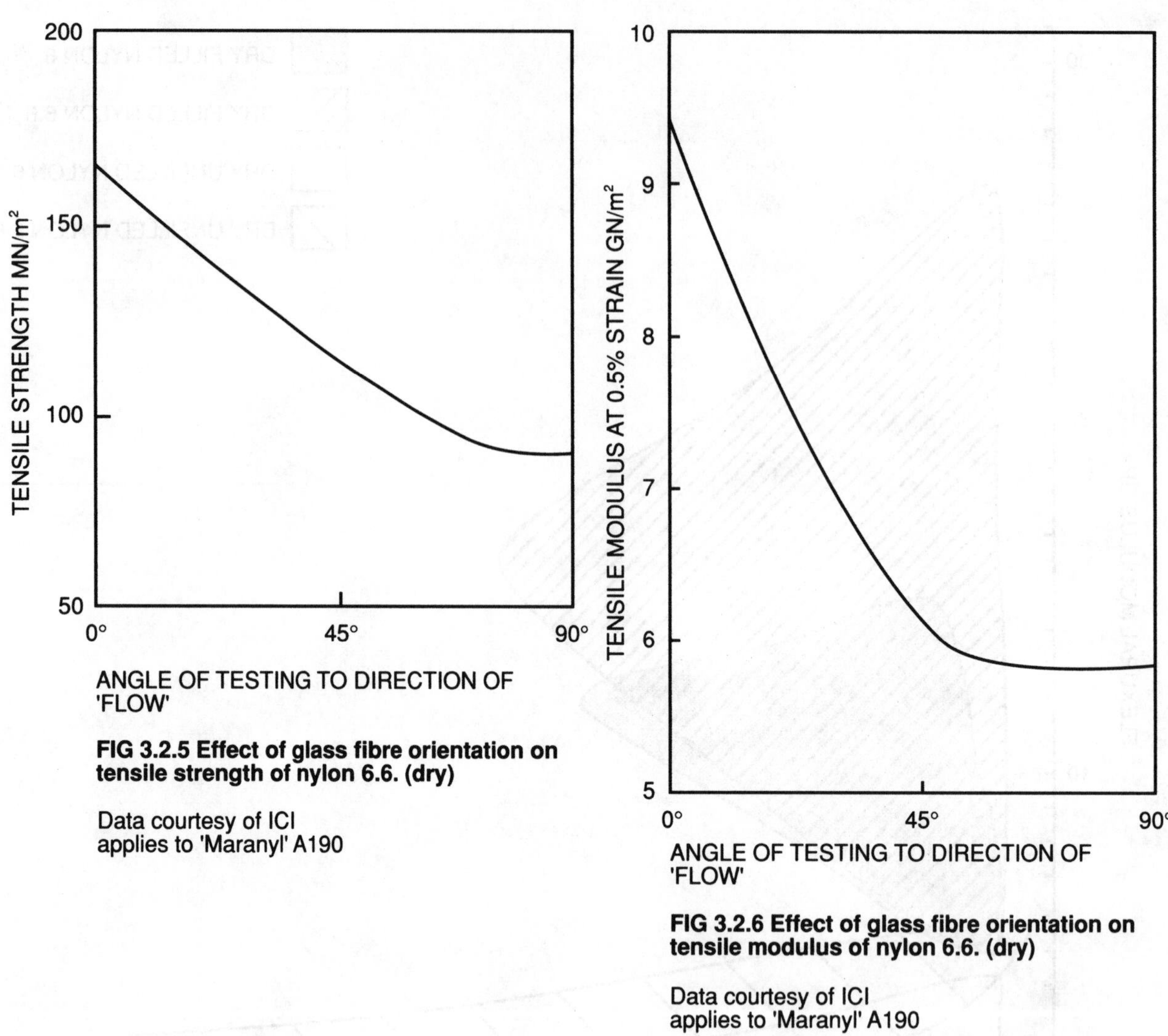

FIG 3.2.5 Effect of glass fibre orientation on tensile strength of nylon 6.6. (dry)

Data courtesy of ICI
applies to 'Maranyl' A190

FIG 3.2.6 Effect of glass fibre orientation on tensile modulus of nylon 6.6. (dry)

Data courtesy of ICI
applies to 'Maranyl' A190

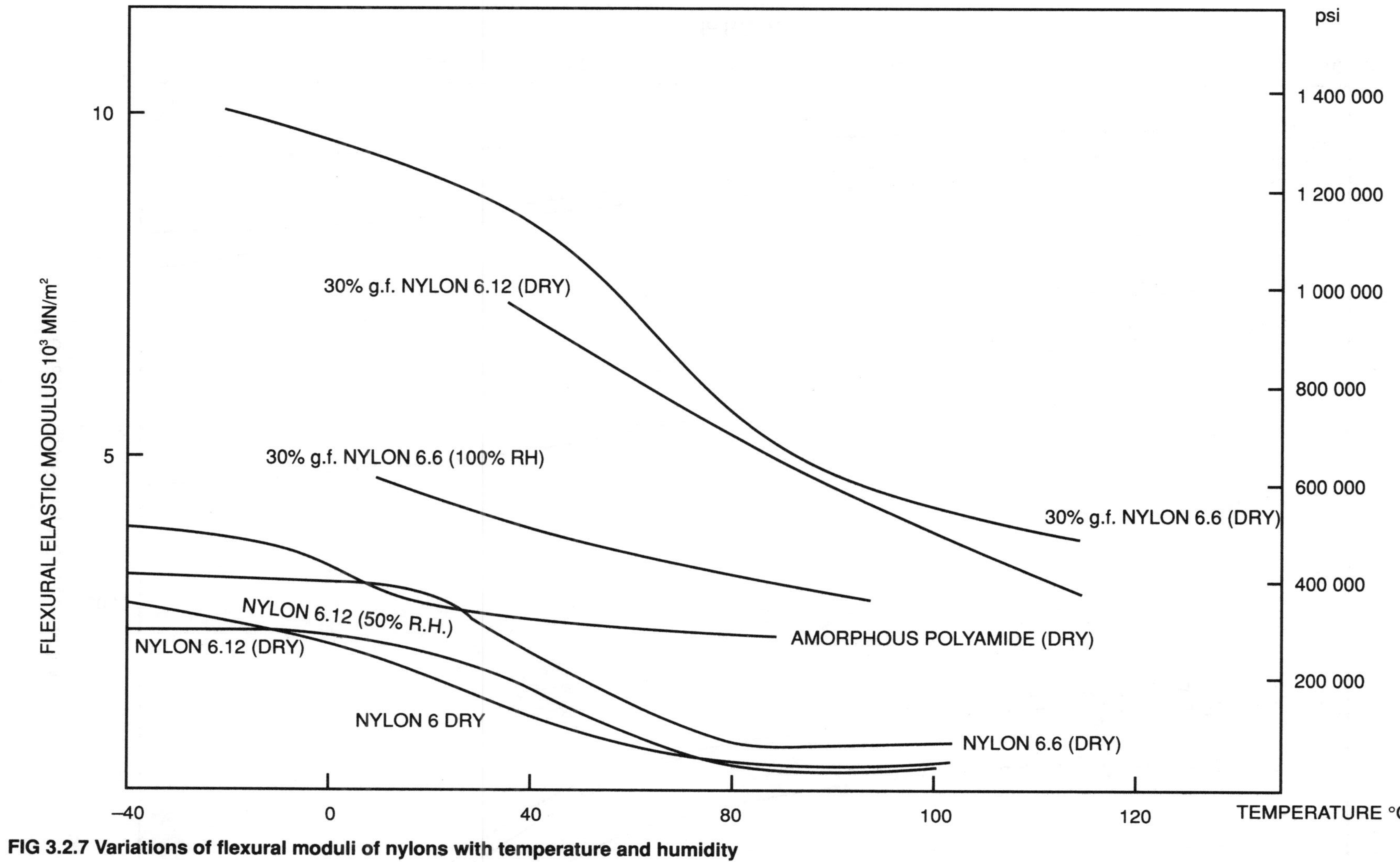

FIG 3.2.7 Variations of flexural moduli of nylons with temperature and humidity

Fig 3.2.7

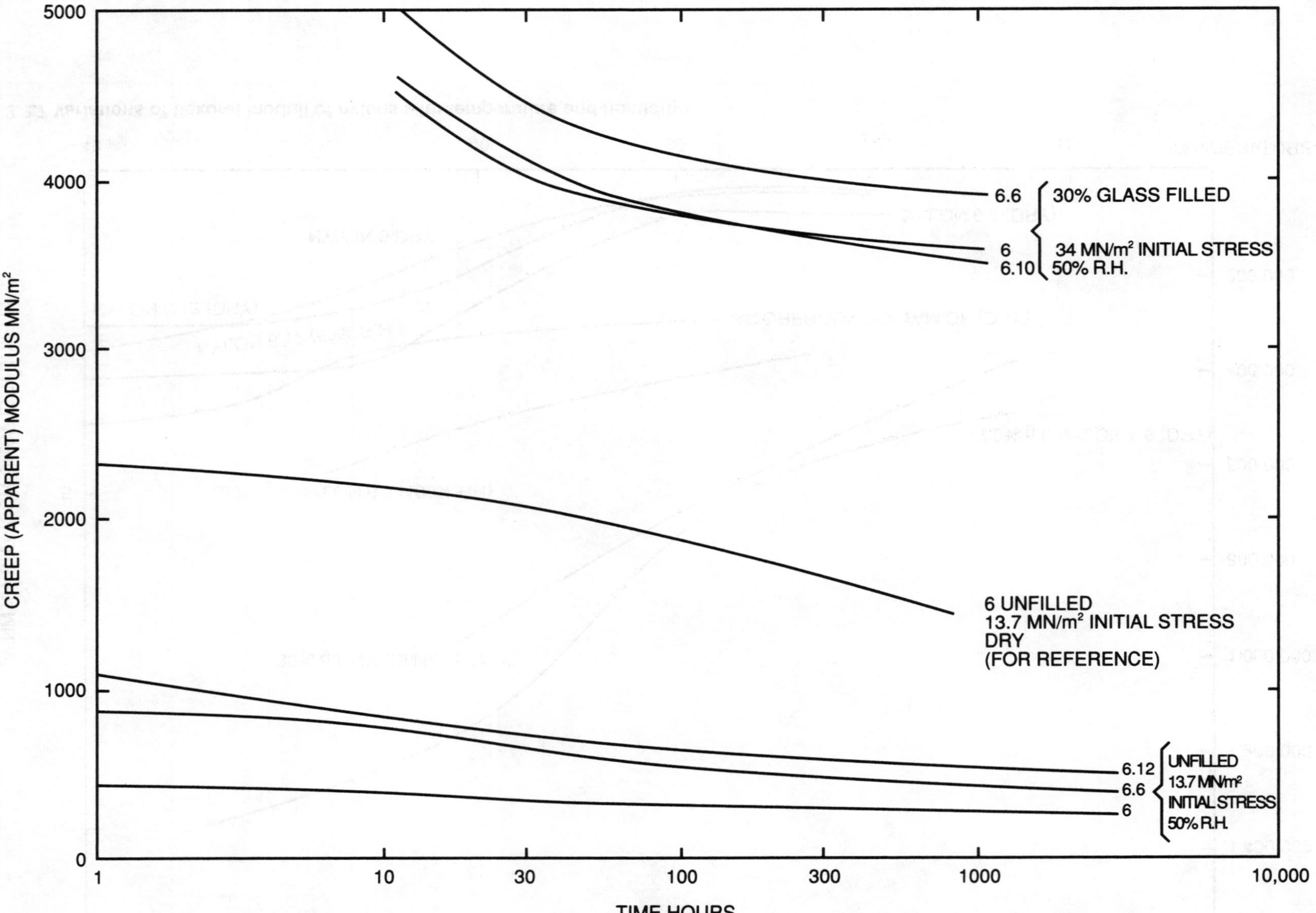

FIG 3.2.8 Effect of glass fibre reinforcement on creep of different types of nylons (room temperature)

Fig 3.2.8

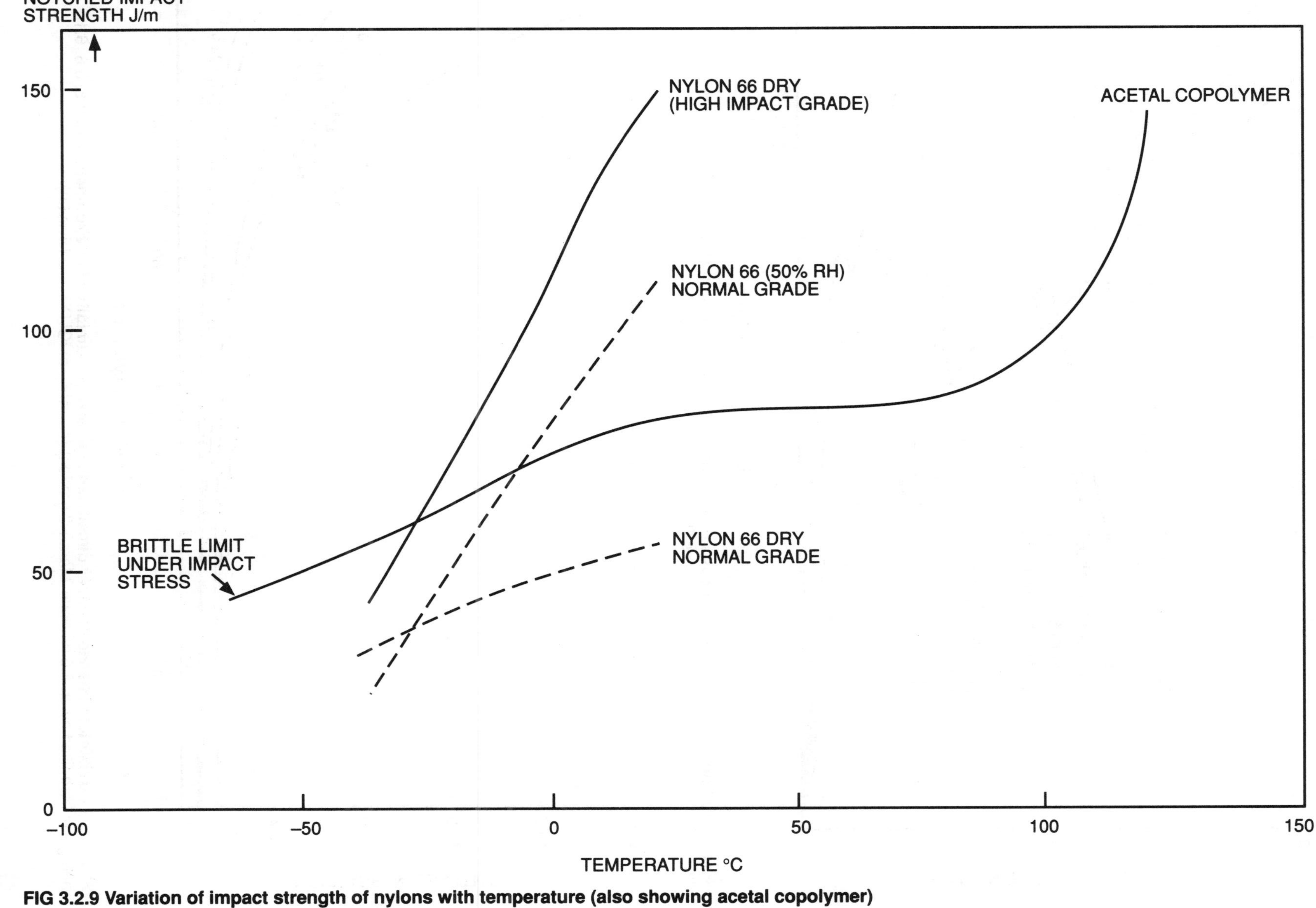

FIG 3.2.9 Variation of impact strength of nylons with temperature (also showing acetal copolymer)

Fig 3.2.9

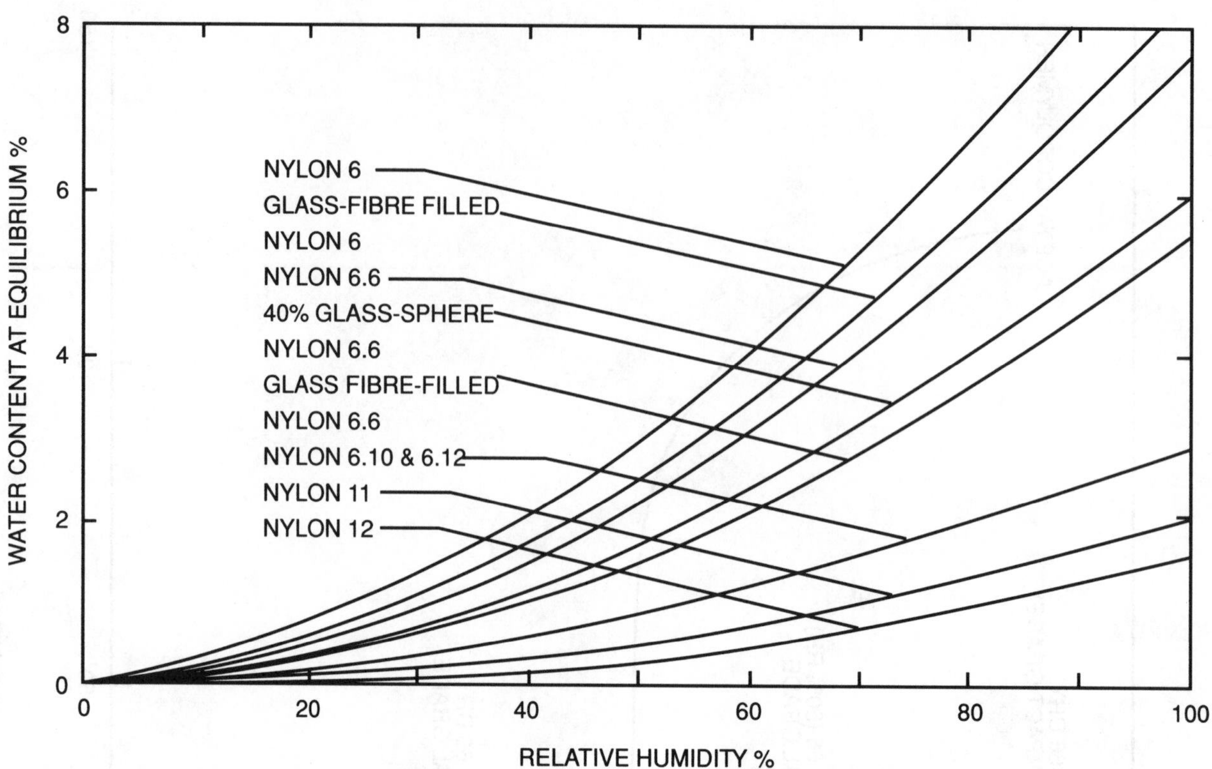

FIG 3.2.10 Equilibrium water content vs relative humidity of environment. Nylon 6, 6.6, 6.10 and glass fibre-filled nylon 6.6 (courtesy of ICI Plastics Division)

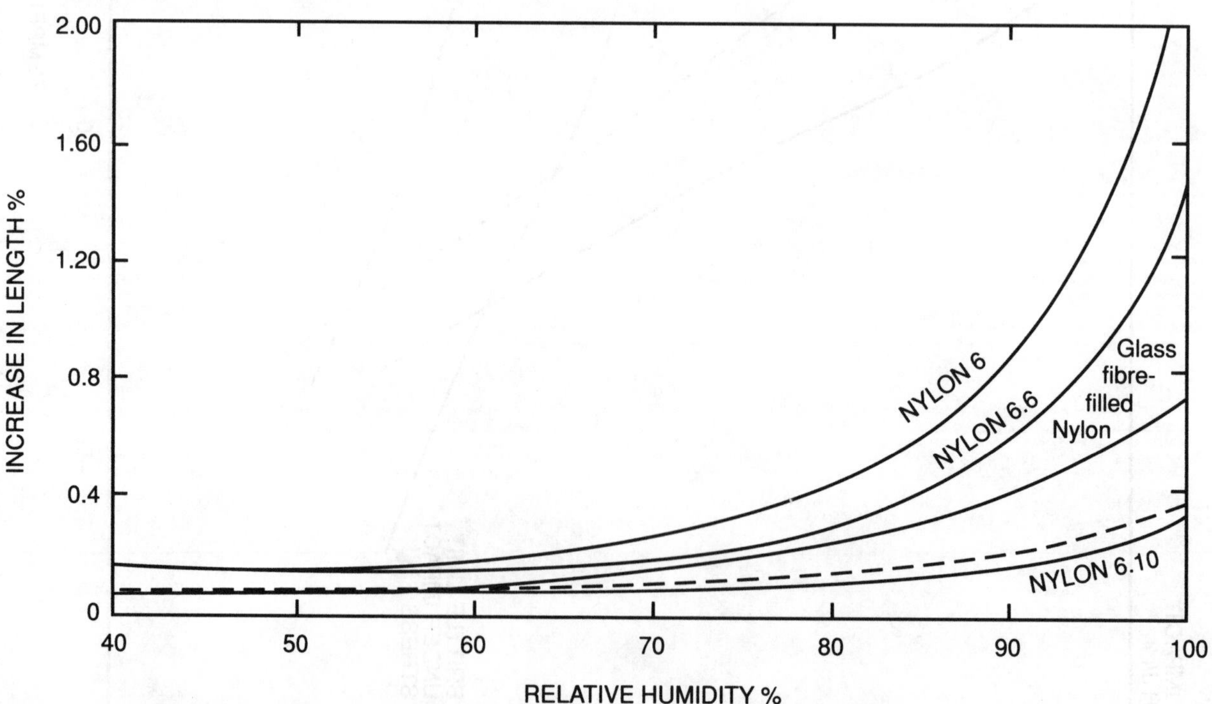

FIG 3.2.11 Equilibrium change in linear dimensions vs relative humidity of environment. Nylon 6, 6.6, 6.10 and glass fibre-filled nylon 6.6 (various 'Maranyl' grades) (courtesy of ICI Plastics Division)

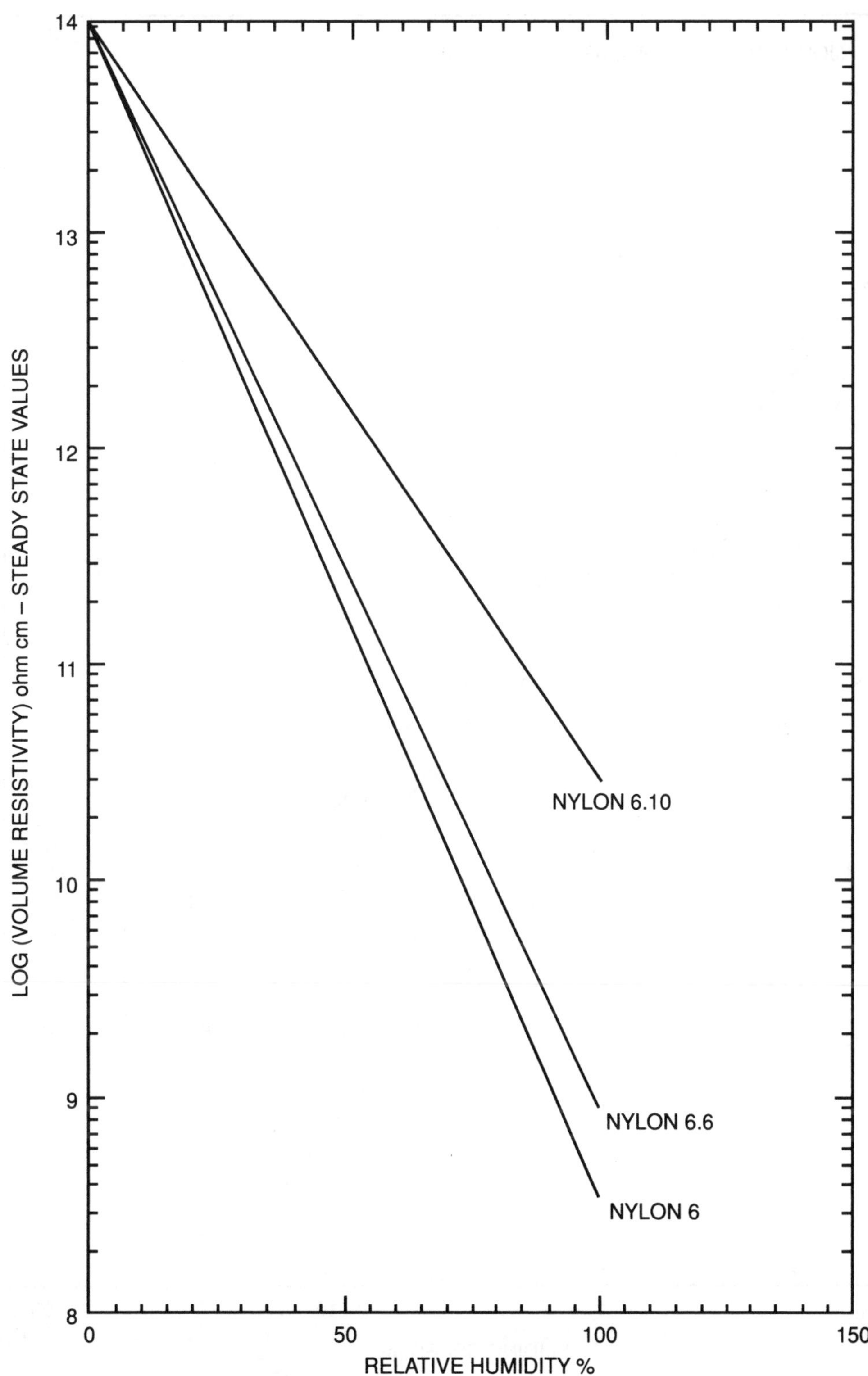

FIG 3.2.12 Steady state volume resistivity vs relative humidity of environment. Equilibrium data. Nylon 6, 6.6, 6.10 ('Maranyl' F103, A100, B100)

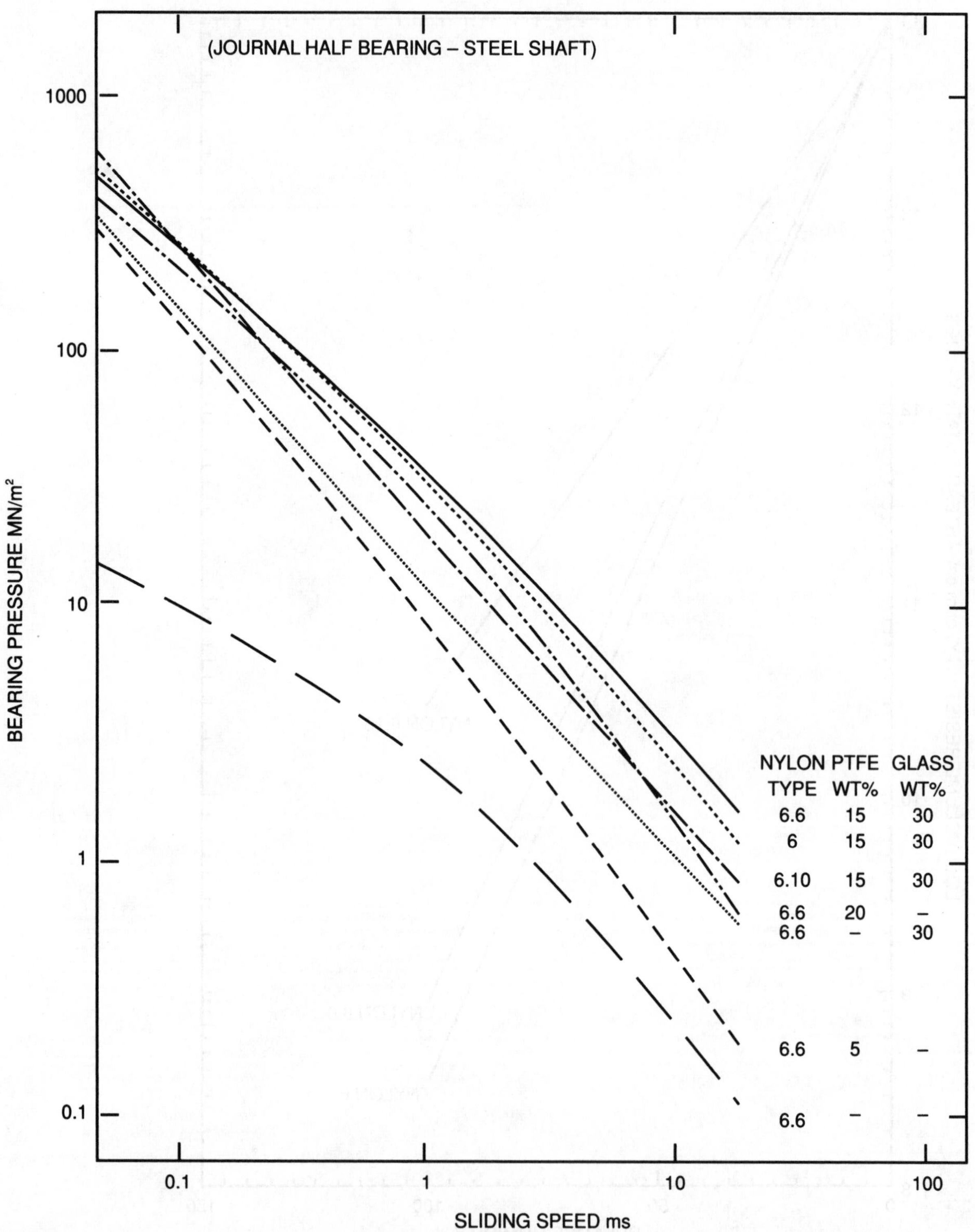

FIG 3.2.13 Limiting P–V curves for different types of nylons showing effect of P.T.F.E. and glass fillers

Fig 3.2.13

3.2.2 Polyacetals

3.2.2.1 General characteristics and properties

The advantages and limitations of polyacetals, compared with other materials, are given in:

Table 3.2.14 *Characteristics of polyacetals*

BASIS AND RANGE OF AVAILABLE TYPES

Acetal resins are based on formaldehyde polymerisation and three basic types are available—homopolymers, copolymers and terpolymers.

Homopolymer	*Copolymer*	*Terpolymer*
Stiffer (~10%), harder and higher resistance to fatigue than copolymer. Highest fatigue endurance of any unfilled commercial thermoplastic at room temperature. Higher tensile and flexural strength (~14%) coupled with lower elongation compared with copolymer. Higher heat deflection temperature compared with copolymer. Not recommended for use with strong acids or bases. Degradation once initiated continues unchecked. If overheated during processing formaldehyde is released. e.g. Delrin	Better stability with long-term high-temperature exposure and exceptional resistance to hot water [a] and hot strong bases. Higher elongation in unfilled condition. Can be processed at wider range of moulding temperatures. Oil impregnated grade available. Lower strength, hardness, stiffness and resilience. Glass-coupled grade available with much improved strength. Highly resistant to thermal and oxidative degradation. e.g. Kematal, Celcon (USA) Hostaform C	Chemical resistance and other properties similar to copolymer except flexural strength and hardness which are highest of acetal types. Mainly employed in blow moulding for such items as aerosol containers. e.g Hostaform T

[a]There is some controversy between major producers as to which acetal is superior in hot water. Homopolymers and copolymers have both been used in contact with water for many years, but the copolymer has had the bulk of hot water applications.

Suitable grades are available for injection, extrusion and blow moulding processing.

The main grades are essentially identified by their molecular weight and relevant flow properties. Lower molecular weight grades are recommended for thin-walled parts and higher molecular weight grades for maximum resilience. By the incorporation of specific additives (e.g. UV stabilisers)—chemical properties can be modified.

Materials designed for specific applications can be obtained by addition of appropriate fillers and/or reinforcements, e.g. glass fibre to improve stiffness and strength (of copolymer) and abrasive wear resistance, and PTFE to improve low friction and adhesive wear characteristics.

Glass fibre-filled copolymers are available which contain a chemical coupling agent which causes greater adhesion between the glass fibre and the copolymer matrix resulting in improved strength and much better impact strength over glass-reinforced (no coupling) homopolymer.

One manufacturer's products are listed in:

Table 3.2.15 *Example of the range of available grades of acetal copolymers*

APPLICATIONS OF POLYACETALS

Polyacetals are used in situations where combinations of resilience, rigidity, dimensional stability, chemical resistance, low friction and wear are required. Some examples are given below.

Gears—spur, helical, bevel or worm.
Cam discs.
Bearing boxes and bushes.
Switch relays, terminal blocks.
Coil formers.
Blower fans and ventilation fans, pump parts.
Parts for office machines and household appliances, washing, machines, bathroom fittings.
† Hinges.
† Springs, snap fittings, screws.
Curtain rail runners.
Aerosol nozzles.
Nuclear engineering applications.
Levers.
Shafts.
Plates.
Valves.
Plumbing fittings,‡ and components for pneumatic technology.
Automotive components—hot water applications such as radiator header taps, water filter bodies, and fuel–contact applications such as tank level mechanisms and carburettor floats.
Irrigation systems in areas of intensive agriculture and low rainfall (e.g. Israel, S. Africa) for compression fittings, jets, sprays, and precision metering valves—preferred due to resistance to pesticides.
Clock and watch parts (dimensional stability, abrasion resistance).

The low-friction filled grades find use in—

(a) platen carriage guides for recorders;
† (b) bearings for computers;
(c) fishing reel components;
(d) gears and cams.

†Where integral spring/hinge properties can be incorporated into one complex moulding acetals are used to replace other thermoplastics, zinc and aluminium die castings, and steel parts.

‡Care must be taken in the selection of acetal for plumbing, especially in components which come into contact with non-ferrous metals or hot water. Special grades are now available which are claimed to pass all necessary technical and water board approvals.

TYPICAL PROPERTIES OF POLYACETALS

A list of properties of polyacetals is given in:
Table 3.2.16 *Typical properties of polyacetals*

The properties given are typical values obtained from standardised tests conducted at room temperature (except where stated otherwise). It should be remembered that properties are affected by processing (see Section 3.2.2.5) and product testing is always advisable.

Before purchasing any grade of material it is necessary to confirm with the supplier that the expected properties can be obtained.

The effect of temperature on mechanical properties is shown in:
Fig 3.2.14 *Temperature variation of yield stress and break elongation for acetal copolymer,* and
Fig 3.2.15 *Comparison of temperature variation of yield stress of acetal copolymer with other common plastics*
Information on creep is provided in:
Fig 3.2.16 *Effect of temperature on the creep of filled and unfilled acetal homo- and copolymers*

3.2.2.2 Chemical and environmental resistance

The resistance of acetals to chemicals is summarised in:
Fig 3.2.17 *Summary of chemical resistance of acetal types*

Immersion in water at temperatures approaching the boiling point can seriously impair the tensile properties of some acetal grades. See:
Fig 3.2.18 *Tensile strength vs time of unreinforced polyacetal resins and compacting materials in 82°C water*

where grades of acetals are compared with competing materials.

3.2.2.3 Friction and wear properties

Polyacetals have high hardness and low friction coefficients which lead to application in bearings. Molybdenum disulphide-filled grades are used where low or very low sliding speeds, high loading pressures and short slide paths are encountered (e.g., swing bearings, joints, guides, etc.). At higher speeds grades with other additives may be employed. PTFE-filled grades are also available.

The effect of PTFE and silicone additions on friction and wear is given in:
Table 3.2.17 *Wear rates and friction of filled and unfilled acetals*

A greater standard of accuracy may be obtained with polyacetal than with nylon and running-in lubrication is not as necessary as for nylon. Polyacetal is an attractive material for fine pitch gears.

ABRASION RESISTANCE

Special grades are available for dry sliding abrasion applications including non-lubricated bearings or bearings given only an initial lubrication.

The abrasion properties can be unfavourably affected by pigmentation—preliminary tests are recommended when using coloured materials.

POLYACETAL BEARINGS

Normally free from stick-slip behaviour because the static and dynamic coefficients of friction are almost equal.

For P–V, P–T values see also Vol. 1, Chapter 1.5.

Running clearances depend upon the operating temperature.

Recommended clearance values for dry or unlubricated bearings in acetal copolymer are:

0.15cm/cm diameter for use above 40°C

0.006–0.0010cm/cm diameter for use up to 40°C (smaller clearances risk bearing seizure).

3.2.2.4 Electrical properties

Properties are typical of those for many thermoplastics used in non-critical applications (see Section 3.1.3).

Acetals cannot be used for applications involving high electric stress at power frequencies at temperatures above 70°C.

Acetals have a useful range of properties from –40°C to +50°C where losses are relatively low and where the dielectric constant changes only slightly with temperature and frequency. At room temperature and 10^8 Hz there are maximum losses which preclude the use of acetals in some applications at high frequencies.

3.2.2.5 Processing

Polyacetals can be injection moulded, blow moulded or extruded on standard equipment and have excellent flow characteristics in the molten state. Different grades are available with special characteristics, e.g. easy flow, faster cycling, etc. Manufacturers of homopolymers and copolymers frequently claim that their own product has better mouldability than their competitors' products. Many leading processors consider that there is no essential difference in mouldability (at least for general purpose grades).

Good temperature control ensures accurately dimensioned mouldings and machines should have at least two heater zones with separate temperature control on the nozzle. Melt temperatures in the range 180–220°C have no marked effect on mechanical properties provided normal moulding cycles are used.

Mould temperature is one of the most important processing parameters for precision injection moulding. In view of the relatively high mould temperatures required (90–125°C), oil is normally used as the thermal control medium. Normal mould temperatures in the region of 55–110°C have no marked effect on ultimate tensile strength, elongation at break, ball indentation hardness and impact strength, but there is a marked *increase* in toughness with increasing mould temperature, see:

Fig 3.2.19 *Influence of mould temperature on impact resistance of polyacetals*

and

Fig 3.2.20 *Influence of residence time in plasticising cylinder on impact resistance of polyacetals*

Thermal degradation is identifiable by a yellow discolouration in the melt or moulding. Homopolymers release formaldehyde if overheated, but if correctly processed, this will not occur. Mouldings can be ejected hot due to good creep resistance of the material.

Thick-walled mouldings require a considerably longer injection cycle, and the moulding costs will increase proportionally. With acetals, the following approximate injection times are needed: with 3mm,35s; 4mm,50s: 5mm,85s.

Normal mouldings can be manufactured with mould tapers around 0.5° but this must be increased up to a maximum of 3° for large mouldings.

MOULD SHRINKAGE

The shrinkage of partially crystalline materials cannot be predetermined with accuracy. It is anisotropic only with glass-filled grades, where distortion may be encountered.

Acetal copolymers (unfilled) have a shrinkage value of between 1.5 and 3.5%. The 20% glass homopolymer exhibits a shrinkage of 0.6–1.5% , and the 25% glass coupled copolymer of 0.4–1.8%. Because of low moisture absorption (0.2% in air at 20°C) no conditioning is required. Dimensional accuracy thus depends upon after shrinkage (post-moulding shrinkage).

POST-MOULDING SHRINKAGE

Post-mould shrinkage is influenced by the mould temperature and the ambient temperature after moulding.

After initial shrinkage, parts maintained at 80°C shrink a further 0.1–0.3% after 3 months, but do not change more than 0.1% after an additional 9 months.

ANNEALING

When processing has been carried out correctly, mouldings do not need to be annealed. In exceptional cases the relationship between annealing time, wall thickness and part geometry must be taken into account. A wall thickness of 2–3mm requires an annealing time of about 20 min. Heating and cooling should be effected slowly.

Mould temperature and injection pressure influence the mould, post-moulding and annealing shrinkage but the absolute values of shrinkage vary with the grade of resin used and suppliers data should be consulted. Shrinkage is a function of the pressure on the melt in the cavity (screw forward rate and injection pressure) and this in turn depends on the viscosity of the resin. Other factors affecting shrinkage are the mould temperature and the degree of crystallinity in the moulding. This applies to all acetals.

DIMENSIONAL TOLERANCES

The following represent average dimensional tolerances that have been achieved on commercial homopolymer mouldings.

	Total tolerance	
Dimension of part (mm)	*Standard control (mm)*	*Fine control (mm)*
0-5	0.064	0.041
10	0.081	0.051
12.7	0.091	0.061
15.2	0.10	0.068
20	0.13	0.084
25.4	0.15	0.10
38	0.23	0.15
51	0.25	0.18
76	0.35	0.23
100	0.40	0.30
127	0.51	0.35
152	0.61	0.40
203	0.72	0.51
254	0.86	0.61

OUTSERT MOULDING

Acetal copolymer is a favoured thermoplastic for outsert moulding of good wear resistance, natural lubricity, fatigue resistance and dimensional stability.

3.2.2.6 Major suppliers and trade names

Suppliers and trade names are listed in:
Table 3.2.18 *Major suppliers and trade names of polyacetals*

TABLE 3.2.14 Characteristics of polyacetals

	Advantages	***Limitations***
Mechanical properties	Excellent combination of toughness, rigidity, fatigue strength and yield strength (spring-like qualities). Very good resistance to stress relaxation. Good creep resistance compared to nylon.	Do not excel in toughness; notch-sensitive.
Wear	Good friction and wear properties. Low coefficient of friction.	
Thermal properties	Maximum service temperature 90–120°C depending on grade, stress and time. High resistance to thermal and oxidative degradation. Good dimensional stability over wide temperature range (–40°C–65°C). Non-toxic combustion products.	Molecular structure is such that non-flammable forms are not practicable. Above about 115°C significant changes in physical properties occur particulary melt flow index and weight loss.
Electrical properties	Good electrical insulation, little affected by moisture. Virtual immunity to arc tracking.	Cannot be used for applications involving high electric stress or power frequencies at temperatures above 70°C. Non-flammable alternatives often preferred.
Environmental	Good moisture and chemical resistance—resistant to nearly all organic solvent and alkali solutions. Low gas and vapour permeability. Little change in mechanical properties with humidity. Copolymer has greater oxidative stability; and resistance to hydrolysis. Proven resistance to pesticidal chemicals.	Attacked by strong mineral acids. Attacked by phenol and aniline. Homopolymers not stable in alkalis. Adversely affected by prolonged exposure to UV in standard unpigmented form.
Food and medicine	Certain grades suitable for contact with foodstuffs, including low friction PTFE lubricated grades of homoploymers.	Materials in moulded form normally contain traces of free formaldehyde (four parts per million) which could be absorbed by foodstuffs.
Fillers	Range of grades available with or without fillers, additives etc. for specific application.	Fillers can reduce strength and elongation.
Processing	Processable by conventional techniques of injection moulding, extrusion and blow moulding with excellent dimensional accuracy and post moulding stability.	Strains in well fabricated components should not exceed 2–5% in long-term applications. High shrinkage causes problems in very thin, flat sections.
Fabrication		Difficult to join to self and other materials. Difficult to electroplate.

TABLE 3.2.15 Example of the range of available grades of acetal copolymers

Grade	*Processing*	*Characteristics*	*Applications*
9010	Injection moulding plunger machines	Lubricated.	As for 9020 R.
9020 R	Injection moulding	Clean running. General purpose moulding grade.	All functional and decorative parts.
9020 G	Injection moulding	Special clean running grade. Continuous working temperatures over 70°C should be avoided.	Components in contact with foodstuffs. Aerosol valves. Camera parts.
9021 R	Injection moulding	Clean running. Higher hardness, stiffness and tensile strength than 9010 and 9020 grades.	As for 9020 R.
13021 R	Injection moulding	Clean running. Fast cycling.	Filling difficult cavities. Multi-impression tools.
27021 R	Injection moulding	Clean running. Fast cycling, easy flowing.	Long flow paths and restricted runner systems. For general use where fast cycling is important.
TVP 1020 R (Terpolymer)	Blow moulding	Clean running. High melt strength. High viscosity.	Containers. Pressure pack bottles.
2520 R	Extrusion injection moulding	High viscosity material without surface lubricant.	Moulded components requiring extra toughness and elongation. Thick section mouldings, extruded rod and tube.
9023 GV	Injection moulding	High strength. Glass-coupled grade.	Parts requiring high tensile strength, stiffness and creep resistance. Low thermal expansion.
9024 M	Injection moulding	Molybdenum disulphide-filled grades.	Parts requiring good slip/stick properties and low friction coefficients in lower speed range.
9024 K	Injection moulding	Special additives	Parts requiring good dry abrasion resistance and low friction coefficients in higher speed range.
VP 9024 TF	Injection moulding	PTFE-filled grade.	Very low friction coefficient. Critical bearings and bushes.
9025	Injection moulding	UV stabilised	Maximum weather resistance.

Courtesy of Farbwerke Hoechst AG.

TABLE 3.2.16 Typical properties of polyacetals

Property		Mechanical properties							Friction and wear properties	
		Tensile yield strength	*Elongation at break*	*Modulus of Elasticity in Tension*	*Flexural strength*	*Flexural modulus*	*Impact strength (IZOD-notched)* (23°C)	*Rockwell hardness*	*Coefficient of Friction (static) dry against steel*	*Abrasion resistance Taber CS-17*
ASTM		D638	D638	D638	D790	D790	D256	D785		D1044
UNITS		MN/m²	%	GN/m²	MN/m²	10³ MN/m²	J/cm OF NOTCH			mg/1000 CYCLES
Material		1	2	3	4	5	6	8	9	10
Standard homopolymer	A	62–70	25–60	3.5	100	2.39–3.02	0.7–1.2	M94	0.1–0.3	14–20
Copolymer Standard	B	62–75	25–60	2.85	90–120	2.60–2.64	0.5–0.75	M80	0.15	14
Copolymer Impact Mod.	B	44–53	75–100	1.7–2.1	50–65	1.40–2.05	0.8–1.25	—	0.9–0.15	17–25
20% Glass reinforced homopolymer	C	59–77	7	6.3	104–117	6.20	0.43–0.70	M90	0.1–0.3	33
25% Glass coupled copolymer	D	130	3	8.7	150	7.4–7.73	0.55–0.95	M99	0.15	40
22% PTFE filled homopolymer	E	48	—	—	71	2.75	0.37	M90	0.05–0.15	9
Terpolymer	F	67	60	2.95	120	2.60	—	M117	0.15	14
Homopolymer UV stabilised		64	35	—		2.78	0.72	M94		
Copolymer UV stabilised		68	60			2.5	0.69	R117		
Copolymer 2% silicone lubricated		55	50			2.1	0.85	R112		
Acetal/Elastomer alloy supertough		45	200			1.38	9.07	R105		

Table 3.2.16

TABLE 3.2.16 Typical properties of polyacetals—*continued*

Property			*Thermal characteristics*						
			Coefficient of thermal expansion	*Heat deflection temperature (0.45 MN/m²)*	*Heat deflection temperature (1.8 MN/m²)*	*Maximum recommended service temperature*	*Flammability*	*Specific gravity*	*Remarks*
ASTM			D696	D648	D648		D635	D742	
UNITS			$10^{-5}\,K^{-1}$	°C	°C	°C	mm/min		
Material			11	12	13	14	15	16	
Standard homopolymer		A	8.1	170	121–124	90 (70–75 at 50% RH)	28	1.39–1.43	
Copolymer	Standard	B	8.4–11.0	153	110–125	104	28	1.41	
	Impact Mod.		—	110–122	77–84	—	—	1.38–1.39	
20% Glass reinforced homopolymer		C	3.8–8.1	174	121–163	90	—	1.56	Used where higher stiffness and dimensional stability required. Intended mainly for injection moulding.
25% Glass coupled copolymer		D	4.0–8.4	166	160–163	104	—	1.50–1.61	
22% PTFE filled homopolymer		E	8.1	166	100	90	—	1.54	Used where low friction and high wear resistance required.
Terpolymer		F	8.4–11.0	153	110	104	—	1.41	
Homopolymer UV stabilised			12	165	130	85	—	1.42	
Copolymer UV stabilised			11	155	105	90		1.41	
Copolymer 2% silicone lubricated			11	148	110	90		1.40	
Acetal/Elastomer alloy supertough			12.3	145	90	80		1.34	

Table 3.2.16—*continued*

TABLE 3.2.16 Typical properties of polyacetals—*continued*

Property		*Physical properties*				*Electrical properties*						
								Dielectric constant		*Dissipation factor*		
		Thermal conductivity	*Specific heat*	*Refractive index*	*Water absorption*	*Volume resistivity*	*Dielectric strength (short time)*	*60 Cycles*	*10^6 Cycles*	*60 Cycles*	*10^6 Cycles*	*Arc resistance*
ASTM		—	—	D542	D570	D257	D149					D495
UNITS		W/m per K	kJ/kg per K		% (24 h)	10^{13} × ohm-m	10^2 kV/m					(s)
Material		1	2	3	4	5	6	7	8	9	10	11
Standard homopolymer (e.g. Delrin)	A	0.224	1.47	All grades opaque	0.25	> 1.0	150	3.7	3.7	0.0048	0.0048	129
Standard copolymer (e.g Hostaform – C Kematal)	B	0.3	1.47		0.22	>1.0	200	3.7	3.7	0.001	0.006	240
Standard terpolymer (e.g. Hostaform T)	C	0.3	1.47		0.22	>1.0	200	3.9	3.9	0.001	0.006	240
20% glass-reinforced homopolymer	D	0.4	—		0.25	0.5	200	4.0	4.0	0.0047	0.036	188
25% glass-reinforced copolymer	E	—	1.21		0.29	1.2	250	4.8	4.8	0.003	0.006	136
22% PTFE homopolymer	F	—	—		0.20	—	—	—	—	—	—	—
Homopolymer UV stabilised					0.25	0.1	180					
Copolymer UV stabilised					0.22	1.4	190					
Copolymer 2% silicone lubricated					0.18	1.5	200					
Acetal/Elastomer alloy supertough					0.27	1.9	189					

Table 3.2.16—*continued*

TABLE 3.2.17 Wear rates and friction of filled and unfilled acetals

Component material	*Wear factor K (10^{-10} in^3 min/ft lb h)*	*Coefficient of Friction*		*Limiting PV (ft lb/in^2 min)*			*Thermal Expansion Coefficient (10^{-5} K^{-1})*	*Thermal conductivity (W/m K)*
		Static	*Dynamic*	*10 fpm*	*100 fpm*	*1000 fpm*		
Unmodified acetal	65	0.14	0.21	4000	3500	<2500	8.5	0.23
Acetal + silicone	20–27	0.08–0.09	0.11–0.12	—	6000–7000	9000	9	0.23–0.24
Acetal 20% PTFE	14	0.07	0.15	10 000	12 500	5500	9.7	—
Acetal 20% PTFE plus silicone	9	0.06	0.11	8000	15 000	12 000	9.7	—

Mating material — steel.

10^{-10} in^3 min/ft lb h ≡ 2.0 × 10^{-17} m^3/J
1 ft lb/in^2 min ≡ 35 J/m^2 s

10 ft/min ≡ 0.05 m/s
100 ft/min ≡ 0.51 m/s
1000 ft/min ≡ 5.1 m/s

Wear (in) = KPTV where T = time (h)
P = pressure (psi)
V = velocity (ft/min)

Nylon–acetal combinations are considered in Section 3.2.1.4.

TABLE 3.2.18 Major suppliers and trade names of polyacetals

Company	*Trade names*	*Grades supplied*	*Forms*
Amcel (UK)[b]	Kematal	U, G, W	Copolymer
Asahi (Japan)[a]	Tenac	—	Homopolymer
BASF	Ultraform	U, G	Copolymer
Celanese Plastics (USA)	Celcon Kematal	U, G, W —	Copolymer —
Du Pont	Delrin	U, G	Homopolymer
Hoechst	Hostaform	G, W	Copolymer and terpolymer
LNP	Fulton	W	Homopolymer and copolymer
	Thermocomp KF	G, S, B, GW	
Polyplastics[b] (Japan)	Duracon	U, G, W	Copolymer
Freeman Chemicals Ltd	Latan	—	—
Vigilant Plastics	RTP 800	—	—
Wilson Fiberfil International SA	Formalafil	—	—

[a] Distributed in the UK by Plastic Products Ltd.
[b] Subsidiary of Celanese Plastics.

U = unfilled; G = glass fibre filled; W = wear resistant grades (PTFE, MoS_2, silicone) S = glass sphere-filled; B = carbon fibre-filled.

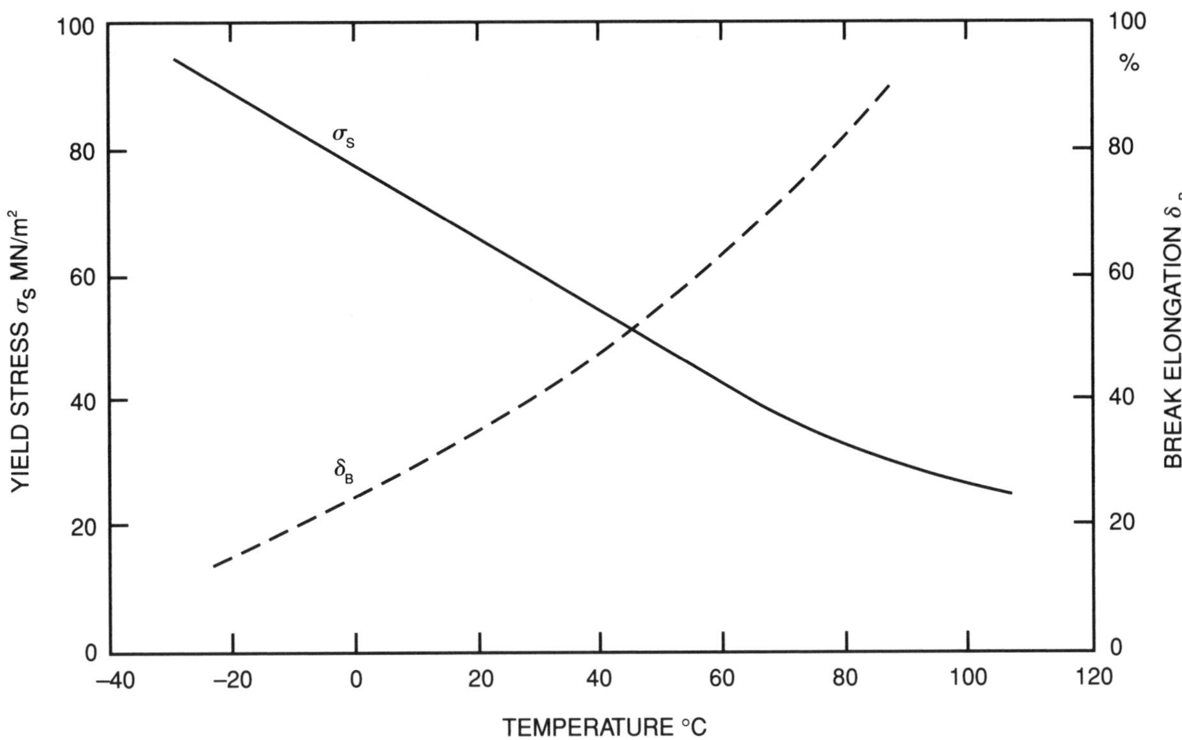

(Strain rate 50 mm/min, test specimen of 4 to 1 proportion, prepared from compression moulded sheet.)

FIG 3.2.14 Temperature variation of yield stress and break elongation for acetal copolymer

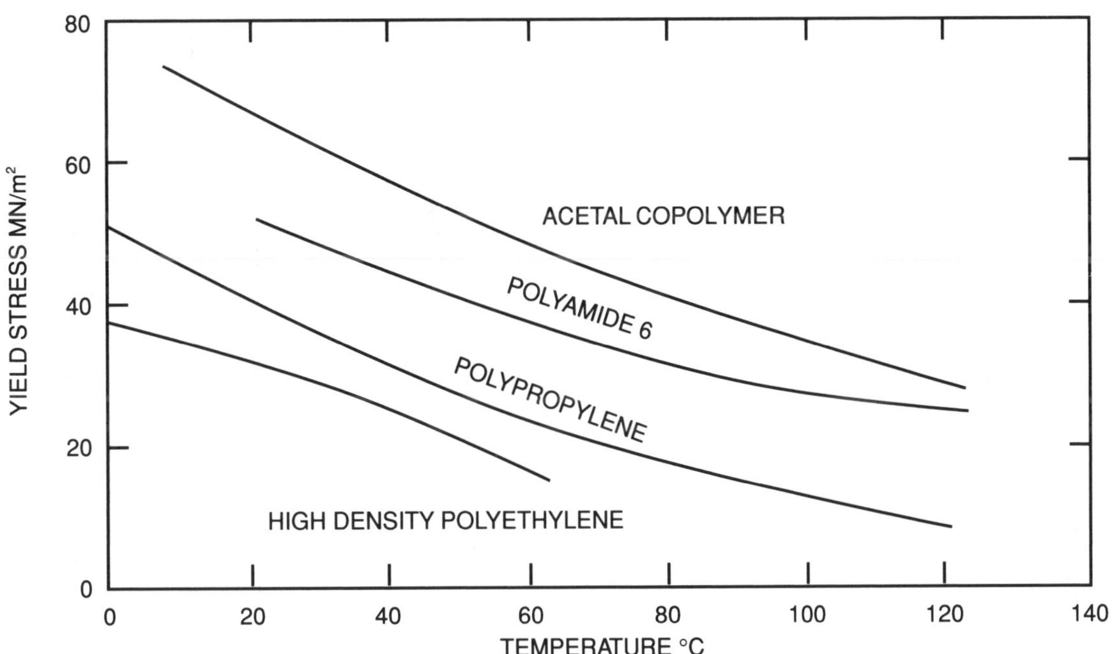

(Strain rate 50 mm/min, test specimen of 4 to 1 proportion, prepared from compression moulded sheet.)

FIG 3.2.15 Comparison of temperature variation of yield stress of acetal copolymer with other common plastics

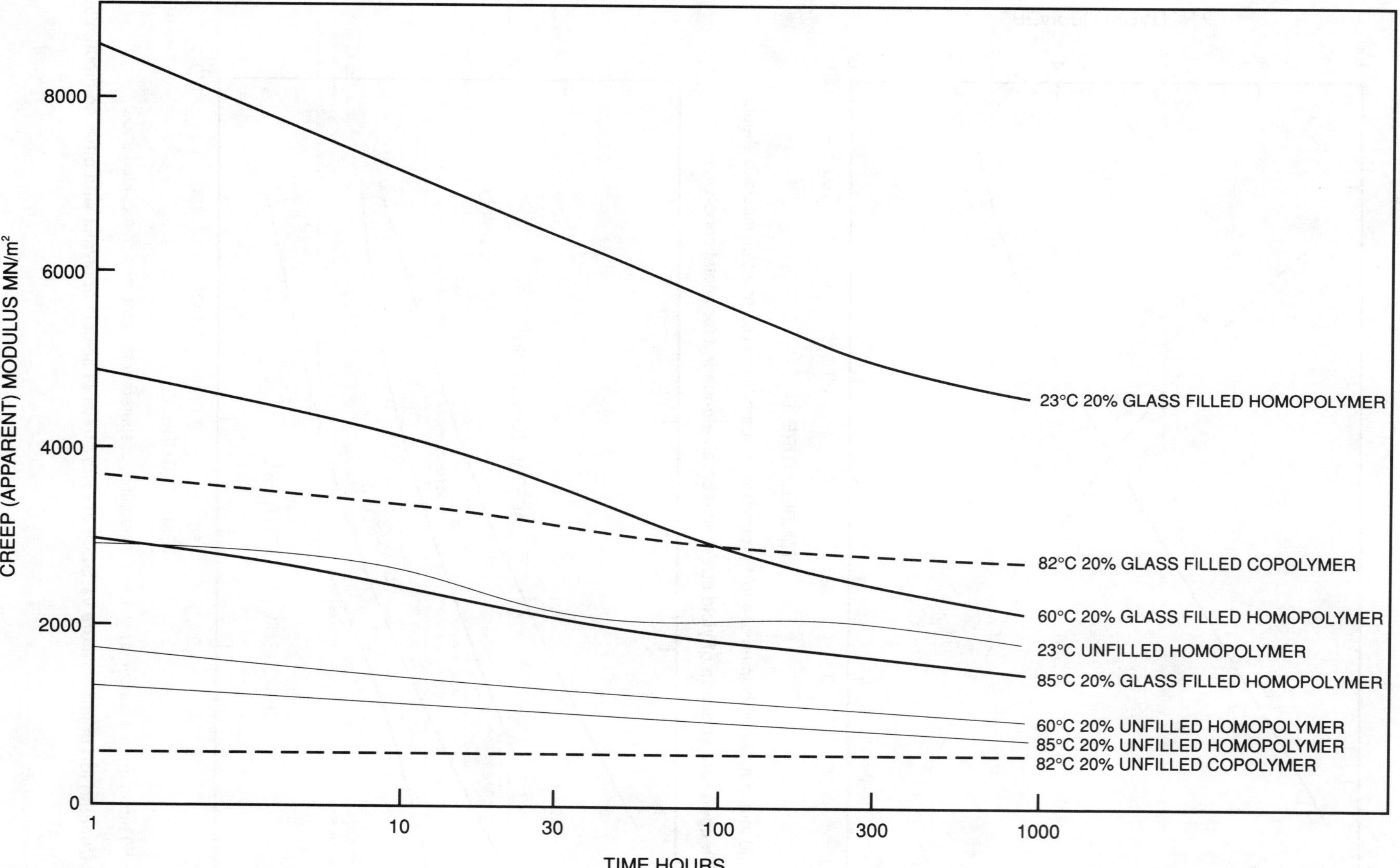

FIG 3.2.16 Effect of temperature on the creep of filled and unfilled acetal homo- and copolymers. Initial load 3.4 MNm^{-2}

Fig 3.2.16

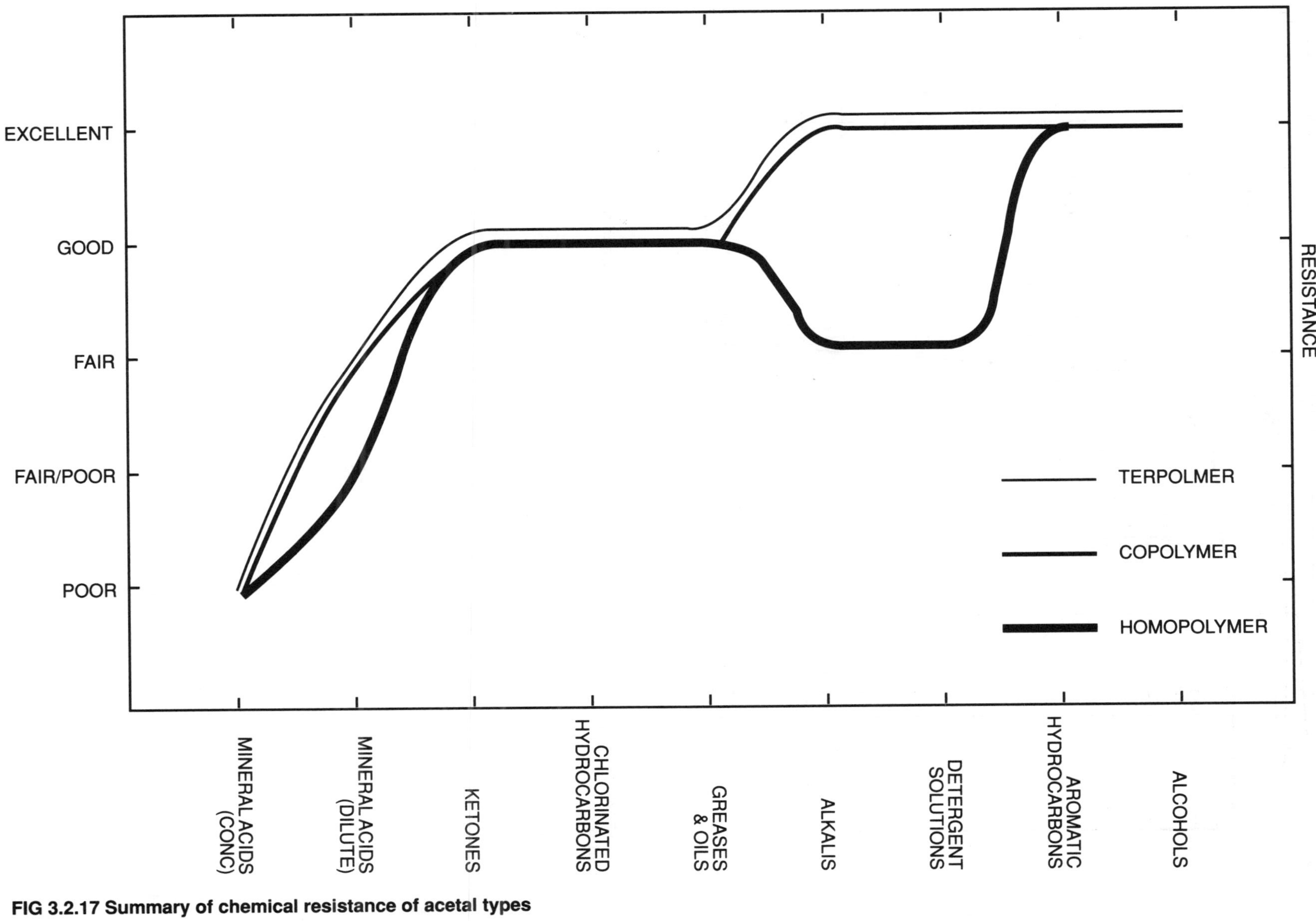

FIG 3.2.17 Summary of chemical resistance of acetal types

Fig 3.2.17

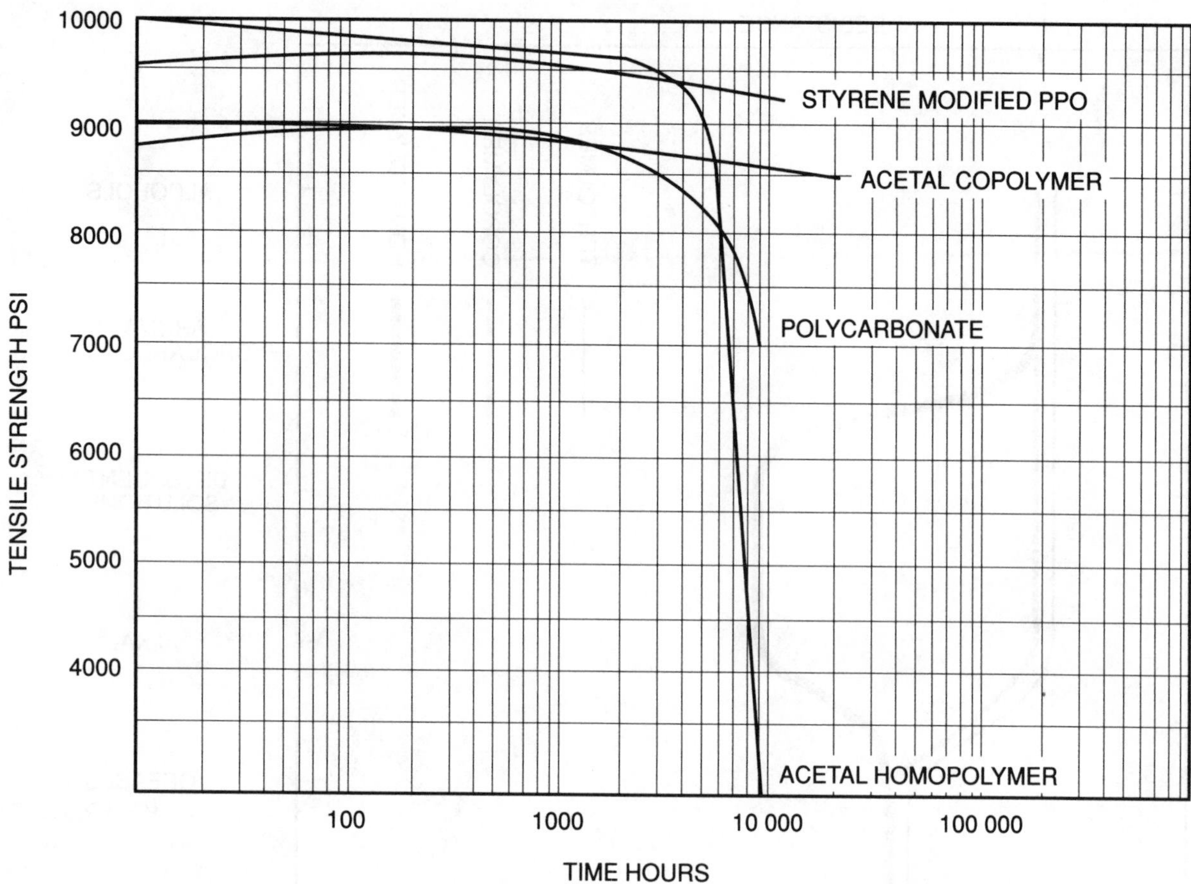

FIG 3.2.18 Tensile strength vs time of unreinforced polyacetal resins and compacting materials in 82°C water
Data courtesy Amcel/Celanese

Fig 3.2.18

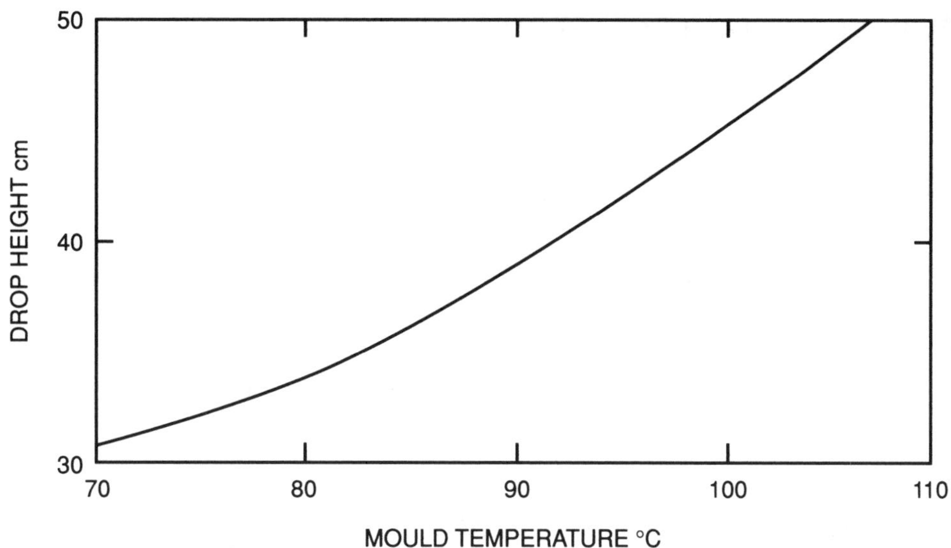

FIG 3.2.19 Influence of mould temperature on impact resistance of polyacetals.
Mean drop height x to achieve 50% failure of test specimens as a function of mould temperature (test specimen: injection moulded sheet measuring 70 x 70 x 4mm, lateral film gate, 1 mm thick)

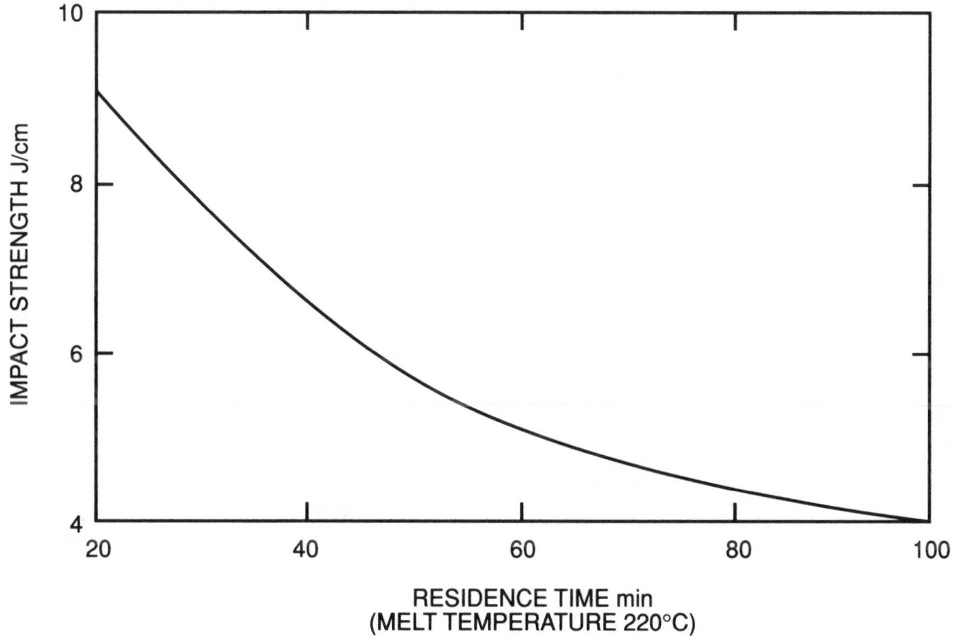

FIG 3.2.20 Influence of residence time in plasticising cylinder on impact resistance of polyacetals

3.2.3 Polyvinyl chloride (PVC) and related polymers

Including vinyl chloride copolymers, chlorinated polyvinyl chloride (CPVC) and polyvinylidene chloride (PVdC)
(For acrylic – PVC alloy see Section 3.2.5;
for ABS – PVC alloys see Section 3.2.8.)

3.2.3.1 General characteristics and properties

PVC - type polymers have many and great advantages, listed in:
Table 3.2.19 *Characteristics of PVC and related polymers*
Their limitations are mainly concerned with elevated temperature or attack by specific reagents.

AVAILABLE TYPES

Polyvinyl chloride (PVC)

Linear polymer of vinyl chloride, with some branching in commercial grades. Fundamental repeating unit.

$$\left[\begin{array}{c} -CH_2-\underset{\displaystyle Cl}{\underset{|}{CH}}- \end{array} \right]_n$$

In commercial polymers n ranges from 500 to 2000, corresponding to a molecular weight range of about 30 000–100 000. End groups (whose chemical nature may vary) and some of the chlorine atoms are weak points in the polymer chain, limiting its heat stability. Stabilisers are necessary in commercial compositions.

Several grades of the basic polymer (PVC 'resin') are available. The main differences among these stem from the method of production (emulsion, suspension or bulk polymerisation) and/or molecular weight. All grades may be compounded with plasticisers and other additives to give a particularly wide range of properties of the resulting compounds, which are still commonly described as PVC; e.g. soft PVC (plasticised compositions) or rigid PVC (unplasticised or lightly plasticised). The most important single factor is the amount and nature of the plasticiser(s), which reduces the inherent hardness and rigidity of the polymer, and imparts flexibility and extensibility (as well as other properties). Other main compounding ingredients which influence the properties are stabilisers and fillers; among the latter some fibrous fillers (e.g. glass fibres) can increase very substantially the modulus and strength of PVC compositions. Processing behaviour and impact properties may also be improved by blending PVC with other polymers, e.g. ABS† or MBS† copolymers, nitrile rubber, chlorinated polyethylene, EVA† copolymers.

Whereas the properties of flexible PVC materials are mainly determined by the nature of the amount of plasticisers, those of rigid PVC are influenced not only by the compounding ingredients but also to a considerable extent by the molecular weight of the polymer and the processing (which may affect the polymer structure and/or the ultimate actual composition of the formulation).

PVC polymer grades and compounds suitable for processing by all main processing techniques are commercially available. Cellular products (foams), flexible and rigid, can be produced.

A special processing advantage of PVC is its ability to form pastes (plastisols) with plasticisers. These may be, for example, applied as coatings and/or expanded into cellular foams, or rotationally moulded to produce hollow articles; in each case 'gelation' by heating converts the fluid paste to a thermoplastic solid.

The other polymers in this general group resemble PVC in most properties, differing usually in one or a few respects. In the following subsections attention is focused on such differences.

Vinyl chloride copolymers

Only those of some practical interest to the product designer are mentioned here. They are copolymers of:

Vinyl chloride/vinyl acetate

Essentially linear copolymers, with the vinyl acetate comonomer units in the main chain. Comonomer content can be up to about 16%. Some of the copolymers may contain a third component in minor proportion.

†M = methacrylate;	B = butadiene;	S = styrene;
A = acrylonitrile;	E = ethylene;	VA = vinyl acetate.

Vinyl chloride/olefin

The olefin component is propylene or ethylene, present in minor proportions (typically up to 10%) in the copolymer chain which is essentially linear.

Vinyl chloride/ester

The common ester components are normally acrylic (e.g. 3-ethyl hexyl acrylate), vinyl fumarate or maleate, present in proportions of up to about 20% within the copolymer chain.

The copolymers are somewhat lower melting, and more tractable in heat processing than PVC homopolymer. They are also more easily soluble in solvents. However, the copolymers are still essentially rigid materials (unlike PVC compositions containing substantial amounts of added external plasticisers). Their mechanical properties are not drastically different from those of rigid PVC compositions.

Chlorinated polyvinyl chloride (CPVC)

This is PVC with additional substitution of hydrogen by chlorine predominantly at the original CH_2 group in the chain. The structure is thus different from that of polyvinylidene chloride

$$\left[\begin{array}{c} -\underset{\displaystyle Cl}{\overset{}{CH}}-\underset{\displaystyle Cl}{\overset{}{CH}}- \end{array}\right]_n$$

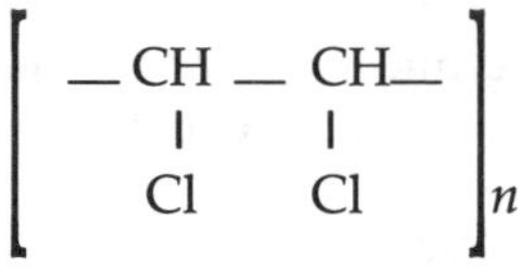

CPVC is similar in most properties to unplasticised PVC, but its softening point is substantially higher, so that its upper service temperature limit is about 100°C. The higher softening point and higher melt viscosity make processing more difficult in comparison with PVC, but the same basic equipment may be used.

Polyvinylidene chloride (PVdC)

Essentially linear polymer of vinylidene chloride, with the fundamental repeating unit

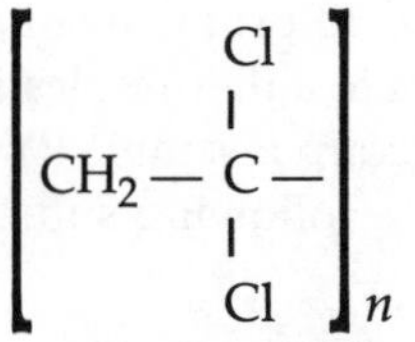

$$\left[CH_2 - \underset{Cl}{\overset{Cl}{\underset{|}{\overset{|}{C}}}} - \right]_n$$

In commercial polymers the value of n is normally above 200 corresponding to molecular weights of 20 000 and over. Unlike PVC, PVdC is normally crystalline, although it can exist in an amorphous form.

Commercial materials are frequently copolymers, e.g with minor proportions (about 10%) of vinyl chloride.

In comparison with PVC, PVdC (especially the crystalline form) is more resistant to organic solvents. This difference is less marked with PVdC containing a comonomer in the chain. The greatest difference, however, is in permeability. The outstanding property of PVdC is its low transmission of permeants, including organic and water vapours. These polymers are listed and compared in:

Table 3.2.20 *Summary of the comparative characteristics of PVC and related polymers*

APPLICATIONS

Polyvinyl chloride

The number and variety of applications of PVC are exceptionally large due to the following technical factors:

(i) The range of properties and property combinations realisable in PVC compositions.
(ii) Special physical forms (cellular PVC) are available.
(iii) A highly versatile fabrication technique peculiar to PVC is available in the use of PVC pastes.
(iv) Cellular PVC and the materials produced from paste may not be the sole constituents of PVC products (as e.g. in cellular toys or artificial leather) but may form important components of composite materials, e.g. flexible PVC foam in motor car and furniture upholstery or vinyl flooring, rigid PVC foam in building panels, PVC coatings (applied in paste form and solidified by heating) on fabrics (tarpaulins, rainwear, protective clothing) and paper (vinyl wallpapers).

Flexible PVC formulations have replaced rubber and other elastomers in a number of applications and given cost savings. Rigid PVC is a cost-effective engineering plastic where extremes of service conditions are not encountered.

The uses of PVC are too numerous and diverse for a unified , comprehensive presentation in a brief section. Only the main or particularly typical uses are therefore mentioned in:

Table 3.2.21 *Typical applications of PVC*

An important application is for 'safety glazing' for which the suitability of the high impact form is demonstrated in:

Table 3.2.22 *Relative advantages of PVC for safety glazing*

Vinyl chloride copolymers

Some of these, e.g. certain grades of vinyl chloride/acetate or vinyl chloride/vinylidene chloride copolymers, find application in adhesive and thin protective coatings (e.g. paints); these are not directly relevant here. (see Vol. 1, Chapter 1.4).

Those vinyl chloride copolymers which are of direct interest to the product designer are used mainly where a rigid polymer with PVC-like service properties is required, but where easier thermal processing (lower melt viscosity, lower processing temperature) is necessary.

A major use has been that of vinyl chloride/acetate copolymer in the moulding of gramophone records, where the grade of copolymer used offers the optimum combination of ease and speed of processing, playing performance, service stability, and cost. This type of copolymer is also the material of vinyl sheeting used for thermoforming, which must combine rigidity with easy heat-softening under fabricating conditions.

Vinyl chloride/acrylic ester copolymer has been used as the material of blow moulded bottles and film, for improved impact strength, and/or heat weldability.

Chlorinated PVC

The principal use for this material is in hot water pipes where its combination of generally PVC-like properties with a much higher softening point makes it particularly suitable.

Polyvinylidene chloride

The most important application of this material is as a coating or laminate component imparting good gas, moisture, vapour, scent and flavour barrier properties to plastics packaging films and containers (e.g. bottles).

Heat-shrinkable and heat-sealable films have been produced from PVdC, as well as heat-sealable layers or coatings on other polymer films.

The solvent resistance of PVdC pipes has been utilised in special applications, as e.g. in chemical installations and apparatus. Use of extruded PVdC products in this and other fields includes also gaskets, valve seats, tape for wrapping joints and conveyor belts. PVdC mouldings are represented by spray-gun handles, acid dippers and components in equipment for rayon manufacture.

3.2.3.2 Mechanical and physical properties

GENERAL CONSIDERATIONS

The range of properties obtainable in PVC compositions through suitable formulation is very wide. In particular the flexibility and softness can be varied over wide limits by plasticisation, which also affects the strength, extensibility and other properties.

Whereas the properties of plasticised PVC materials are mainly determined by the nature and amount of the plasticisers, those of rigid PVC are influenced not only by the compounding ingredients but also to a considerable extent by the molecular weight of the polymer and the processing (which may affect the polymer structure and/or the ultimate actual composition of the formulation). Thus it should be particularly noted that where data have been quoted other than as a range of rigid PVC compositions (and especially the creep data) they should be regarded as illustrating the general scale of values and type of behaviour; they would not be directly and accurately representative of every individual rigid PVC material.

ASTM test data are given for the mechanical and physical properties in:

Table 3.2.23 *Mechanical and physical properties of PVC and related polymer compounds,* and

Fig 3.2.21 *Tensile strength vs elongation for PVCs*

Variation with temperature is shown in:

Fig 3.2.22 *Typical variation of mechanical properties with temperature in PVC rigid extrusion grade.*

From the designer's point of view it is particularly worth noting that even for rigid PVC the practical upper temperature limit in service is about 60°C and the acceptable design stress level at room temperature will vary from about 4.1 MN/m^2 (600psi) for some vinyl chloride copolymers, or unplasticised PVC modified by blending with polymeric modifiers to about 6.89 MN/m^2 (1000 psi) for rigid PVC polymer (and certain copolymer) compositions. The modulus figure recommended for use in design considerations for PVC and vinyl chloride copolymers is about 1030 MN/m^2 (150 000 lbf/in^2). This is comparatively low, because of the creep behaviour. As with other amorphous polymers the strength of rigid PVC is higher at low temperatures: other things being equal the value at –50°C is approximately double that at room temperature. The fall in strength of unplasticised PVC with increasing temperature is marked: at 60°C the strength is reduced to approximately half the room temperature value or less.

The variation of creep with temperature for one grade of rigid PVC is illustrated in:

Fig 3.2.23 *Variation in creep of PVC high impact rigid injection moulding compound with temperature*

CPVC has the best creep resistance of unreinforced PVC types and the level of performance is indicated in:

Fig 3.2.24 *Stress rupture curves for CPVC hot water pipes*

Creep curves comparing PVC with other materials are in Section 3.1.2.

As with all data of this type it should be used for comparison purposes only. It is imperative to check with material suppliers that the chosen material can meet the performance requirements of the components to be manufactured.

3.2.3.3 Chemical and environmental resistance

Temperature limitation parameters have been linked with environmental resistance, although some of the parameters (such as softening and cold flex limitations) should strictly be classed as mechanical property limitations. Temperature limits are listed in:

Table 3.2.24 *Recommended service temperature limits for PVC and related polymer compounds*

The resistance of PVC and related types to most of the environments encountered in service is listed in:

Table 3.2.25 *Environmental properties of PVC and related polymer types*

3.2.3.4 Electrical properties

Generally vinyls are electrically non-conductive but can be formulated to vary from very good insulators to low conductivity materials. Their electrical properties are tabulated in:

Table 3.2.26 *Electrical properties of PVC and related polymer compounds*

3.2.3.5 Processing

The information in this section is intended to indicate to a product designer without specialist knowledge of plastics processing what major techniques are available for processing PVC and related polymers, and what types of products may be produced thereby. The most important general points relevant to the relationship between the material, processing and design in PVC and other chlorine-containing polymers, are also indicated.

FORMULATION FOR HEAT-STABILITY IN PROCESSING

The industrial techniques used in the manufacture of products, as well as semi-products (e.g. sheeting, pipe), and for further processing the latter, are listed in:

Table 3.2.27 *Processing of PVC and related polymers*

All of these techniques involve heat treatment at comparatively high temperatures. PVC polymer is easily decomposed by heat and the most important single consideration is that the composition used for any particular product must be formulated not only for the desired service properties, but also for satisfactory heat processability; this in turn may affect performance characteristics. Similar considerations apply to other chlorine-containing polymers, some of which (e.g. vinyl chloride/acetate copolymer) are used instead of PVC in applications where easier thermal processing is required. Thermal degradation shows as discoloration (from yellow to dark brown) and deterioration in properties.

MACHINING

Like most other plastics PVC and other chlorine-containing polymers can be machined. This method does not involve thermal processing. However, it is economical for mass production and—in extreme cases—local degradation can still occur due to frictional heat. In industrial practice machining is a secondary operation, confined largely to fabrication of PVC sheeting (e.g. sawing, drilling).

GENERAL POINTS RELEVANT TO MOULDED PRODUCT DESIGN

Moulding shrinkage and tolerance

These will vary with the formulation and moulding conditions. In general the mould shrinkage of PVC and related materials will increase with mould temperature and decrease with moulding pressure. Rigid PVC compositions have the lowest shrinkage, but even in those it may be up to 0.8%. Mould shrinkage is approximately as follows:

Rigid PVC and PVC copolymers	: 0.1–0.8%	
Flexible PVC	: 1.0–5.0%	(varies with plasticiser)
CPVC	: 0.3–0.7%	
PVdC	: 0.5–2.5%	

Section thickness and shape

In injection moulding unplasticised PVC the normal practical lower thickness limit is about 2 mm. In special cases (where very short flow paths and large gates are possible) this may be reduced by about half. The highest readily producible section thickness is about 15 mm. Abrupt changes in thickness and sharp corners should be avoided : both provide stress concentration points.

RE-USE OF SCRAP

This is commonly practised in production (extrusion, moulding), and acceptable, provided that the formulation is suitable and the material to be re-worked is not significantly degraded. However, where product quality is of particular importance even 'good' scrap percentage should be kept low (below about 20%).

3.2.3.6 Trade names and suppliers

Trade names and suppliers are listed in:
Table 3.2.28 *Major suppliers and familiar trade names*

TABLE 3.2.19 Characteristics of PVC and related polymers

	Advantages	***Limitations***
Mechanical properties	Good combination of stiffness and impact strength (rigid formulation), toughness, extensibility, high ratio of strength to weight (flexible formulations). Acrylic–PVC alloy with very high impact strength available (see Section 3.2.5).	Recovery from bending or stretching damped compared with rubber. Comparatively low heat distortion and softening temperature (even rigid compositions—except CPVC).
Physical properties	Good electrical insulation (enhanced by suitable formulation).	
Environmental resistance/flammability	Non-flammability (reduced or enhanced by some plasticisers).	Tendency to progressive degradation at elevated temperatures—in extreme situations HCl gas is given off. Stiffening and embrittlement at low temperatures (rigid compositions) and flexible unless specially formulated. Susceptible to staining by sulphides (flexible formulations—unless specially modified).
Resistance to solvents, etc.	Good resistance to acids, alkalis, oils and many corrosive inorganic chemicals, oxygen, ozone; good water barrier; PVC polymer is resistant to alcohols, aliphatic hydrocarbons and detergent solutions but these reagents may extract plasticisers from flexible formulations. Good weathering properties (can be enhanced or impaired by additives).	Susceptible to microbiological attack (especially flexible formulations unless fungicides or bactericides, which are expensive, are added). Attacked by ketones, some grades swollen or attacked by chlorinated and aromatic hydrocarbons, esters, some aromatic ethers and amines, and nitro-compounds.
Application to food and medicine	PVC and PVdC used in packaging applications. Blown bottles and containers also made from PVC (non-toxic grades).	
Formulation and processing	Formulation versatility giving materials ranging from elastomers to rigid engineering thermoplastics and processing possibilities from paste coatings to injection moulding.	Adhesion to many substrates limited in the absence of a primer.
Miscellaneous	Dimensional stability at room temperature for rigid formulations. Almost unlimited range of colours—transparent to opaque according to formulation. Low cost.	Relatively high specific gravity for polymeric material.

TABLE 3.2.20 Summary of the comparative characteristics of PVC and related polymers

Material		*Advantages*	*Limitations*
PVC	(Flexible)	Wide range of flexibility possible—competes for rubber applications at lower cost. Ability to form pastes (plastisols) by addition of plasticisers for coating and casting.	Flexible grades most susceptible to staining, chemical and microbiological attack due to presence of plasticisers. Creep properties inferior to rigid PVC.
PVC	(Rigid)	Lower coefficient of friction and better abrasive wear resistance than flexible grades. Easier to process and cheaper than CPVC. Can be glass fibre-reinforced to give improved strength and stiffness and lower coefficient of thermal expansion.	Attacked by strong acids and some solvents. Swollen by aromatic hydrocarbons. Lower maximum service temperature than CPVC.
PVC copolymers	(Less rigid than rigid PVC)	Processing characteristics and impact strength improved by blending with copolymers.	Creep properties generally inferior to rigid PVC. Lower tensile strength.
CPVC	(Rigid)	Considerably higher upper service temperature than rigid PVC. Best creep properties of PVC types. Resistant to acids, alkalis, most organic solvents, oil and grease.	More difficult to process than PVC due to higher softening point and melt viscosity. Higher cost.
PVdC	(Flexible)	Best resistance to all acids and most common alkalis. Unaffected by aromatic and aliphatic hydrocarbons, alcohols and esters. Best resistance to permeation by organic and aqueous vapours.	Higher cost Thermal processing more difficult.

TABLE 3.2.21 Typical applications of PVC

Product and/or application	*PVC type*	*PVC properties important in the application*	*Remarks*
Building applications			
Piping: potable water conduits, water mains, drainage, soil pipe systems, rain water systems and guttering, gas conduits, venting.	Rigid	Chemical and corrosion resistance, good barrier properties, lightness (in comparison with traditional materials), non-flammability, ease of jointing and installation, strength.	
Structural and cladding elements: panelling and partitioning (interior), wall cladding (exterior), building panels, glazing, roofing, window frames, hollow bricks.	Rigid	Lightness with rigidity, corrosion resistance, no need for protective coatings, permanent decorative finish possible, good thermal insulation (in cellular form).	Cellular, rigid PVC is used in some applications, e.g. composite panels.
Miscellaneous products: flexible 'see-through' doors, folding doors, roof underlay, greenhouse glazing, suspended ceilings, flooring (continuous and tiles).	Flexible (plasticised)	Toughness, clarity (with suitable formulations), non-flammability, chemical and moisture resistance, wear resistance.	
Electrical applications			
Insulation and sheathing (wire and cable):	Flexible (plasticised)	Abrasion resistance, resistance to moisture and chemicals, non-flammability, good ageing characteristics, ease of processing, acceptable electrical properties.	Rubber and the polyolefins have somewhat better electrical properties, but PVC is a better all-round insulation and sheathing material, superior in the other properties listed.
Terminal boxes and conduit:	Rigid	Lightness, ease of handling and installation, moisture and corrosion resistance, non-flammability, good finish.	
Battery separators:	Rigid porous	Stiffness, controllable pore size, resistance to chemicals, corrosion and water.	
Packaging applications			
Rigid foil (thin sheeting): may be fabricated *inter alia* into containers, blister and skin packs, nesting trays (for confectionery or biscuits).	Rigid	Rigidity and strength, suitability for thermoforming, clarity (in transparent formulations), non-flammability, dimensional stability, right balance of permeability to O_2 and moisture (for food wrapping).	PVC rigid foils are normally produced by extrusion casting.
Flexible film: used directly for wrapping food and other products *inter alia* as 'shrink'- or 'cling'- wrap.	Flexible (plasticised)	Clarity, non-fogging (e.g. when wrapping moist food), orientability (for built-in shrinkage properties), wrinkle resistance, non-flammability, right balance of permeability and barrier properties (for food wrapping).	Flexible PVC films are usually produced by extrusion blowing.
Blown bottles and containers: used for oil, wine, beverages, shampoos, certain cosmetics.	Rigid	Rigidity, toughness, clarity (in transparent formulations), low weight (compared to glass), chemical resistance, non-tainting, resistance to environmental stress cracking, non-toxicity (special formulations).	

Table 3.2.21

TABLE 3.2.21 Typical applications of PVC—*continued*

Product and/or application	*PVC type*	*PVC properties important in the application*	*Remarks*
Coated fabric and paper applications Leathercloth: used in upholstery, clothing, travel and fancy goods. Protective and foul-weather clothing: tarpaulins, life-rafts, hovercraft skirts, conveyor belts. Vinyl wallpapers	Flexible (plasticised)	Applicability in paste form, water barrier properties, wear resistance, flexibility, toughness, adhesion to substrates.	
Foam applications Rigid foam: used as core in sandwich structure (building and boat building) buoyancy blocks, fishing floats, insulation (thermal and acoustic), shock absorbent materials.	Rigid	Good insulation properties, rigidity, moisture and chemical resistance, non-flammability.	
Flexible foam: upholstery (especially motor cars), leathercloth and fancy goods, carpet backing, underlays, embossed wallpapers, foam flooring, shoe soles, soft toys.	Flexible (plasticised)	Production via paste, heat-bondability and embossability, non-flammability, good resilience properties.	In upholstery uses PVC foam can be heat sealed or bonded to itself and to PVC top layers.
Safety glazing	Transparent sheet	Combination of impact strength, rigidity and surface hardness.	Transparent PVC sheet, along with acrylic and poly-carbonate have been used for many years as safety and vandal resistant glazing.

Table 3.2.21—*continued*

TABLE 3.2.22 Relative advantages of PVC for safety glazing

Property	*Units*	*High impact PVC 'Pacton'*	*Clear PVC 'Darvic'*	*Polycarbonate*	*Acrylic 'Perspex'*
Sheet thickness	mm	3	3	3	6.35
Impact strength Falling dart	% Pass	100	70	100	0
Flexural Modulus 60 s, 23 °C	GN/m^2	2.8	3.4	2.4	2.9
Surface hardness Pencil	—	6H	6H	2H	9H

Table 3.2.22

TABLE 3.2.23 Mechanical and physical properties of PVC and related polymer compounds

Properties		*ASTM test method*	*Units*	*Rigid PVC and PVC copolymers*	*CPVC*	*PVDC*	*PVC + 20% glass fibre*
Tensile strength		D638	MN/m^2	24–62	35–62	21–34	96
Modulus of elasticity in tension		D638	10^3 MN/m^2	2.4–4.1	2.5–3.2	0.3–0.5	7.6
Elongation		D638	%	2.0–40	4.5–65	Up to 250	3
Impact strength (Izod notched)		D256	J/cm of notch	0.2–10.6	0.5–3.0	0.2–0.5	0.8
Flexural strength		D790	MN/m^2	69–110	60–117	29–43	145
Modulus of elasticity in flexure		D790	10^3 MN/m^2	2.0–3.4	2.6–3.1	2.4–3.4	6.9
Compressive yield strength		D695	MN/m^2	6.8–7.6	—	1.4–1.9	83
Hardness	Rockwell	D785	—	R106–120	R117–122	M50–65	
	Shore	D676		D70–85	—	>A95	D90
Specific gravity		D792	no units	1.4–1.54	1.38–1.58	1.65–1.75	1.58
Thermal expansion coefficient		D696	10^{-5}/°C	7.0–25.0	4.4–8.0	19.0	4.1
Thermal conductivity		C177	10^{-2} W/m°C	0.125–0.24	0.13	0.125	—
Specific heat		—	10^3 J/kg°C	0.8–1.2	1.4	1.3	—
Refractive index		D542	—	1.52–1.55	—	1.60–1.63	—
Clarity		—	—	Transparent to opaque			Opaque

Table 3.2.23

TABLE 3.2.23 Mechanical and physical properties of PVC and related polymer compounds—*continued*

UV stabilised PVC	*UPVC*	*High impact UPVC*	*Plasticised PVC*			*Cross-linked PVC*	*Structural foam*
			0–100% elongation	*100–300% elongation*	*>300% elongation*		
45	51	42	20	40	11	28	24
50	60	150	95	280	450	150	30
0.6	0.8	6	10.6 +	10.6 +	10.6 +	0.8	0.3
2.9	3.0	2.5	0.03	0.007	0.003	2.8	9
R108	R110	D76					R95
			A85	A65	A60	A99	
1.4	1.4	1.38	1.34	1.3	1.24	1.35	0.7
6	6	7	10	14	16	6	0.6

Table 3.2.23—*continued*

TABLE 3.2.24 Recommended service temperature limits for PVC and related polymer compounds

Limiting parameter		*ASTM test method*	*Units*	*Rigid PVC and PVC copolymers*	*CPVC*	*PVdC*	*PVC + 20% glass fibre*	*UV stabilised PVC*	*UPVC*	*High impact UPVC*	*Plasticised PVC*			*Cross-linked PVC*	*Structural foam*
											0–100% elongation	*100–300% elongation*	*>300% elongation*		
Maximum recommended service temperature (continuous—no stress)		—	°C	66–80	100–110	77–100	66	50	50	50	50	50	50	95	50
Deflection temperature (°C)	1.8 MN/m^2	D648	°C	54–80	72–112	54–66	82	67	67	63	—	—	—	120	55
	0.45 MN/m^2			56–82	102–117	88–100	—	70	70	67	—	—	—	130	60
Flammability		D635	—	Non-burning to self ext.	Non-burning	Self ext.	—	VO	VO	VO	V2 >3	HB	HB	VO >2	VO
Cold flex temperature		D1043	°C	—	—	—	—								
Cold bend temperature		—	°C	—	—	—	—								

Table 3.2.24

TABLE 3.2.25 Environmental properties of PVC and related polymer types

PVC type	*Flexible PVC*	*Rigid PVC*	*CPVC*	*PVdC*
Weatherability	Performance can be obtained by correct choice of additive. White, black, silver and pastel shades (heavily pigmented with titanium dioxide) compounds give best results. Acrylic laminate on top of PVC recommended for severe conditions.			
Chemical resistance	Basic PVC resins highly resistant. However, plasticisers are attacked or removed by certain solvents, chemicals and oils. All grades are dissolved by some ketones.			
	Attacked or swollen by aromatic solvents, aldehydes, naphthalenes and certain chloride, acetate and acrylate esters. Can be attacked by strong acids.		Resistant to acids, common alkalis, oil and grease and most organic solvents.	Excellent resistance to all acids and most common alkalis, aliphatic and aromatic hydrocarbons alcohols and esters, oils and grease.
	Sulphur reacts with metal stabilisers used with some grades to produce staining.	Some impact modifiers used in rigid PVC can reduce chemical resistance.		
	Can be printed on and marked by solvent inks (ball-point).	Normal impact grades have better chemical resistance than high impact.		
		Resistant to staining.		
	All grades resistant to common alkalis and weak acids.			
Water absorption (24-hr test ASTM D570)	Moisture absorption characteristics are formulation-sensitive.			
	0.2–1.0%	0.03–0.4%	0.02–0.15%	0.1%
Thermal degradation	Discolouration can occur by exposure to heat with inadequate stabilisation.			
Maximum recommended service temperature °C (no stress)	66–80	66–80	100–110	77–100
Fire and flame resistance	Chlorine constituents give general flame and fire resistance. Most formulations rated slow burning to self-extinguishing. Rigid formulations non-dripping. HCl gas can be given off. Chloride and phosphate plasticisers used to improve fire and flame resistance—antimony oxide gives best results.			
Fungus and bacteria resistance	High-cost additives necessary for tropical conditions, earth burial, refrigerator gaskets, etc.	As plasticisation of PVC increases risk of microbiological attack, less flexible types of PVC are less susceptible. However, special additives may be necessary for some applications, e.g. hospital flooring.		
Foodstuff	Grades with specially chosen plasticisers, stabilisers, etc., are available for use with food and drink. Evidence that alcohol can release free PVC monomer (believed to be carcinogenic) led to withdrawal of PVC miniature spirit bottles.			

Table 3.2.26

TABLE 3.2.26 Electrical properties of PVC and related polymer compounds

Electrical properties	*ASTM test method*	*Units*	*Flexible PVC*	*Rigid PVC and PVC copolymers*	*CPVC*	*PVdC*	*PVC + 20% glass fibre*	*UV stabilised PVC*	*UPVC*	*Plasticised PVC 100–300% elong.*	*Cross-linked PVC*	*Structural foam*
[a]Dissipation (power) factor 60 Hz 10^3 Hz 10^6 Hz	D150	—	0.08–0.15 0.07–0.16 0.04–0.14	0.007–0.02 0.009–0.017 0.006–0.019	0.019–0.021 0.09–0.011 0.02	0.03–0.045 0.06–0.075	0.009–0.016	0.025	0.025	0.1	0.01	0.025
Dielectric constant 60 Hz 10^3 Hz 10^6 Hz	D150	—	5.0–9.0 4.0–8.0 3.3–4.5	3.2–3.6 3.0–3.3 2.8–3.1	3.08 2.8–3.6 3.2–3.6	4.5–6.0 3.5–5.0 3.0–4.0	3.8 3.4	3.1	3.1	6	3.3	2.5
	This property is frequency-sensitive and also affected by filler and plasticiser content.											
Volume resistivity 50% RH and 23°C	D257	ohm-m	10^9–10^{13}	—	$1–8 \times 10^{13}$	10^{12}–10^{14}	10^{11}					
Dielectric strength Short time ⅛ in (0.3 cm) thickness	D149	$10^2 \times$ kV/m	120–160	140–200	480–600	160–240	165					

[a] Dissipation and loss factor. Current frequency and formulation affect loss factor.
Peaking frequency is shifted by the type of filler; plasticiser content and type affect peak height.

TABLE 3.2.27 Processing of PVC and related polymers

Process	*Polymer*[a]	*Typical products*	*Remarks*
Calendering	PVC (rigid or flexible); VC/Ac	Sheet; film (flexible or rigid); flooring and floor tiles (flexible).	Typical thickness tolerance in good production of sheet: 15–20% on 0.0025 in.
Extrusion	PVC (rigid or flexible); VC/Ac	Sheet (flexible or rigid). Film (flexible or rigid). Pipe (flexible or rigid). Profiles (flexible or rigid). Coated wire and cable (flexible coating).	Film may be produced via a flat die ('casting') or in tubular form (blown film). Cellular products (e.g. sheet) can be produced by extrusion.
Compression moulding	PVC (rigid or flexible VC/Ac	Thick sheeting or laminates. Cellular articles. Gramophone records.	
Injection moulding	PVC (rigid or flexible) VC/Ac VC/VdC PVdC	Pipe fittings, electrical fittings, gaskets, footwear (including microcellular soles).	
Blow moulding	PVC (rigid)	Bottles and other containers.	In this process a tube, the 'parison', is blown up into conformity with a mould. The parison is usually extruded (sometimes injection moulded).
Paste processing (i) Coating (by dipping, spreading or transfer techniques) (ii) Moulding (compression, rotational, transfer)	PVC paste (suspension of polymer particles in plasticiser, with other additives).	 Coated fabrics, vinyl wall-papers, PVC-backed carpets, coated work-gloves, coated metal sheet and articles. Skins for car armrests, cellular toys, pads, sports goods.	Paste products are mainly flexible, but semi-rigid and even rigid ones can be made with special formulations. Cellular products (e.g. foam for upholstery, toys, artificial leather with cellular layer) may be made.
Powder processing (i) Coating (by dripping or spraying) (ii) Spreading (on temporary support)	PVC (flexible or semi-rigid) V/Ac	 Coated metal sheet, wire trays. Sintered porous sheets for battery separators.	
Thermoforming	VC/Ac or PVC (less usual) in the form of rigid sheet (mainly) or plasticised sheet (less common)	Blister packs, nesting trays, containers, ceiling panels.	
Jointing (by various heat welding methods, friction welding, high frequency welding and cementing)	PVC and related materials	Joints and laminates of many kinds.	This is a secondary process, which may be classified as a fabrication technique.

[a] VC = vinyl chloride, Ac = acetate, VdC = vinylidene chloride.

TABLE 3.2.28 Major suppliers and familiar trade names

Company	*Trade names*	*Types*
PVC		
Eni Chemical (UK) Ltd	Ravinl (and others)	Rigid, UPVC, flexible. structural foam
Atochem (UK)	Vinoflex	Flexible
	Lucorex, Resinoplast (and others)	UPVC, UV stab., flexible
	Vinuran	Modified
BASF	Vinoflex	Flexible, UPVC
	Vinidur	Rigid and elastomer-toughened
	Vinuran	Modified
BIP	Beetle	Flexible
	Cobex; Fromopas	Rigid
	Velbex	Industrial Black
	Vybak	Rigid sheet
BP	Breon	Rigid
Blane	Blane	Plasticised (clay)
Borden	Borden	—
Chemische Werke (Huls)	Vestolit	Flexible, UPVC, structural foam
Colorite	Unichem	Flexible
Conoco	Conoco	Flexible and rigid
Cornelius	Komadur	Rigid
Courtaulds Acetate[a]	Celvin	Flexible sheet
DSM	Varlan	UPVC
Diamond Shamrock	Dacovin	Rigid and flexible
Dynamit Nobel	Astraglass	Flexible
	Astralit	Rigid
	Astralon	Rigid
	Trosiplast	Rigid and flexible, UPVC, UV stab.
	Trovidur	Foam
Ethyl	Ethyl	Flexible

[a]Formerly British Celanese.

Table 3.2.28

TABLE 3.2.28 Major suppliers and familiar trade names—*continued*

Company	*Trade names*	*Types*
PVC—*continued*		
Firestone	FPC	Flexible
B.F. Goodrich	Geon	Rigid and flexible, UPVC
Hoechst	Hostalit	UPVC, UV stab.
	Genotherm, Genopak, Genolon	Rigid film
Hooker	Rucodur TM	Rigid, impact mod. with EPDM
	Rucoblend	Rigid
ICI	Corvic	—
	Darvic	Rigid
	Pacton	Rigid, high impact sheet
	Vynalast	Laminate TP blend
	Welvic	Rigid, UPVC, UV stab.
Lanza	Lanza	Rigid, flexible, V/A copolymer
Phillips	Ultryl	—
Solvay	Benvic	Rigid and flexible
	Solvic	—
Tenneco	Tenneco	Flexible
Ugine Kuhlmann	Ekavyl	—
Norsk Hydro Polymers Ltd	Hyvin	UPVC, flexible, cross-linked, structural foam
PVDC		
ICI	Viclan	
Solvay/Laporte	Ixan (coatings)	
CPVC		
Goodrich	Geon	
Dynamit Nobel UK	Rhenoflex	
Atochem (UK)	Lucolor	

Table 3.2.28—*continued*

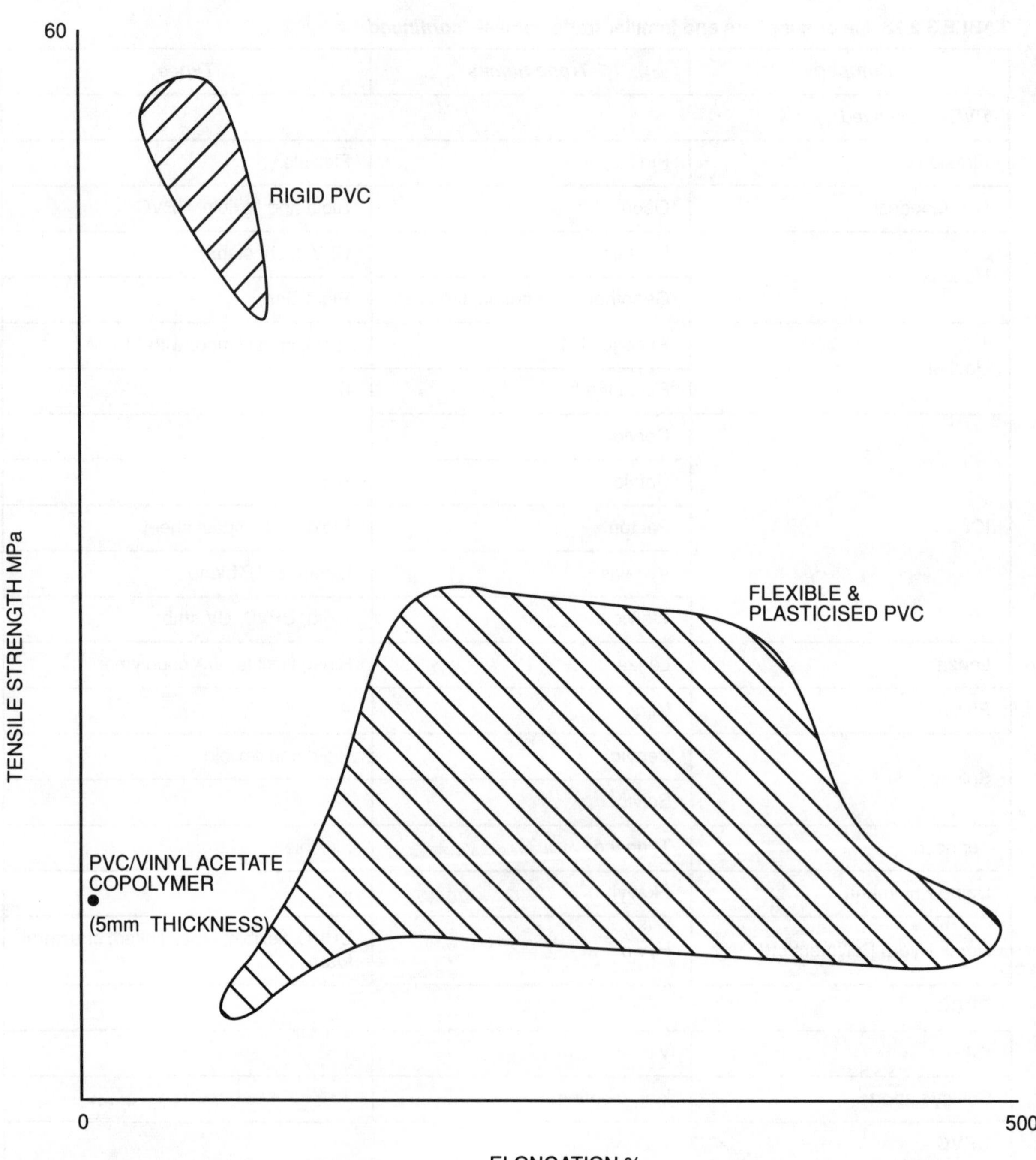

FIG 3.2.21 Tensile strength vs elongation for PVCs

Fig 3.2.21

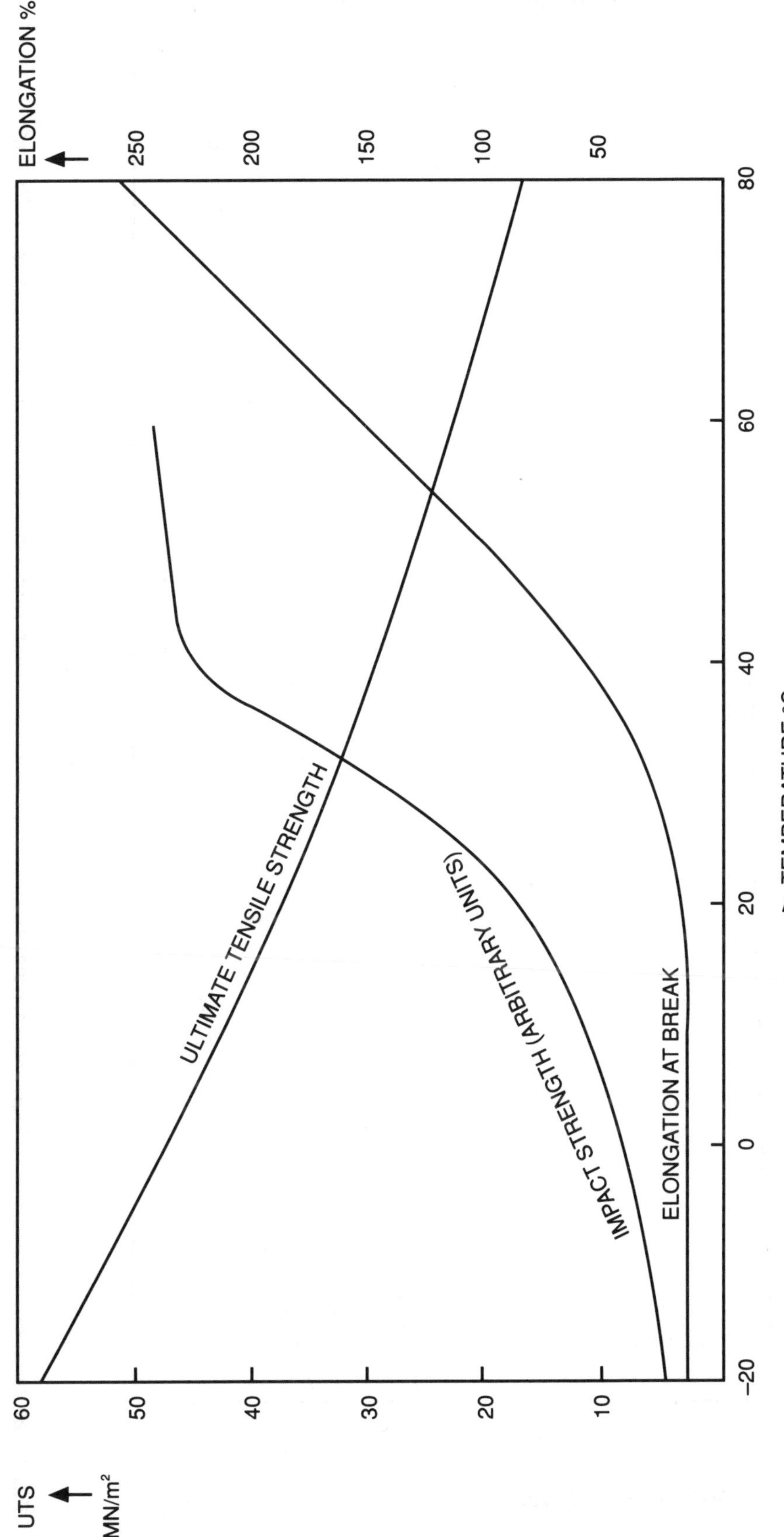

FIG 3.2.22 Typical variation of mechanical properties with temperature in PVC rigid extrusion grade

Fig 3.2.22

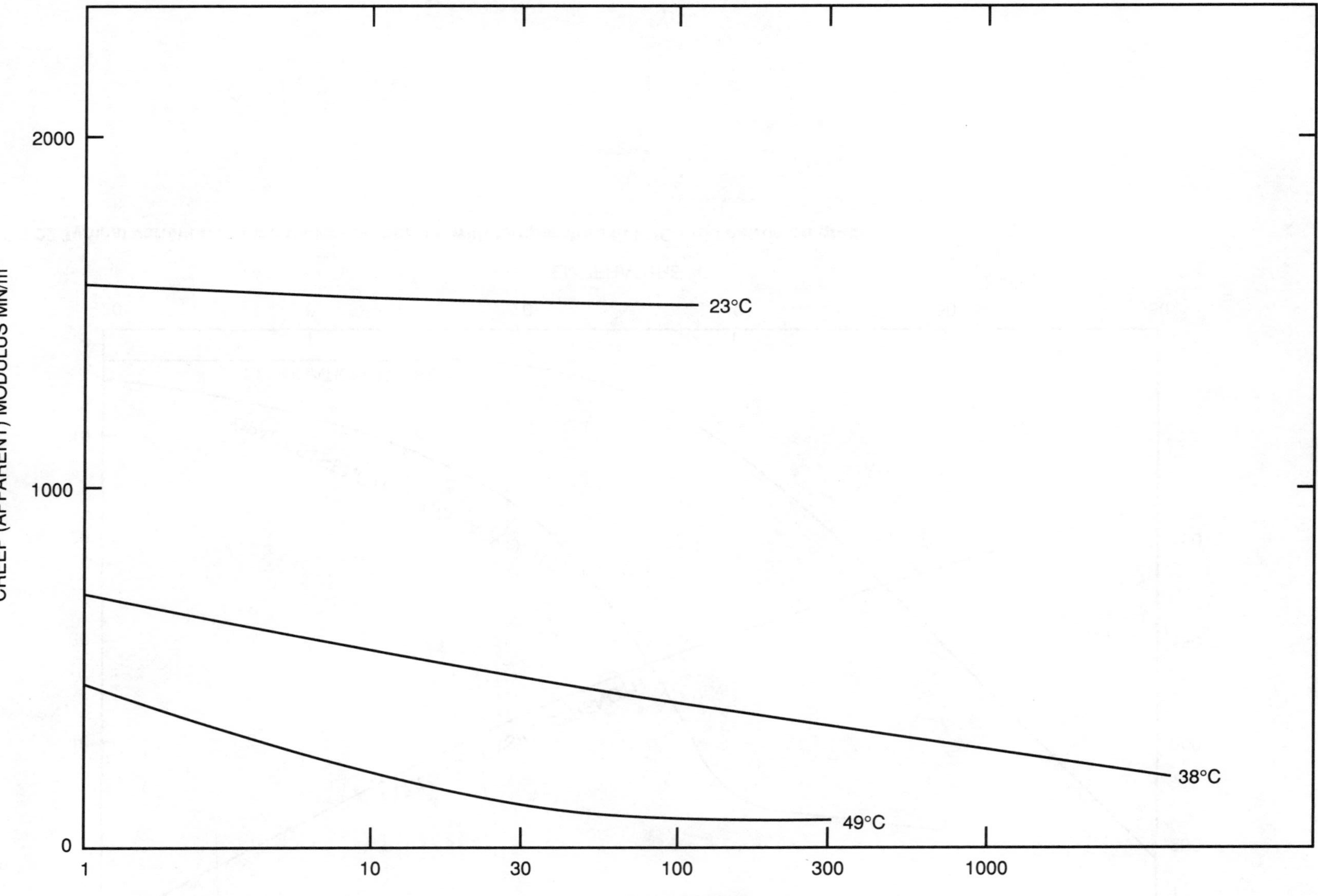

FIG 3.2.23 Variation in creep of PVC high impact, rigid, injection moulding compound with temperature. Initial applied stress 3.4 MN/m²

Fig 3.2.23

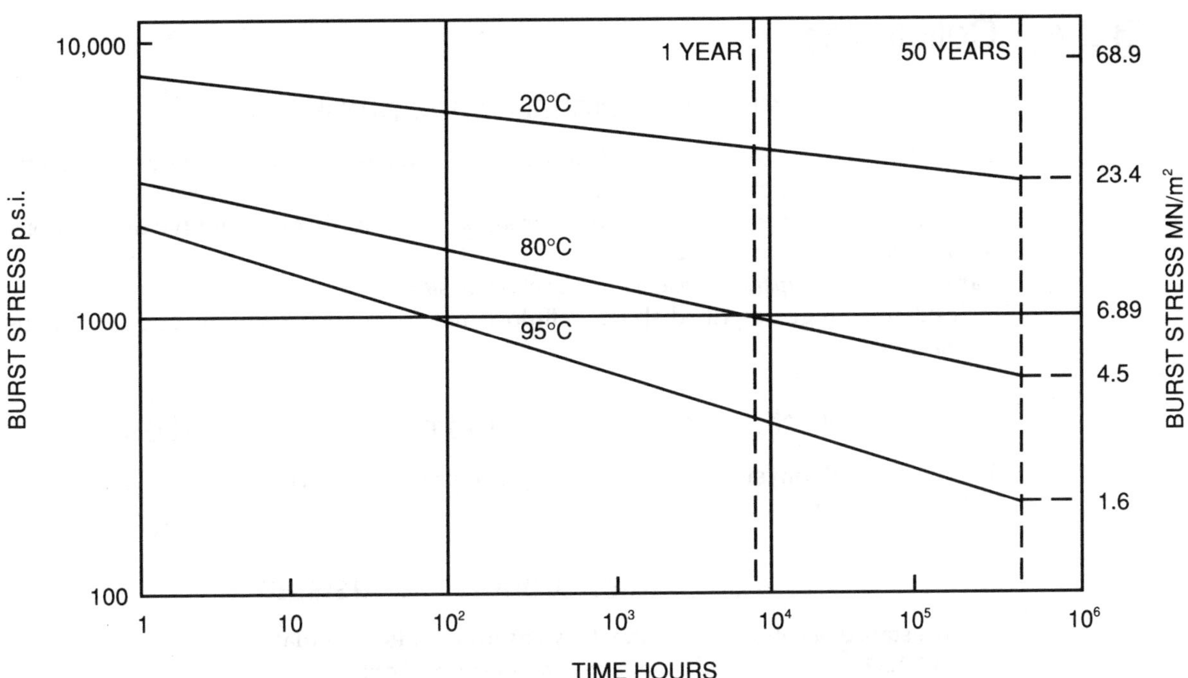

FIG 3.2.24 Stress rupture curves for CPVC hot water pipes

3.2.4 Cellulosics

3.2.4.1 General characteristics and properties

The general characteristics of cellulosics, their advantages and limitations are listed in:
Table 3.2.29 *Characteristics of cellulosics*

The available types, together with their advantages, limitations and the applications of the several types are listed in:
Table 3.2.30 *Comparative characteristics of cellulosics*

Beside the range described above, cellulose may form the basis of the so-called 'natural plastics'.

3.2.4.2 Mechanical, physical and electrical properties

The mechanical, physical, electrical and optical properties of cellulosics are listed in:
Table 3.2.31 *Properties of cellulosics*

3.2.4.3 Chemical and environmental resistance

The resistance of cellulosics to attack by environments is tabulated in:
Table 3.2.32 *Chemical and environmental properties of cellulosics*

3.2.4.4 Processing

FORMING METHODS

Cellulosics can be cast into sheet or film, injection moulded, extruded, blow and rotational moulded into structural components and thermoformed from sheet. In addition, grades of butyrate and propionate are available for fluidised bed and electrospray coatings. Cellulose acetate has the lowest processing rate with consequent increase in processing costs. Impact strength varies with processing temperatures and can be doubled by increasing the processing temperature from 150 to 210°C.

SHRINKAGE

Mould shrinkage is approximately 0.4% in the flow direction and 0.7% transverse to the flow direction. Average shrinkage of 0.4–0.5% can generally be expected with section thicknesses of 2–3 mm and 0.15–0.25% with 1 mm section thicknesses. Post-mould shrinkage is negligible.

FABRICATION

Cellulosics may be bonded to themselves using certain solvents (see Vol. 1, Chapter 1.7). Different cellulosics may be joined using two component adhesives which are also employed for joining to other plastics and materials.
Bonding by solvents is preferred to welding.
All cellulosics are suitable for normal cutting and machining operations.

3.2.4.5 Trade names and suppliers

Trade names and suppliers are listed in:
Table 3.2.33 *Trade names and suppliers of cellulosics*

TABLE 3.2.29 Characteristics of cellulosics

	Advantages	*Limitations*
Mechanical properties	Wide range of available properties to suit specific applications by varying plasticiser content. Unique combination of toughness and transparency with relatively low cost. Excellent toughness even at low temperatures. Not susceptible to stress cracking.	Natural polymer base causes greater variations in properties than truly synthetic polymer. Tendency to creep under load.
Optical properties	Unique combination of toughness and transparency with relatively low cost. Almost unlimited colour range—transparent, translucent and opaque.	Optical properties inferior to acrylics.
Thermal properties	Low specific heat and thermal conductivity, which is responsible for pleasant feel with consequent applications for handles, pen barrels, typewriter keys, etc.	All types will burn slowly (self-extinguishing grades of cellulose acetate are available).
Electrical properties	High dielectric constant, good dielectric strength and volume resistivity. No dust attraction as a result of electrostatic pick-up.	High electrical dissipation (power) factor.
Environment	Moisture does not greatly affect mechanical properties except hardness or stiffness which drop slightly. Good grease resistance. Excellent dimensional stability (except cellulose acetate).	Attacked by strong acids and alkalis. Dissolved or swollen by alcohols, esters, ketones, aromatic and chlorinated hydrocarbons. Permeable to water vapour and gases in varying degrees. Staining can occur on contact with some other plastics (either material can be affected).
Food and medicine	Non-toxic grades suitable for use with foodstuffs are available.	
Processing	Some grades are easy to process (excellent flow properties giving fine detail reproduction) and fabricate by conventional techniques—also available as foils, films (can be cast from solution or extruded) and coatings. High gloss finish possible—electrostatic pick-up quickly dispersed—no dust patterns.	
Derivation	Basic polymers derived from natural products (cellulose) which is basis of cotton, wood, etc.	

TABLE 3.2.30 Comparative characteristics of cellulosics

Type	*Advantages*	*Limitations*	*Applications*
Cellulose acetate	Harder, stiffer and stronger grades available than other cellulosics. Best resistance to organic solvents. Cheapest raw material. Self-extinguishing grades available. Film grades can be printed.	Poorest dimensional stability of the cellulosics. Plasticisers not so stable as in other cellulosics. Least resistance to aqueous solutions. Not recommended for outdoor use. Lower toughness than cellulose acetate butyrate and cellulose propionate. Slowest and therefore most expensive to process. Highest specific gravity.	Extruded tape. Film and sheet used in packaging. Blister packs. Skins, rigid transparent containers. Premium toys, tools handles, electrical appliance housings, electrical insulation, shields, lenses, eyeglass frames.
Cellulose acetate butyrate	Softer flows (higher plasticiser content) than propionate available. Better low temperature impact strength than cellulose propionate.	Unpleasant odour during processing and very slight odour in finished products which may be masked by additives.	Pen and pencil barrels, tubular packages. Used in metallised form for signs and decorative purposes. Heavy gauge sheet is used for outdoor signs, small weather shelters and skylights. Film and packaging applications similar to acetate. Automotive components, blister packs, toothbrush handles, safety goggles, lighting fixtures.
Cellulose propionate	Available as fluidised bed and electrostatic spray coatings. Weather stabilised grades available. Slightly better toughness than cellulose acetate.	Raw material more expensive than cellulose acetate. Not flame retardant.	
Ethyl cellulose	Outstanding toughness at low temperature. Lowest water absorption and hence highest dimensional stability. Best dielectric properties. Best resistance to alkalis. Excellent processing properties. Lightest of cellulosics.	Most expensive of cellulosics. Dissolved by most organic solvents. Oxidises progressively. Not flame retardant.	Flashlight cases, fire extinguisher components, electrical appliance parts.
Cellulose triacetate	Highly resistant to grease and oils, most common solvents and moisture. Good dimensional stability. Good resistance to distortion under heat. High dielectric strength. Easily punched, creased, folded and pressure formed.	Does not vacuum form well. Must be processed in solution. Applications usually restricted to thin sheeting.	Thin sheeting used for visual aids, graphic arts, photographic albums, photographic film base, motion picture sound tape; electrical insulation applications.
Cellulose nitrate	Outstanding dimensional stability. Low water absorption. Good toughness.	Low flammability characteristics. Lack of stability to heat and sunlight. Cannot be used in standard moulding and extrusion equipment; formed into hollow articles by pressing between heated sheets.	Toilet articles and industrial items fabricated from sheet, rod and tube.

Table 3.2.30

TABLE 3.2.31 Properties of cellulosics

Properties	*Units*	*ASTM test method*	*Cellulose acetate*	*Cellulose acetate butyrate*	*Cellulose propionate*	*Ethyl cellulose*	*Cellulose nitrate*
Tensile yield strength	MN/m^2	D638	29–77	20–57	20–60	20–28	35–70
Elongation	%	D638	6–70	50–100	29–100	5–40	10–40
Modulus of Elasticity in Tension	$10^3 MN/m^2$	D658	0.65–4.0	0.5–2.0	0.6–2.15	2.2–2.5	1.4–2.8
Flexural yield strength	MN/m^2	D790	20–76	17–59	15–66	28–84	56–70
Flexural modulus	$10^3 MN/m^2$	D747	0.8–2.48	0.4–2.0	0.6–2.21	—	—
Compressive yield strength	MN/m^2	D695	27–57	19–51	18–65	69–240	—
Impact strength IZOD notched (23°C)	J/cm of Notch	D256	0.2–3.2 (very good −20 → 60°C)	0.3–5.2	0.25–5.7	1.0–4.25	—
Hardness	Rockwell	D785	R34–125	R30–115	R41–122	R50–115	M25–50
Maximum continuous service temperature	°C	—	55–95	55–110	60–110	60–82	60
Heat deflection temperature 1.81 MN/m^2	°C	D648	48–86	56–185	54–80	49	—
Heat deflection temperature 0.445 MN/m^2	°C	D648	52–105	60–104	58–101	65	—
Specific gravity	—	D792	1.22–1.34	1.15–1.22	1.17–1.24	1.09–1.17	1.35–1.40
Thermal coefficient of linear expansion	10^{-5} °C^{-1}	D696	8–18	11–17	11–17	10–20	12–16
Thermal conductivity	W/m per K	—	0.16–0.32	0.16–0.32	0.16–0.32	0.16–0.28	0.13–0.21
Specific heat (RT)	J/kg per K	—	1250–1750	1250–1700	1250–1700	1250–3150	1386–1596
Volume resistivity (50% RH and 23°C)	10^{-2} ohm.m	D257	10^{12}–10^{13}	1–5 × 10^{13}	10^{13}–10^{14}	10^{12}–10^{14}	—
Dielectric strength 0.3cm (1/8in) thickness	10^2V/mm	D149	100–240	100–160	120–180	140–200	120–240
Dielectric constant 60 Hz	—	D150	3.5–7.5	3.7–4.5	3.7–4.3	2.7	6.7–7.3
Dielectric constant 10^3 Hz	—	D150	3.4–7.0	4.0–5.0	3.6–4.3	2.5–3.5	—
Dielectric constant 10^6 Hz	—	D150	3.2–5.0	3.2–6.2	3.3–3.8	2.0–3.0	6.2
Dissipation power factor 60 Hz	—	D150	0.01–0.06	0.01–0.02	0.01–0.04	0.005–0.020	0.06–0.15
Dissipation power factor 10^3 Hz	—	D150	0.01–0.07	0.01–0.03	0.01–0.04	0.002–0.020	—
Dissipation power factor 10^6 Hz	—	D150	0.01–0.10	0.02–0.04	0.01–0.05	0.010–0.060	0.07–0.10
Arc resistance	s	D495	50–310	—	175–190	60–80	—
Refractive index	—	D542	1.46–1.50	1.46–1.49	1.46–1.49	1.47	1.50
Clarity	—	—	Transparent, translucent or opaque				
Transmittance	%	—	88	88	88	—	—
Haze (film, thin sheet)	%	—	<1	<1	<1	—	—

TABLE 3.2.32 Chemical and environmental properties of cellulosics

Environment \ Cellulosic type	Cellulose acetate	Cellulose acetate butyrate	Cellulose propionate	Ethyl cellulose
Weak acids	Slight attack on all types.			
Strong acids	All types decompose.			
Weak alkalis	Slight attack.			Resistant.
Strong alkalis	Decompose			Resistant or very slight attack.
Solvents	All types resistant to oils and greases with slight variation in respect of nature of oil, i.e. animal, vegetable, mineral.			
	More resistant than other types, e.g. resists carbon tetrachloride, not resistant to some alcohols, acetone, phenols, ketones and esters.	Resistant to aliphatic hydrocarbons and ethers. Swelled or dissolved by alcohols, esters, ketones, aromatic and chlorinated hydrocarbons.		Dissolves in most common solvents.
Water absorption 24 h %	1.9–4.5	0.9–2.2	1.2–2.8	0.5–1.8
Weathering	Not recommended for outdoor use.	Stabilised grades available.		Better than acetate but not as good as stabilised grades of CP and CAB.
Stability	Not as good as other types.	Excellent — plasticisers used in these materials do not evaporate and are not prone to extraction by water.		
Flammability ASTM D635 mm/min	27–32	27–32	27–32	

Table 3.2.32

TABLE 3.2.33 Trade names and suppliers of cellulosics

Company	*Trade name*	*Forms*
Bayer (Albis (UK) Ltd)	Cellidor A	Cellulose acetate
	Cellidor B	Cellulose acetate butyrate
	Cellidor P	Cellulose propionate
BIP	Bexoid	Cellulose acetate (sheet)
	Xylonite	Cellulose nitrate
Dynamit Nobel	Cellonex	Cellulose acetate
Eastman	Tenite A	Cellulose acetate
	Tenite B	Cellulose acetate butyrate
	Tenite P	Cellulose propionate
Emser Werke	Setilithe, Tubiceta	Cellulose acetate
Hercules	Hercules	Ethyl cellulose
		Cellulose acetate
Mazzuchelli	Sicalit	Cellulose acetate
		Cellulose nitrate
Courtaulds	Dexcel	Cellulose acetate
EMS Grilon UK	Setilithe	Cellulose acetate

3.2.5 Acrylics (including PMMA, SMMA and acrylic casting dispersion)

3.2.5.1 General characteristics and properties

The advantages and limitations of acrylic plastics are listed in:
Table 3.2.34 *Characteristics of acrylics*

AVAILABLE TYPES

In addition to the monomers the range of acrylic plastics includes an acrylic–PVC alloy of very high toughness, an acrylic casting dispersion containing 50% by volume silica and a copolymer, styrene methyl methacrylate (SMMA), which combines the important properties of PMMA, polystyrene and SAN.

SMMA has the sparkling clarity and resistance to abuse, temperature and scratching of PMMA acrylics, and the mouldability of polystyrene at a price competitive with SAN. SMMA is slightly less tolerant of UV exposure than acrylics. Some products formerly made from acrylic are now made in SMMA, for reasons of economy.

Because of the structural similarity of the various grades of PMMA, acrylics are more conveniently classified according to product form, as in:
Table 3.2.35 *Basis and range of available types of acrylic plastics*

APPLICATIONS

Acrylic monomers and SMMA

Transparent items such as automotive rear light lenses, signs, lighting fittings, motorcycle windscreens, meter cases.
Films, both coloured and colourless, are used to protect metal, wood and plastics from outdoor or UV exposure.
Fibre optics.

Acrylic modified PVC sheet

Corrosion resistant parts, seating, machine housings enclosures and as a durable wall-covering material.

Acrylic casting dispersion

Replaces ceramics and cast iron for kitchen sinks, sanitary ware (up to 180°C) and bathroom furniture.

TYPICAL PROPERTIES

The properties of the available grades of PMMA, acrylic modified PVC, acrylic casting dispersion and SMMA are listed in:
Table 3.2.36 *Typical properties of acrylics (PMMA)*, and
Table 3.2.37 *Typical properties of styrene methyl methacrylate (SMMA)*

3.2.5.2 Chemical and environmental resistance

The resistance of acrylics to a variety of environments is tabulated in:
Table 3.2.38 *Chemical and environmental resistance of acrylics*

3.2.5.3 Processing

Acrylic sheet may be shaped by air, vacuum, mechanical pressure or by combinations of these methods. Cast acrylic sheet has a high molecular weight and cannot be vacuum formed to the complex shapes possible with extruded sheet. Both types of sheet are machined on ordinary woodworking machines, but special tool shapes and rake angles are recommended for optimum results.

Granular acrylics may be processed by extrusion or injection moulding, but because of the high melt viscosity of the material, higher than average pressures are required.

Processing temperature	—	Extrusion	160 – 200°C
		Injection moulding	210 – 240°C
		Mould temperature	50 – 70°C
		Mould shrinkage	0.1 – 0.8%

3.2.5.4 Trade names and suppliers

Major suppliers and trade names are given in:
Table 3.2.39 *Trade names and suppliers of acrylics*

TABLE 3.2.34 Characteristics of acrylics

	Advantages	*Limitations*
Mechanical properties	High impact resistance—breakage resistance 6–17 times greater than window glass. Stronger and lighter than cellulosics. Rigid.	Lower impact strength and formability than cellulose acetate butyrate and rigid PVC. Low scratch resistance compared with glass. Flexible grades not available.
Optical properties	Excellent optical clarity and complete transparency–equivalent to the best optical glass.	
Thermal properties	Good low temperature characteristics.	Combustible—self-extinguishing grades available in sheet form only. Heat resistant grades generate smoke when ignited.
Environmental properties	Excellent long-term resistance to weathering and sunlight. Resistant to weak acids and alkalis, aliphatic hydrocarbons and most detergents and cleaning fluids. Low water absorption. Can be coated with abrasion and solvent resistant coating (fluoroplastic/silica).	Attacked by strong oxidising acids, chlorinated hydrocarbons, esters and ketones, toluene and lacquer thinners. Susceptible to cracking and crazing if long term tensile stresses exceed 1500 psi (10 MN/m^2).
Processing and finishing	Can be cast as sheet and thermoformed; injection moulding, extrusion, high impact and heat resistant grades available. Toughness can be improved by molecular orientation induced during processing (stretch forming). Excellent dimensional stability and close dimensional tolerance mouldings. Able to accept decorative paints, lacquers and metallic deposits. Can be solvent cemented.	Polystyrene is easier to process than acrylics.
Costs	Cheaper than cellulose propionate and cellulose acetate butyrate. High impact acrylics claimed to be cheaper than polycarbonate.	Higher cost than rigid transparent PVC and polystyrene.
Food	Some grades suitable for contact with food.	

TABLE 3.2.35 Basis and range of available types of acrylic plastics

Basic forms	*Available grades*	*Remarks*
Cast or extruded sheets	Standard. High impact. Self-extinguishing. Mirrored. Super thermoforming.	Available in transparent, translucent and opaque colours with variety of surface textures. Thermoformed sheet can exhibit elastic memory and revert to former shape at elevated temperature. Greater resistance to solvent stress crazing than moulding components.
	Acrylic–PVC alloy sheet for thermoforming 'modified compound'.	Opaque material, 2–3 times tougher than ABS, more likely to dent than fracture.
Injection moulding and extrusion compounds	Standard (easy flow). High impact. High molecular weight (tougher and more robust when ejected from mould). Heat resisting.	No self-extinguishing grade currently available. Grades with superior stress cracking resistance have been developed.
Acrylic casting dispersion	70% SiO_2 in 30% methyl methacrylate resin by weight (50:50 by volume; density of 1.8 gcm^{-3}). Greater Polymerised product is prepared by a method resembling RIM.	Hard, rigid with very high quality surface.

TABLE 3.2.36 Typical properties of acrylics (PMMA)

	Property		ASTM test	Units	Cast sheet	
					Standard	High impact
Mechanical	Tensile yield strength		D638	MN/m^2	55–85	45
	Flexural strength		D790	MN/m^2	80–130	60
	Elongation		—	%	5	40–140
	Modulus of elasticity in tension		D638	MN/m^2	2400–3300	1855
	Flexural modulus		D790	MN/m^2	2400–3800	2070
	Impact strength		D638 (IZOD)	J/cm of notch	0.2	1.0
	Hardness		D789	Rockwell	M99	M61
Thermal	Max. continuous service temperature (no load)		—	°C	80–90	60–65
	Deflection temperature	0.45 MN/m^2	D648	°C	105	90
		1.81 MN/m^2	D648	°C	90	85
	Thermal conductivity		—	W/M per K	0.16–0.25	—
	Thermal expansion		D696	$10^{-6}\ K^{-1}$	70	85
Electrical	Volume resistivity		D257	ohm–m	$>10^{13}$	—
	Dielectric strength		D149	10^2 kV/m	170–200	215
	Dielectric constant	60 Hz	D150	—	3.3–4.5	3.9
		10^6 Hz			2.1–3.2	
	Power factor	60 Hz	D150	—	0.05	0.06
		10^6 Hz			0.015–0.03	
	Arc resistance		D495		No tracking	
Miscellaneous	Specific gravity		D792	—	1.19	1.20
	Flammability		D635	mm/min	33	33
	Water absorption		D570	% in 24 hr	0.4	—
	Mould shrinkage		—	%	—	—
Optical	Refractive index		D542	—	1.49	1.50
	Transmittance 3 mm thick		D1003–61	%	92	92
	Haze		—	%	1	2.6

Table 3.2.36

TABLE 3.2.36 Typical properties of acrylics (PMMA)—*continued*

Sheet	*Moulded parts*			*Acrylic casting dispersion*
Acrylic–modified PVC	*Standard*	*High impact*	*High temperature*	
45	65–78	40–63	70	70
74	90–140	50–110	80	—
100	6	20–70	3	4
2300	2600–3300	1400–2800	3240	—
2760	2900–3200	1450–2600	3450	3000
8.0	0.2	1.1	0.2	0.2
R105	M92	M110	M100	M100
—	80–90	80	90–110	50
81	100	85	115	103
71	90	75	105	90
0.15	0.16–0.25	0.16–0.2	—	—
63	65–80	70	55	60
—	10^{13}	10^{13}	—	10^{15}
170	200–300	150–800	200	200
3.9	3.3–3.9	3.5–4.0	3.0	—
3.4	2.2–3.2	2.5–3.1		
0.002	0.05	0.05	0.04	—
0.03	0.02–0.03		0.004–0.040	
—	—	—	160	—
1.35	1.19	1.15	1.16	1.19
—	18	15	33	—
0.1	0.3	0.4	0.2	0.3
—	0.1–0.8	0.4–0.8	—	—
Opaque	1.49	1.49	1.52	Opaque
—	9	90	92	—
—	<3	5	—	—

Table 3.2.36—*continued*

TABLE 3.2.37 Typical properties of styrene methyl methacrylate (SMMA)

Property	*Units*	*NAS–P–154 (modified flow)*	*NOAN*
Tensile strength (break)	MN/m^2	59	56
Elongation	%	2.1	2.1
Tensile elastic modulus	GN/m^2	3.3	3.4
Izod impact strength	J/m notch	16	16
Specific gravity	—	1.09	1.08
Hardness	Rockwell M	72	73
Heat deflection temperature (1.8 MPa)	°C	96	99
Water absorption (24 h)	%	0.15	0.11
Light transmission	%	90	90
Refractive index	—	1.56	1.57
Mould shrinkage	%	0.2–0.6	0.2–0.6
Burning rate	mm/min	<3.8	<3.8

Data supplied by Richardson Company/Normandy Plastics.

TABLE 3.2.38 Chemical and environmental resistance of acrylics

Environment	*Effect*
Weak acids Weak alkalis	No effect except hydrofluoric and hydrocyanic acids which will cause attack.
Strong acids	Attacked by mineral acids.
Strong alkalis	Resistant.
Solvents	Soluble in chloroform, trichlorethylene, toluene, ethylene dichloride. Attacked by esters and ketones. A number of organic fluids whilst not solvents may cause cracking and crazing, e.g. aliphatic alcohols and detergents, especially under stress.
Weather	Transparency, gloss and dimensional shape are virtually unaffected by years of exposure to salt spray, normal or industrial atmospheres. Resistant to exposure to light from fluorescent lamps—no darkening or deterioration. Degraded by high intensity UV light below wavelength of 265 nm. Special formulations resist UV emission from mercury vapour light sources. SMMA is marginally less resistant to UV light.
Flame	Burning rate of acrylics (except the self-extinguishing grades which are available as sheet) vary from 12 to 32 mm/min, which classifies them as slow burning (ASTM D635).

TABLE 3.2.39 Trade names and suppliers of acrylics

PMMA		
Company	***Trade name***	***Form***
Cdf Chemie	Altulite	
CYRO	Acrylite, Cyrolite	
Cornelius Chemical	Plexit 55	
Degussa	Degalon	
Du Pont	Lucite	
Eni Chemical	Vedril	
B F Goodrich	Carboset	
ICI	Diakon	Moulding powder
	Perspex	Sheet
Mitsubishi	Shinkolite	
Rohm & Haas	Plexiglass	
	Implex	
	Oroglass	Sheet
Stanley Plastics	Transpalite	Mouldings

Acrylic/PVC alloy		
Company	***Trade name***	***Form***
Du Pont	DKE	
Rohm & Haas	Kydex	

Acrylic casting dispersion		
Company	***Trade name***	***Form***
ICI	Asterite	Moulding compound

SMMA		
Company	***Trade name***	***Form***
Richardson Polymer Corp	SMMA	

Table 3.2.39

3.2.6 Polyurethane thermoplastics

There is a very wide range of polyurethanes, most of which are thermosetting and many of which are elastomeric.

Thermoplastic polyurethanes have associated processing advantages and can be formulated to meet a wide range of hardness specifications as follows:

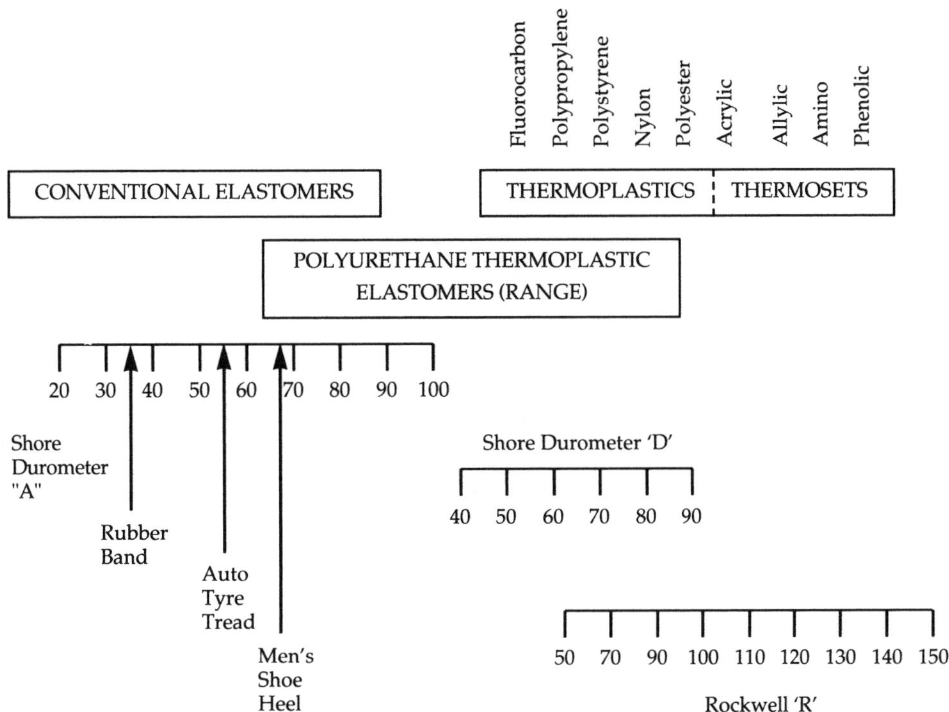

Polyurethanes also include castable versions (including RIM), millable gums, hard coating materials and rigid and flexible foams.

All solid polyurethanes are dealt with in 'Polyurethane Elastomers', Section 3.4.14 and foams in 'Foamed Plastics', Chapter 3.6.

3.2.7 Fluoroplastics

3.2.7.1 General characteristics and properties

The advantages and limitations of fluoroplastics are listed in:
Table 3.2.40 *Characteristics of fluoroplastics*

Fluoroplastics are linear polymers, varying in degree of fluorination from a fully fluorinated carbon such as PTFE to partially fluorinated polymers such as PVF or copolymers with non-fluorinated monomers (e.g. with ethylene).

The high chemical and thermal stability of fluoroplastics generally decreases with lower degrees of fluorination. In addition, the copolymers contain non-fluorinated segments and are therefore less resistant to chemicals and heat than fully fluorinated plastics. However, partially fluorinated plastics have advantages such as better processibility and in some cases improved mechanical properties.

Fillers, such as glass fibres, carbon and graphite, metal and metal oxide powders, mica and molybdenum disulphide are added to fluoroplastics to improve wear and creep resistance, thermal conductivity, load-bearing capacity or electrical properties.

AVAILABLE TYPES AND THEIR APPLICATIONS

There are a very large number of fluoroplastics, the most important of which are listed, together with their advantages, limitations, available forms, applications and relative cost, in:
Table 3.2.41 *Comparison of commercially available fluoroplastics*

TYPICAL PROPERTIES

The properties of fluoroplastics vary according to the fluorine content, as illustrated in:
Fig 3.2.25 *General property trends in fluoroplastics*

Properties of commercial fluoroplastics are also significantly influenced by the content of filler. Typical fillers employed, and the areas of application of the resultant plastics, are given in:
Table 3.2.42 *Use of fillers in PTFE*

The mechanical , thermal, electrical and other properties of fluoroplastics are tabulated in:
Table 3.2.43 *Typical properties of fluoroplastics*

The strength of fluoroplastics, in common with other materials, decreases with increasing temperature.
That of PTFCE is:

Temperature	Tensile strength (MN/m^2)
23°C	32
100°C	11
200°C	2

3.2.7.2 Chemical and environmental resistance

CHEMICAL RESISTANCE

Fluoroplastics are inert to most chemicals and solvents, but all types are attacked by molten alkali-metals, elemental fluorine, oxygen difluoride and chlorine trifluoride. In addition, halogenated solvents at high temperatures and pressure can cause degradation. Some absorption can occur, e.g. exposure to carbon tetrachloride at 150°C and 75 psi causes a weight increase of between 1 and 3%. Weight changes are reversible on removal of the absorbed material.

PVdF (PVF_2) is attacked by fuming sulphuric acid.
ECTFE is attacked by chlorinated solvents above 150°C.

WEATHERING

Fluoroplastics (with the exception of PVdF which shows slight chalking after long exposure) are not affected by ultraviolet light.

FUNGI AND BACTERIA

No effect.

FLAMMABILITY

PTFE, FEP, PFA, PVdF and PTCFE are self-extinguishing (Group 1).
ETFE and ECTFE are rated SE - O., ETFE drips and ECTFE chars.

WATER ABSORPTION (% in 24h ASTM D570)

PTFE	FEP	PVdF	PFA	PTFCE	ETFE	ECTFE
0.01	0.01	0.03	0.03	0.00	0.1	0.01

RADIATION

Low doses of radiation greatly affect properties of PTFE especially in the presence of oxygen: PTFE has only limited use after exposure beyond 10^4 rad. Mechanical and electrical properties are affected.

STATIC BUILD-UP

PTFE tubing is used to transport aggressive chemicals, solvents and gases, both as tubing and as lining in steel pipe. One problem is its tendency to hold and generate static electrical charges; this prevents the safe use of PTFE where a spark caused by static electricity could cause an explosion or fire.

Grades of PTFE are available from some suppliers with sufficient surface conductivity to prevent static build-up; static can occur if the surface resistivity exceeds 10^{10}ohms. The normal surface resistivity for PTFE is about 4 x 10^{12}ohms but static-free grades have SRs less than 10^9ohms.

FILLERS

The environmental resistance of fluoroplastics is affected by filler materials.
The resistance of filled PTFE is shown in:
Table 3.2.44 *Resistance of filled PTFEs to aggressive environments*

3.2.7.3 Wear

The wear factors for various filled PTFE compounds are:

Filler	*% by Volume*	*Wear factor K [a] (10-8 cm3 s/cm kg h)*	
No filler		7000–1400	*P*=0.47 kg/cm^2; *V*=75m/s
Glass fibre	22	5.5	*P*=2.33 kg/cm^2; *V*=75 cm/s.
Glass fibre/MoS_2	12.2/2.3	5	
Carbon/graphite	26.3	4	
Bronze/MoS_2	23.8/2.3	4	

[a] 10^{-8}cm^3 s/cm kg h ≡ 2.8 × 10^{-17} m^3/J

3.2.7.4 Processing

PROCESSING OF PTFE

PTFE is obtained in the form of aqueous dispersion, fine powder (coagulated dispersion) or pre-sintered extrusion powder.

The dispersion can be used to impregnate asbestos, glass cloth, graphite, etc., which after sintering give composite materials with some of the characteristic properties of PTFE, for example, low friction non-stick surfaces.

Powders are used for moulding and extrusion, fine powders for thin extrusions (for example tape and thin-walled tubing), and pre-sintered powders for thicker sections, e.g. rods, sheet mouldings, etc.

The granular polymer is moulded for cold compaction of a pre-form followed by sintering to form a homogeneous mass. The technique requires careful control of preforming pressure and its rate of application as well as the sintering conditions. Rods, tubes, stepped diameter mouldings or sheet can be produced but the method does not permit continuous production.

Isostatic moulding consists of pressurising the powder uniformly by fluid. Pre-forms thus produced are then sintered as before. Fairly complex shapes, such as beakers, joints, T-pieces, plugs, etc. can be produced by this method. This process is very similar to that described for metals and ceramics (see Vol. 1, Chapter 1.6). Extrusion coating of iron is possible with PTFE and some commercial iron coating grades specially designed for this process are available.

PROCESSING OF ECTFE

ECTFE extrudes and moulds well in the range 260–300°C. Extrusion and injection moulding (except for temperature) are comparable to HDPE.

PROCESSING OF EFTE

EFTE has good moulding characteristics. Melt flow allows filling of thin sections. Cycle times are equivalent to other engineering thermoplastics.

PROCESSING OF PFA

PFA has good mechanical properties at melt viscosities much lower than PTFE: it combines fluoroplastic performance with ease and economies of thermoplastic processing. Film and sheet grades are available.

PROCESSING OF FEP

FEP can be handled by extrusion and injection moulding.

PROCESSING OF PVdF

Extrusion of PVdF pellets is particularly easy; pipes and sections, film and sheet, filaments and wire coatings are all made this way. Non-orientated films can be blow extruded, being more rigid than film made by flat die extrusion—since the material picks up static electricity the usual steps must be taken.

Injection moulding is possible using screw or ram machines; a temperature of 220°C is recommended. It is not advisable to exceed 250°C. The use of light alloy moulds is not recommended because the material sticks to the metal (this problem can be overcome by using a mould release agent for each cycle).

PVdF can be compression, transfer or blow moulded and is well suited to centrifugal moulding. Articles made by the first two methods, however, are not generally free from internal stresses, but may, if necessary, be annealed.

3.2.7.5 Trade names and suppliers

Trade names and suppliers of fluoroplastics are given in:

Table 3.2.45 *Trade names and suppliers of fluoroplastics*

TABLE 3.2.40 Characteristics of fluoroplastics

	Advantages	*Limitations*
Environmental	Inert to attack by acids, alkalis and almost all organic solvents. Excellent weatherability. Do not degrade in UV light and are not attacked by fungi or bacteria.	Attacked by molten alkali metals, elemental fluorine, oxygen difluoride and chlorine trifluoride. Some grades permeable. Filled grades have inferior resistance because of fillers.
Wear	Anti-stick and low friction characteristics (friction decreases with increasing loads: anti-seizing). Filled grades have best bearing properties of any plastic material.	Sliding speed has marked effect on wear rate of unfilled PTFE.
Dimensional stability		Linear expansion increases at room temperature (16–20°C) with an accompanying change of volume of between 1 and 1.8%.
Thermal properties	Stable at high temperatures (up to 260°C continuous exposure). Tough at low temperatures (down to –160°C). Self-extinguishing.	Stiffer at low temperatures (but not brittle). Not satisfactory for high loading at elevated temperatures. Toxic products upon decomposition. High thermal expansion—difficult to machine to close tolerances.
Mechanical properties		Unfilled grades have low creep resistance. Some grades have low strength.
Electrical properties	Stable electrical properties over wide range of frequency, environmental conditions, and temperature. Low dielectric constant and loss. Non-tracking.	
Processing and fabrication	Can be bonded after thermal etching. Easily machined on metalworking tools.	High melt viscosity—PTFE cannot be processed by conventional melt-extrusion and moulding techniques.
Miscellaneous		High density for a plastic material. High cost. Presence of voids in material affects mechanical strength and electrical properties: optimum S.G. is 2.15–2.23.

TABLE 3.2.41 Comparison of commercially available fluoroplastics

Type	*Advantages*	*Limitations*	*Available forms*	*Typical applications*	*Relative cost*
PTFE Polytetrafluoroethylene	Cheapest of fluoroplastics. With FEP best chemical resistance. High service temperature (250°C).	Cannot be injection moulded or extruded—processed by powder metallurgy technique, high processing temperature. Coatings are porous. Mechanical and electrical properties are affected by exposure to radiation.	Preformed rods, plates, tubes. Compression and isostatic mouldings. As composite in dry bearings. Powders and aqueous dispersion. Coatings.	High and low temperature electrical and electronic insulation (increasingly specified for computers and aircraft where thin insulation is a must); bearings; wear resistant, low friction surfaces; chemical processing equipment; valves; pump impellers; liners; pipes; gaskets; anti-adhesive and anti-icing coatings.	100–150
PTFE-filled	Retention of chemical and heat resistance of PTFE with improvement of the following properties: 1. Wear resistance by factor of 100. 2. Creep resistance by factor of 10. 3. Thermal conductivity by factor of 5. 4. Stiffness by factor of 12. 5. Thermal expansion by factor of 0.3–0.5.	Some glass-filled PTFEs prone to discolouration.	Pre-formed components (bearings, etc). Compression mouldings.	As for PTFE but parts requiring better creep or wear resistance. PPS/PTFE alloy used as coating.	100–150
FEP Tetrafluoroethylene hexafluoropropylene copolymer	Chemical resistance equivalent to PTFE but able to be extruded and injection moulded with low mould shrinkage. Non-porous coatings possible. Low water absorption. Films approved by FDA. Does not support combustion.	More expensive than PTFE. Lower maximum service temperature (200°C). High thermal expansion coefficient. Low dielectric constant. Not recommended in radioactive environment.	Injection mouldings. Compression mouldings. Extrusions. Film. Coatings.	As for PTFE where processing advantages outweigh extra cost. Recommended for fluid handling systems, electrical and mechanical systems where excellent performance under extreme conditions is required. Films used for conveyor belt surfacing, heat shrinkable industrial wall covers and high temperature release sheeting for reinforced plastics. Electrical applications include flexible cables and capacitors. Coatings.	160–200
PFA/TFA Tetrafluoroethylene tetrafluoromethyl vinyl ether copolymer	Easier to process than FEP and with better mechanical properties and higher service temperature—40% stronger than PTFE at 250°C. Chemical resistance similar to PTFE.	More expensive than PTFE and FEP Not recommended in radioactive environments.	Injection mouldings. Extrusions. Blow mouldings. Film and Sheet.	As for PTFE and FEP where better processing characteristics justify extra cost. Film and sheet can be used as liners for chemical process vessels, electrical insulation and laminates, release surfaces.	330
PTFCE (CTFE) Polytrifluorochloro-ethylene	Lowest permeability to water vapour of any thermoplastic. Harder and more creep resistant than PTFE or FEP. Excellent low temperature properties, inherent flexibility.	Higher coefficient of friction than FEP and PTFE. Less resistant to swelling by solvents than FEP and PTFE. Lower melting point but	Injection mouldings. Extrusions. Compression mouldings. Coatings. Films.	Seals, diaphragms, chemical process parts and linings (competes with stainless steel). Cryogenic applications—valve seats, packings, electrical insulation. High frequency electrical insulation	330

Table 3.2.41

TABLE 3.2.41 Comparison of commercially available fluoroplastics—*continued*

	Good radiation resistance. Wide range of grades available, soft and flexible, glass fibre reinforced with higher heat deflection temperature, transparent film, pore-free coatings. High impact strength at 200°C. Good electrical properties to 160°C. Films pass NASA liquid oxygen compatibility test MSFC-106A.	more difficult to mould or extrude than FEP. More expensive than PTFE or FEP. Service temperature range –200°C to +200°C, although tensile strength diminishes rapidly with only moderate increases in temperature (see Section 3.2.7.1).		up to 160°C. Films used in packaging of aerospace hardware, key components in transparent, laminated, moisture barrier constructions, cap liners, gaskets, pressure sensitive tapes.	
PVdF (PVF_2) Polyvinylidene fluoride	Higher mechanical strength, better creep resistance and abrasion resistance than PTFE. Good insulating properties, although higher dielectric constant and loss factor than PTFE and FEP. Transparent under 100μm thickness.	Less chemically inert than PTFE. Attacked by fuming sulphuric acid and slight bleaching on long-term exposure to UV light. Lower maximum service temperature than PFTE, FEP and PTFCE. Will decompose, releasing HF, if heated above 350°C.	Injection mouldings. Extrusions. Compression mouldings. Coatings.	Electrical insulation. Chemical process equipment. Corrosion resistant metal coatings in chemical plant, industrial and commercial buildings. Heat shrinkable tubing (selected for abrasion resistance). Some types used as piezoelectric film.	160
PVF Polyvinylfluoride	Good weathering properties, abrasion resistance, colour retention, chemical inertness and stain resistance. Can be laminated to range of materials.		Film.	Aircraft cabin interiors, lighting panels, wall coverings and many construction applications.	
EFTE and ECTFE Ethylene trifluorochloroethylene copolymer	Better tensile strength, toughness and processability than other fluoroplastics. Chemical resistance superior to PVF_2. ETFE stiffer above 150°C. ECTFE stiffer below 100°C. Good radiation resistance. With EFTE, higher temperature properties can be improved by cross-linking by irradiation, or by glass fibre reinforcement. EFTE not attacked by solvents to 220°C.	Lower heat resistance. ECTFE attacked by chlorinated solvents above 150°C.	Injection mouldings. Extrusions. Film and Sheet.	Wire and cable insulation. Chemical resistant linings and coatings, laboratory ware. Moulded electrostructural parts. Films used in demanding high temperature, electrical applications e.g. insulation of computer wire, oil well logging cable, aircraft hook-up wire and nuclear power cable.	220–230
PPS/PTFE alloy Alloy of polyphenylene sulphide and polytetrafluoroethylene	Better high-temperature mechanical properties and tougher than conventionally reinforced PTFE. Wear and chemical resistance of PTFE retained. Less susceptible than PTFE to creep and cold flow. Less porous, more dimensionally stable.	Limited availability, especially ex-USA. Premium priced.		Valve seals in gas and liquid chromatography. Wrist pins in I/C engines. Various packings, bushings and seals.	150–230

Table 3.2.41—*continued*

TABLE 3.2.42 Use of fillers in PTFE

Filler	*General Description*	*Areas of application*
Glass	Up to 36% by volume, optimum about 22%. Increasing glass content improves creep resistance and chemical resistance, particularly to mineral acids. Inert gas sintering improves porosity and further improves creep resistance.	Recommended for valve seats, gaskets, seals and components requiring resistance to creep and chemical attack. Suitable as a bearing material for low PV values. At high loads and speeds the wear rate increases and there is the risk of scoring shafts.
Bronze	About 25–30% by volume (NB—Filled PTFE compositions containing bronze are NOT suitable for use in applications involving contact with foodstuffs.)	Good bearing material and preferable to glass either at higher speeds, where its greater thermal conductivity is used, or against softer mating materials.
Bronze/graphite	About 40% by volume bronze and graphite filled. Very good dry and wet wear resistance.	Bearing applications involving operation in water and solvents, or dry operation at low speeds. Lower wear rate than bronze filled PTFE, especially in water.
Powdered coke	28% by volume. Very good corrosion resistance with reasonable wear and creep resistance.	Used for applications involving hydrofluoric acid. Higher wear rate than glass-filled grade, but less likely to score shafts. Piston ring material in oil-free compressors.
Graphite	15% by volume. Lower coefficient of friction than other filled grades.	Recommended where reasonably good deformation resistance and/or wear resistance are required together with a very low coefficient of friction. Some specialised end-uses arise from the relatively low volume resistivity.

Table 3.2.42

TABLE 3.2.43 Typical properties of fluoroplastics

	Property		*Units*	*PFTE unfilled*	*PTFE 30% glass-filled*	*PTFE 40% bronze/ graphite*	*PTFE powdered coke*
	Specific gravity		—	2.1–2.2	2.25	3.2	2.1
Mechanical	Tensile strength (RT)		MN/m^2	10–44[b]	12–20[b]	5–8[b]	12–16[b]
	Tensile Elastic Modulus		GN/m^2	0.41–0.55	—	—	—
	Elongation (RT)		%	250–400[b]	200–300[b]	10–15[b]	100–150[b]
	Flexural Modulus		GN/m^2	0.35–0.70	1.65	—	—
	Izod impact strength		J/cm notch	1.3–2.1	2.2	—	—
	Hardness		Shore	D50–D55	D60–D75	—	D60–D65
	Coefficient of friction		—	0.06–0.25	0.12–0.15	0.15	0.13
Thermal	Cont. service temperature		°C	+ 260	+ 260	+ 260	—
	Heat distortion temperature	0.45 MN/m^2	°C	120	—	—	—
		1.8 MN/m^2	°C	—	—	—	—
	Coefficient of thermal expansion		$10^{-5}\,K^{-1}$	9.9	6.0–8.0	5.2–5.4	6.4–6.9
	Thermal conductivity		W/mK	0.24	0.36	0.76	—
Electrical	Volume resistivity		ohm–m	$>10^{16}$	$>10^{13}$	—	—
	Dielectric constant	60 Hz	—	2.1	—	—	—
		10^3 Hz	—	—	—	—	—
	95% RH	10^6 Hz	—	2.05	2.35	—	—
	Dry	10^6 Hz	—	2.05	2.35	—	—
	Loss factor	60 Hz	—	0.0002	—	—	—
	Dry	10^3 Hz	—	0.0001	—	—	—
	95% RH	10^3 Hz	—	0.0001	—	—	—
		10^6 Hz	—	0.0002–0.0004	—	—	—
	Dielectric strength		10^2 kV/m	155–200	—	—	—
Misc.	Arc resistance		s	>300	—	—	—
	Water absorption 24 h		%	0.01	—	—	—
	Clarity[a]		—	Opaque	—	—	—
	Mould shrinkage		%	0.5–3.0	2.0	—	—
	Refractive index		—	—	—	—	—

[a] TP = transparent. TL = translucent.
[b] NB: PTFE, filled or unfilled, should never be used at strains beyond the yield point. Furthermore, tensile breaking stress and strain, though used for *quality control*, are unsatisfactory for design purposes.

Table 3.2.43

TABLE 3.2.43 Typical properties of fluoroplastics—*continued*

PTFE ceramic reinforced	*FEP*	*FEP 20% Glass fibre-coupled*	*EPE*	*PFA*	*PFA 20% Glass fibre-reinforced*
—	2.14–2.17	2.2	2.12–2.17	2.12–2.19	2.24
5–17[b]	17–24	40	24–28	29	33
—	0.35–0.48	—	—	—	—
—	250–330	2.5	300	200	200
0.32	0.63–0.67	5.5	0.65–0.70	0.69	0.7
—	No break	2.0	—	No break	7.0
—	D58	R65	D55	D60	D68
—	0.08	—	—	—	—
—	204	150	Near 260	260	170
—	70	—	—	—	—
—	—	158	—	—	82
3.06–3.6	8.5–18.9	5	—	12–19.9	13.6
—	0.20	—	—	—	—
10^{13}	2×10^{16}	10^{12}	—	10^{16}	10^{16}
2.9–3.6	2.1	—	2.1	2.1	—
—	2.1	2.5	—	2.1	2.9
—	—	—	—	—	—
2.9–3.6	2.1	—	2.1	2.1	—
0.0005–0.0015	0.0002	—	0.0001–0.0002	0.00002	—
0.0002	0.0002	0.0005	—	0.00002	0.001
—	—	—	—	—	—
0.0005–0.0015	0.0007	—	0.0003	0.00008	—
120–160	195–240	130	—	200	400
—	165–300	—	—	—	—
—	<0.01	0.01	0.03	0.03	0.04
—	TP–TL	—	—	TP–TL	—
—	3.0–6.0	0.4	—	4.0	0.8
—	1.34	—	—	—	—

Table 3.2.43—*continued*

TABLE 3.2.43 Typical properties of fluoroplastics—*continued*

	Property		*Units*	*PVF*	*PVDF (PVF_2)*	*PVDF 20% C fibre-reinforced*	*PTFCE (CTFE)*
	Specific gravity		—	1.45	1.76–1.78	1.76	2.05–2.16
Mechanical	Tensile strength (RT)		MN/m²	40	42–58	100	31–39
	Tensile Elastic Modulus		GN/m²	—	0.84–1.75	—	1.0–1.26
	Elongation (RT)		%	150	50–300	6	100–175
	Flexural Modulus		GN/m²	1.4	1.4–2.25	5.5	1.5–2.0
	Izod impact strength		J/cm notch	1.8	1.9–2.1	1.2	1.3–1.4
	Hardness		Shore	D80	R110	D90	R110–R115
	Coefficient of friction		—	—	—	—	0.08
Thermal	Cont. service temperature		°C	150	148	130	176–198
	Heat distortion temperature	0.45 MN/m²	°C	121	149	150	126
		1.8 MN/m²	°C	82	85–148	140	—
	Coefficient of thermal expansion		10^{-5} K^{-1}	9	12–15.5	6	6.9–8.5
	Thermal conductivity		W/mK	—	0.15–0.24	—	0.25
Electrical	Volume resistivity		ohm–m	10^{11}	5×10^{12}	10^{3}	10^{16}
	Dielectric constant	60 Hz	—	—	8.4–10	—	2.6–2.7
		10^3 Hz	—	8.0	7.72	—	—
	95% RH	10^6 Hz	—	—	—	—	—
	Dry	10^6 Hz	—	—	6.43	—	2.3–2.37
	Loss factor	60 Hz	—	—	0.05	—	0.02
	Dry	10^3 Hz	—	0.5	0.018	—	—
	95% RH	10^3 Hz	—	—	—	—	—
		10^6 Hz	—	—	0.184	—	0.007–0.01
	Dielectric strength		10^2 kV/m	200	100	—	200–240
Misc.	Arc resistance		s	—	50–60	—	360
	Water absorption 24 h		%	0.05	0.04	0.04	—
	Clarity[a]		—	—	TP–TL	—	—
	Mould shrinkage		%	—	2.5–3.0	0.7	0.5–3.0
	Refractive index		—	—	1.42	—	1.43

Table 3.2.43—*continued*

TABLE 3.2.43 Typical properties of fluoroplastics—*continued*

ETFE			*ECTFE*	*ECTFE Glass fibre-filled*	*PTFE/PPS alloy*	*PTFE 15% Graphite*
Modified	***30% Carbon fibre***	***30% Glass fibre***				
1.7	1.73	1.89	1.80	1.85	2.14	2.11
44	1.03	96	48	82	45	27
1.65	—	—	1.65	—	—	—
100–400	2–3	4–5	200	10	6 (yield)	240
1.38	11.4	7.2	1.65	6.5	1.17	1.03
No break	2.4	4.0	No break	3.7	2.21	1.4
R50–D75	—	R74	R95	R100	D70	D63
—	—	—	—	—	0.08	—
148–198	—	—	165–179	130	246–260	180
150–180	—	—	165–180	—	—	120
70	241	238	75	200	>260	95
4.2	1.45	3.1	8	2.9	7.2	12
—	—	—	0.16	—	0.28	—
10^{14}	—	—	10^{13}	10^{13}	—	—
2.6	—	—	2.3	—	—	—
2.6	—	—	—	2.9	—	—
—	—	—	—	—	—	—
2.6	—	—	2.3	—	—	—
—	—	—	0.0005	—	—	—
0.0008	—	—	—	0.004	—	0.0007
—	—	—	—	—	—	—
0.005	—	—	0.015	—	—	—
155–160	—	—	190–195	350	—	—
72	—	—	—	—	—	—
0.03	—	<0.02	0.01	0.01	<0.03	0.07
TP in thin sec.	—	Opaque	TP–TL	—	—	—
3.0–4.0	0.15–0.25	0.2–0.3	—	0.04	—	—
1.40	—	—	2.0–2.5	—	—	—

Table 3.2.43—*continued*

TABLE 3.2.44 Resistance of filled PTFEs to aggressive environments

	Glass-filled	*Bronze–graphite*	*Powdered coke*[a]	*Graphite*[a]	*Bronze*
50% H_2SO_4	S	U	PS	PS	U
Conc. HCl	PU	U	PS	PS	U
Conc. HNO_3	PU	U	PS	PS	U
40% NaOH	PU	PS	PS	PS	PS
0.880 NH_4OH	S	U	PS	PS	U
Benzene	S	PS	PS	PS	PS
Phenol	S	PS	PS	PS	PS
Trichloroethylene	S	PS	PS	PS	PS
Ethanol	S	PS	PS	PS	PS
Fluorine	PU	—	PS	PS	—
Chlorine	PS	PS (dry)	PS	PS	PS (dry)
Bromine	PS	—	PS	PS	—
HF	U	PS[b]	PS	PS	PS[b]
SO_2	PS	—	PS	PS	—
Mercury	PS	U	PS	PS	U

S = Satisfactory.
U = Unsatisfactory.
PS = Not tested, but probably satisfactory.
PU = Possibly unsatisfactory.

[a] Attacked only by oxidation, notwithstanding a slight reduction in weight due to attack or impurities in filler.

[b] Up to 70°C.

Data courtesy of ICI.

TABLE 3.2.45 Trade names and suppliers of fluoroplastics

Polymer	*Abbreviation*	*Company*	*Trade names*	*Types*
Polytetrafluoroethylene	PTFE	Allied Chemicals	Halon	U, G, GP, GM, P, M
		Du Pont	Teflon	U, G Graphite
		Hoechst	Hostaflon	U, G Graphite
		ICI	Fluon	U, M, G, K, P Graphite
		LNP	Flourocomp	M
		Montedison	Algaflon	—
		Atochem UK	Soreflon, Foraflon	
Polytrifluorochloroethylene or Polychlorotrifluoroethylene	PTFCE or CTFE	Allied Chemicals	Plaskon CTFE	
		Firestone	FPC	PVC-filled
		3M	KEL-F	U, G
		Atochem UK	Voltalef	—
Copolymer of tetraflouroethylene and hexafluoropropylene	FEP	Du Pont	Teflon FEP	U
		LNP	Thermocomp LF	G
	EPE	Du Pont	Teflon	U
Copolymer of tetrafluoroethylene with tetrafluoromethyl vinyl ether	PFA	Du Pont	Teflon PFA	U, G
		Hoechst	Hostaflon TFA (similar to PFA)	
Copolymer of ethylene with trifluorochloroethylene	ECTFE or PECTFE	Allied Chemicals (Montedison UK)	Halar	U
Copolymer of ethylene with tetrafluoroethylene	ETFE or PETFE	Du Pont	Tefzel	U, G
		Hoechst	Hostaflon ET	
		LNP	Thermocomp	C, G
Polyvinylidene fluoride	PVdF or PVF_2	Courtaulds	Grafil PVDF	C
		Dynamit Nobel	Dyflor	
		Pennwalt	Kynar	U, S, Carbon Black
		Solvay	Solef	U, A
		Suddeutsche Kalkstickstoff	Vidar	
		Atochem UK	Floraflon	
Polyvinylfluoride	PVF	Diamond Shamrock	Dalvor	—
		Du Pont	Tedlar	
Copolymer of tetrafluoroethylene and vinylidene fluoride	TFE/VDF	Pennwalt	Kynar	
Alloy of polytetrafluorethylene and polyphenylene sulphide	PTFE/PPS	International Polymer Corp.	Alton	U
Copolymer of hexafluoropropylene and vinylidene fluoride	FKM	Du Pont	Vitron	*Elastomers*
		3M	Fluorel	
		Montedison	Technoflon	*See*
Perfluoroelastomer	PFE	Du Pont	Kalrez	*Section 3.4.9*
Copolymer of vinylidene fluoride and chlorotrifluoroethylene	CTFE/VDF	3M	KEL-F Elastomer	

U = unfilled, G = glass fibre, P = graphite, M = bronze, K = coke flour, C = carbon fibre, S= glass bead, A = asbestos fibre.

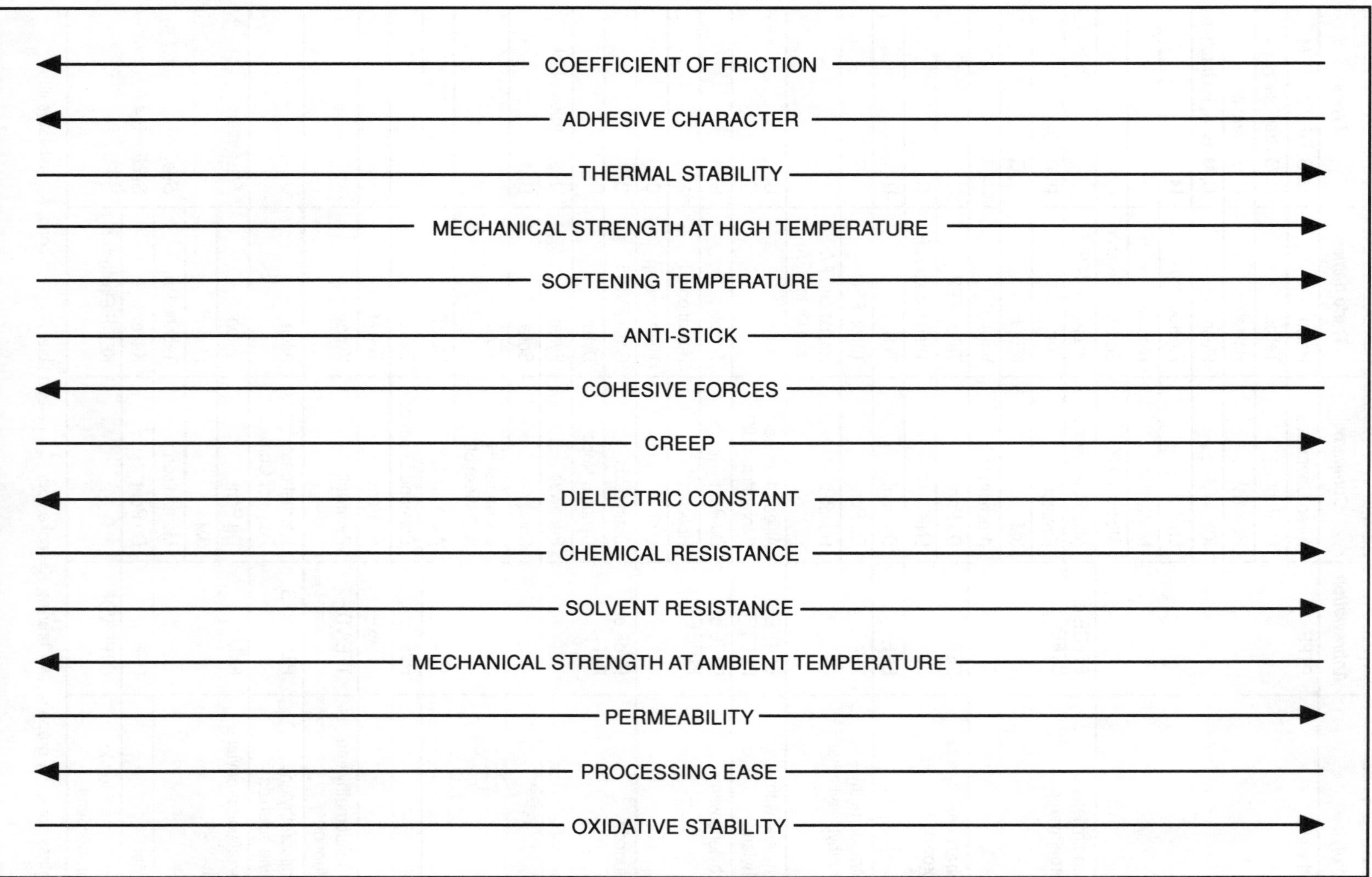

FIG 3.2.25 General property trends in fluoroplastics

Fig 3.2.25

3.2.8 Polystyrene, SAN, ABS, styrene maleic anhydride, ASA (styrene based polymers) (for SAN/polysulphone, see section 3.2.16.)

3.2.8.1 General characteristics and properties

The advantages and limitations of styrene based polymers are set out in:
Table 3.2.46 *Characteristics of styrene based polymers*

AVAILABLE TYPES

The homopolymer polystyrene (PS) is the simplest of this class, being the product of the direct polymerisation of monomeric styrene. PS is atactic and is amorphous since it is incapable of crystallisation. Its simple repeating structure may be represented:

$$\left[-CH_2 - \underset{\underset{C_6H_5}{|}}{CH} - \right]_n$$

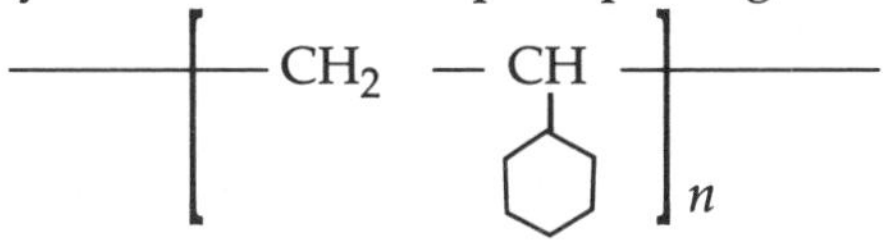

Expanded polystyrene (EPS) is physically entirely different from the parent dense polymer, but chemically its constitution is the same.

High impact polystyrene (HIPS), or toughened polystyrene as it is often called, is produced by modifying polystyrene with a rubbery additive. The additive is generally either a styrene—butadiene rubber (SBR), with a styrene content of 25–30%—or polybutadiene. This is generally achieved by polymerising styrene monomer containing 5–20% dissolved rubber. It is not possible to represent the structure of such a product simply but an important feature is that discrete rubber particles exist in the matrix. These particles interfere with the crack propagation process.

The styrene—acrylonitrile materials (SAN) are true copolymers said to contain 20–30% acrylonitrile. In simple terms the structure may be represented thus:

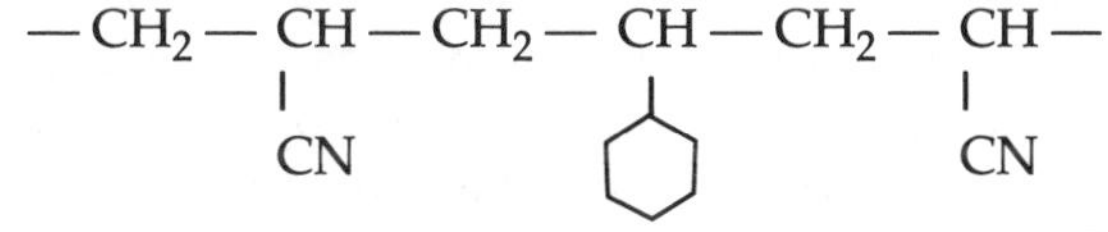

$$-CH_2-\underset{\underset{CN}{|}}{CH}-CH_2-\underset{\underset{C_6H_5}{|}}{CH}-CH_2-\underset{\underset{CN}{|}}{CH}-$$

Like polystyrene, SAN can also be modified by rubber additions to give improved impact strength. Modification by the addition of acrylic rubbers is of significance but a much more widely used class of materials is produced by polybutadiene modification of SAN. These products are known as ABS. ABS materials essentially consist of a continuous phase of styrene– acrylonitrile copolymer in which an elastomeric phase, having a degree of compatibility with the copolymer, is dispersed. There are a variety of routes used in preparation of these polymers, and by varying the ratio of the three monomers and also the means of preparation there is a wide range of products available. Structurally these compositions are not capable of simple diagrammatic representation.

A recently introduced class of compounds is the butadiene—styrene copolymer series (BDS) under the name of 'K-resin', two types of which are now available. It is not intended that these copolymers should be substitutes for polystyrene itself, which in some respects they resemble, but rather they should be used where polystyrene falls short of the desired properties, notably impact toughness. It is claimed that K-resin combines good impact toughness with clarity.

ASA is styrene–acrylonitrile copolymer modified with butadiene-free acrylate elastomers.

Glass fibre-reinforced PS, SAN and ABS are available and the reinforced grades are processed almost exclusively by injection moulding. The glass-reinforced grades are available in natural colour and a whole variety of pigmented colours. The reinforced materials are used because they offer specific advantages over unreinforced grades for some applications. Some of these advantages are increased stiffness, strength and creep resistance, improved temperature performance, dimensional stability, and lower mould shrinkage.

Typical properties of some glass fibre-reinforced materials are tabulated in Section 3.2.8.2. Other levels of reinforcement are also available.

TYPICAL APPLICATIONS

Applications of polystyrene and other styrene based polymers are given in:
Table 3.2.47 *Typical applications of polystyrene*
Table 3.2.48 *Typical applications of styrene based polymers*

TYPICAL PROPERTIES

The properties of several grades of styrene are compared in:
Table 3.2.49 *Comparative characteristics of styrene based polymers;*
Fig 3.2.26 *Tensile strength vs elongation for polystyrene and related polymers,* and
Fig 3.2.27 *Tensile strength vs impact strength for polystyrene and related polymers.*

3.2.8.2 Mechanical and physical properties

The mechanical, physical, electrical and optical properties of styrenes are tabulated in:
Table 3.2.50 *Typical properties of common grades of styrene based polymers*

These properties are significantly modified by reinforcement and filling.
The effect of glass reinforcement on a variety of properties, and the effect of glass bead filling and asbestos reinforcement, are compared in:
Table 3.2.51 *Typical properties of glass fibre reinforced styrene based polymers compared with the unreinforced form,* and
Table 3.2.52 *Typical properties of commonly filled grades of styrene based polymers*

The mechanical properties of styrene based polymers vary with temperature, formulation and moulding parameters. The properties most significant to the designer are toughness and creep. The effect of temperature on toughness is shown in:
Fig 3.2.28 *Influence of temperature on Izod impact toughness of styrene based polymers*

The impact properties are also very sensitive to moulding procedure. Compression moulded specimens have very much higher impact strengths than injection moulded specimens. Moulding temperature also influences impact strength but to a very much smaller extent. Directionality of impact strength can be very strongly influenced by moulding temperature. At low temperatures, properties in the direction of flow may be better by several hundred per cent than properties across the direction of flow, while at higher temperatures the properties tend to an average.

Unfortunately, formulation to increase toughness usually impairs other desirable properties and it may prove necessary to compromise on one or other property. For instance, as the impact strength of ABS increases, cost, creep strain and opacity also tend to increase while tensile strength, rigidity, hardness and heat resistance tend to decrease.

In a similar way to toughness (though usually by adopting contrary procedures) creep resistance can be improved by formulation. The property values which can be achieved are shown in:
Fig 3.2.29 *Tensile creep at 23°C (general purpose ABS);*
Fig 3.2.30 *Tensile creep at 70°C (high temperature ABS);*
Fig 3.2.31 *Tensile creep at 23°C (creep resistant ABS);* and
Fig 3.2.32 *Tensile creep at 70°C (high temperature creep resistant ABS)*

3.2.8.3 Environmental resistance

The main environmental parameter to be taken into consideration with styrene based polymers is temperature, most styrenes having maximum service temperatures at, or below, 100°C. The main temperature parameters, and the water absorption, are listed in:

Table 3.2.53 *Environmental behaviour of styrene based polymers.*

3.2.8.4 Processing

Most styrene based polymers can be injection moulded.

POLYSTYRENE

PS may be processed by injection moulding, extrusion and blow moulding. Biaxially oriented sheet can also be vacuum formed. The material is heat stable and can be processed over a wide range of conditions. All the usual finishing processes may be carried out in PS, and painting, decorating, printing and metallising are particulary effective. In injection moulding, mould shrinkage for PS is low (0.2–0.6 % unreinforced and 0.05–0.1% glass reinforced), and is little affected by processing conditions.

PS is processed by extrusion generally in single screw machines using a single stage or two stage screw.

Polystyrene containing flame retardant additive is compatible with other polystyrene resins, but not with other, different, thermoplastic resins.

Dust generated from machining operations can be hazardous due to toxicity and/or flammability. Polystyrene may, subject to this caution, be machined, solvent welded, hot stamped, vacuum metallised or painted.

HIGH IMPACT POLYSTYRENES

The processability of medium impact PS is similar to that of the homopolymer, general purpose grades. Mould shrinkage and dimensional stability are also similar. As with general purpose polystyrene, orientation in fabricated articles may give rise to anisotropy.

The high impact grades are used on a large scale in injection moulding, extrusion or blow moulding and vacuum forming. Printing and decoration needs no special pretreatment. In processing these materials it is necessary to avoid overheating or long residence times which cause yellowing.

STYRENE–ACRYLONITRILE COPOLYMER

SAN polymers may be processed by injection moulding, extrusion or blow moulding. Sheet can also be thermoformed with a small draw ratio. These materials are conventional in their moulding behaviour.

ACRYLONITRILE–BUTADIENE–STYRENE

ABS can be injection moulded, extruded, calendered, blow moulded and vacuum formed. It is possible to vacuum form large parts, currently 2.3–3.3m (8–10 ft) square due to the excellent hot strength character of ABS at the processing temperature. The resins show good flow properties and can be utilised in the fabrication of intricate components.

In injection moulding of components subsequently to be electroplated, special precautions must be taken in order to achieve a satisfactory finished article. Critical care must be given to the moulding conditions to avoid surface blemishes, internal strain or sink marks. Failure to observe and adhere to the optimum conditions results in poor plating appearance and adhesion difficulties. Mould release agents may affect plating performance.

Mould shrinkage is generally between 0.4 and 0.7% for unfilled grades, and with glass fibre reinforcement these values reduce to around 0.1–0.15%. ABS may be thermally welded by hot-plate, gas, induction, resistance or ultrasonic methods, Self-tapping screws, rivets and even nails are available, suitable for use with ABS. Painting, vacuum metallising, electroplating, printing, hot-foil stamping and flocking are all successfully used for decorative purposes. Good adhesion is obtained without special surface preparation (see Vol. 1, Chapter 1.7).

BUTADIENE–STYRENE COPOLYMER

Only two grades of 'K-resin' are available. One is satisfactory for sheet extrusion, thermoforming and injection moulding applications and the other is formulated to provide the higher melt strength necessary for blow moulding usage.

For extrusion use the makers recommended either single-stage screw or two-stage screw proportioned to minimise the shear heat input of the melt. Melt degradation occurs with excessive heat or long residence times. The makers claim that, in thermoforming thin-walled items, this material behaves like impact polystyrene. In injection moulding it has been found that optimum impact strength with good clarity is achieved with a melt temperature of 232°C and a mould temperature of 38°C.

3.2.8.5 Cross-linked 'thermosetting' polystyrene

This material is a rigid polystyrene plastic consisting of a cross-linked copolymer of styrene and divinylbenzene. The material has the low dissipation factor and stable dielectric constant of polystyrene at ultra-high frequencies with improved rigidity, higher temperature resistance and excellent creep resistance. It is flammable (burns without melting), not soluble (though softens) in ketones, chlorinated and aromatic hydrocarbons.

It is comparable with fluoroplastics, polystyrene, polyethylene oxide as a low loss dielectric and is particulary useful for its UHF insulating properties. It has good radiation resistance and its dielectric constant is not affected by 1000h immersion in boiling water.

Temperature °C	*Dissipation Factor*			*Dielectric constant*		
	1MHz	*10MHz*	*8.5 GHz*	*1MHz*	*10 MHz*	*8.5 GHz*
25	0.00012	0.00025	0.00066	2.531	2.531	2.529
40	0.00011	0.00020		2.53	2.53	
60	0.00009	0.00016		2.53	2.53	
80	0.00008	0.00021	0.00072	2.52	2.52	2.518
100	0.00012	0.00031		2.52	2.52	
120	0.00021	0.00042	0.00076	2.51	2.51	2.508
140	0.00030	0.00054		2.50	2.50	
155			0.00080			2.47

3.2.8.6 Styrenic polymer alloys (see also Section 3.2.16 for ABS/polysulphone, SAN/polysulphone.)

Key properties of these alloys can be varied with composition. Typical properties are shown in:

Table 3.2.54 *Typical properties of ABS/polymer alloys*

The following is a general summary of some styrene polymer alloys:

ABS/POLYCARBONATE

Comparative characteristics

These alloys have characteristics intermediate between ABS and Polycarbonate (see Section 3.2.13).

Principal advantageous characteristics include:

High impact strength and hardness.
Improved heat distortion temperature compared to ABS.
Dimensional stability.
Glass-filled grade has exceptional rigidity.
Electrical properties independent of moisture and temperature.

Processing and fabrication

ABS/Polycarbonate can be processed on any of the commonly available injection moulding machines and extruders. A slight moisture content will not harm the materials, although for large area parts quality can be impaired by processing damp material.

Melt temperatures range from 220 to 280°C.

Shrinkage is uniform ranging between 0.4 and 0.7%.

Parts may be solvent bonded by methylene chloride or ethylene chloride, or solvent-cemented with ethylene chloride—8% polycarbonate.

Electroplatable grade available from some suppliers.

Typical Applications

Electrical and electronic appliances.
Automotive parts.

ABS/PVC

Generally flame retardant, high impact materials. Applications include electrical appliance housings, and skins for automotive instrument panels.

SEMI-FLEXIBLE ABS ALLOY (ABS PLUS UNSPECIFIED CONSTITUENT)

Intended to compete with EPDM/PP, PC, R-RIM and SMC as resilient bumpers and front ends for cars.

ABS–POLYURETHANE

Very high impact strength. Polyurethane contributes lubricity, wear resistance, toughness, low temperature impact strength and chemical resistance; ABS provides increased stiffness and higher initial modulus plus improved injection moulding properties. The melt flow of the alloy is almost as high as that of general purpose ABS grades.

POLYSTYRENE/EPDM

Properties typical of polystyrene but with the additional advantage of weather/light stability. Applications broadly similar to the above and structural foam technology is under review.

SAN/OLEFIN RUBBER

Alloy has impact strength close to polycarbonate and ABS, but retains its toughness over long periods of outdoor exposure.

3.2.8.7 Trade names and suppliers

Major suppliers and trade names of polystyrenes and styrene based polymers are listed in:
Table 3.2.55 *Trade names and suppliers of styrenes*

TABLE 3.2.46 Characteristics of styrene based polymers

Advantages	***Limitations***
Wide range of properties. Low cost of basic material.	Brittle unless modified.
Low cost of processing because of relatively low processing temperature.	Low elevated temperature strength.
Low mould shrinkage.	
Low water absorption.	Degraded by UV radiation, Attacked by some solvents.
Clear, transparent.	High impact materials not transparent.
Easily produced as foam.	
Excellent dielectric properties.	

TABLE 3.2.47 Typical applications of polystyrene

Material	*Comments*	*Typical applications*
Polystyrene General purpose types	For mouldings of brilliance, clarity, rigidity, surface hardness where good moulding flow is needed for large items with fairly thick sections.	Refrigerator parts and accessories, e.g. trays, boxes, compartment doors. Packaging, e.g. cosmetics, tablet phials, costume jewellery. Homewares. Advertising novelties. Light fittings. Prismatic reflectors. Diffusers. Toys.
Easy flow general purpose types.	Low melt viscosity for moulding intricate shapes with high clarity. Also for very thin-walled products. When moulding with multiple cavity tools or producing insert mouldings, fast mould filling is possible and good flow round inserts is obtained avoiding weld lines.	Thin-wall packages, e.g. cosmetics, table phials. Accessories for refrigerators and other items similar to the above under general purpose types.
Heat resistant general purpose types	Higher heat distortion temperature, low monomer content, low odour (making it suitable for foodstuffs). Good clarity and mould filling as other grades.	Packaging, food products, vending machine products. Housewares, e.g. kitchenware in particular.
Heat resistant high molecular weight types	For applications where good resistance to distortion is essential. Good rigidity and hardness.	Kitchenware. Hot food and drink containers.
High impact polystyrene Medium impact grades	For injection moulding. Properties balanced to give good flow character with toughness, rigidity and good gloss in products.	Thin-wall products for food packaging. Closures and container lids. Toys and novelties.
High impact grades for injection moulding		Refrigerator fittings. Kitchen/bathroom cabinets. Toilet seats and tanks. Closures. Instrument control knobs and housings.
High impact grades for extrusion	For extrusion of sheet to be used for vacuum forming.	Refrigerator liners and interiors. Furniture drawers. Extruded panelling.
Heat resistant, high impact grade	Generally for injection moulding	Housings for electrical components. Components of appliances such as fans. Radio and television cabinets.
Extra high impact grades	For highest impact resistance	As for high impact grades. e.g. refrigerator door liners subject to impact by bottles.
Ultra-high impact strength grades	Generally for injection moulding of parts for heavy duty use.	Shoe heels. Tote boxes. Machine housings. Appliance components.

Table 3.2.47

TABLE 3.2.48 Typical applications of styrene based polymers

Material	*Comments*	*Typical applications*
Styrene–acrylonitrile copolymers (SAN)	Processed by injection moulding and extrusion.	Kitchen and picnicware Cups, tumblers, trays Tooth-brush handles Refrigerator components Radio knobs and scales Instrument lenses Cosmetic packaging
Acrylonitrile–butadiene styrene (ABS)	Injection moulding, extrusion and vacuum forming.	Telephone casings Housings for domestic appliances, e.g. vacuum cleaners Luggage cases Safety helmets Car fascia and instrument panels, body panels Toys Complete small boats (vacuum formed) Furniture Food mixer housings Automobile radiator grilles (often chromium plated) Refrigerator door and tank liners Margarine tubs
Semi-flexible ABS		Resilient bumpers Automotive front end assemblies
Coextruded ABS/acrylic sheet	Combines impact strength (ABS) with gloss/surface finish (acrylic).	Sanitary ware Automotive and caravan construction
Butadiene–styrene polymers	High clarity and good impact strength and toughness.	Blister packs Fabricated boxes and clear food containers Cassette holders Medical tubing, blood filter housing. Photographic parts—flip flash, photo displays
ASA	High resistance to outdoor ageing.	Traffic lights Road signs Lawn-mower housings Hot-water drainage pipes Caravan parts Garden furniture Boat hulls
Styrene maleic anhydride terpolymer	Developmental	Possible uses: Wheel hub cover Radio speaker grill Car instrument cluster

TABLE 3.2.49 Comparative characteristics of styrene based polymers

	Polystyrene (PS)	***HIPS***	***SAN***
Mechanical properties	Available in a wide range of different grades. Generally hard and rigid but brittle.	Impact strength may be up to seven times greater than polystyrene. Tensile stength and hardness reduced.	Low mechanical properties may be enhanced by glass reinforcement or modification with acrylic ester. Generally greater rigidity and superior to polystyrene.
Thermal properties	Low thermal capability. Low softening temperature.	Heat stability and softening temperature inferior to polystyrene.	Low thermal capability.
Environmental	Low moisture absorption. Susceptible to stress cracking and crazing coupled with poor resistance to attack by oils and certain chemicals, notably esters, higher alcohols, aromatics, chlorinated hydrocarbons and ketones. Generally resistant to dilute acids and unaffected by concentrated or dilute alkalis. Unsuitable for outdoor use. Burns—products of combustion can be toxic. Best X-ray resistance except for PEEK.	As PS. Flame retardant grades available.	Enchanced resistance to oils, greases and certain chemicals compared with polystyrene. Soluble in ketones, esters and some chlorinated hydrocarbons. Better resistance to stress cracking and crazing. Modification with acrylic ester avoids age yellowing. Water absorption is ten times greater than PS.
Food applications # Special control of solvents, additives, etc., may be required.	Used in refrigerator parts and accessories. Heat resistant grades used extensively in contact with food›#. Free from odour and taste.	As polystyrene but used in applications where higher impact strength required#.	As polystyrene but used in applications where superior properties required#.
Electrical properties	Good electrical insulation properties. Very low power factor. Very low permeability in high frequency electric fields.		
Optical properties	Transparent grades available.	Stability to light reduced compared to polystyrene.	Transparent grades available.
Processing	Good mouldability. Suitable for injection moulding, extrusion or vacuum forming.	As PS	Injection moulding and extrusion.
Miscellaneous	Low cost.		
Relative cost/unit weight	100	110	130

TABLE 3.2.49 Comparative characteristics of styrene based polymers—*continued*

ABS	*BDS*	*Styrene/Maleic Anhydride Terpolymer*	*ASA*
Very good combination of mechanical properties, heat and chemical resistance at a cost lower than most engineering thermoplastics. Low elongation compared to elastomers. High impact strength. Glass reinforced grades available for increased stiffness. Grades with highest impact strength have lowest tensile strength and modulus.	Good impact strength in combination with clarity.	Inferior to ABS in strength and impact strength.	Similar to ABS.
See above. Some self-extinguishing grades have low heat stability. Low max. continuous service temp. High thermal expansion for a plastic.		Slightly better than high-heat ABS.	
See above. Limited weathering resistance. Poor solvent resistance. Low water absorption. Moulded parts have good ageing resistance. Stain resistant.	Tendency to suffer stress cracking and crazing in contact with some food products (better than PS). Unsuitable for long term outdoor exposure. Softening by mineral and vegetable oils, glycols. Soluble in hydrocarbons, alcohols, ketones, esters, ethers. Not affected by methanol, ethanol or aqueous solutions of these.	Known resistance to oils, naphtha and gasoline.	Very high resistance to ageing.
#	#Recommended in many packaging applications including food containers but see left. Can be sterilised by ethylene oxide or Co60 radiation.	No data	
Electrical properties are generally good with low water absorption but not outstanding for any specific application.		No data.	
Opaque and clear.	Good impact strength combined with clarity.	No data.	
Injection moulding, extrusion and vacuum forming. Special grades available for electroplating applications. Glass reinforced grades show considerably reduced moulding shrinkage. May be machined, drilled and welded.	Suitable for extrusion, injection moulding, blow moulding or thermoforming depending on grade.	Good processability. Platable grades available.	
Metal coatings show excellent adhesion. Lightweight.			
150–200	180	176–206 (developmental)	

Table 3.2.49—*continued*

TABLE 3.2.50 Typical properties of common grades of styrene based polymers (polystyrene, SAN)

	Property		***Units***	***Polystyrene general purpose***	***High impact polystyrene***	***High temperatue polystyrene***	***Thermosetting polystyrene***
Mechanical	Tensile strength	−40°C	MN/m^2	—	—	—	—
		20°C	MN/m^2	35–84	12–62	34–84	48
		70°C	MN/m^2	—	—	—	—
	Tensile elongation		%	1.0–4.5	7–60	1.5–20	—
	Tensile modulus		GN/m^2	2.8–3.5	0.9–3.5	—	—
	Compressive strength		MN/m^2	77–112	27–62	77–12	—
	Flexural strength	−40°C	MN/m^2	—	—	—	—
		20°C	MN/m^2	83–118	27–69	58–120	79
		70°C	MN/m^2	—	—	—	—
	Flexural modulus	−40°C	GN/m^2	—	—	—	—
		20°C	GN/m^2	2.8–3.2	1.0–3.4	—	—
		70°C	GN/m^2	—	—	—	—
	Impact strength (IZOD)	−40°C	J/cm notch	—	—	—	—
		−25°C	J/cm notch	—	—	—	—
		20°C	J/cm notch	0.13–0.34	0.27–4.0	0.13–0.9	—
	Hardness		Rockwell R	M65–M90	M35–M70	M40–M85	—
Physical	Specific gravity		—	1.04–1.11	0.98–1.10	1.04–1.1	1.04–1.06
	Coefficient of thermal expansion		10^{-5} K^{-1}	6–8	3.4–21	6–8	—
	Thermal conductivity		W/mK	0.08–0.13	0.04–0.12	0.08–0.12	—
	Volume resistivity		ohm-m	10^{11}–>10^{15}	>10^{14}	10^{11}–10^{-15}	>10^{14}
Electrical	Dielectric constant	60Hz	—	2.45–3.1	2.4–4.8	2.4–3.4	—
		10^3 Hz	—	2.4–2.65	2.4–4.5	2.4–3.2	—
		10^6 Hz	—	2.4–2.7	2.4–3.8	2.4–3.1	2.53
	Power factor	60 Hz	—	0.0001–0.0006	0.0004–0.0020	0.0005–0.0030	—
		10^3 Hz	—	0.0001–0.0003	0.0004–0.0020	0.0005–0.0030	—
		10^6 Hz	—	0.0001–0.0004	0.0004–0.0020	0.0005–0.0050	0.0004
	Dielectric strength		10^2 kV/m	200–280	120–240	160–240	200
	Arc resistance		s	60–140	20–140	60–135	—
Optical	Refractive index		—	1.59–1.60	Translucent/ opaque	1.57–1.60	—
Processing	Linear mould shrinkage		%	0.1–0.6	0.2–0.8	0.1–0.8	—

Table 3.2.50

TABLE 3.2.50 Typical properties of common grades of styrene based polymers (polystyrene, SAN)—*continued*

20–30% glass-filled polystyrene	*SAN unfilled*	*SAN 20%–30% glass-filled*	*UV Stabilised*	*2% Silicone lubricated*	*Structural foam*	*UV stabilised HIPS*	*Fire retardant HIPS*
—	—	—					
63–105	63–84	60–104	32	30	22	40	28
—	—	—					
0.75–1.3	1.5–3.7	1.1–3.8	3	2	2	5	2.5
5.9–9.0	2.8–3.8	2.8–9.8					
94–126	98–120	154					
—	—	—					
74–140	91–133	154–182					
—	—	—					
—	—	—					
5.6–7.0	Up to 3.8	5.6–12.6	3	3	1.6	2.1	2
—	—	—					
—	—	—					
—	—	—					
0.2–2.25	0.17–0.25	0.2–0.2	0.2	0.2	0.3	0.8	0.6
M70–M95	—	M100, E60	M80	M70	M30	M30	M30
1.2–1.33	1.07–1.10	1.2–1.46	1.1	1.07	0.8	1.1	1.13
1.8–4.5	3.6–3.8	2.7–3.8	7	7	10	7	12
—	0.12	—					
$3.2–10^{14}$	$>10^{14}$	$1.4–2.1 \times 10^{15}$	10^{13}	10^{13}	10^{14}	10^{13}	10^{13}
—	2.6–3.4	3.6					
—	2.6–3.3	3.5	2.6	2.6	3	2.8	2.8
—	2.6–3.1	3.4					
0.004–0.014	0.006–0.008	0.006					
0.001–0.004	0.007–0.012	0.006	0.0003	0.0002	0.0005	0.0006	0.0006
0.001–0.003	0.007–0.010	0.006					
140–170	160–200	200	180	200	100	150	150
25–40	100-150	16					
Translucent/ opaque	1.56–1.57	Translucent/ opaque					
0.1–0.2	0.2–0.7	0.1–0.2	0.5	0.5	0.6	0.5	0.5

Table 3.2.50—*continued*

TABLE 3.2.50 Typical properties of common grades of styrene based polymers (ABS, BDS)—*continued*

	Property		*Units*	*ABS general purpose and high strength*	*ABS maximum impact*	*ABS heat resistant*	*ABS semi-flexible*	*ABS 20–40% glass-filled*	*ABS transparent*
Mechanical	Tensile strength	–40°C	MN/m²	68–80	52	80	—	—	—
		20°C	MN/m²	17–62	32–45	42–51	—	60–133	39–44
		70°C	MN/m²	24-28	17	32	—	—	—
	Tensile elongation (break)		%	10–140	5–70	3–20	—	2.5–3.0	25–75
	Tensile modulus		GN/m²	0.9–2.8	1.6–2.4	2.2–2.4	—	4.2–7.0	1.3–2.5
	Compressive strength		MN/m²	17–84	39	50–70	—	84–154	49–70
	Flexural strength	–40°C	MN/m²	105	82	106–120	—	—	—
		20°C	MN/m²	68	33–56	68–88	—	112–190	67–77
		70°C	MN/m²	40	29	39–54	—	—	—
	Flexural modulus	–40°C	GN/m²	2.5	2.0	2.5–2.8	—	—	—
		20°C	GN/m²	2.1–3.1	1.8–2.5	1.4–2.4	1.0–1.6	5.6–9.8	2.1–2.5
		70°C	GN/m²	1.6	1.3	1.5–2.0	—	—	—
	Impact strength (IZOD)	–40°C	J/cm notch	0.9	0.75–1.3	0.5	—	—	—
		–25°C	J/cm notch	1.05	1.4	0.35–0.8	2.5	—	—
		20°C	J/cm notch	0.95–2.00	2.6–5.1	1.0–3.25	6.00	0.5–1.2	1.25–2.5
	Hardness		Rockwell R	R100	R88, R85–105	R100–115	—	M65–100	R98–107
Physical	Specific gravity		—	1.04–1.07	1.02	1.04–1.07	—	1.10–1.38	1.07
	Coefficient of thermal expansion		10^{-5} K^{-1}	6–13	5.8–11.0	3.3–9.3	—	2.9–3.6	9.0–9.5
	Thermal conductivity		W/mK	0.18–0.32	0.18–0.32	0.18–0.32	—	—	—
	Volume resistivity		ohm-m	1.2×10^{14}	$1.0–4.8 \times 10^{14}$	1.2×-10^{14}	—	—	2.5×10^{13}
Electrical	Dielectric constant	60Hz	—	2.4–5.0	2.87	2.91	—	—	3.7
		10^3 Hz	—	2.4–4.5	2.86	2.91	—	—	—
		10^6 Hz	—	2.4–3.8	2.76	2.44	—	—	3.2
	Power factor	60 Hz	—	0.003–0.008	0.005	0.005	—	—	0.015
		10^3 Hz	—	0.004–0.007	0.006	0.006	—	—	—
		10^6 Hz	—	0.007–0.015	0.009	0.008	—	—	0.015
	Dielectric strength		10^2 kV/m	140–185	160–180	140–185	—	—	160
	Arc resistance		s	50–85	50–85	50–85	—	—	125–130
Optical	Refractive index		—	Translucent to opaque	Translucent/ to opaque	Translucent to opaque	Opaque	Translucent to opaque	Transparent (1.536) to opaque
Processing	Linear mould shrinkage		%	0.4–0.9	0.4–0.9	0.4–0.9	—	0.1–0.2	0.5–0.8

Table 3.2.50—*continued*

TABLE 3.2.50 Typical properties of common grades of styrene based polymers (ABS, BDS)—*continued*

BDS	*Styrene/ Maleic anhydride terpolymer*	*ASA*	*ABS Fire retardant*	*ABS High impact (UV stabilised)*	*ABS (low gloss)*	*ABS (plating)*	*ABS (structural foam)*	*SMA copolymer*	*SMA copolymer 30% glass fibre*
—	—	—	—						
25–28	34–36	47–66	34	35	30	42	25	52	75
—	—	—	—						
15–100	40	10–20	6	6	8	8	4	1.8	1.4
1.2–1.4	2.07–2.14	2.2–2.9	—						
—	—	—	—						
—	—	—	—						
47	55–58	—	—						
—	—	—	—						
—	—	—	—						
—	2.21–2.35	2.5	2.1	2.3	1.7	2.4	1.5	3	6
—	—	—	—						
—	—	—	—						
—	—	—	—						
0.2	1.55–1.70	1.0	1.8	3.0	4.0	4.0	0.7	0.3	1.4
—	—	—	R96	R103	R80	R107	R60	L105	R101
1.01	1.07	1.07	1.20	1.06	1.03	1.06	0.85	1.08	1.22
—	—	8–11	7	9	10	8	10	7	3.6
—	—	0.17	—						
—	—	10^{14}	10^{11}	10^{12}	10^{14}	10^{14}	10^{14}	10^{13}	10^{13}
2.5	—	—	—						
—	—	—	3.26	3.0	2.8	2.7	2.6	2.5	2.6
—	—	3.3–3.5	—						
0.0004–0.001	—	—	—						
—	—	—	0.008	0.007	0.007	0.008	0.008	0.001	0.01
—	—	0.015–0.030	—						
120	—	220	140	200	200	200	100	120	120
—	—	—	—						
Transparent 1.57–1.60	—	—	—						
0.4–1.0	0.4–0.6	0.5	0.7	0.6	0.7	0.6	0.8	0.4	0.3

Table 3.2.50—*continued*

TABLE 3.2.51 Typical properties of glass fibre reinforced styrene based polymers compared with the unreinforced form

Polymer			*Unreinforced*			*Glass fibre-reinforced compositions*								
						20% wt			*30% wt*			*40% wt*		
Property	***Units***	***ASTM***	***PS***	***SAN***	***ABS***	***PS***	***SAN***	***ABS***	***PS***	***SAN***	***ABS***	***PS***	***SAN***	***ABS***
Tensile strength at 23 °C	MN/m^2	D638	44	68	40	72	108	93	86	119	99	99	128	110
Tensile Modulus at 23 °C	GN/m^2	D638	3.1	3.4	2.0	7.2	7.9	4.2	8.9	11.0	—	11.3	13.4	7.2
Flexural strength	MN/m^2	D790	58	68	68	103	136	120	111	151	127	120	159	137
Flexural modulus	GN/m^2	D790	3.1	3.7	2.2	4.5	7.5	5.8	8.2	10.3	7.5	10.3	12.7	9.6
Izod impact strength (notched ½" × ¼")	J/cm	D256	0.18	0.21	3.15	0.48	0.53	0.80	0.53	0.58	0.74	0.64	0.58	0.69
Specific gravity	no units	D792	1.05	1.08	1.04	1.20	1.22	1.20	1.28	1.31	1.28	1.38	1.40	1.38
Coefficient of thermal expansion	$10^{-5} K^{-1}$	D696	7.2	6.4	7.7	3.9	3.7	3.6	3.4	3.4	3.2	2.8	2.7	2.1
Heat deflection temperature at 0.45 MN/m^2	°C	D648	85	102	87	93	96	101	101	101	104	104	104	107
Water absorption 24 h	%	D570	0.05	0.28	0.30	0.07	0.15	0.15	0.05	0.10	0.14	0.05	0.08	0.10
Mould shrinkage	%	D955	0.003	0.005	0.005	0.001	0.001	0.0015	0.0005	0.0005	0.001	0.0005	0.0005	0.001

Table 3.2.51

TABLE 3.2.52 Typical properties of commonly filled grades of styrene based polymers

Property	*Units*	*ABS 30% glass bead*	*ABS 15% glass bead 15% glass fibre*	*SAN 30% glass bead*	*SAN 40% asbestos fibre*[a]	*Polystyrene 30% glass bead*	*SAN UV stabilised*	*High impact SAN*	*Fire retardant SAN*	*SMA copolymer 30% glass fibre*
Tensile strength at 23 °C	MN/m^2	38	69	48	76	34	70	48	70	75
Tensile Modulus	GN/m^2	—	—	—	14.3	—	—	—	—	—
Elongation	%	2.5–3.5	2–3	2–3	—	2–3	2.5	3.5	2.5	1.4
Flexural strength	MN/m^2	55	96	69	126	48	—	—	—	—
Flexural modulus	GN/m^2	3.8	5.1	4.8	11.0	3.8	3.5	2.5	3.5	6.0
Izod impact strength (notched) ½" × ¼"	J/cm notch	0.38–0.54	0.65–0.86	0.27–0.40	0.28	0.27–0.40	0.25	0.6	0.2	1.4
Specific gravity	no units	1.28	1.28	1.31	1.51	1.28	1.08	1.07	1.2	1.22
Coefficient of thermal expansion	10^{-5} K^{-1}	—	—	—	2.1	—	7	9	7	3.6
Heat deflection temperature 0.45 MN/m^2	°C	110	112	99	105	88	96	95	90	125
Heat deflection temperature 1.8 MN/m^2	°C	87	101	76	101	71	84	82	84	112
Flammability	UL94	—	—	—	VO at >3 mm	—	HB	HB	VO >2	HB
Water absorption (24 h)	%	—	—	—	0.3	—	0.25	0.5	0.3	0.1
Mould shrinkage	%	—	—	—	0.2	—	0.6	0.5	0.5	0.3

[a] Asbestos-reinforced SAN is flame retardant and exhibits extremely high levels of moulding accuracy and dimensional stability. The high modulus makes the material competitive with pressure die-casting alloys.

Table 3.2.53

TABLE 3.2.53 Environmental behaviour of styrene based polymers

Property	*Units*	*Polystyrene general purpose*	*High impact polystyrene*	*High temperature polystyrene*	*Thermosetting polystyrene*	*20%–30% glass-filled polystyrene*	*SAN unfilled*	*SAN 20–30% glass-filled*	*UV stabilised SAN*	*High impact SAN*	*Fire retardant SAN*
Max. cont service temperature	°C	105	90	80–115	—	90–105	90–105	90–110	50	55	55
Heat deflection temperature at 1.8 MN/m^2	°C	65–113	64-93	85–115	100	100–110	87–193	100–115	84	82	84
Vicat softening point	°C	82–103	78-100	93–103	—	—	92–110	—			
Flammability	—	SB	SB	SB	Burns	—	SB	SB	HB	HB	VO>2
Water absorption (24h)	%	0.03–0.4	0.1–0.6	0.05–0.40	0.05	0.05–0.10	0.2–0.35	0.08–0.22	0.25	0.5	0.3

Property		*Units*	*ABS general purpose and high strength*	*ABS maximum impact*	*ABS heat resistant*	*ABS semi-flexible*	*ABS 20–40% glass-filled*	*ABS transparent*	*BDS*	*Styrene/ maleic anhydride terpolymer*	*ASA*	*ABS Fire retardant*	*ABS high impact UV stabilised*	*ABS (low gloss)*	*ABS (plating)*	*ABS (structural) foam*
Max. cont service temperature		°C	75–85	60–80	90–100	>80	95–110	50	—	—	85–95	70	70	70	70	70
Heat distortion temperature at 1.8 MN/m^2	Unannealed	°C	83–87	83–89	91–109	—	98–116	75–85	70	102–107	—	82	89	121	89	82
	Annealed	°C	—	95–101	104–120	—	—	—	—	115–118	95–101					—
Vicat softening point	1 kg load	°C	—	—	—	—	—	—	—	123–129	—					
	5 kg load	°C	83–97	83–97	95–101	—	—	—	93	112–117	93–107					
Flammability		—	SB	SB	SB	—	SB	—	SB	—	—	VO >1.5	HB	HB	HB	HB
Water absorption (24h)		%	0.35–0.40	0.42	0.40	—	0.18–0.40	0.4	0.08–0.09	—	0.45	0.4	0.3	0.3	0.3	0.6

TABLE 3.2.54 Typical properties of ABS/polymer alloys

Property		*Units*	*ABS/ polycarbonate unfilled*	*ABS/ polycarbonate 20% glass-filled*
Tensile yield strength		MN/m^2	45–56	85
Elongation at break		%	—	2.5
Tensile elastic modulus		GN/m^2	1.95–2.5	6.0
Flexural strength		MN/m^2	75–96.5	130
Flexural modulus		GN/m^2	2.76	—
Impact strength Izod notched	23 °C	J/m notch	270–550	—
	−40 °C	J/m notch	80	—
Max. continuous service temperature		°C	—	—
Heat deflection temperature 1.8 MN/m^2		°C	100–122	—
Expansion coefficient		10^{-5} K^{-1}	6.25–8.5	—
Thermal conductivity		W/m per K	0.18–0.20	—
Volume resistivity	dry	ohm–m	$>10^{14}$	—
	24 h water	ohm–m	$>10^{14}$	—
Dielectric constant dry	60 Hz	—	2.9	—
	10^3 HZ	—	2.9	—
	10^6 HZ	—	2.9	—
Power factor dry	60 Hz	—	0.004	—
	10^3 HZ	—	0.004	—
	10^6 HZ	—	0.007–0.008	—
Dielectric strength 24 h	dry	10^2 kV/m	240	—
	water	10^2 kV/m	240	—
Specific gravity		—	1.12–1.16	1.26
Linear mould shrinkage		%	0.4–0.7	—
Water absorption 24 h		%	0.6–0.7	—

TABLE 3.2.54 Typical properties of ABS/polymer alloys—*continued*

ABS/ rigid PVC	*ABS/ flexible PVC*	*ABS/ polysulphone unfilled*	*ABS/ semi-flexible alloy*	*Polystyrene/ EPDM*	*ABS/ polyurethane*
37–41	21	51	—	14–35	30
—	—	30	—	14–55	—
1.93–2.76	0.7	2.38	—	—	1.5
34–80	—	92	—	33–73	—
1.72–2.65	—	2.52	1.0–1.6	1.8–2.4	—
675	810	513	6.0	—	432
70	60	—	2.3 at −30 °C	—	124
95	—	—	85 (Vicat)	72–91 (Vicat)	—
70–99	74	150	—	—	83
8.3	—	6.48	—	—	—
0.27	—	—	—	—	—
3–5 × 10^{12}	—	3.64 × 10^{13}	—	—	—
—	—	—	—	—	—
3.9–5.0	—	3.14	—	—	—
2.8–4.5	—	3.13	—	—	—
2.8–3.8	—	3.26	—	—	—
0.004–0.034	—	—	—	—	—
0.002–0.012	—	—	—	—	—
0.006–0.011	—	—	—	—	—
150–240	—	170	—	—	—
—	—	—	—	—	—
1.16–1.22	1.13	1.13	—	1.01–1.04	1.04
0.3–0.55	—	0.7	—	—	—
0.02–0.3	—	—	—	—	—

Table 3.2.54—*continued*

TABLE 3.2.55 Trade names and suppliers of styrenes

<table>
<tr><th>Company</th><th colspan="2">Trade names</th><th>Fillers</th></tr>
<tr><td colspan="4">Polystyrene (PS) and high impact polystyrene (HIPS)</td></tr>
<tr><td>Amoco</td><td colspan="2">Amoco</td><td>U</td></tr>
<tr><td>ARCO</td><td colspan="2">Dylark, Dylene</td><td>U, G</td></tr>
<tr><td>Atochem</td><td colspan="2">Lacqrene</td><td>U, V, F</td></tr>
<tr><td>BASF</td><td colspan="2">Polystyrol</td><td>U, Butadiene Rubber, V, G, R</td></tr>
<tr><td>BP Chemicals</td><td colspan="2"></td><td>U, V</td></tr>
<tr><td>CdF Chemie</td><td colspan="2">Gedex</td><td>—</td></tr>
<tr><td>Chemische Werke (Huls)</td><td colspan="2">Vestypor, Vestyron</td><td>U, V, F</td></tr>
<tr><td>Cole</td><td colspan="2">Novalite, Stilletex</td><td>—</td></tr>
<tr><td>Dow</td><td colspan="2">Styron</td><td>U, V</td></tr>
<tr><td>Fiberfil/Dart</td><td colspan="2">Styrafil</td><td>G</td></tr>
<tr><td>Fiberite (Vigilant)</td><td colspan="2">RTP</td><td>G</td></tr>
<tr><td rowspan="2">Foster Grant</td><td colspan="2">Fostalite, Fostarene,</td><td>U</td></tr>
<tr><td colspan="2">Tuf-Flex</td><td>—</td></tr>
<tr><td>Freeman Chemicals</td><td colspan="2">Lastirol</td><td>G</td></tr>
<tr><td rowspan="3">Hammond</td><td rowspan="3">Gordon</td><td>Superdense</td><td>U</td></tr>
<tr><td>Superflex</td><td>U, Copolymer, Rubber/Styrene</td></tr>
<tr><td>Superflow</td><td>U</td></tr>
<tr><td>Hoechst</td><td colspan="2">Hostyren</td><td>U, V</td></tr>
<tr><td>Huntsman Chemical</td><td colspan="2">Huntsman</td><td>U, V</td></tr>
<tr><td>LNP</td><td colspan="2">Thermocomp</td><td>G, PTFE, Silicon, S</td></tr>
<tr><td>Monsanto</td><td colspan="2">Lustrex</td><td>U, V</td></tr>
<tr><td>Montedison</td><td colspan="2">Edistir</td><td>U, V</td></tr>
<tr><td>Rexene</td><td colspan="2">Rexene</td><td>U</td></tr>
<tr><td>Rhone Poulenc</td><td colspan="2">Afcolene E</td><td>—</td></tr>
<tr><td>Shell</td><td colspan="2">Shell DP, Carinex</td><td>U</td></tr>
<tr><td>Sterling</td><td colspan="2">Sternite</td><td>U</td></tr>
<tr><td>Sumitomo</td><td colspan="2">Eslorite</td><td>U</td></tr>
</table>

U = unfilled; G = glass fibre-filled; S = glass bead/sphere-filled; V = UV stabilised; R = silicone lubricated; F = structural foam.

Table 3.2.55

TABLE 3.2.55 Trade names and suppliers of styrenes—*continued*

Company	***Trade names***	***Fillers***
Polystyrene (PS) and high impact polystyrene (HIPS)—continued		
TBA	Arpylene	G
Thermofil	—	G
Ugine Kuhlmann	Styropor	—
Wilson Fibrefil International SA	Styrofil	G, R, F
Styrene–butadiene or butadiene styrene (BDS)		
Phillips	K–Resin	—
BASF UK	Styrolux	—
Styrene/maleic anhydride		
Monsanto	Cadon	Terpolymer
Arco	Dylark	Copolymer, glass-filled
Acrylonitrile butadiene styrene (ABS)		
Altulor	Altuchoc	—
Eni Chemical	Ravikral	U
BASF	Terluran	U
Bayer AG	Novadur	G, Clay, U
Borg Warner	Cycolac	U
DSM	Ronfalin	U
Filberfil/Wilson	Abasafil	G
Hammond	Gordon	U
JSR	JSR	U, Elastomeric
LNP	Thermocomp	G, PFFE, Silicon
Mitsubishi Rayon	Shinko–Lac	U
Mobay/Abtec	Abson	U
Monsanto	Lustran	U, G
Montedison	Urtal	U, G
Polymon	Formid	U
Rexene	Rexene	U

U = unfilled; G = glass fibre-filled; S = glass bead/sphere-filled; V = UV stabilised; R = silicone lubricated; F = structural foam.

Table 3.2.55—*continued*

TABLE 3.2.55 Trade names and suppliers of styrenes—*continued*

Company	***Trade names***	***Fillers***
Acrylonitrile butadiene styrene (ABS)—continued		
SIR (UK)	Restiran	—
Sterling	Sternite	U
Thermofil	—	G
Ugine Kuhlmann	Kapronet, Ugikal,	—
	Metacrylene BS	—
Uniroyal	Royalite	—
Eni Chemical	Ravikral	—
	Urtal	—
Dow Chemical Co.	Dow ABS	—
Freeman Chemicals Ltd	Lostilac	—
Ferro (GB) Ltd	Starflam ABS	—
Vigilant Plastics	RTP 600	—
Styrene–acrylonitrile copolymer (SAN)		
BASF	Luran	U, V, H, F, G
Dow	Tyril	U, V, H, F
Fiberfil/Dart	Acrylafil	G
Fiberite (Vigilant)	RTP	G
LNP	Thermocomp	G, S, PTFE, Silicon, Graphite
Monsanto	Lustran	Copolymer
TBA	Arpylene	A, G
Thermofil	—	G
SIR (UK)	Restil	—
Freeman Chemicals Ltd	Lastil	U, V, F, G
Montedison	Kostil	U, V, H, F
Wilson Fibrefil International SA	Acrylafil	G
Styrene–acrylonitrile acrylate (ASA)		
BASF	Luran S	—

U = unfilled; G = glass fibre-filled; S = glass bead/sphere-filled; A = asbestos fibre-filled; V = UV stabilised; H = high impact; F = fire retardant.

Table 3.2.55—*continued*

TABLE 3.2.55 Trade names and suppliers of styrenes—*continued*

Company	*Trade names*	*Fillers*
Styrenic polymer alloys ABS/polycarbonate		
Bayer AG	Bayblend	
BASF (UK)	Terblend B	
Borg-Warner	Cycoloy (US)	
	Cycolac (UK)	
Mobay	Bayblend (US)	
ABS/PVC		
Borg-Warner	Cycovin	
JSR	JSR	
Schulman	Polyman	
Uniroyal	Kralastic	
Abtec	Abson	
DSM	Ronfalin V	
Dynamit Nobel UK	Troisplast PVC/ABS	
ABS/polysulphone		
Amoco Chemicals (UK) Ltd	Mindel A	
BASF UK	Terluran S	
Semi-flexible alloy (ABS + unspecified constituent)		
Borg-Warner	Cycolac X–340	
SAN/Olefin rubber		
Uniroyal	Rovel	
Polystyrene/EPDM		
Hoechst	Hostyren XS	

Table 3.2.55—*continued*

Fig 3.2.26 and Fig 3.2.27

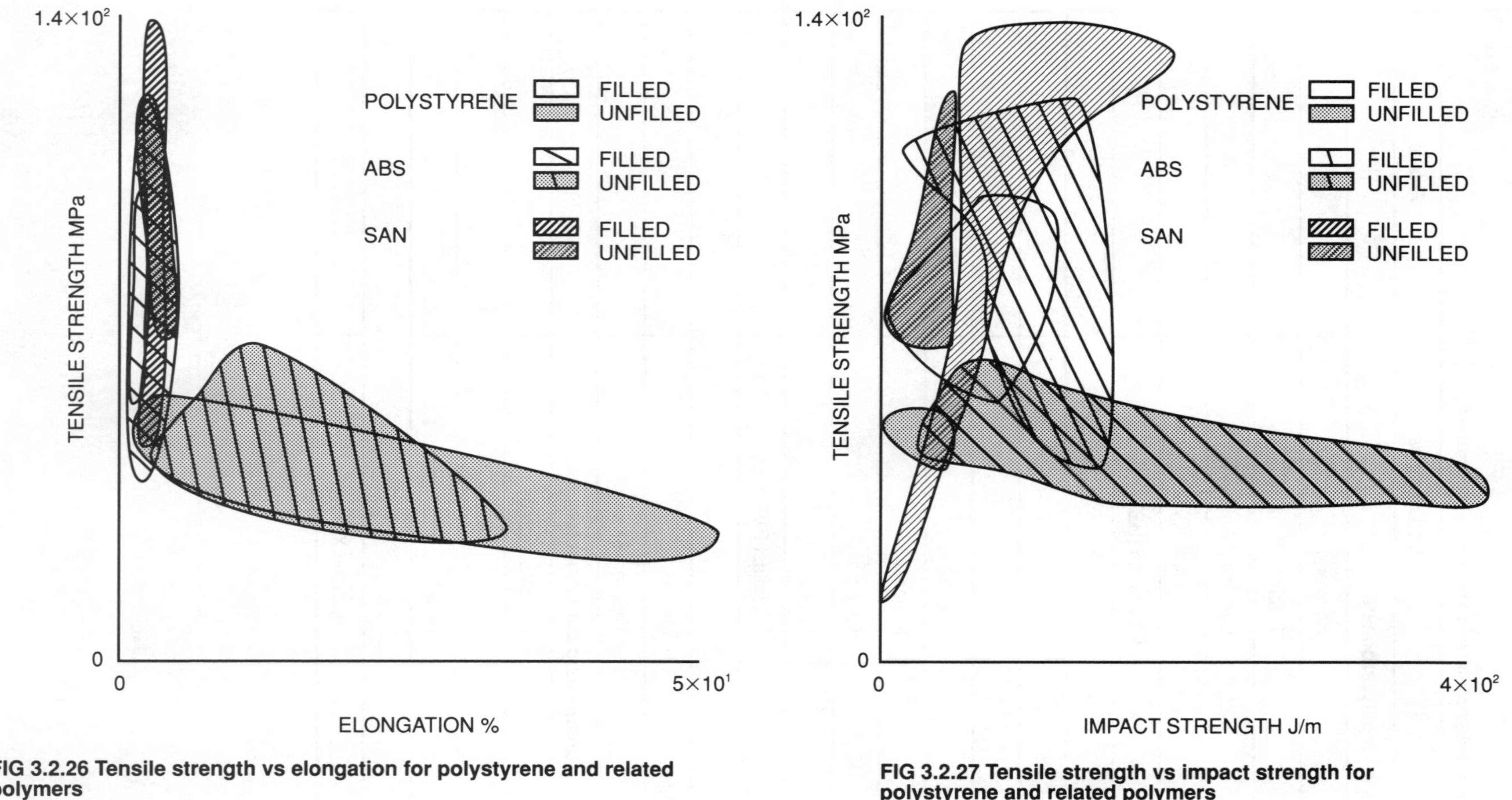

FIG 3.2.26 Tensile strength vs elongation for polystyrene and related polymers

FIG 3.2.27 Tensile strength vs impact strength for polystyrene and related polymers

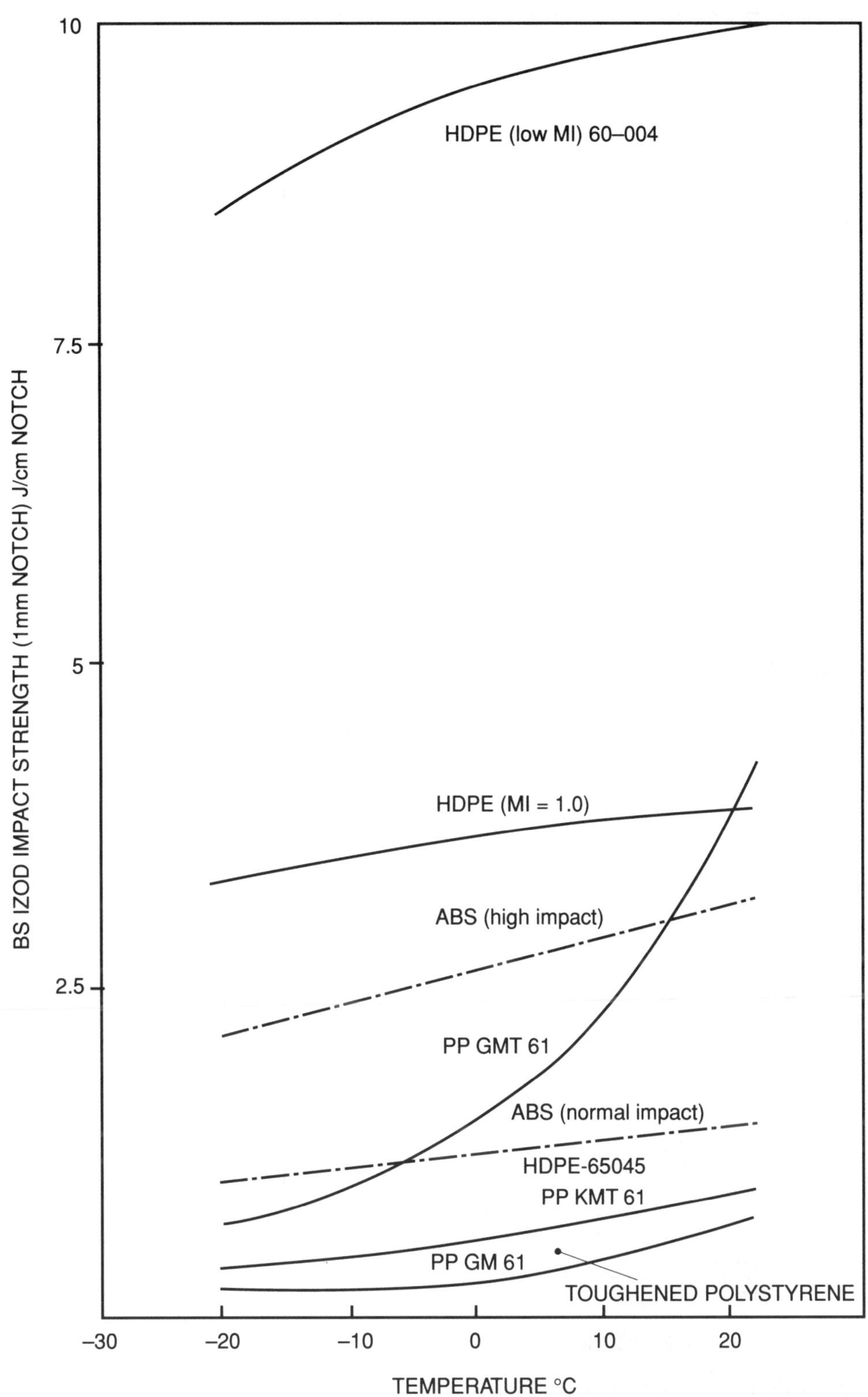

FIG 3.2.28 Influence of temperature on Izod impact toughness of styrene based polymers

Fig 3.2.28

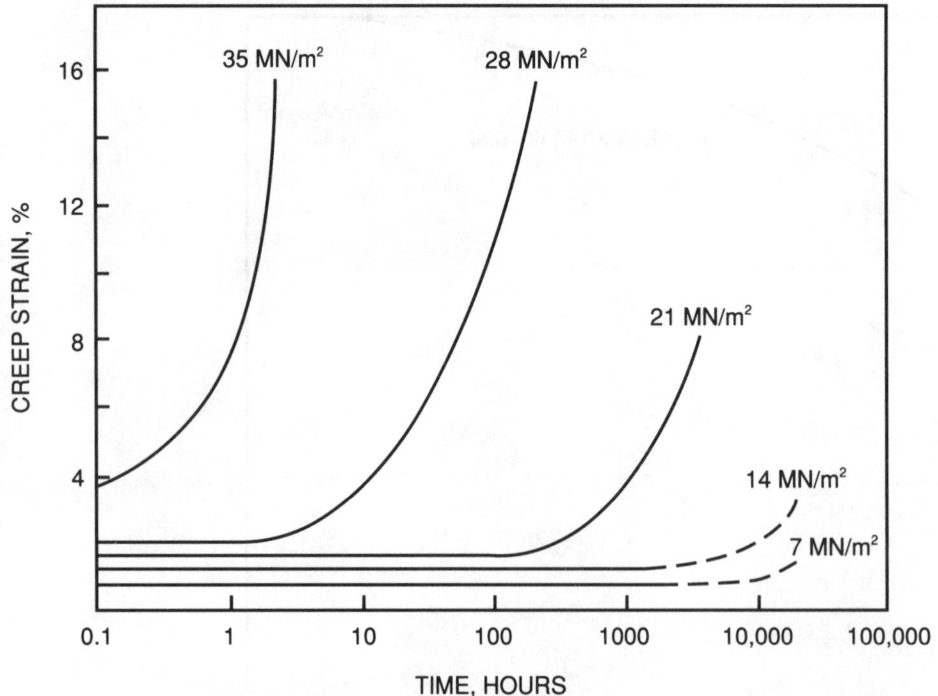

FIG 3.2.29 Tensile creep at 23°C (general purpose ABS)

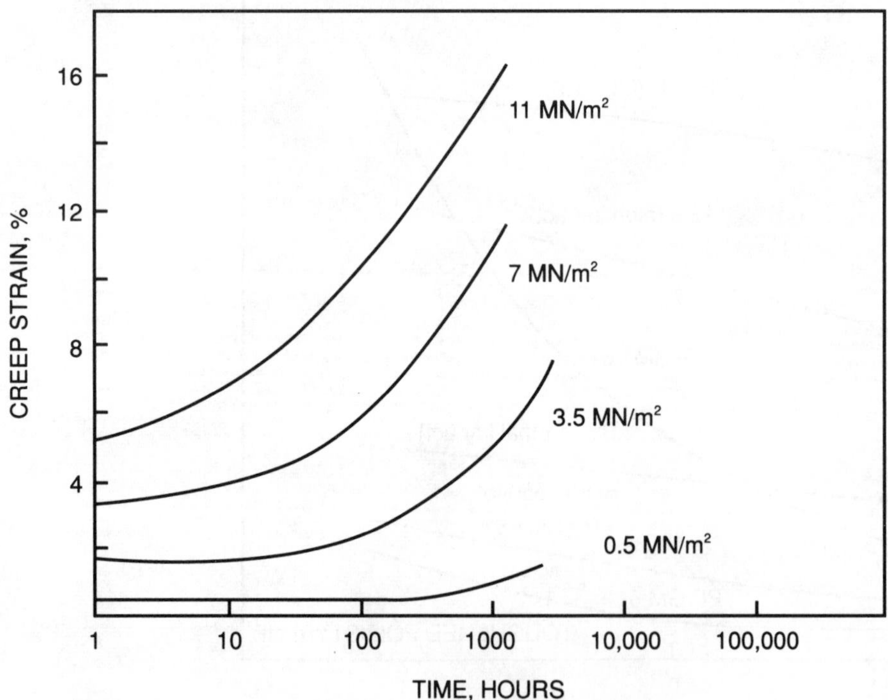

FIG 3.2.30 Tensile creep at 70°C (high temperature ABS)

Fig 3.2.29 and Fig 3.2.30

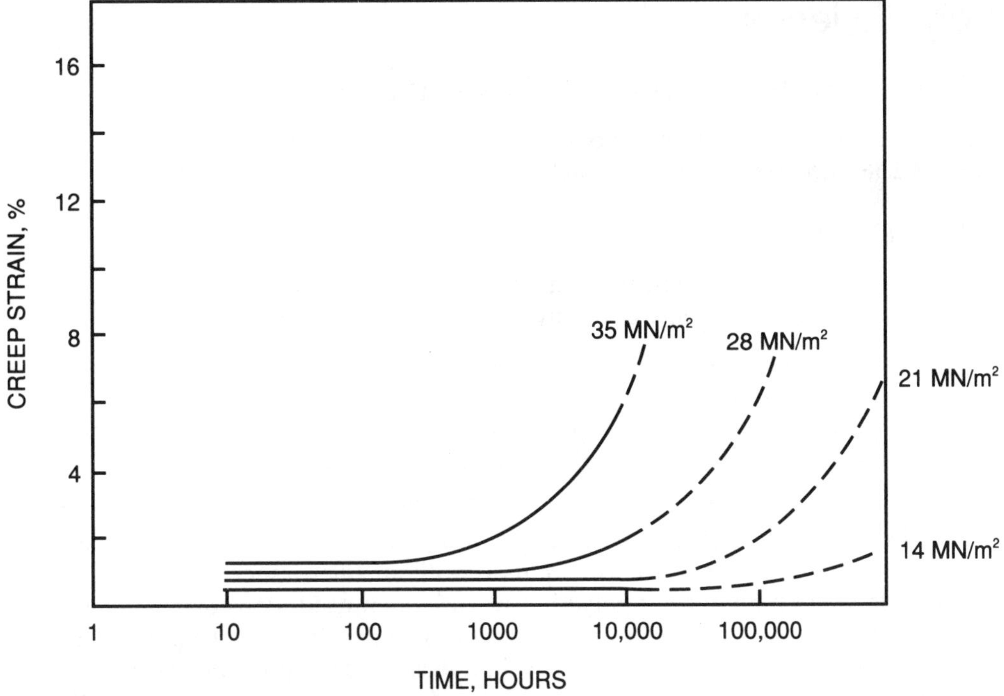

FIG 3.2.31 Tensile creep at 23°C (creep resistant ABS)

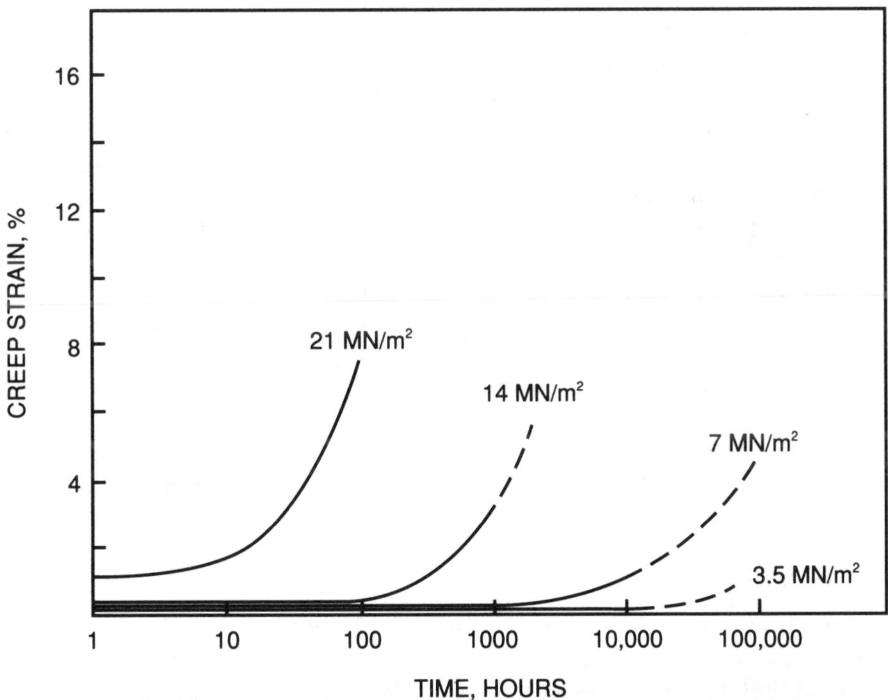

FIG 3.2.32 Tensile creep at 70°C (high temperature creep resistant ABS)

3.2.9 Polyethylenes

3.2.9.1 General characteristics and properties

The advantages and limitations of polyethylenes are tabulated in:
Table 3.2.56 *Characteristics of polyethylenes*

AVAILABLE TYPES

Polyethylene is a thermoplastic material produced by the polymerisation of the olefinic monomer ethylene, and is now generally divided into two main groups:

(a) Low density (0.918–0.93) polyethylene (LDPE) produced by a high pressure and temperature process;
(b) High density (0.950–0.965) polyethylene (HDPE) produced usually by a low pressure process with special metallic derivatives as catalysts.

Polyethylenes with densities in the range 0.930–0.945 are not normally made directly but can be produced by blending high and low density components. Unlike other thermoplastics, polyethylenes are normally used without fillers, such as glass, although blends of low and high density polyethylenes are often used. Polyethylene grades are normally characterised by *density* and *melt index*. As the melt index of a polyethylene of a given density is increased (molecular weight decreased) improvements are found in the processability and gloss, accompanied by deterioration in impact strength and stress-cracking resistance. Consequently the choice of melt index is normally a compromise between ease of processing and appearance, and performance.

Ultra-high molecular weight polyethylene (UHMWPE) has a molecular weight in the range $2–6 \times 10^6$ compared to $3–5 \times 10^5$ for conventional HDPE. Properties of such resins include outstanding abrasion resistance, exceptional impact resistance at cryogenic temperatures, low friction, fatigue resistance and vibration damping.

For a given melt index an increase in density gives improvements in modulus of rigidity, surface hardness, tensile yield strength, softening point and solvent resistance, but decreases impact strength and permeability rate.

A further basic polymer variation in grades, especially with high density polyethylenes, is *molecular weight distribution*. Widening of the distribution improves flow, particularly during extrusion, but a narrower distribution is preferred for injection moulding since cooling and shrinkage are more even, resulting in less distortion and internal strain which could give rise to stress-cracking.

Also available are polyethylene copolymers which contain a small proportion of comonomer, such as propylene.

Linear low density polyethylenes (LLDPE), unlike most low density PEs, have a linear structure, more typical of a HDPE. They exhibit marked differences in properties and processability.

Cross-linked polyethylene

Uncontrolled cross-linking of polyethylene is undesirable, but intentionally carried out by high-energy radiation or chemical additives, it provides a material with enhanced creep resistance at high temperatures and stress-cracking resistance. But the material is no longer thermoplastic and cannot be subsequently reprocessed once cured.

Low and high density polyethylenes are compared in:
Table 3.2.57 *Comparative characteristics of low and high density polyethylenes*, and
Fig 3.2.33 *Tensile strength vs elongation for polyethylenes*

Their structures and densities are illustrated in:
Fig 3.2.34 *Density distribution and structures of the three forms of polyethylene*

APPLICATIONS

LDPE	Injection mouldings:	Kitchen utility ware, toys, process tank liners, closures, packages, sealing rings, battery parts.
	Blow mouldings:	Squeeze bottles for packaging, containers for drugs.
	Film:	Wrapping materials for food, clothes, etc.
	Wire and cable:	High frequency insulation, jacketing.
	Pipe:	Chemicals handling, irrigation systems, natural gas transmission.
HDPE	Refrigerator parts, packaging, structural housing panels, pipe, defroster and heater ducts, sterilisable houseware and hospital equipment, hoops, battery parts, blow moulded containers including automotive petrol tanks, film wrapping materials, wire cable and insulation, and chemical resistant pipe.	
Glass-Filled HDPE	Automotive fender liners, tote boxes, blower and fan casings, chemical pipe fittings and pump components.	
UHMWPE	Applications requiring abrasion resistant sheet, plate or mouldings, e.g. pumps, filtration of coarse media, chemical processing, food cutting boards.	
LLDPE	Film resins aim at markets for LDPE, EVA. For mouldings, applications include rubbish bins and quality kitchen ware.	

TYPICAL PROPERTIES

The mechanical, physical and environmental properties and mould shrinkage of polyethylenes are listed in:
Table 3.2.58 *Typical properties of polyethylenes*, and
Fig 3.2.35 *Tensile strength vs impact strength for HDPE*
A comparison with EVA polyethylene copolymer is provided in Section 3.2.12

3.2.9.2 Mechanical properties and creep

The more significant mechanical properties of polyethylenes are listed in Table 3.2.58. The impact properties are outstanding. The most significant property as far as the designer is concerned is creep. The relationship between creep, density and temperature is illustrated in:
Fig 3.2.36 *Interpolated stress–temperature curves for polyethylenes (2% total strain at 1000h)*

It can be seen that at room temperature there is a linear relationship between creep and density and this is true irrespective of the melt index. However, melt index will affect rupture behaviour. Thus for a given density, for long-term applications, the higher the melt index, the lower will be the permissible strain.

The creep performance of PE is enhanced by radiation cross-linking, frequently of blends with butyl, or conjugated diene butyl rubber.
Typical properties of these blends are shown in:
Table 3.2.59 *Irradiation cross-linked elastomer/polyethylene blends*

3.2.9.3 Chemical and environmental resistance

Some data on the temperature limitations and environmental resistance of polyethylene are given in Table 3.2.58. The effects of the more important environments are shown in:
Table 3.2.60 *Chemical resistance and environmental properties of low and high density polyethylene*

3.2.9.4 Processing

GENERAL

Normal plastic design rules apply to polyethylene. All stress raisers such as sharp corners and rapid changes in section should be avoided because of the dangers of premature impact failures or stress cracking. Provided care is taken with ejection from the mould, slight undercuts can be tolerated as the material is flexible enough to be 'jumped' off the mould (especially LDPE).

Since they are crystalline polymers, polyethylenes show a fairly high level of mould shrinkage and care is necessary in moulding to close tolerances. Most shrinkage is dependent on molecular orientation and is greater along the line of flow than across it. This can give rise to warping of large flat areas, which should therefore be avoided if possible. Sinking and voiding are essentially the same problem and, like warping, result from high volumetric shrinkage on cooling from the melt.

PROCESSING TECHNIQUES

Polyethylenes may be processed by normal plastics processing techniques; by extrusion, injection moulding, blow moulding, compression moulding, extrusion moulding, calendering, vacuum forming and by rotational powder casting. Three aspects have to be considered in selecting the grade of polyethylene for a given finished article: the ease of processing, the product, the most suitable processing technique for the article concerned, and the demands imposed on the finished article. In general, polyethylenes with a low melt index are suitable for extrusion; and with a high melt index, for injection moulding. The hardness and rigidity of the finished articles depend on the density of the polyethylene concerned.

Generally the techniques for polypropylene are applicable (see Section 3.2.10)

EXTRUSION

For smoothest extrusions at highest output rates, materials of wide molecular weight distribution and of the lowest molecular weight consistent with form retention should be used, otherwise surface defects obtrude.

INJECTION MOULDING

With polyethylenes there is a large specific volume change with temperature, the effect being more pronounced the greater the molecular weight distribution, and this can make moulding to size difficult. There is also considerable orientation which can cause distortion and warping. At higher injection cylinder temperatures this effect is reduced. It may be further reduced, particularly on large flat pieces, by changing the gate design. On flat rectangular surfaces, the internal stresses caused by such orientation may be reduced by moving gates from the centre to one corner of the area, by providing multiple pinpoint gates, by introducing ribbed surfaces and by the use of film gates. On centre-gated circular or cylindrical objects, such as dishes and buckets, the flat areas should be dished or made to vary in thickness to distribute the moulded-in stresses.

Generally, the heavier the wall section, the slower the cooling and the greater the mould shrinkage. Hence, mould temperatures of 30–60°C and wall thicknesses of 0.080–0.100 in are used where possible in moulding large, flat-surfaced pieces. Use of material of as high melt index as is consistent with the end use of the object also helps to control shrinkage by lowering the melt temperatures required to fill the mould cavities having thin walls. The use of screw-injection machines permits the use of material with lower melt indices and thus gives a greater choice of grades.

Impact strength varies with injection moulding temperature. Injection moulded samples produced at high temperatures, and particularly in thicker sections, have similar impact strength to that of virtually unoriented compression moulded samples.

ROTATIONAL CASTING

This technique is particularly suited to the production of large mouldings. In this method polyethylene powders are charged into the mould which is then heated by air or hot liquid whilst being rotated about two mutually perpendicular axes. The powder flows over all the inside surfaces of the mould, melts and is solidified when the mould is cooled. This process is fairly expensive as cycle times are quite long. However, for very large mouldings this method may prove the most economic.

3.2.9.5 Trade names and suppliers

Trade names and suppliers of polyethylenes are listed in:
Table 3.2.61 *Trade names and suppliers of polyethylenes*

TABLE 3.2.56 Characteristics of polyethylenes

	Advantages	***Limitations***
Mechanical properties	Good resistance to impact over wide temperature range (–40–90°C).	Lower creep resistance and inferior fatigue properties to polypropylene (no 'integral' hinge effect).
Environmental	Resistant to wide range of chemicals, most inorganic acids and alkalis at room temperature. Good moisture resistance. Insoluble in organic solvents below 60°C but may be swollen.	Susceptible to weathering, UV light and environmental stress-cracking. Alloying of PE with butyl rubber and EPDM improves environmental stress-cracking resistance. Low resistance to permeation by gases and vapours (although HDPE is superior to polypropylene). Attacked by strongly oxidising acids. Dissolves in common solvents above 60°C.
Thermal properties		Very high thermal expansion. Flammability rating 2.5 cm/min. Poor temperature capability.
Food and medicine	Non-toxic grades available for use with foodstuffs.	
Electrical properties	Good electrical properties—not appreciably affected by humidity due to low water absorption. Power factor virtually unaffected by frequency or temperature.	Power factor very sensitive to slight degrees of oxidation and to polar additives.
Processing	Ease of processing by all thermoplastic methods.	High mould shrinkage (typical for crystalline polymers). Difficult to bond or print on.
Cost	Low cost.	

TABLE 3.2.57 Comparative characteristics of low and high density polyethylenes

Low density polyethylene (LDPE)	***High density polyethylene (HDPE)***
Tough at much lower temperature (embrittles at –60°C). Translucent. Flexible. Upper service temperature limit of 88°C. *Linear low density (LLDPE) exhibits:* Lower warpage. Improved environmental stress-crack resistance. Improved heat resistance. Improved toughness, puncture resistance and tear strength in films.	Better chemical resistance. Better creep resistance. Five times stiffness of LDPE at room temperature. Higher temperature service (up to 130°C). More resistant to permeation. Notch sensitivity little affected by temperature but deteriorates with increase in molecular weight (melt index) and moulding temperature. Can withstand hot water sterilisation but not steam.

TABLE 3.2.58 Typical properties of polyethylenes

	Property		ASTM Test	Units	Low-density polyethylene (LDPE)	High-density polyethylene (HDPE)	HDPE 30% glass fibre
Mechanical	Tensile yield strength		D638	MN/m^2	0.4–16	21–38	70
	Flexural strength		D790	MN/m^2	—	35–50	80
	Elongation at break		—	%	90–650	600–1500	2–3
	Modulus of Elasticity in Tension		D638	MN/m^2	120–240	420–1400	—
	Flexural elastic modulus		D790	MN/m^2	55–410	690–1800	6300
	Impact strength		D638 (IZOD)	J/cm of notch	No break	0.6–No break	0.55
	Hardness		D785	Shore	D41–50	D60–70	R85[b]
Physical	Specific gravity		D792	—	0.918–0.940	0.950–0.965	1.17
	Thermal conductivity		—	W/m per K	0.33	0.46–0.52	0.37
	Thermal expansion		D696	10^{-6} K^{-1}	100–200	110–130	50
Electrical	Volume resistivity		D257	ohm-m	1.2×10^{13}–$>10^{14}$	$>10^{14}$	—
	Dielectric strength		D149	10^2 kV/m	180–400	180–200	—
	Dielectric constant	60Hz	D150	—	2.25–2.35	2.30–2.35	—
		10^6 Hz			2.25–2.35	2.30–2.35	—
	Power factor	60 Hz	D150	—	<0.0005	<0.0005	—
		10^6 Hz			<0.0005–0.0009	<0.0005	—
	Arc resistance		D495	s	135–160	—	—
Thermal/Environmental	Max. continuous service temperature (no load)		—	°C	85–100	120–130	—
	Deflection temperature	0.45 MN/m^2	D648	°C	38–50	60–95	132
		1.81 MN/m^2	D648	°C	32–40	43–55	127
	Water absorption		D570	%	<0.01–0.03	<0.01	0.2
Processing	Mould shrinkage		—	%	1.5–5	2–5	0.3–0.4

[a] Ultra high molecular weight polyethylene.
[b] Rockwell.
[c] Values highly dependent on strain rate
[d] With inorganic fillers.

Table 3.2.58

TABLE 3.2.58 Typical properties of polyethylenes—*continued*

UHMWPE	*UV stabilised LDPE*	*Linear LDPE*	*UV stabilised HDPE*	*Chlorinated PE*	*Cross-linked PE*	*PE foam*	*Ethylene/ propylene copolymer*
17–24	9	20	30	12.5	18	0.8	26
28							
300–500	300	500	100	700	350	130	500
140–760							
900–960	250	350	125	2	600	20	600
No break	10.6+	10.6+	1.50	10.6+	10.6+	10.6+	1.5
D60–70	D48	D48	D68	SA70	SD58	SA30	R75
0.937–0.960	0.92	0.92	0.96	1.16	0.93	0.1	0.9
0.35–0.44							
200[d] (130)	200	200	120	180	200	200	160
$>10^{14}$	10^{13}	10^{14}	10^{15}	10^{11}	10^{14}	10^{10}	10^{13}
280	250	250	210	120	210	200	300
2.3							
2.3							
0.0002							
0.0002							
—							
90	50	50	55	60	90	50	60
95	50	45	75	35	60	—	93
68–82	35	37	46	25	60	—	54
<0.01	0.01	0.01	0.02	0.01	0.01	0.01	0.01
8–10	3	3	3	3	3	—	1.5

Table 3.2.58—*continued*

TABLE 3.2.59 Irradiation cross-linked elastomer/polyethylene blends

Plastic blend ratio	***50/50 (CDB/HDPE)***				***50/50 (Reg. Butyl/HDPE)***			
Radiation dosage (Mrad)	0	2.5	7.5	15	0	2.5	7.5	15
Stress/strain properties at RT								
Modulus at 100%, MPa	6.6	7.6	7.3	7.8	6.2	6.2	6.2	6.1
Modulus at 300%, MPa	—	8.1	8.5	9.9	6.4	6.5	—	—
Tensile strength, MPa	6.9	12.5	11.7	11.5	7.8	6.3	6.3	6.2
Elongation at break (%)	300	630	500	400	600	253	230	163
Stress/strain properties at 150°C								
Modulus at 100%, MPa	—	1.6	—	0.9	(Not measurable)	—	—	0.2
Modulus at 300%, MPa	—	—	—	—		—	—	0.4
Tensile strength, MPa	—	1.6	—	1.7		—	—	0.6
Elongation at break (%)	500	260	—	200		600	—	490

(Data supplied by Exxon Chem. Co.)

TABLE 3.2.60 Chemical resistance and environmental properties of low and high density polyethylene

Environment	*Effect*
General	The lower the melt index the better the resistance to chemical and solvent attack and to environmental stress-cracking. HDPE is generally superior to LDPE. Polyethylene copolymers are in general more susceptible to solvent and chemical attack but considerably more resistant to environmental stress-cracking than homopolymers of the same melt index.
Acids and alkalis	Resistant to most acids and alkalis. Attacked by highly oxidising acids such as concentrated nitric, glacial acetic, fuming sulphuric and hydrogen peroxide (above 20°C).
Solvents	Insoluble in most organic solvents below 60°C but some absorption, softening or embrittlement may occur in alcohol, esters, amines and phenols. Above 70°C dissolves in benzene, xylene, toluene, amyl acetate, trichlorethylene, paraffin and turpentine, but not in glycerine, ether, carbon disulphide, acetone or linseed oil. LDPE is relatively poor in aromatic and chlorinated hydrocarbons.
Oils and greases	Mineral oils and petrol cause marked swelling but effect of animal and vegetable oils is much less.
Other media	Attacked by liquid and gaseous halogens, although satisfactory life may be achieved at room temperature in presence of gaseous chlorine and fluorine.
Foodstuffs	Inert—can be used in contact with foodstuffs.
Weathering and UV light exposure	Degradation occurs through photo-oxidation—can be prevented by adding 2% of carbon black.
Heat	Oxidises when heated in air to above 60°C, hence processing time should be as short as possible. Anti-oxidants are normally added (~ 0.1%) to reduce oxidation during processing.
Water	Low absorption—equilibrum at 20°C HDPE 0.01% LDPE 0.15%
Environment stress-cracking	Very susceptible especially when in contact with polar compounds, surface active agents (e.g. detergents) and silicone compounds. Increased by increasing molecular weight (decreasing melt index) and increasing density. Decreased by decreasing stress (eliminated below threshold level) by decreasing moulding stresses (frozen-in-stains).
Flammability	2.5 cm/min

TABLE 3.2.61 Trade names and suppliers of polyethylenes

Company	*Trade names*	*Type*	*Fillers*
Allied Chem.	Paxon	H	U
	Suprel	M/UHMW	U
Amoco	Amoco	H	U
Eni Chemical (UK) Ltd	Eraclene	L, H	U, V
	Fertene	L	V
	Riblene	L	U, V
ARCO	Dylan, Super Dylan	L, H,	U
Atochem	Lacqtene	L, LLDPE, H	U, V
BASF	Neopolen	W, F	U
	Lupolen	H, M, L, LLDPE	U, V
	Lucobit	Bitumen ethylene copolymer	
BIP	Bexal	Sheet	
Bayer	Baylon	L	—
CdF Chemie	Lotrene	L	U, V
Chemische Werke (Huls UK)	Vestolen	L, M, H	U, V
Chemplast	Ertalene	—	—
Chemplex	Chemplex	L, H	U
Cole	Trade names: Hilo Blend, Playrite, Figure Compound, Dollymix, Tip Top		
DSM	Stamylan, Stamylax	H, L, M, LLDPE, W	U, V
Dow	Dow PE	M	U
	Dowlex	LLDPE	U
	Dow	Chlorinated	Polyethylene
Du Pont	Alathon	H	U
Dynamit Nobel	Trofil	—	—
Eastman	Tenite	Ethylene/Propylene Copolymer	
Exxon/Esso	Exxon	L	—
	Escorene	LLDPE, L	U, V

Key: U = unfilled; G = glass fibre-filled; C = carbon fibre-filled; H = high density; M = medium density; L = low density; LLDPE = linear low density; UHMW = ultra-high molecular weight; V = UV stabilised; W = chlorinated; X = cross-linked; F = foam.

Table 3.2.61

TABLE 3.2.61 Trade names and suppliers of polyethylenes—*continued*

Company	*Trade names*	*Type*	*Fillers*
Fiberfil/Dart	Ethofil	H	G
Fiberite	RTP	H	G
Gulf	Poly-eth	L, H	U
Hoechst	Hostalen	L, H, UHMW	U, V
ICI	Alkathene	L	—
LNP	Thermocomp	H	G, PTFE
Mitsui	Novatec	H	U
	Yukalon	L	U, C
	Hi-zec	H	U
	Neo-zec	L, M	U
Montedison	—	LLDPE	U
Northern Petrochem.	Norchem	L	U
Phillips	Marlex	H, L	U, copolymers
Polypenco	—	UHMW	U
Rhone Poulenc	Natene, Mandene	M, H	U
SIR	Sirtene	—	—
Shell	Carlona	L	U, V
Solidur	—	UHMW	—
Solvay/Laporte	Eltex	H, L	U, V
	Fortiflex	H, M	U
Sumitomo	Sumikathene	L, M	U
Thermofil	—	H	G
Union Carbide	Bakelite	L, H	U, C
	—	LL	—
US Ind. Chem.	Microthene	L	U
	Petrothene	H, L, M	U, C, Clay
Statoil (UK) Ltd	Statoil	L, H	U, V
Neste Chemicals (UK) Ltd	Neste LDPE	L, LLDPE	U, V

Key: U = unfilled; G = glass fibre-filled; C = carbon fibre-filled; H = high density; M = medium density; L = low density; LLDPE = linear low density; UHMW = ultra-high molecular weight; V = UV stabilised; W = chlorinated; X = cross-linked; F = foam.

Table 3.2.61—*continued*

TABLE 3.2.61 Trade names and suppliers of polyethylenes—*continued*

Company	***Trade names***	***Type***	***Fillers***
BP	Alkathene	L	V
	Rigidex	M, H	U, V
Croxton & Gary	Dow	W	U
AEI Cables Ltd	Sioplas	W	U
BXL Plastics	Plastizote	F	U
Freeman Chemicals Ltd	Latene	Ethylene Propylene Copolymer	

Key: U = unfilled; G = glass fibre-filled; C = carbon fibre-filled; H = high density; M = medium density; L = low density; LLDPE = linear low density; UHMW = ultra-high molecular weight; V = UV stabilised; W = chlorinated; X = cross-linked; F = foam.

Table 3.2.61—*continued*

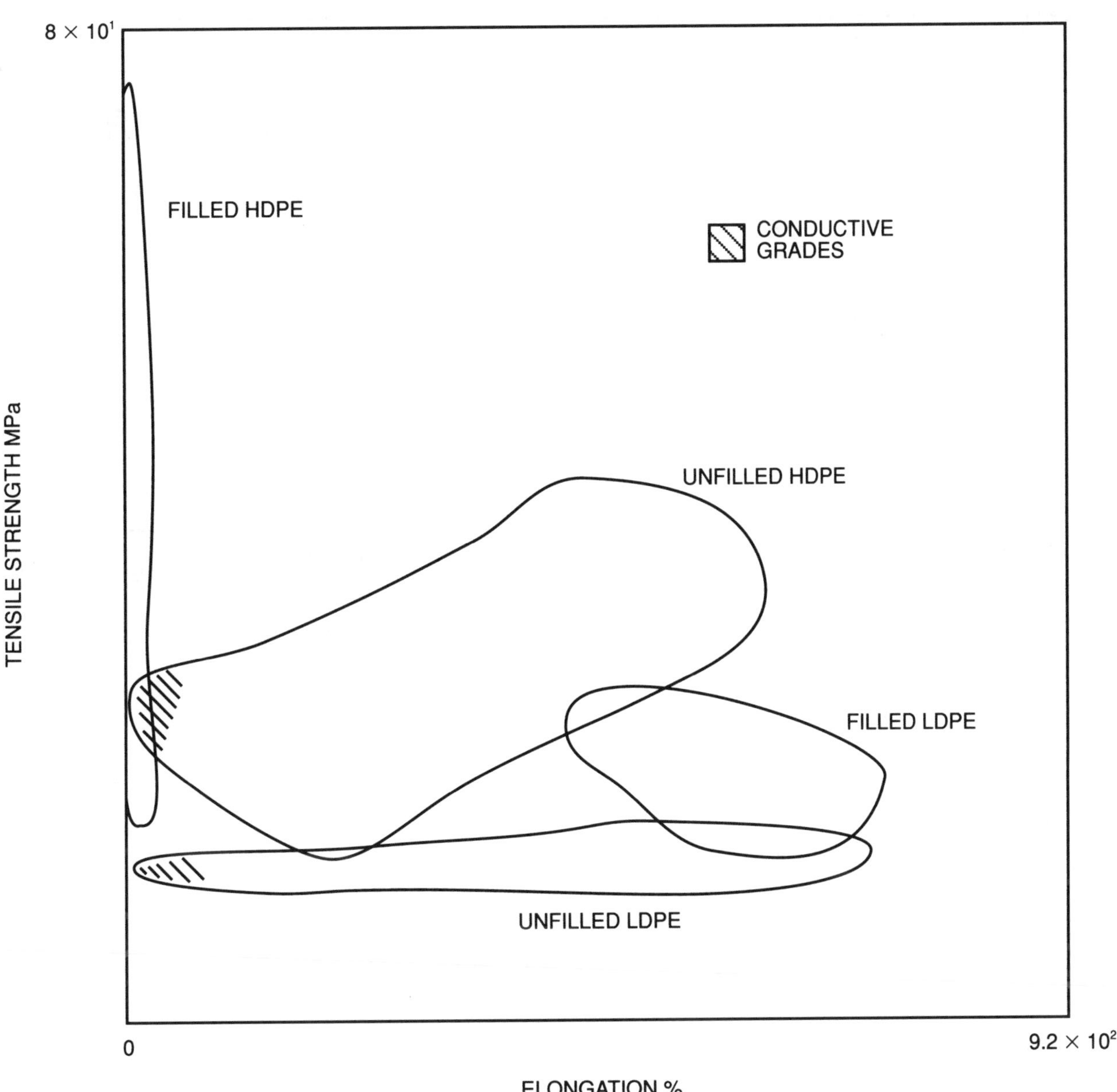

FIG 3.2.33 Tensile strength vs elongation for polyethylenes

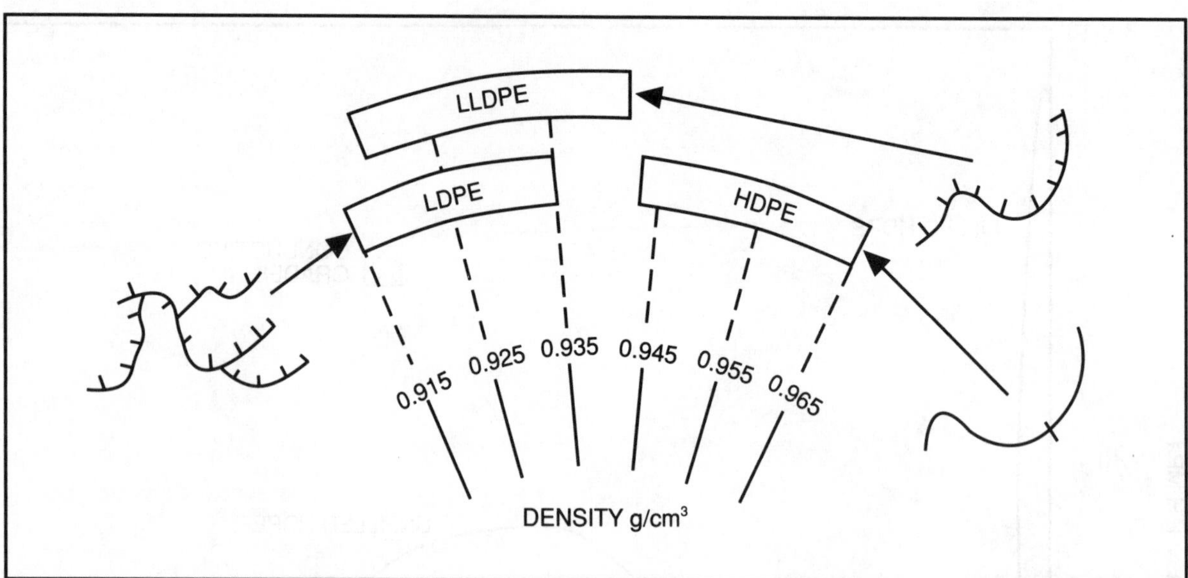

FIG 3.2.34 Density distribution and structures of the three forms of polyethylene

Fig 3.2.34

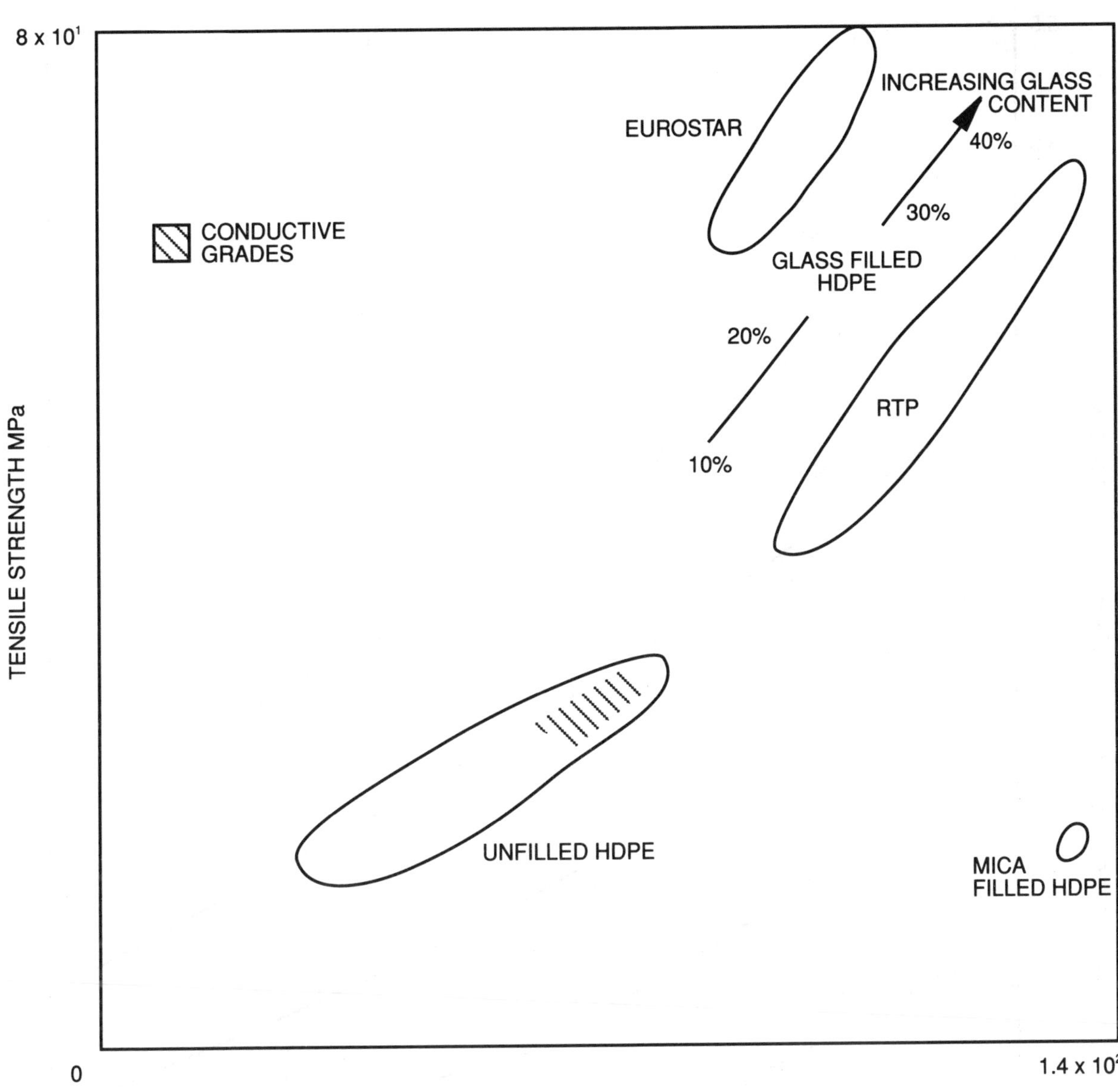

FIG 3.2.35 Tensile strength vs impact strength for HDPE

Fig 3.2.35

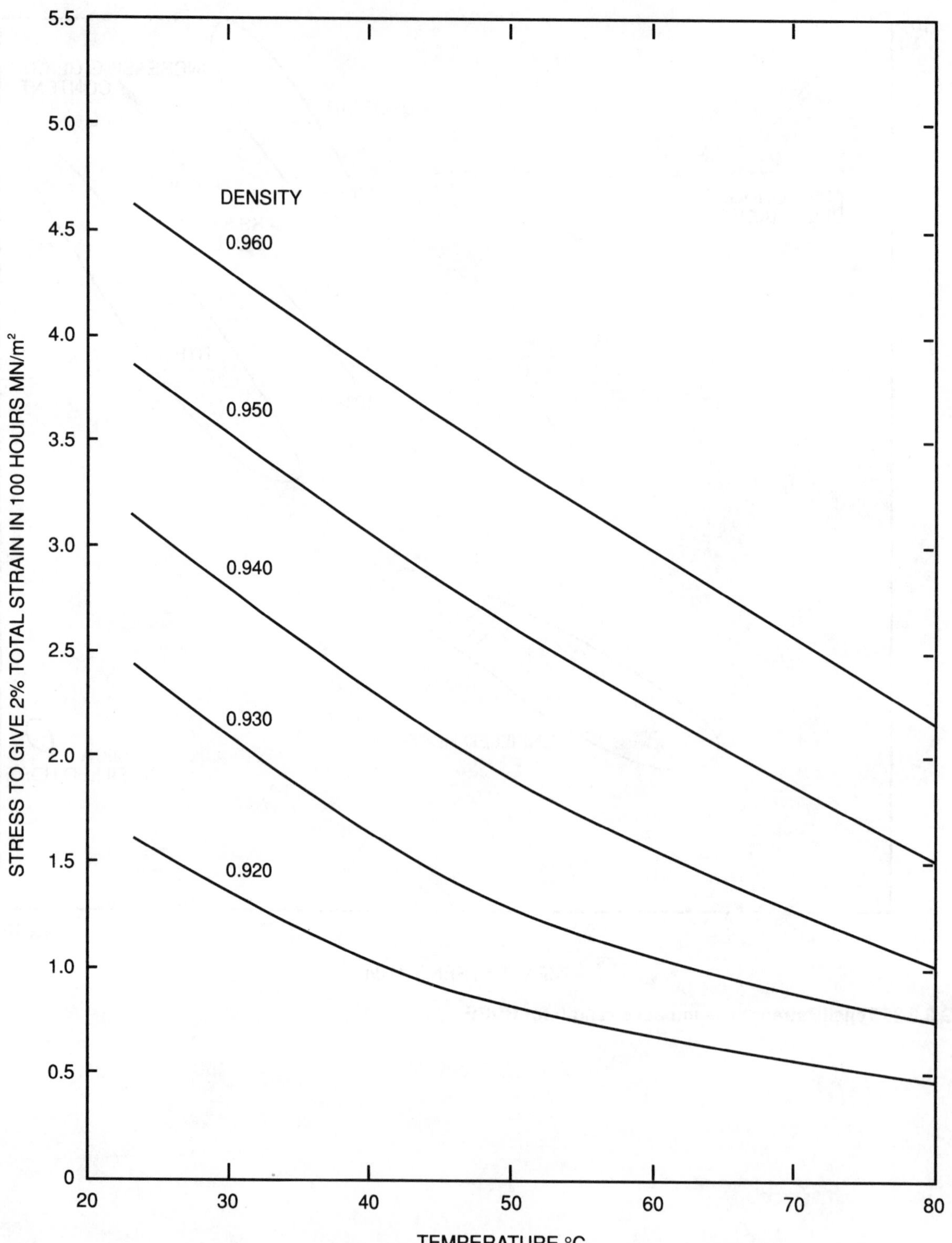

FIG 3.2.36 Interpolated stress–temperature curves for polyethylenes (2% total strain at 1000 h)
(Data courtesy of Shell Chemicals)

Fig 3.2.36

3.2.10 Polypropylenes

3.2.10.1 General characteristics and properties

The advantages and limitations of polypropylene are listed in:
Table 3.2.62 *Characteristics of polypropylene*

AVAILABLE TYPES

Polypropylene is a thermoplastic material produced by the polymerisation of the olefinic monomer propylene in the presence of a catalyst.

$$\left[-CH_2- \overset{\displaystyle CH_3}{\underset{\displaystyle H}{\overset{|}{\underset{|}{C}}}} - \right]_n$$

The spatial arrangement of the methyl groups attached to each second carbon atom in the polymer chain may vary in three ways to give isotactic, syndiotactic or atactic polypropylene. The regularity of the isotactic and syndiotactic forms allows crystallisation, whereas the atactic form, being irregular, cannot crystallise. In commercially available polypropylene the molecules are substantially isotactic, although some syndiotacticity may be present to a small extent. No significant commercial use has been found for atactic polypropylene. The commercial isotactic polypropylenes crystallise to a degree of 60–70%.

Polypropylene is available commercially as either homo- or copolymers and in numerous grades and formulations. Moulding and extrusion grades are available in low, medium and high melt flows. Also available are grades acceptable by the USFDA, medium- or high-impact types, heat and ultraviolet-radiation-stable formulations. Copolymerisation with relatively small amounts of other monomers, e.g. 5% ethylene, modifies the polypropylene structure, giving products with improved impact properties (between 10 and 30 times better than the homopolymer with the higher value applying to low melt index copolymers) but reduced creep strength. Filled grades are also available, typical fillers being glass, and talc.

A special type of reinforced polypropylene where the glass reinforcement is chemically coupled to the polymer is an important class of engineering plastics. Long fibre reinforced grades also offer significant improvement in mechanical properties.

APPLICATIONS

General

Polypropylene combines strength, rigidity, temperature resistance, excellent chemical resistance and electrical insulation properties. Copolymers offer better impact behaviour at low temperatures.

Three other characteristics can be exploited to considerable advantage: the ability to form an integral hinge, the ability to accept a textured finish and the ability to provide snap fits. The possibility of using snap-fit assemblies is a consequence of the toughness and resilience of polypropylene, and the unique combination of this feature with the integral hinge and textured finish often leads to a reduction in the number of components and assembly operations required, as well as providing aesthetic advantages. Many of the successful applications listed below incorporate one or more of these features.

Mouldings

Car components, e.g. accelerator pedals with integral hinge, fascia, glove boxes, trim panels, radiator fans, heater ducting, steering wheel covers, mats, grilles, parcel shelves, door handles, air cleaners, water pumps, air-intake noise suppressors, lead-acid battery cases and bumpers.
Household goods, e.g. washing machine parts—agitators, taps, tubes, filters, pump housings, tubs; dishwasher parts—tubs, doors, spray arms, pump housings, detergent dispensers; vacuum cleaner housings, nozzles and parts; hairdrier casings, and hoods; suitcases with textured finishes; refrigerator liners; refuse bins; picnicware.
Electrical goods, e.g. radio and TV housings and cabinets, capacitors, coil formers, control knobs, transformer housings, industrial lights, insulators for field electric fencing, aerial parts, cable and wire coatings, electrical switch gear, cable connectors and fittings.
Kitchen equipment, e.g. mixing machine components, electric kettles, baby feeding bottle warmers, children's plates, buckets, thermos flask cases, strainers, sink units, wash-basins and waste-pipe fittings.
Containers, e.g. food packaging (boil-in-the-bag, etc.), goods wrappings, cosmetic compacts, contact lens cases, first aid cases, bottles, drums, jerrycans, containers for food, pharmaceuticals and cosmetics, tool boxes, cigarette boxes.

Microsphere-filled polypropylene

Replacement of glass-filled ABS dashboard components; chemical pump rotor/stator sets; vacuum-filter elements; vacuum drums.

Talc-filled polypropylene

Moulded tops for automotive batteries. General components.

Wood-flour-filled polypropylene

Intended for press-forming in matched metal dies to make simple, panel-like parts, either plain or clad with PVC sheet or textiles. Typical cycle times are 30s, working with pre-heated (160°C) material and cold tools. Typically used for automotive door and roof liners and vehicular interior panels. Starting product is extruded sheet.

TYPICAL PROPERTIES

Typical properties of filled and unfilled grades of polypropylene are listed in:
Table 3.2.63 *Typical properties of polypropylene grades*, and
Fig 3.2.37 *Tensile strength vs elongation for polypropylenes*

The Modulus of Elasticity, one of the more important mechanical properties, is significantly increased by aging if a considerable period elapses between moulding and loading. The relationship between modulus and time of aging over a range of temperature and the permissible load for a given deformation may be determined by constructing a nomogram for the polypropylene under consideration.

The effect of temperature on toughness is shown in:
Fig 3.2.38 *Influence of temperature on Izod impact toughness of polypropylenes*

The influence of fillers and reinforcements on some of the mechanical properties of polypropylene homopolymer is shown in:
Fig 3.2.39 *Variation of flexural modulus with filler type and loading of polypropylene homopolymer*, and
Fig 3.2.40 *Variation of tensile strength with filler type and loading of polypropylene*

Longitudinally oriented polypropylene is extremely resistant to fatigue in bending. This has enabled the concept of the integral hinge to be realised commercially. Most unfilled grades of polypropylene are suitable, the quality of the hinge being mainly dependent upon the design.

The phenomenon of high resistance to repeated flexing is based on a strong tendency to orientation in one direction when the material is pressed through a narrow opening. Orientation, a property which tends to be undesirable in many injection moulding applications, is in this case advantageous.

Integral hinges give the following design benefits:

Very tough and flexible, better than most conventional hinges.
Do not tear when stresses are exerted.
Low production costs, the hinge being built-in.
Very light in weight, only 33% of the weight of aluminium.
Silent.
Resistant to many environments, and no galvanic problems.

3.2.10.2 Chemical and environmental resistance

The effects of the environments to which polypropylenes are normally exposed are listed in:
Table 3.2.64 *Chemical and environmental characteristics of polypropylenes*

3.2.10.3 Processing

Polypropylene can be injection moulded, blow moulded or extruded. In addition, preprocessed sheet, film, pipe, wire and cable covering are available.

Mould shrinkage is usually between 1 and 2%. Compared with other crystalline thermoplastics polypropylene exhibits a much lower degree of anisotropy and this gives mouldings which are less prone to distortion.

POST-MOULDING PROCESSING

Welding

Polypropylene sheets can be welded by the hot-gas filler-rod technique normally used for other thermoplastic sheet materials. By this technique butt joints, T joints and right-angled joints can be made, and weld strengths up to 90% of the parent sheet can be achieved.

Polypropylene sheets can also be joined by ultrasonic welding at 20 kHz. Because of the mechanical loss characteristics of polypropylene, this method is presently suitable only for thin sections.

Other welding techniques which can be used satisfactorily for polypropylene are butt welding without the use of a filler rod, and spin or friction welding. High frequency welding is not feasible with polypropylene due to its very low dielectric loss factor.

Adhesives

Because polypropylene is chemically inert it is difficult to find really good adhesives for the material, and their use should be avoided whenever possible. If adhesives must be used it is often possible to find a contact adhesive adequate for the requirement, but this will be very inferior in strength to the rest of the material.

Machining

Polypropylene can be cut, sawn, milled, drilled, tapped, turned and routed by using normal techniques applied to thermoplastics. Standard metalworking tools and lubricants are satisfactory. High cutting speeds and low rates of feed should be used.

Pipe manipulation

Extruded polypropylene pipe can be manipulated by the methods used for other plastic pipes. The pipe can be bent by heating and using a bending spring or sand to maintain its shape, and it can be welded using polypropylene filler rod. Although jointing can be accomplished by tapping the pipe for insertion into polypropylene or metal fittings, this technique is to be avoided because polypropylene is notch-sensitive.

Decoration

Products manufactured from polypropylene may be decorated in several ways, although the inertness and chemical resistance of polypropylene make it difficult to obtain good adhesion of the various coatings without some form of pre-treatment of the surface. Coloured products are best produced by using pigmented polymer at the fabricating stage. Suitable pre-treatments include: 1. application of a special primer; 2. immersion in potassium permanganate solution; 3. etching the surface with trichloroethylene vapour; 4. immersion in chromic acid solution; 5. flaming in a hot blue gas flame and 6. electrical treatment.

Painting and lacquering

Paints and lacquers can be applied to primed polypropylene articles by spraying, dipping or spreading. Most types of commercially available air drying and thermosetting finishes give satisfactory results.

Printing

Polypropylene articles can be printed by the usual techniques, such as silk screen, flexographic, offset and photogravure, with specially formulated inks. The surface to be printed should be pre-treated by an electronic process or by flash-heating the surface with a naked flame.

With film an electronic process is normally used. As slip and antistatic additives tend to migrate to the surface it is usual to treat the film in line with the extrusion process.

Metallisation

Two methods are normally used, either vacuum metallisation or electroplating. Vacuum metallisation is the process normally used, and conventional units can be used. The primed polypropylene surface must be lacquered and thoroughly dried before metallisation, when a highly reflective surface is not required. After metallising, an appropriate top coat may be applied as either a protective coating, or as a tint to give gold, brass or other metallic colour effects.

Metallising polypropylene by electroplating is becoming more popular. The process involves non-electric chemical deposition of a metal, usually nickel, on to chromic acid etched polypropylene, followed by electro-plating with the selected metal. The bond strengths are higher than obtained on ABS.

Hot foil stamping

Hot foil stamping is commonly used. However the temperature required to obtain good adhesion of the design is higher than for many other plastics. Hot foil stamping may be carried out on untreated polypropylene, although use of a special primer (applied to either the polypropylene or the back of the foil) gives better adhesion. Since pre-treatment may be avoided and as there is no printing ink to be cleaned off, the process is often the most economical.

Decals

Decals may be used for decoration. These comprise pre-printed polypropylene foils which are placed within the mould cavity and onto which the melt from the injection moulding machine is directly applied.

Antistatic treatments

Grades of polypropylene are available which contain antistatic additives. Post-treatments with antistatic agent, applied by dipping, spraying or wetting with a sponge may also be used. Such antistatic post-treatments, however, lose their effectiveness after contact with water.

3.2.10.4 Trade names and suppliers

Trade names and suppliers are listed in:
Table 3.2.65 *Major suppliers and familiar trade names of polypropylenes*

TABLE 3.2.62 Characteristics of polypropylene

	Advantages	*Limitations*
Mechanical properties	Very good fatigue resistance—possesses integral hinge property. Filled grades available with improved stiffness—comparable with unfilled nylon, and lower price. Good abrasion resistance.	
Environmental	Chemical resistance similar to polyethylenes (i.e. resistant to most inorganic acids, alkalis and salts) but without susceptibility to environmental stress cracking when in contact with alcohols, esters, detergents or polar hydrocarbons. Mechanical and electrical properties unaffected by submersion in water. Very low water absorption.	Heat ageing stability adversely affected by contact with metals. Attacked by highly oxidising acids such as fuming sulphuric, liquid and gaseous halogens and swells rapidly in chlorinated solvents and aromatics. 2% carbon black needed to prevent degradation by ultraviolet light.
Thermal properties	More rigid than polyethylenes and retains mechanical properties to elevated temperatures due to high melting point.	Embrittles below −17°C. Upper service temperature 90–120°C depending on grade and anti-oxidant content (upper limit filled grades, lower limit flame retardant grades).
Electrical properties	Excellent dielectric properties.	
Food and medicine	Non-toxic grades available for use with foodstuffs. Can be steam sterilised.	
Fabrication and processing	Can be joined by hot gas, hot tool, induction or friction welding. Available as pipe sheet rod and injection mouldings.	Not readily thermoformed because of low melt strength and hence requires close control of processing parameters. Joined more easily to other materials, e.g. wood and aluminium, than to itself.
Miscellaneous	Low density (0.9 g/cm^3) Low cost.	Colour purity adversely affected by reinforcing materials. More expensive than polyethylene.

TABLE 3.2.63 Typical properties of polypropylene grades

<table>
<tr><th colspan="3">Property</th><th>ASTM Test</th><th>Units</th><th>General purpose unfilled homopolymer</th><th>High impact unfilled copolymer</th></tr>
<tr><td rowspan="7">Mechanical</td><td colspan="2">Tensile yield strength</td><td>D638</td><td>MN/m²</td><td>27–40</td><td>19–35</td></tr>
<tr><td colspan="2">Flexural strength</td><td>D790</td><td>MN/m²</td><td>40–50</td><td>28</td></tr>
<tr><td colspan="2">Elongation</td><td>—</td><td>%</td><td>30–>200</td><td>30–>200</td></tr>
<tr><td colspan="2">Tensile Elastic Modulus</td><td>D638</td><td>GN/m²</td><td>0.5–1.9</td><td>0.8–1.3</td></tr>
<tr><td colspan="2">Flexural Modulus</td><td>D790</td><td>GN/m²</td><td>1.2–1.7</td><td>0.7–1.4</td></tr>
<tr><td colspan="2">Impact strength Izod</td><td>D638 (IZOD)</td><td>J/cm notch</td><td>0.2–1.2</td><td>0.8–6.4</td></tr>
<tr><td colspan="2">Hardness</td><td>D789</td><td>Rockwell</td><td>R80–105</td><td>R28–95</td></tr>
<tr><td rowspan="3">Physical</td><td colspan="2">Specific gravity</td><td>D792</td><td>—</td><td>0.9–0.91</td><td>0.9–0.91</td></tr>
<tr><td colspan="2">Thermal conductivity</td><td>—</td><td>W/m per K</td><td>0.17–0.20</td><td>0.25</td></tr>
<tr><td colspan="2">Thermal expansion</td><td>D696</td><td>10^{-5} K^{-1}</td><td>6.8–10.4</td><td>7.2–10.6</td></tr>
<tr><td rowspan="7">Electrical</td><td colspan="2">Volume resistivity</td><td>D257</td><td>ohm-m</td><td>>10^{14}</td><td>10^{14}</td></tr>
<tr><td colspan="2">Dielectric strength</td><td>D149</td><td>10^2 kV/m</td><td>240–280</td><td>180–255</td></tr>
<tr><td rowspan="2">Dielectric constant</td><td>60Hz</td><td rowspan="2">D150</td><td rowspan="2">—</td><td>2.2–2.6</td><td>2.2–2.3</td></tr>
<tr><td>10^6 Hz</td><td>2.1–2.6</td><td>—</td></tr>
<tr><td rowspan="2">Power factor</td><td>60 Hz</td><td rowspan="2">D150</td><td rowspan="2">—</td><td>0.0005–0.0007</td><td><0.0016</td></tr>
<tr><td>10^6 Hz</td><td>0.0002–0.0020</td><td>—</td></tr>
<tr><td colspan="2">Arc resistance</td><td>D495</td><td>s</td><td>125–136</td><td>123–140</td></tr>
<tr><td rowspan="4">Thermal/Environmental</td><td colspan="2">Maximum service temperature</td><td>—</td><td>°C</td><td>110–120[a]</td><td>—</td></tr>
<tr><td rowspan="2">Deflection temperature</td><td>0.45 MN/m²</td><td>D648</td><td>°C</td><td>95–135[a]</td><td>90–115</td></tr>
<tr><td>1.81 MN/m²</td><td>D648</td><td>°C</td><td>46–67</td><td>49–65</td></tr>
<tr><td colspan="2">Water absorption 24h</td><td>D570</td><td>%</td><td><0.01–0.3</td><td><0.01–0.02</td></tr>
<tr><td>Processing</td><td colspan="2">Mould shrinkage</td><td>—</td><td>%</td><td>1.0–2.5</td><td>0.9–2.0</td></tr>
<tr><td></td><td colspan="2">Remarks</td><td></td><td></td><td>[a] Heat resistant grades</td><td></td></tr>
</table>

Table 3.2.63

TABLE 3.2.63 Typical properties of polypropylene grades—*continued*

30% glass fibre-filled homopolymer	*40% long glass fibre reinforced homopolymer*	*25% Talc filled homopolymer*	*20% glass microsphere filled homopolymer*	*Wood flour-filled grade*
47–103	110	28–35	45	—
63–137	—	48–50	70	—
2–4	—	12–40	4–5	—
4.0–6.9	—	2.4–2.8	—	4.0
4.2–6.55	7.5	2.8–3.2	2.7–3.8	3.65
0.3–1.6	2	0.6–4.3	0.35–0.6	—
R90–115	—	—	D71 (Shore)	—
1.12–1.13	1.22	1.08	0.99	1.04
0.33	—	—	0.3	—
3.1–1.13	—	6–8	4.3	3–7
10^{13}– 1.5 $\times 10^{14}$	—	10^{14}	1.7 $\times$ 10^{14}	—
150–205	—	195–200	150	—
2.4	—	—	—	—
2.2–2.9	—	2.4	—	—
0.001	—	—	—	—
0.001–0.003	—	0.004–0.006	—	—
70–80	—	—	—	—
90–120	130	—	118	—
146–163	—	115–135	121	—
130–150	156	90	113	106
0.03–0.04	—	0.05	0.5	0.5
0.4–0.6	—	1.0–1.4	1.5	—
Chemically coupled grades are available to improve strength, stiffness, dimensional stability and creep resistance.		Easy flow characteristics. High degree of fire retardance possible. Up to 40% loadings are commonplace.	Low mould shrinkage, low cycle times at lower cost with slight increase in machine wear rates.	

Table 3.2.63—*continued*

TABLE 3.2.63 Typical properties of polypropylene grades—*continued*

	Property		ASTM Test	Units	UV stabilised homopolymer	UV stabilised copolymer	Fire retardant polypropylene
Mechanical	Tensile yield strength		D638	MN/m²	32	24	25
	Flexural strength		D790	MN/m²	—	—	—
	Elongation		—	%	150	300	15
	Tensile Elastic Modulus		D638	GN/m²	—	—	—
	Flexural Modulus		D790	GN/m²	1.5	1.2	1.8
	Impact strength Izod		D638 (IZOD)	J/cm of notch	0.4	0.8	0.5
	Hardness		D789	Rockwell	D72	R80	R97
Physical	Specific gravity		D792	—	0.908	0.91	1.2
	Thermal conductivity		—	W/m per K	—	—	—
	Thermal expansion		D696	10^{-5} K^{-1}	10	10	9
Electrical	Volume resistivity		D257	ohm-m	10^{14}	10^{15}	10^{12}
	Dielectric strength		D149	10^2 kV/m	280	280	230
	Dielectric constant	60Hz	D150	—			
		10^6 Hz					
	Power factor	60 Hz	D150	—			
		10^6 Hz					
	Arc resistance		D495	s			
Thermal/Environmental	Maximum service temperature		—	°C	100	90	100
	Deflection temperature	0.45 MN/m²	D648	°C	110	100	110
		1.81 MN/m²	D648	°C	57	60	70
	Water absorption 24h		D570	%	0.02	0.03	0.01
Processing	Mould shrinkage		—	%	1.5	2	1.1
	Remarks						

Table 3.2.63—*continued*

TABLE 3.2.63 Typical properties of polypropylene grades—*continued*

40% talc-filled homopolymer	*20% $CaCO_3$ filled homopolymer*	*40% $CaCO_3$ filled homopolymer*	*Elastomer modified polypropylene*	*Structural foam*
30	26	24	23	15.8
—	—	—	—	—
8	80	60	350	40
—	—	—	—	—
3.2	2	2.8	0.75	0.8
0.3	0.5	0.4	5	0.5
R85	R85	R90	R70	R80
1.25	1.03	1.23	0.904	0.6
—	—	—	—	—
5	6	4	11	16
10^{14}	10^{13}	10^{13}	10^{13}	10^{14}
200	180	180	280	260
—	—	—	—	—
—	—	—	—	—
—	—	—	—	—
—	—	—	—	—
—	—	—	—	—
100	100	100	90	95
130	105	110	82	72
80	68	71	55	46
0.02	0.04	0.04	0.02	0.02
0.7	1.2	1.0	2	2.3

Table 3.2.63—*continued*

TABLE 3.2.64 Chemical and environmental characteristics of polypropylenes

Environment	*Effect*
Acids and alkalis	Resistant to most acids and alkalis up to 60°C or higher. Attacked by highly oxidising agents such as fuming nitric or sulphuric acids.
Organic solvents	Resistant to higher aliphatic solvents and polar substances, (ethanol, acetone, etc.) but aromatics and chlorinated solvents cause swelling with reduction in strength and increase in flexibility. No solvent exists at room temperature but above 80°C toluene, xylene and chlorinated hydrocarbons dissolve polypropylene.
Other chemicals	Attacked by liquid and gaseous halogens.
Environmental stress-cracking	High resistance under stress to solutions of detergents up to 100°C and brittle fracture is rare.
Water	Very low water absorption, 0.03% in 24 h, 0.5% over 6 months and 2% after 6 months at 50°C. Properties and product dimensions are virtually independent of humidity changes. Permeability is very low and responsible for use of polypropylene film as moisture barrier.
Foodstuffs	Non-toxic grades available for use with foodstuffs.
Weather and light stability	Polypropylene is adversely affected by the combined action of air and UV radiation. The chemical basis of such photodegradation consists of chain fission accompanied by oxidation. This results in deterioration of the mechanical and electrical properties, and of the surface appearance of the polymer. Polypropylene is available in light-stabilised grades of three types; opaque pigments which act as a screen to the radiation, or ones which inhibit the chain reaction initiated by the UV light, however the highest degree of stabilisation is obtained with Carbon Black which screens the polymer from the radiation. Care has to be exercised in the choice of grade of Carbon Black and its concentration, to ensure maximum protection while avoiding any side-effects on other polymer characteristics. In considering the possible choices of either pigments or UV stabiliser systems for sunlight protection, it must be remembered that all polypropylenes contain thermal antioxidants and interactions may occur between them. Such interactions may be either synergistic or antagonistic in terms of weathering life of the polymer.
High energy radiation	Mouldings made from polypropylene should not be used in situations where they are exposed to a total radiation dose of more than 0.1 megarad. At higher dosage levels material damage occurs at an increasing rate. Thus, for example in a radiation steriliser with 2.5 megarad gamma-rays after exposure for some months it is very probable that the sterilised parts will have suffered some embrittlement. Some grades are available which have been stabilised to withstand such irradiation.
Flammability	Polypropylene, being a high molecular weight paraffinic hydrocarbon, is combustible. In contact with flame the plastic melts and then ignites. Polypropylene is not self-extinguishing. Grades of polypropylene are available which are flame resistant, such that the material will not burn. Some slight sacrifice in other properties occurs.

TABLE 3.2.64 Chemical and environmental characteristics of polypropylenes—*continued*

Environment	*Effect*
Heat	High softening temperature offers the possibility of use in applications closed to other low-cost thermoplastics. However, the presence in the structure of the polymer of tertiary carbon atoms in the molecule, renders it particulary liable to oxidation at elevated temperatures. Antioxidants do not affect the physical properties of the material with the exception of colour and electric properties, dielectric loss behaviour varying slightly with the antioxidant system used. Generally the heat resistant grades of polypropylene are less suitable than the other grades for high frequency work. The heat ageing stability of polypropylene is adversely affected by contact with metals. The most deleterious effect on the heat ageing life of polypropylene is given by copper and brass, reducing the heat ageing life by between 40 and 10 fold. Nickel and chromium are next in order of severity giving around a two- to three fold decrease in life, while iron, lead and zinc are somewhat less severe. Tin and aluminium have barely any effect and may be considered virtually inert. Typical ageing lives are given below. Effect of metal contact on life of polypropylene at 100°C Environment — Life at 100°C (h) Air — 87 600 (10 years) Aluminium — 82 000 Tin — 82 000 Zinc — 65 000 Lead — 55 000 Iron — 48 000 Chromium — 40 000 Nickel — 30 000 Brass — 8 000 Copper — 2 500 Special stabilisation systems are incorporated in various grades to give improvement in the ageing life in contact with such metals, but ideally such contact should be avoided. If metal contact cannot be avoided then aluminium or tin should be used, although chromium plated, nickel plated or cadmium plated metal inserts have proven satisfactory. In polypropylene applications such as wire-covering where contact with copper is unavoidable the subsequent service temperatures should not exceed about 60°C.

Table 3.2.64—*continued*

TABLE 3.2.65 Major suppliers and familiar trade names of polypropylenes

Company	*Trade names*	*Fillers*
Adell	Adell	U, S, G
Allied Chem.	Plaskon	U
Alloy Polymers	Caprez DPP	Electroplatable grade
Amoco	Amoco, Torlon PP	U, T, C
Anic	Kastilene	U, A, T, C
Atochem	Lacqtene P	U, V, C
BASF	Novolen	U, V, F, T, E, C
BIP	Bexel	Sheet
Chemische Werke (Huls)	Vestolen P	U, G, T, V, F, R, C
DSM	Stamylan P	U, V, E, C
Eastman	Tenite	U, T
Exxon/Esso	Exxon	U, T
Ferro (GB) Ltd.	Ferrolene, Starpylen	T, B, G
Fiberfil/Dart	Plaslode, Profil	B, G, I
Fiberite (Vigilant)	RTP	G, T, B, R
Foster Grant	Hostalen	U
Freeman Chemicals	Latene	F, T, B, G
Gulf	Gulf	U
Hercules (Himont Ltd)	Pro-Fax	U, G, M, T, V, C
Hoechst	Hostalen	U, V, F, T, G, E
	Trespaphan	Film
ICI	Propathene	U, G, T, V, F, B, F, R, C
LNP	Thermocomp	G, PTFE
Mitsui	Noblen, Polypro	U, G, C
Montedison	Moplen	—
Phillips	Marlex	U, C
Plastichem	Ballotini	S
Polymon	Formid	T

KEY: All grades homopolymer except where C = copolymer indicated available.
U = unfilled; G = glass fibre-filled; S = glass sphere; T = talc; A = asbestos; B = calcium carbonate; L = limestone; M = mineral; I = mica. V = UV stabilised; F = fire retardant; E = elastomer modified. R = structural foam.

Table 3.2.65

TABLE 3.2.65 Major suppliers and familiar trade names of polypropylenes—*continued*

Company	*Trade names*	*Fillers*
Rexene	Rexene	U, C
Rhone Poulenc	Napryl	U, C
SIR	Sirtene P	—
Shell	Shell	U, V, C
Shulman	Polyflam	U, M, C
Solvay/Laporte	Eltex P	U, V, C
Statoil	Statoil	U, V, C
Sumitomo	Noblen	U
TBA	Arpylene	A, T, L, G
Thermofil	—	G, T, B
Tokuyama Soda	Polypro	U, G
Wilson Fibrefil International SA	Profil	T, B, G, R

KEY: All grades homopolymer except where C = copolymer indicated available.
U = unfilled; G = glass fibre-filled; S = glass sphere; T = talc; A = asbestos; B = calcium carbonate; L = limestone; M = mineral; I = mica. V = UV stabilised; F = fire retardant; E = elastomer modified. R = structural foam.

Table 3.2.65—*continued*

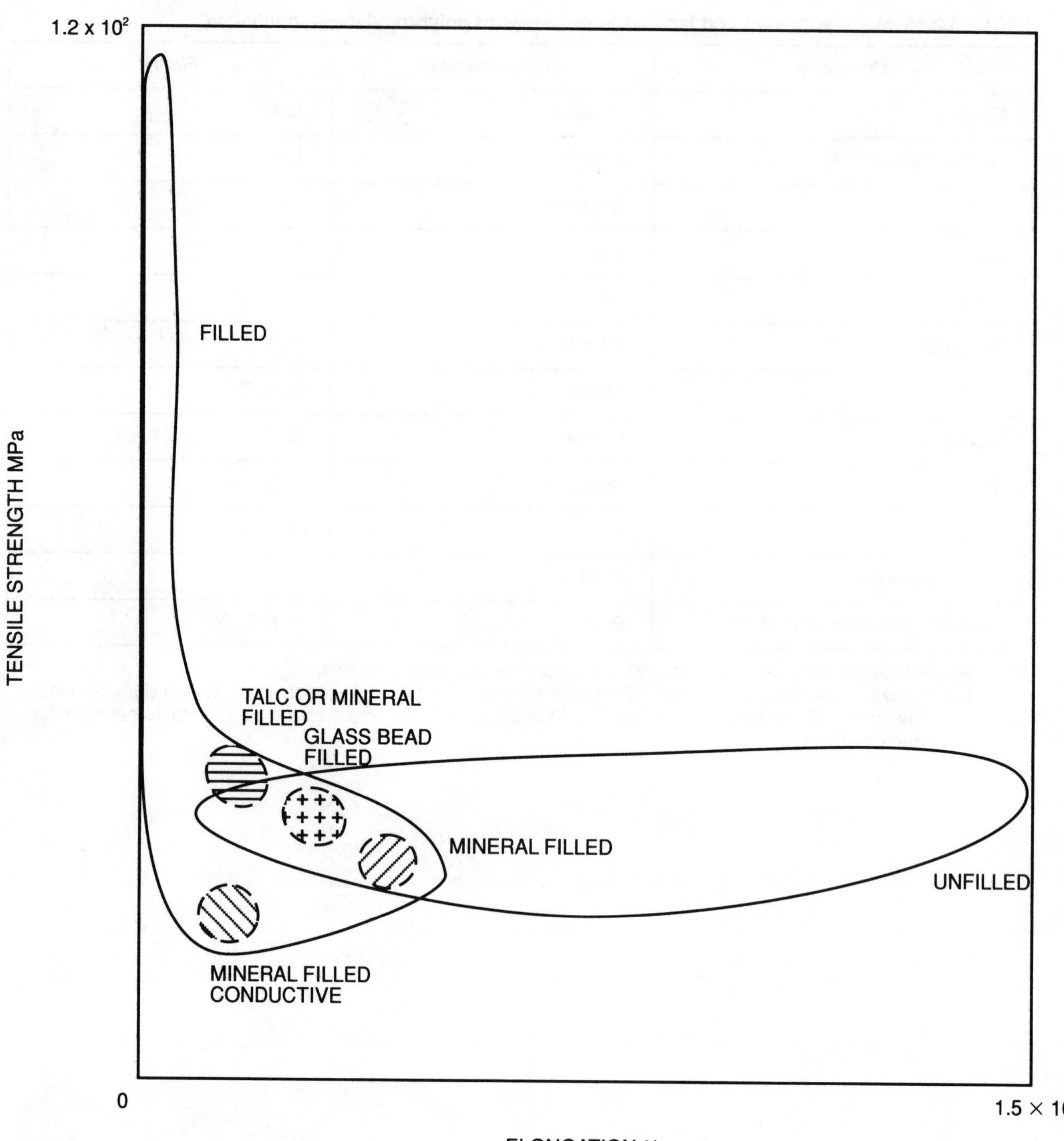

FIG 3.2.37 Tensile strength vs elongation for polypropylenes

Fig 3.2.37

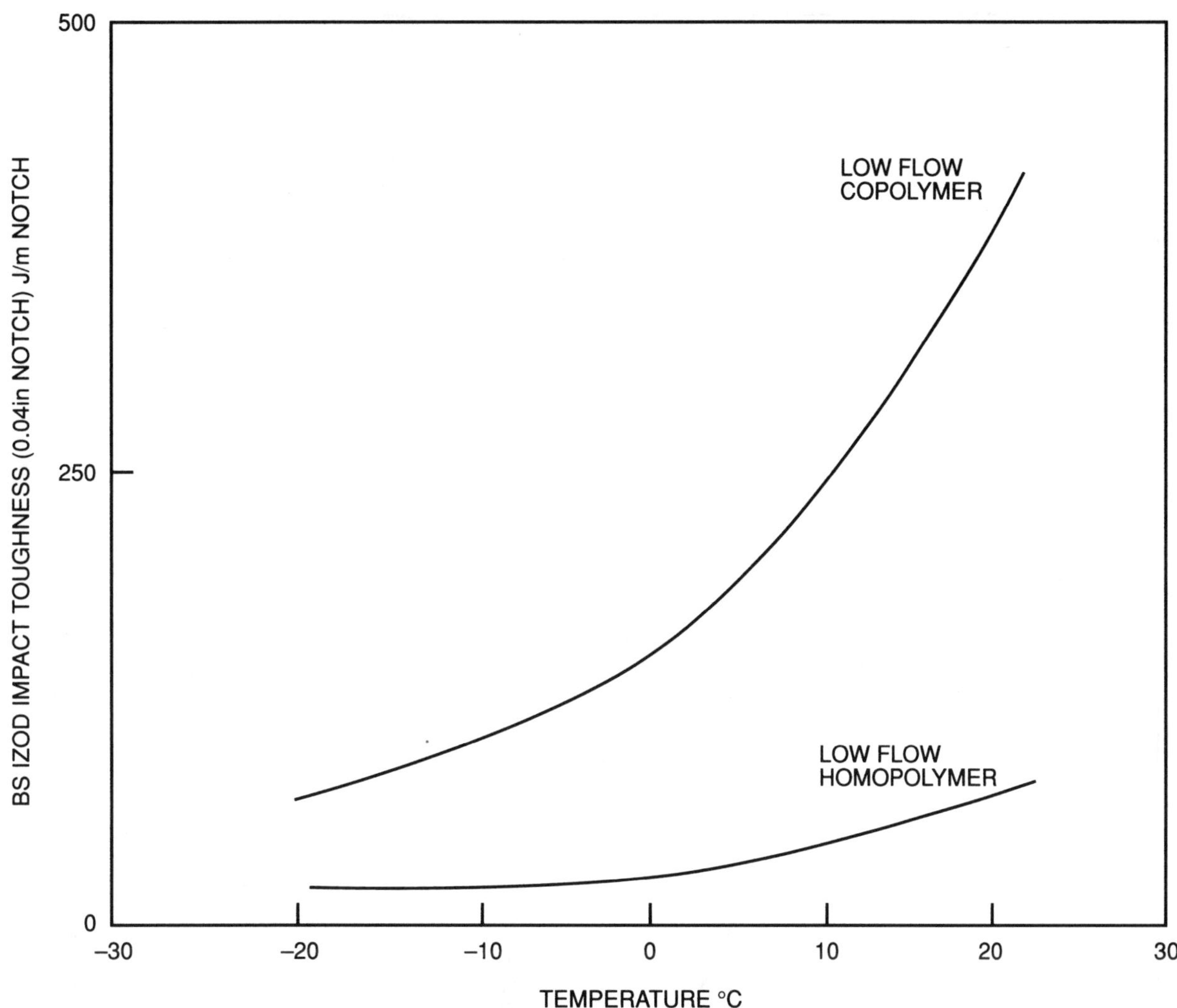

FIG 3.2.38 Influence of temperature on Izod impact toughness of polypropylenes

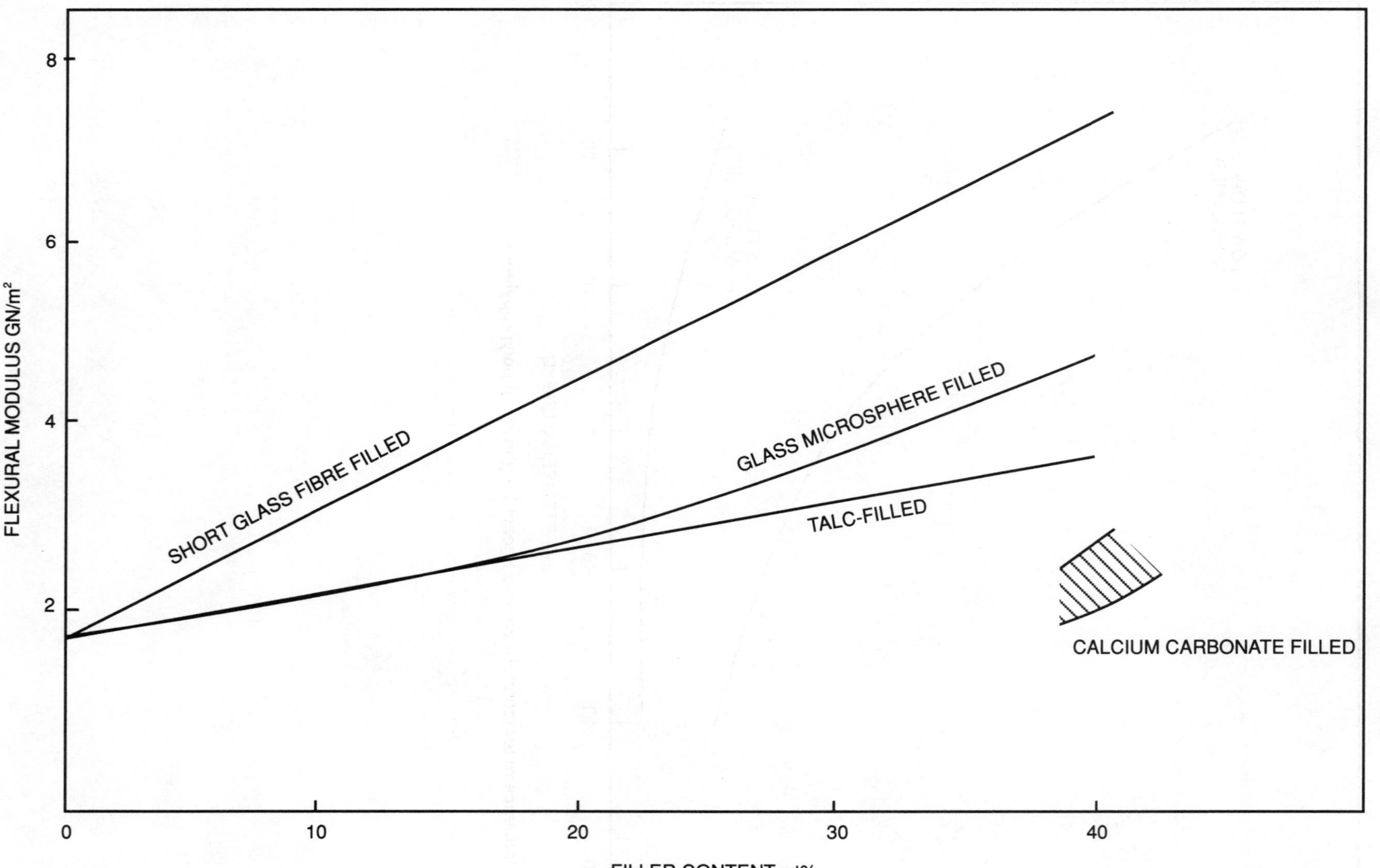

FIG 3.2.39 Variation of flexural modulus with filler type and loading of polypropylene homopolymer

Fig 3.2.39

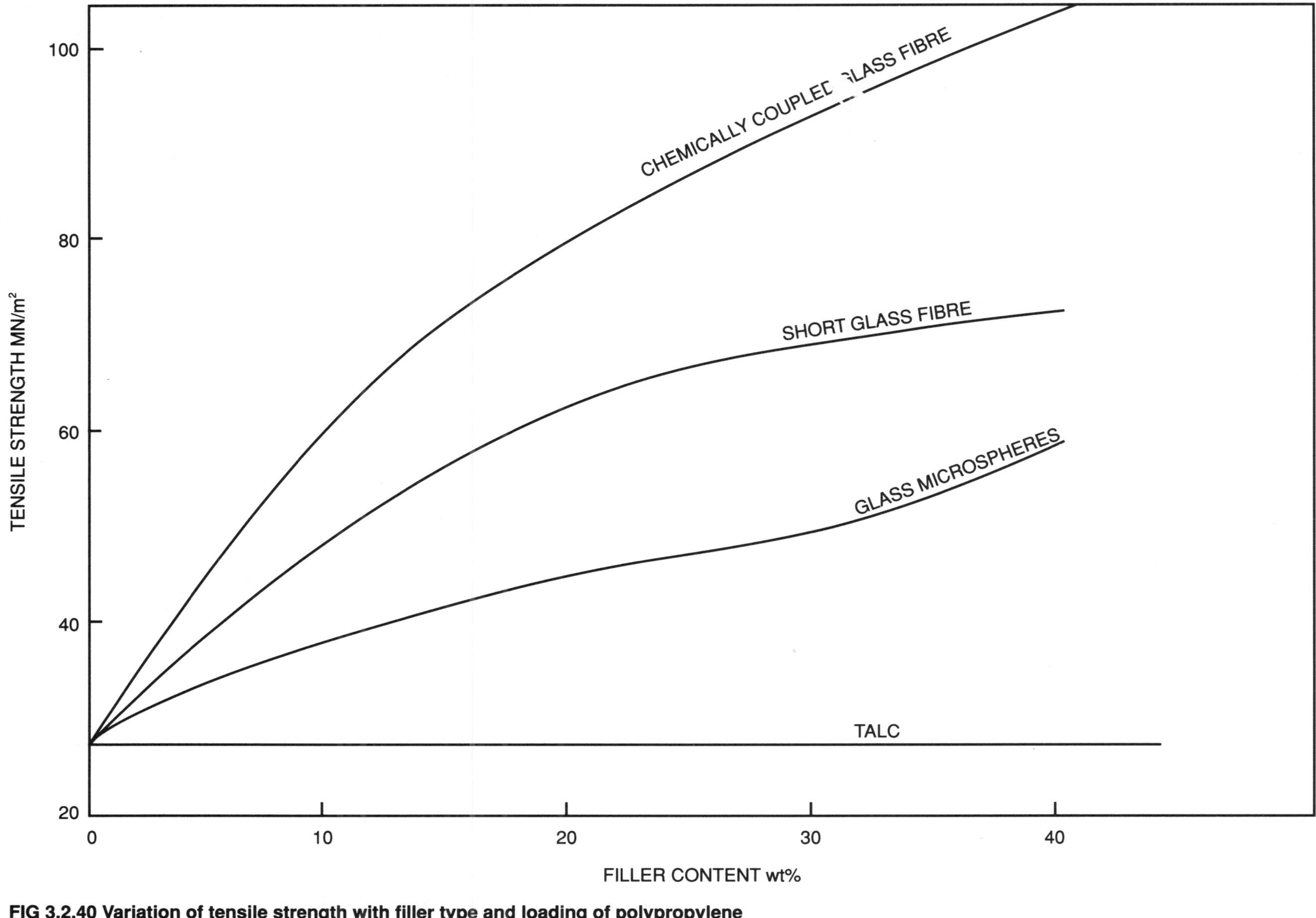

FIG 3.2.40 Variation of tensile strength with filler type and loading of polypropylene

Fig 3.2.40

3.2.11 Polybutylene

3.2.11.1 General characteristics and properties

The advantages and limitations of polybutylene resins are listed in:
Table 3.2.66 *Comparative characteristics of polybutylene*

AVAILABLE TYPES AND APPLICATIONS

Polybutylene resins are available in both the homo- and copolymer form in a number of grades to fit a variety of applications.

Homopolymer

Better low temperature toughness than copolymer. Available as general purpose grades.

Copolymer

Available for blending and general extrusion. Greater flexibility than homopolymer, impaired low temperature brittleness, better creep resistance. General purpose grades. Most are approved for food packaging.

Film

Weatherability and UV inhibited grades for outdoor applications. Industrial and food packaging including 'boil in bag'.

Pipes

General purpose and high-temperature grades. Stress-cracking, chemical and abrasion resistance combined with coilability and heat resistance lead to application in industrial piping. Suitable to 1000 psi at 20°C and 500 psi at 85°C. Pipes are competitive with copper, coated steel and other plastics on cost/performance basis. Pipes can be joined by heat fusion, compression and cold forming.

Other applications

High-wear areas (chute linings, slurry pipelines, guard rails), electrical insulation, pressure vessels, protective coatings and foams.

Elastic and high-tenacity fibres.

TYPICAL PROPERTIES

The typical properties of a number of grades of polybutylene (marketed by Shell) are listed in :
Table 3.2.67 *Typical properties of polybutylene*

3.2.11.2 Environmental properties

Polybutylenes have good resistance to most domestic and industrial environments, with the exception of hydrocarbons, as follows:

Water	—	very good, 24-h absorption 0.03%.
Dilute mineral acids	—	excellent.
Concentrated mineral acids	—	excellent.
Oxidising acids	—	fair.

Alkalis	—	excellent.
Alcohols	—	excellent.
Ketones	—	excellent.
Aliphatic hydrocarbons	—	good.
Aromatic hydrocarbons	—	fair.
Halogenated hydrocarbons	—	fair.
Detergents	—	excellent.
Fats and oils	—	good.
Stress cracking	—	very good.

Weatherability depends on colour:

Natural	—	poor.
Pigmented	—	fair.
Carbon black filled	—	excellent.

3.2.11.3 Processing

Polybutylene can be injection moulded and extruded.

Extrusion temperatures,	190 – 290°C.
Injection temperature,	240 – 290°C
Mould temperature,	40 – 80°C
Mould shrinkage,	1.5 – 3.0%

Immediately after processing polybutylene is soft and flexible and care is required handling parts to avoid deformation. Changes in characteristics including hardness, density, rigidity and strength take place during 2–5 days (normalizing or crystallisation period).

3.2.11.4 Trade names and suppliers

Suppliers of polybutylene and familiar trade names are listed in:
Table 3.2.68 *Trade names and suppliers of polybutylene*

TABLE 3.2.66 Comparative characteristics of polybutylene

	Advantages	*Limitations*
Mechanical properties	Unusual combination of flexibility, toughness and abrasion resistance. Film grades have greater impact and puncture resistance than LDPE and rubber-modified HDPE.	
	Resistant to 110°C. Good creep resistance.	Films not recommended for use below –35°C.
Environmental properties	Good barrier properties. Good resistance to environmental and mechanical stress-cracking.	
	Resistant to soaps, detergents, acids and bases up to 90°C. Resistant to aliphatic solvents at room temperatures.	Partially soluble in aromatic and chlorinated hydrocarbons above 60°C. Absorbs some hydrocarbons at room temperature.
Electrical properties	Good electrical insulating characteristics.	
Optical properties		Optical properties of film intermediate between LDPE and HDPE.
Processing	High filler loadings (up to 85%) accepted. Processed on same equipment as LDPE. Film grades available and FDA approved. Films can be hot-filled or used for boil-in-bag applications.	Poor machinability.

TABLE 3.2.67 Typical properties of polybutylene

	Property	Unit	ASTM test method	General purpose grades		
				0100	0200	0300
Mechanical	Tensile yield strength	MN/m^2	D–638	13	14	14
	Ultimate tensile strength	MN/m^2	D–638	34	31	31
	Tensile elongation	%	D–638	350	350	350
	Tensile Modulus	GN/m^2	D–638	0.20	0.24	0.24
	Modulus of Elasticity	GN/m^2	D–638	—	—	—
	Izod impact strength	—	D–256	No break	No break	No break
	Hardness, shore	D scale	D–2240	50	55	55
	Design stress C rating	MN/m^2, 20 °C MN/m^2, 82 °C	D–2837	—	—	—
Physical	Specific gravity	—	D–1505	0.915	0.915	0.915
	Coefficient of thermal expansion[a]	10^{-5} K^{-1}	D–696	—	—	—
	Thermal conductivity[a]	W/m per K	C–177	—	—	—
Thermal and environmental	Environmental stress-crack resistance	Hours	D–1693	No failure[b]	No failure[b]	No failure[b]
	Brittleness temperature[a]	°C	D–746	–17	–17	–17
	Melting point range[a]	°C	DTA	124–126	124–126	124–126
	Vicat softening point[a]	°C	D–1525	—	—	—
	Water absorption	% day^{-1}	D–570	—	—	—
Miscellaneous	Melt index	g/10 min	D–1238	0.4	1.8	4.0
	Colour	—	—	Natural	Natural	Natural
	End use	—	—	Blending	Blending	Blending

[a]Compression moulded specimen conditioned 10 days, 73 °F, 50% RH.
[b]15 000 h @ 50 °C, 10% Igepal CO630.

Data supplied by Shell Chemical Company.

TABLE 3.2.67 Typical properties of polybutylene—*continued*

General purpose grades			*Pipe grades*				
0400	*0700*	*8640*	*4101*	*4103*	*4121*	*4127*	*4128*
14	14	12	16	16	16	16	16
29	29	31	33	33	33	33	33
350	350	350	280	280	280	280	280
0.24	0.24	0.19	—	—	—	—	—
—	—	—	0.26	0.26	0.26	0.26	0.26
No break	No break	No break	—	—	—	—	—
55	55	50	60	60	60	60	60
—	—	—	7	7	7 3.5	7 3.5	7 3.5
0.915	0.915	0.908	0.925	0.930	0.925	0.930	0.930
14	—	—	113	13	13	13	13
—	—	—	0.21	0.21	0.21	0.21	0.21
No failure[b]	No failure[b]	No failure[b]	—	—	—	—	—
−17	−17	−17	—	—	—	—	—
124–126	124–126	124–126	124–126	124–126	124–126	124–126	124–126
—	—	—	113	113	113	113	113
0.015	—	0.03	—	—	—	—	—
20.0	10.0	1.0	0.4	0.4	0.4	0.4	0.4
Natural	Natural	Natural	Black	Blue	Black	Grey	Beige
Blending	Blending	Blending	Cold water pipe	Cold water pipe	Hot water pipe	Hot water pipe	Hot water pipe

Table 3.2.67—*continued*

TABLE 3.2.68 Trade names and suppliers of polybutylene

Supplier	*Trade names*	*Notes*
BASF	Oppanol B	Polyisobutylene
BP	Hyvis	
Chemische Werke	Vestalon BT	
Shell	—	Referred to as 'Polybutylene'
Witco	Witron	

3.2.12 Polyolefin copolymers (EVA, ionomer, polyallomers, EVA resins, EEA)

3.2.12.1 General characteristics and properties

The advantages and limitations of polyolefin polymers are given in:
Table 3.2.69 *Characteristics of polyolefin copolymers*

AVAILABLE TYPES

So-called polyolefin copolymers are all copolymers of ethylene and some other polymer. They include ethylene–vinyl acetate (EVA) which contains up to 40% vinyl acetate and ethylene–ethyl acrylate (EEA).

Also included under this heading are polyallomers. These are propylene ethylene copolymers which, because of their block polymer structure, differ essentially from ethylene and propylene blends. Copolymers formulated by different techniques and ionomers which, because of ionic linkages, have very special properties, are also included in this section.

It is difficult to generalise on the characteristics of polyolefin copolymers because of their very wide range of compositions and properties. They are all, however, very flexible, and easy to process, but have very low moduli and low service temperatures.

Ethylene—vinyl acetates (EVA)

Copolymers of ethylene with a minor proportion (up to 18%) of vinyl acetate. The properties can vary between materials similar to LDPE to non cross-linked elastomers depending on the ethylene/vinyl acetate ratio. The crystalline melting points, and hence the softening points and service temperatures of EVA copolymers, are largely related to the vinyl acetate content, and are all lower than those of LDPE. Hence there is a limit imposed on the upper service temperature, varying from 65°C for low VA grades, to about 45°C for the highest VA grades normally recommended for extrusion and moulding applications.

EVA copolymers are very useful as blend additives as they may confer increased flexibility, low temperature toughness, chemical inertness, etc. When blended with LDPE environmental stress crack resistance may be significantly improved. A blend of a suitable EVA grade with polypropylene or HDPE will have greater flexibility and a lower softening point. The low crystallinity of EVA copolymers enables them to accept high loadings of inert fillers without embrittlement. A range of semi-conductive thermoplastics can therefore be produced by loading with finely divided Carbon Black.

As the VA content increases the product becomes progressively more translucent.

Ethylene–vinyl acetate resins

Compositions of ethylene–vinyl acetate resins having 18–40% by weight of vinyl acetate are less crystalline, more rubbery materials than EVA. Whilst they are used for moulding in place of rubbers and thermoplastics, their main uses are in blends with fillers, waxes, resins and polymers.

Ethylene–ethyl acrylate (EEA)

When compared to EVA, EEA is more thermally stable, has better low and high temperature properties and faster moulding cycle times.

Polyallomers

Highly crystalline polymers prepared from two monomers — propylene and ethylene, which when polymerised have the following structure:

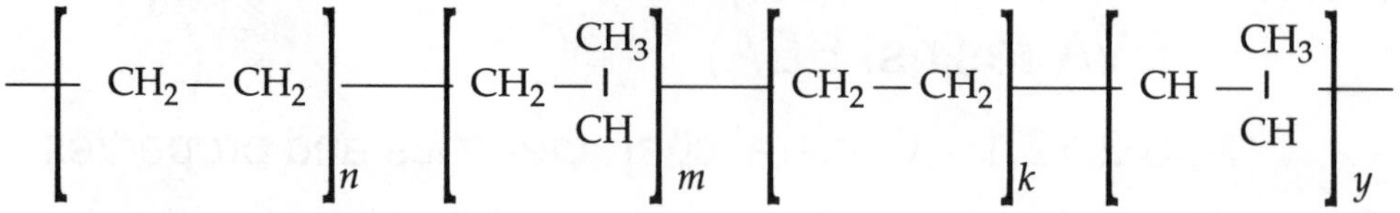

$$-\left[CH_2-CH_2\right]_n-\left[CH_2-\underset{CH}{\overset{CH_3}{|}}\right]_m-\left[CH_2-CH_2\right]_k-\left[CH-\underset{CH}{\overset{CH_3}{|}}\right]_y-$$

This degree of crystallinity is normally only found in polymers prepared from one monomer.

Various polyallomer formulations are available with combinations of high stiffness and medium impact strength and vice versa, also heat and light stabilised grades and others suitable for use with foodstuffs.

Polyallomers crystallise slowly and the process continues after ejection from the melt, thus rapid cooling is not necessary to ensure toughness as is essential with polypropylene and high density polyethylene.

Ionomers

Copolymers of ethylene with a minor proportion of acid monomer such as methacrylic acid, with metal ions (such as Na, K, Mg or Zn) forming ionic cross-links between polymer chains. Ionomers have the stiffness in the cold of a cross-linked material but the cross-links are heat fugitive so the polymer processes as a thermoplastic.

The ionic interchain cross-links give ionomers increased strength, stiffness and toughness without destroying melt processability. In the melt, residual ionic linkages impart outstanding melt strength, and the drawing characteristics of ionomers are superior to other thermoplastics. Thus performance in vacuum forming and skin packaging is outstanding and extrusion coating is assisted. Ionic polymers exhibit a sharply reduced permeability to oily materials in comparison with conventional polyethylene.

Ionomers will accept high filler loadings, yet still retain a high proportion of their physical properties. At the loading levels normally used for pigmentation (i.e. 1–2%) there is very little effect on physical properties. Even a composition containing 60% by weight of an inert filler will retain over 70% of its tensile strength.

APPLICATIONS

Ethylene–vinyl acetate (EVA)

EVA copolymers' inherent flexibility and fast processing characteristics make them alternatives to both natural and synthetic rubber in some applications. Similarly EVA copolymers can be used instead of plasticised PVC particulary where plasticiser migration would be a problem. The low density and the short production cycle times required for EVA must be taken into account when comparing costs with those of rubber and plasticised PVC. (The flexibility is retained at low temperatures, down to –70°C.)

Existing applications for EVA copolymers containing low vinyl acetate contents (less than 15%) include:

Record turntable mats.
Ice-cube trays.
Film for fresh meat wrap.
Disposable gloves.
Beer tubing.
Spirit bottling lines.
Corrugated hose — swimming pool cleaner hose.
— vacuum cleaner hose.
— anaesthetic hose.

Protective strip — car door buffer.
— draught excluder.
Gaskets, lightly loaded and where no special high temperatures occur, e.g. freezer doors.
Disposable teats for baby bottles.
Base pads for staplers, portable radios and other light electrical equipment.
WC pan connectors.
Ski-stick baskets.
Closures and closure wads.
Car door bumpers.
Bumper surrounds for domestic vacuum cleaners.
Cable sheathing.

Ethylene–vinyl acetate resins

Uses of EVA copolymers containing high VA contents (18–40%) are mainly in blends with tack resins and waxes for products in the hot melt class of adhesives. Such hot melt adhesives have the advantage of the absence of solvent which allows them to act almost instantaneously.

Ethylene–Ethyl Acrylate

EEA copolymers have been used successfully for tubing, gaskets, toys and furniture parts. When blended with high density polyethylene, a material with high impact resistance is produced. This blend has been used for drums and containers, dustbins, vanes for snow blowers and car bumpers. EEA film products have good tear strength, puncture resistance and weathering resistance and can be used for agricultural applications, flexible gloves, hospital pads and heavy duty packaging.

Polyallomers

Polyallomers are used in moulded parts requiring toughness and good hinge properties. Examples are containers with integrally moulded hinges and threaded holes; frozen food and heat sterilisable food containers; shoe lasts. Embossed sheet is used for luggage shells.

Ionomers

Ionomers find applications as film, sheet, and tube and for wire and cable coating, injection moulded articles and shape extruded articles. Typical applications include:

Film and thin sheeting for:

Skin packaging.
Vacuum packaging.
Heat seal packaging.
Wrappings for fatty foods.
Vacuum formed articles requiring a deep draw.

Extrusion coatings of:

Paper and paperboard.
Cotton and wood.
Metals (e.g. aluminium foil).
Ceramics.
Wire and cable.

Bottles for:

- Vegetable oils.
- Cosmetics.
- Liquid shortenings.
- Table syrups.

Tubing for:

- Liquid foods
- Beer
- Aerosols
- Hypodermic syringes

Automotive and electrical appliances
Closures
Housewares
Safety equipment
Sports goods
Toys
Tools and Tool handles
Hammer heads
Appliance housings and feet
Furniture and cabinet trim
Shoe heel pieces.

TYPICAL PROPERTIES

The more important properties of polyolefins are listed in:
Table 3.2.70 *Typical properties of polyolefin copolymers*

3.2.12.2 Chemical and environmental resistance

The influence of chemical and other environments on EVA are listed in:
Table 3.2.71 *Chemical and environmental resistance of EVA*

The resistance of polyallomers is listed in:
Table 3.2.72 *Chemical and environmental resistance of polyallomers*

3.2.12.3 Processing

ETHYLENE—VINYL ACETATE (EVA)

The range of relatively low vinyl acetate content EVA copolymers is primarily intended for processing by conventional thermoplastic processing techniques such as extrusion and injection moulding.

Moulds for EVA copolymers should be made from high-quality chrome–nickel steel and should be hardened. For mouldings with a high gloss finish, highly polished chrome-plated mould surfaces are required, whereas vapour-honed or sand-blasted surfaces give a matt finish but with the advantage of easier mould release.

In general, EVA copolymers will flow for a longer path length than a low-density polyethylene of the same melt flow index, thus allowing larger and thinner sections in the component. As EVA copolymers have a rubbery nature it is possible to mould severe undercuts in EVA, though to prevent damage in removal undercuts should have rounded edges. On thin sections knock-out pins, if used, should be of generous cross-section if punch-through is to be avoided. Ejector plates, rings etc. are preferred to pins whenever practicable.

A taper on mouldings is desirable to aid ejection, a taper of 2° usually being sufficient although high VA content copolymer may need greater tapers.

The mould shrinkage is dependent on the type of gate used and its position, the characteristics of the polymer and the processing conditions used. However, for mould design purposes a shrinkage allowance of 0.012cm/cm may be taken as a guide.

EVA copolymers may be extruded in the form of simple profiles, complex sections being difficult to produce due to the limp and sticky nature of the product. To maintain constant dimensions in profiles other than tube, it is necessary to use a post-forming device.

Wide sheet may be extruded in EVA copolymers by means of conventional sheet extrusion equipment. Lower extrusion temperatures are normally required than for LDPE. As EVA copolymers have relatively low strength in the semi-molten state care must be taken to avoid excessive tension during haul off.

EVA copolymers are well suited to blow moulding techniques offering excellent bottle bottom welds, excellent flexibility, excellent environmental stress crack resistance and good transparency. As transparency is a function of the cooling rate it is much improved when moulds are made of beryllium/copper alloys which have a high thermal conductivity.

POLYALLOMERS

Polyallomers are well suited to a wide variety of injection moulding, extrusion and thermoforming processing. These materials possess excellent flow characteristics, good mould finish, good plastic memory and low mould shrinkage making them useful materials for injection moulding. Crystallisation of polyallomer from the melt occurs more slowly than with highly crystalline homopolymer, most of the crystallisation occurring after solidification and consequently the processing conditions necessary to produce tough products are less critical for polyallomer than for most formulations of polypropylene and high-density polyethylene. The slow crystallisation of polyallomers also causes solidification of the melt to be slightly delayed thereby permitting polyallomer to flow into long, thin mould sections more readily than polypropylene.

After moulding, polyallomers normally take between 24 and 48h to complete crystallisation, and to develop maximum physical properties. This is most noticeable in stiffness. Moulding shrinkage is usually complete after this period and is almost equal in different directions. This moulding shrinkage is generally of less magnitude than with polyethylene.

Polyallomers' excellent mouldability permits small parts with a wall thickness of 0.010in (0.25mm) to be moulded, as their sharp decrease in viscosity at their melting point allows them to flow into mould cavities more readily than do most other thermoplastics. This allows them to be used when the mould design is difficult to fill and when thin sections are necessary.

IONOMERS

Ionomers may be readily processed on all conventional thermoplastic equipment.

For extrusion applications ionomers may be processed using conventional equipment, similar to that used for polyethylene; however, due to their higher melt viscosity, ionomers require 10—40% more extruder horsepower than polyethylene. Grades are available for extrusion as sheet, film, tube and for extrusion coating onto wire, and onto various substrates and laminates.

Extruded sheet may be thermoformed on normal vacuum forming equipment. Draw ratios of up to 6 : 1 are possible with accurate control of temperature and forming rate.

POST-MOULD PROCESSING

Separate parts made from ionomer may be joined by heat sealing, spin welding or by epoxy or rubber based adhesives after suitable pretreatment. Improved adhesion is obtained if the surface is oxidised by flaming or chemical treatment prior to applying the adhesive.

High frequency welding and ultrasonic welding are not suitable.

3.2.12.4 Trade names and suppliers

Major suppliers and trade names of polyolefin copolymers are given in:
Table 3.2.73 *Trade names and suppliers of polyolefin copolymers*

TABLE 3.2.69 Characteristics of polyolefin copolymers

Material	*Advantages*	*Limitations*	*Relative cost/unit wt*
Ethylene–vinyl acetate EVA	Elastomeric-like qualities of softness and flexibility (without the use of plasticisers) with processability of thermoplastics (injection, blow, compression, transfer, rotational moulding and extrusion). Alternative materials to elastomers or flexible PVC where plasticisers would give problems. Greater resilience than flexible PVC. Processability, toughness and flexibility superior to LDPE (flexibility retained to −70°). Good optical clarity and gloss possible in finished parts. Good chemical resistance as compared with inexpensive flexible PVC. Can be used with foodstuffs and sterilised by low-level atomic radiation or chemical sterilisants. Better ozone resistance than rubbers. Can accept high proportion of inert fillers without embrittlement. High friction coefficient (non-skid).	Low resistance to heat (45–65°C) (hence cannot be steam sterilised) and solvents—attacked by chlorinated hydrocarbons, benzene and its derivatives. Electrical properties inferior to LD polyethylene but competitive with rubber and vinyl components. Dielectric properties frequency dependent. Degraded by ultraviolet light unless protected by 2–3% carbon black filler. High tack at moderate to high temperatures. Inferior elastomeric behaviour to true rubbers. Abrasion resistance inferior to plasticised PVC or elastomers. Surface mars easily.	140–200
Ethylene–ethyl acrylate (EEA)	More thermally stable, better low and high temperature properties, ability to accept higher filler loadings and a higher crystallisation temperature (faster moulding times), compared to *EVA*.		
Polyallomers	Combination of many of the best properties of both HDPE and crystalline polypropylene. Better fatigue properties (hingeability) than polypropylene. Better processability than polyethylene. Remains ductile down to −40°C. Impact strength up to 6.4 J/cm of notch. Lightest of solid polyolefins (SG 0.91). Superior to rubber modified polypropylene in colour, clarity, mouldability and electrical properties. Heat and UV light stabilised grades available. Resistant to strong alkalis and weak acids.	Slow burning. Attacked by strong oxidising acids. Swollen by ethylene dichloride and xylene. Bleached, swollen and warped by carbon disulphide, carbon tetrachloride, chloroform, benzene and its derivatives.	100–150
Ionomers	Highly transparent (80–90% in visible region) grades available. Cross-linked structure gives good toughness but characteristic thermoplastic processing possibilities (injection and blow moulding, extrusion) are retained. Flexible material without the use of plasticisers. Resistant to oils and greases and may be used in contact with foodstuffs. Better environment stress-cracking resistance than polyethylene. Flexible at low temperatures (at least down to −70 to −105°C). Drawing characteristics (superior to other thermoplastics) responsible for film and extrusion coating applications. Better resistance to corona discharge than polyethylene. Abrasion resistance comparable with nylons.	Low stiffness and creep resistance. Maximum service temperature 71°C. Flammability rating (ASTM D635) 23–28 mm/min. Attacked by acids. Poor resistance to UV light, unless stabilisers are used, which destroy transparency.	200–290

Table 3.2.69

TABLE 3.2.70 Typical properties of polyolefin copolymers

	Property		*ASTM test/ BS test*	*Units*	*Ethylene–vinyl acetate (EVA)*	
					Vinyl acetate (%)	
					7.5	*18*
Mechanical	Tensile strength (break)		D638	MN/m^2	16.2	12.3–17.2
	Elongation		D638	%	700	650–750
	100 s Modulus (0.2% strain)		—	MN/m^2	156	45–47
	Flexural Modulus		D790	MN/m^2	—	—
	Impact strength Izod		D256	J/cm notch	No break	
	Hardness		—	Shore/ Rockwell	D46	D36–38
	Compression set		BS903	%	—	37
	Rebound resilience		BS903	—	—	54
	Softness		BS2782	—	6	15–16
Physical	Specific gravity		D792	—	0.926	0.937
Electrical	Volume resistivity		D257	ohm–m	10^{13}	10^{13}
	Dielectric strength		D149	10^2 kV/m	240–310	
	Dielectric constant	60 Hz	—	—	2.5–3.2	
	Power factor	60 Hz	—	—	0.003–0.020	
Optical	Refractive index		D542	—	—	—
	Clarity			—	Transparent to opaque	
Thermal and environmental	Maximum service temperature		—	°C	55–65	
	Heat deflection temperature	1.8 MN/m^2	D648	°C	33	
		0.45 MN/m^2	D648	°C	62	
	Cold bond temperature		BS2782	°C	—	<–70
	Cold flex temperature		BS2782	°C	—	–33 to –39
	Charpy embrittlement temperature		BS2782	°C	–20 to –30	–30 to –35
	Water absorption (24 h)		D570	%	0.05–0.13	
	Flammability		D635	mm/min	very slow	
	Stress-crack resistance		D1693	F50-h	>24	>300
Production	Mould shrinkage		—	%	0.7–1.1	
	Melt flow index		D1238	dg/min	2	2–33

[a]Brittleness temperature (D746–57T). [b]At yield.

Table 3.2.70

TABLE 3.2.70 Typical properties of polyolefin copolymers—*continued*

EVA resins 28% vinyl acetate		*Ethylene–ethyl acrylate (EEA)*		*Polyallomer*	*Ionomer (moulding grade)*	*LD polyethylene (for comparison purposes)*
Melt flow index		*Ethyl acrylate (%)*				
Low	*High*	*15*	*13*			
11.0	1.6	20[b]	18[b]	21–28	28	12.8
800	950	>600	>600	400–500	450	600
24	—	45	36	—	—	240
—	—	—	—	500–800	180	—
—	—	D32	D27	2.7	4.5–7.0	No break
D25 (A79)	A67	—	—	R50–85	D56–68	D50
39	—	—	—	—	—	—
65	—	—	—	—	—	—
29	—	—	—	—	—	4.5
0.95	0.95	0.93		0.896–0.899	0.93–0.96	0.918
—	—	—		10^{13}	mainly $>10^{14}$	See Section 3.2.9
—	—	180–280		345	255–433	
—	—	2.7–2.9		2.3	2.4–2.5	
—	—	0.003–0.02		0.0005	0.001–0.005	
—	—	—		1.492	1.51	—
—	—	Translucent		Transparent/ translucent	Transparent	—
—	—	85–90		110	70	85–100
—	—	—		49–51	—	32–40
—	—	—		93	41	38–50
<–70 –40 –30 to –35	— — —	— — —		— — —	— — –107[a]	— — –10 to –20
—	—	0.04		<0.01	0.1–0.4	<0.01–0.03
—	—	—		Slow	23–28	—
>300	—	—		—	—	1.2
—	—	1.0–3.5		1.0–2.0	0.7–1.9	1.5–5.0
5	400	1.3	6	2	2–10.0	2

Table 3.2.70—*continued*

TABLE 3.2.71 Chemical and environmental resistance of EVA

Environment	***Effect***
Acids and alkalis	Resistant to weak acids, attacked by strong acids. Resistant to weak and strong alkalis.
Solvents	Resistant to alcohols, glycols and most oils and greases. Attacked by chlorinated and aromatic hydrocarbons and ketones.
Environmental stress-cracking	Excellent resistance compared with LDPE.
Foodstuffs	Inert—can be sterilised by radiation (dose must be restricted to 2.5 megarads or discolouration and chemical breakdown will occur) disinfectants or ethylene oxide but cannot be steam sterilised.
Water absorption	ASTM D570, 0.07–0.13% in 24 h.
UV light	Degradation occurs unless 2–3% Carbon Black is added. Useful life unprotected varies between 1 and 5 years depending on climate.

TABLE 3.2.72 Chemical and environmental resistance of polyallomers

Environment	*Period*	*Test*	*Effect*
Water absorption 24 h, ⅛" thick		D570	0.01%
Burning rate		D635	Slow
Sunlight			Crazes; can be protected
Weak acids		D543	None
Strong acids		D543	Attacked slowly by oxidising acids
Weak alkali		D543	None
Strong alkali		D543	Very resistant
Organic solvents		D543	Resistant below 80 °C
10% Hydrochloric acid 30% Sulphuric acid Oleic acid 1% Sodium hydroxide Soap solution Olive oil Wessen oil Ethylene glycol Glycerine Formaldehyde Sodium hypochlorite	1 year at 23 °C		None
5% Acetic acid 10% Ammonium hydroxide 10% Sodium hydroxide 2.5% Calcium chloride 50 and 95% Ethanol 5% Methanol 5% Phenol 10% Sodium chloride 2% Sodium ethanol carbonate Water	1 year at 23 °C		Slight yellowing
Acetone Butyl acetate Ethyl acetate 100% Methanol Nitric acid Toluene Turpentine	1 year at 23 °C		Slightly bleached and slightly swollen
Ethylene dichloride Xylene	1 year at 23 °C		Swollen
Carbon disulphide Carbon tetrachloride Chloroform Benzene and derivatives Heptane	1 year at 23 °C		Bleached, swollen and warped

TABLE 3.2.73 Trade names and suppliers of polyolefin copolymers

Supplier	*Trade name*	*Note*
EVA		
Atochem UK	Evatane	
ARCO	Dylan	
BASF	Lupolen	
Bayer	Levasint, Levapren	
Chemplex	Chemplex	
Du Pont	Alathan	
Hoechst	Hostalen	
Monsanto	Montothene	
Pantasote	Pantalast	Copolymer with PVC
Sumitomo	Evatate	
Union Carbide	Bakelite	
U.S. Ind. Chem.	Ultrathene	
EVA resin		
Du Pont	Elvax	
ICI	Evatane	
EEA		
B.P. Chemicals	DPDM	15% ethyl acrylate
Polyallomer		
Eastman	Tenite PAL	
Ionomer		
Du Pont	Surlyn A	

3.2.13 Polycarbonates

3.2.13.1 General characteristics and properties

The advantages and limitations of polycarbonates are listed in:
Table 3.2.74 *Characteristics of polycarbonates*

AVAILABLE TYPES

Commercially available polycarbonate resins are derived from a reaction between bisphenol A and phosgene. The structural formula is given below and consists of polar carbonate groups separated by aromatic hydrocarbon groups.

$$\left[\bigcirc - \underset{CH_3}{\overset{CH_3}{\underset{|}{\overset{|}{C}}}} - \bigcirc - O - \overset{O}{\overset{\|}{C}} - O \right]_n$$

The presence of benzene rings restricts the flexibility of the molecule, the repeat unit of which is quite long. In spite of the regularity of the polymer, conventionally fabricated samples do not exhibit a high degree of crystallinity and the largely amorphous structure ensures the characteristic toughness of polycarbonate components. The rigid nature of the molecule leads to high heat distortion and glass transition temperatures.

Mechanical properties are affected by molecular weight but above 30 000 (typical of commercial materials with extrusion grades having slightly higher values than injection moulding grades) the effect is small.

Additives, fillers, etc., are used to produce grades with special application properties. The grades include: heat and UV stabilised; glass-filled, for improved strength, stiffness and heat deflection resistance at the expense of impact strength; hydrolysis stabilisers for food contact applications; and grades with special flame retardance. Components may be clear or opaque in a wide range of colours. Low melt viscosity grades are available for injection moulding and are a necessity for hard-to-fill moulds, and it is usual to incorporate mould release agents.

Moulding materials are supplied in granule form and may be converted to finished components by injection moulding, extrusion or casting, Pre-formed sheets may be thermoformed, blow formed or embossed.

Recent grades, which are probably block copolymers, claim to have overcome the limitation of notch sensitivity, and processing problems associated with the incorporation of additives. Other mechanical, optical and processing characteristics appear unchanged.

APPLICATIONS

Electronics and electrical engineering

Housings for computers, table top calculating machines and magnetic disc packs
Relays and power tools
Components for electrical calculating machines
Tape take-up discs
TV sweep units and line transformers
Winding supports
Coil formers and housings
Telephone housings for mining operations
Telephone dials
Plug and socket terminal blocks

Electronics and electrical engineering—continued

Guard panels
Time switch chassis
Covers and bases for switch boxes
Housings and chassis for current meters
Bases, mounting and starter enclosures for fluorescent lamps
Switch plates
Dust hoods and lids
Protected switchgear
Special and power plugs
Fuse boxes
Insulators and insulating panels
Battery covers
Push-button controls, single and pilot lamp covers
Plugs and connectors
Components for street lamps
Doors for light standards
Terminal boxes
Shielding for electrical equipment (antistatic grade)

Photographic equipment

Housings forminiature and narrow-gauge cameras and flashes as well as:
Slide and narrow-gauge film projectors
Lens holders
Camera shutter trims
Film and slide cassettes
Film advance spools
Viewers
Apertures
Exposure meters
Camera lens grade (developmental)

Lighting and optical equipment

Lamp covers
Switch enclosures
Light screens, panels and control equipment
Lamp sockets
Binocular enclosures
Microscope components

Mechanical and process engineering, precision mechanics

Components for pneumatic controls and multi-stage liquid pumps
Filter cups
Inspection glasses
Protective hoods
Housings
Filter gates and panels
Cold water pumps
Control cams
Valves
Levers
Chassis
Push Buttons
Fan wheels
Components for sewing machines

Traffic and Automotive

Housings and coloured lenses for traffic signals
Traffic signs
Directional signs
Vehicle covers (snowmobile)
Headlight lenses
Reflectors
Tail lamps
Trafficators
Warning lamps and instrument housings
Heating grilles
Ventilation and radiator grilles
Overriders
Fuse boxes and covers
Instrument covers
Interior lamps
Housing for automobile aerial motors
Covers for ship's lamps
Bumpers

Office supplies and stationery

Components for typewriters and calculating machines
Housings for ball-point and fountain pens
Stencils
Topographic film
Cores for paper rolls
Ink ducts
Slide rule components
Rulers, triangles
Lettershoot buckets

Household and domestic ware

Tableware
Baby feeding bottles
Coffee filters
Shaver housings
Shaving blade support
Lighters
Components for kitchen appliances
Vacuum cleaner housings
Hair curlers
Water containers
Cutlery
Mixers and blenders
Coffee grinders and percolator components
Inch rules
Chocolate moulds

Safety equipment

Machine guards, safety helmets, glasses, goggles and shields
Safety glazing
Miners' battery cases (acid resistance)

Miscellaneous

Greenhouse sheeting†
Building panels†
Medical supplies
Housings for blood cleaning filters
Vending machines (sweet dispensers)
Components for fishing equipment

†Double-walled sheeting grades available (2–10mm thickness)

Bearings

Special lubricated polycarbonates find applications in severe frictional environments.

TYPICAL PROPERTIES

Typical properties of polycarbonates are listed in:
Table 3.2.75 *Typical properties of polycarbonates,*
Fig 3.2.41 *Tensile strength vs elongation for polycarbonates*, and
Fig 3.2.42 *Tensile strength vs elongation for filled polycarbonates*

3.2.13.2 Mechanical properties

The allowable working stress levels for unfilled and glass-reinforced polycarbonates taking into account creep over a series of temperatures and fatigue at room temperature are listed in:
Table 3.2.76 *Allowable working stress levels for polycarbonate resins*

Impact toughness (by the Falling Dart method) is compared with that of acrylics and cellulose acetate over a range of temperatures in:
Fig 3.2.43 *Comparison of impact strength of polycarbonate with other polymers over a range of temperatures*

3.2.13.3 Friction and wear properties

The wear factor, friction and limiting PV for polycarbonates filled with a variety of friction reducing additives are listed in:
Table 3.2.77 *Friction and wear properties of polycarbonates*

3.2.13.4 Processing

Polycarbonates can be injection moulded and extruded, but materials need to be kept dry or dried before use to avoid surface marks. Usually, with adequately polished moulds high gloss surfaces can be achieved. Colours are stable to normally-used moulding conditions. Care needs to be taken to employ such processing conditions in both injection and extrusion as will minimise internal strains, especially if contact with stress-cracking or crazing agents is a possibility in use.

Mould shrinkage is largely uniform and generally 0.7–0.8% for unreinforced grades; glass-reinforced materials demonstrate a degree of anisotropy of shrinkage within the approximate limits of 0.25 and 0.50%, being smaller in the flow directions.

When using flame retardant grades, it being essential to hold material temperature to less than 300°C, care must be exercised in respect of moulding flow path length. With standard materials slightly higher temperatures can be used to improve flow—this cannot be done with flame retardant grades.

Fabricating processes such as machining and welding present little difficulty. For machining, cooling by air or clean water is adequate. Excellent results can be achieved by polishing, but alkaline pastes should be avoided as there is a risk with these of chemical surface attack. All forms of welding, including solvent and adhesive methods are feasible. In the case of solvent welding it has to be borne in mind that removal of the last traces of solvent is difficult.

It is frequently advisable (as with many plastics) to consider annealing mouldings and extrusions before, say, solvent welding or if there is a risk of contact with active agents.

3.2.13.5 Trade names and suppliers

The major suppliers of polycarbonates and trade names are listed in:
Table 3.2.78 *Major suppliers and familiar trade names of polycarbonates*

TABLE 3.2.74 Characteristics of polycarbonates

	Advantages	*Limitations*
Mechanical properties	Very high resistance to impact damage even at subzero temperatures (remains flexible down to –150°C); 8–11 times acetal, nylon or polypropylene and three times ABS. Good creep resistance in dry conditions up to 115°C.	Low scratch resistance (scratch resistant coatings available). [a]Notch-sensitive and susceptible to crazing under strain. Glass-filled grades have lower impact strength.
Environmental properties	Resistant to weak acids, aliphatic hydrocarbons, paraffins, alcohols (except methanol) and some animal and vegetable oils and greases. UV stabilised grades show no degradation of mechanical properties after five years exposure; colours show slight change. Low moisture absorption and good dimensional stability.	Although polycarbonate can be repeatedly cleaned in hot water, unlimited long-term exposure to hot water above 60°C is not advised. Similar remarks apply to steam sterilisation. Attacked by strong oxidising acids, alkalis, ammonia, methanol, aromatic, chlorinated and perchlorinated hydrocarbons, aldehydes, ketones, ethers and esters. Unstabilised grades degrade on exposure to UV light; mechanical properties lowered and surface cracks develop; clear materials yellow, craze and lose gloss; opaque materials lose colour and surface gloss.
Thermal properties	Non-flammable, self-extinguishing and low smoke grades available. Good creep resistance and high heat deflection temperature lead to applications previously limited to opaque thermoplastics.	Upper service temperature (air) 145°C. Short-term water/moisture/steam 140°C. Long-term water 60°C.
Electrical properties	Very good dielectric properties stable over a wide range of temperature, frequency and humidity. Antistatic (low surface resistivity) grade available.	Poor resistance to tracking.
Optical properties	Grades available with transparency equivalent to glass but much tougher.	
Food and medicine	Approved grades available for use with foodstuffs. Free from taste, toxicity and odour. Can be steam sterilised with little loss of dimensions.	
Processing	Can be injection moulded, compression, transfer and blow moulded and extruded. Can be formed by punching and rolling; easily machinable. Can be solvent cemented, adhesive bonded, hot and ultrasonically welded. Good dimensional stability — can be moulded to much closer tolerances than nylon or acetals.	Special care required to avoid moisture pick-up in moulding. Selection of adhesives and paints must be undertaken with care owing to poor solvent resistance.
Miscellaneous	Wide range of colours available. Glass-filled foamed grades obtainable.	

[a] Newly developed grades are claimed to have overcome this.

TABLE 3.2.75 Typical properties of polycarbonates

	Property		ASTM Test	Units	Low-to medium melt viscosity unfilled grades	Glass-filled grades			
						10%	20%	30%	40%
Mechanical	Tensile yield strength		D638	MN/m^2	58–70	65	140	110	—
Mechanical	Tensile break strength		D638	MN/m^2	>70	50	90	—	120
Mechanical	Elongation		D638	%	80–120 (30[a])	10–20	3–4	4	2–3
Mechanical	Tensile Elastic Modulus		D638	GN/m^2	2.2–2.4	3.5	6.0	—	10.0
Mechanical	Flexural strength		D790	MN/m^2	100–105	110	145	—	190
Mechanical	Flexural Modulus		D790	GN/m^2	2.5	3.7	5.6	8	10.0
Mechanical	Impact strength (Izod)		D256	J/cm	600–950 (80[a])	200–300	100	100	100
Mechanical	Hardness		D789	Rockwell	70–75M (93M[a])	M85	M118	M92	M93
Physical	Specific gravity		D792	—	1.2	1.27	1.35	1.42	1.52
Physical	Thermal conductivity		—	W/m K	0.18–0.21	0.21	0.21	0.22	0.22
Physical	Thermal expansion		D696	$10^{-6} K^{-1}$	60–70	40	30	—	20
Electrical	Volume resistivity		D257	ohm-m	10^{15}	10^{15}	10^{13}	10^{13}	10^{15}
Electrical	Surface resistivity		—	ohm	10^{15} (10^{5}[b])	—	—	—	—
Electrical	Dielectric strength (short time)		D149	10^2 kV/m	150–160	180	220	230	200
Electrical	Dielectric constant	60Hz	D150	—	3.0–3.17	3.1	3.2	3.3	3.4
Electrical	Dielectric constant	10^6 Hz	D150	—	2.96–3.0	3.0	3.2	3.3	3.4
Electrical	Power factor	60 Hz	D150	—	0.0009 (0.0027[a])	0.0008	0.0009	0.0009	0.0013
Electrical	Power factor	10^6 Hz	D150	—	0.0010 (0.0055[a])	0.0075	0.0073	0.0070	0.0067
Electrical	Arc resistance (St. steel)		D495	s	10–11	5–10	5	—	5
Thermal/Environmental	Max. cont. service temperature	Electrical	UL	°C	100–125	125	120	—	120
Thermal/Environmental	Max. cont. service temperature	With impact	UL	°C	100–115	115	120	—	120
Thermal/Environmental	Max. cont. service temperature	Without impact	UL	°C	100–125	125	130	+15	130
Thermal/Environmental	Heat deflection temperature	0.45 MN/m^2	D648	°C	138–145 (166[a])		210	152	—
Thermal/Environmental	Heat deflection temperature	1.8 MN/m^2	D648	°C	127–140 (149[a])	140–145	209	146	145–150
Thermal/Environmental	Flammability		UL94	—	V–2 (V–0[c])	V–0	V–1	—	V–1
Thermal/Environmental	Water absorption (24h)		D570	%	0.12–0.19	0.12	0.26	0.12	—
Production	Mould shrinkage		—	%	0.4–0.7	0.1–0.5	0.2–0.5	0.2	0.1–0.3

[a] High HDT grade. [b] Antistatic grade. [c] Some FR grades.

Table 3.2.75

TABLE 3.2.75 Typical properties of polycarbonates—*continued*

UV stabilised	*Fire retardant*	*Structural foam*	*Polycarbonate PBT alloy*	*15% PTFE-filled*
55	60	40	50	50
90	85	70	5	85
2.4	2.8	2.2	2.1	2.12
650	640	100	800	136
M70	M70	M40	M75	M65
1.2	1.25	0.85	1.22	1.28
70	70	80	90	70
10^{14}	10^{14}	10^{13}	10^{13}	10^{15}
230	200	100	220	230
115	115	115	115	115
145	143	130	113	140
137	137	125	103	132
0.15	0.15	0.18	0.1	0.13
0.6	0.6	0.7	0.8	0.6

Table 3.2.75—*continued*

TABLE 3.2.76 Allowable working stress levels for polycarbonate resins

	Temperature (°C)	*Unfilled (MN/m^2)*	*10% Glass (MN/m^2)*	*20% Glass (MN/m^2)*	*40% Glass (MN/m^2)*
Continuous load 6 h +	–54	30	39	42	49
	–18	16	21	27	39
	23	14	19	25	35
	71	7	10	12	21
	91	4	5	7	14
	120	0	1.5	4	7
Intermittent load < 6 h followed by an equal recovery period before reloading	–54	36	42	67	84
	–18	29	34	53	76
	23	28	32	49	70
	71	22	27	39	53
	91	21	25	34	43
	120	18	21	26	36
Fatigue endurance limit at 2.5×10^6 cycles	23	7	25	39	53

TABLE 3.2.77 Friction and wear properties of polycarbonates

Component material	*Wear factor K (10^{-10} in^3 min/ ft lb h)*	*Coefficient of friction*		*Limiting PV (ft lb/in^2 min)*			*Thermal expansion[a] coefficient (10^{-5} K^{-1})*
		Static	*Dynamic*	*10 fpm*	*100 fpm*	*1000 fpm*	
Unmodified polycarbonate	2500	0.31	0.38	750	500	500	6.8
Polycarbonate 15% PTFE	75	0.09	0.15	15000	20000	10500	7.0
Polycarbonate 15% PTFE/silicone	40	0.06	0.09	14000	23000	13000	7.0
Polycarbonate/ silicone	65	0.09	0.10	10000	18000	12000	6.8

[a] At 0.3 MN/m^2.

10^{-10} in^3 min/ft lb h ≡ 2.0×10^{-17} m^3 J^{-1} 10 ft/min ≡ 0.05 m/s;

1 ft lb/in^2 min ≡ 35 J/m^2 s 100 ft/min ≡ 0.51 m/s;

1000 ft/min ≡ 5 m/s

Wear (in) ≡ *KPTV* where *T* = time (h).

Table 3.2.76 and Table 3.2.77

TABLE 3.2.78 Major suppliers and familiar trade names of polycarbonates

Company	*Trade names*	*Fillers*
Adell	Adell	G, U
ATO	Orgalan	U, G
Altulor	Altulex PC	—
EniChemical	Sinvet	U, V, G
Bayer AG	Makrolon	U, G, P, V, F
	Makroblend	PBT alloy
BASF	Ultradur	PBT alloy
Courtaulds	Grafil	C
Dow	Calibre	U, V, F, G
Ferro (GB) Ltd.	Starglas PC	G, C
	Star	
Fiberfil/Dart	Polycarbafil	G, PTFE
Fiberite (Vigilant)	RTP	G, C
Freeman Chemicals Ltd	Latilon	U, V, F, G, C
General Electric (Engineering Polymers UK)	Lexan	U, Polyethylene, G, P, V, F, S
	Xenoy CL100	PBT alloy
LNP	Thermocomp	G, PTFE, Silicon, C
Mitsui	Novarex	—
Mobay (US)	Merlon	—
Teijin	Multilon, Panlite	U, G
Thermofil	—	G, PTFE
Wilson Fibrefil International SA	Polycarbafil	G, S

U = unfilled. G = glass fibre. C = carbon fibre. P = Alloy with PBT for special automotive use.
V = UV stabilised. F = fire retardant. S = structural foam.

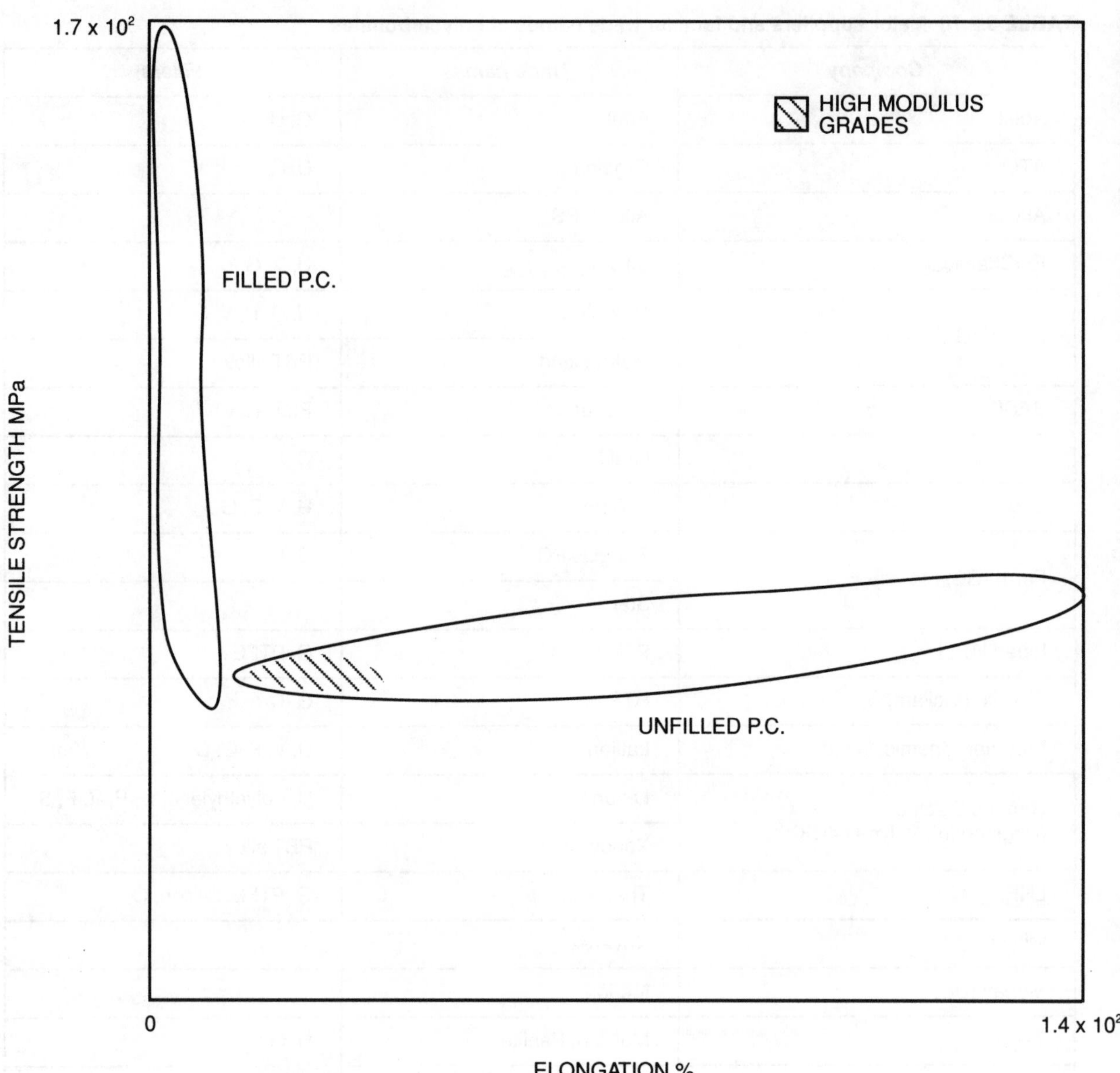

FIG 3.2.41 Tensile strength vs elongation for polycarbonates.

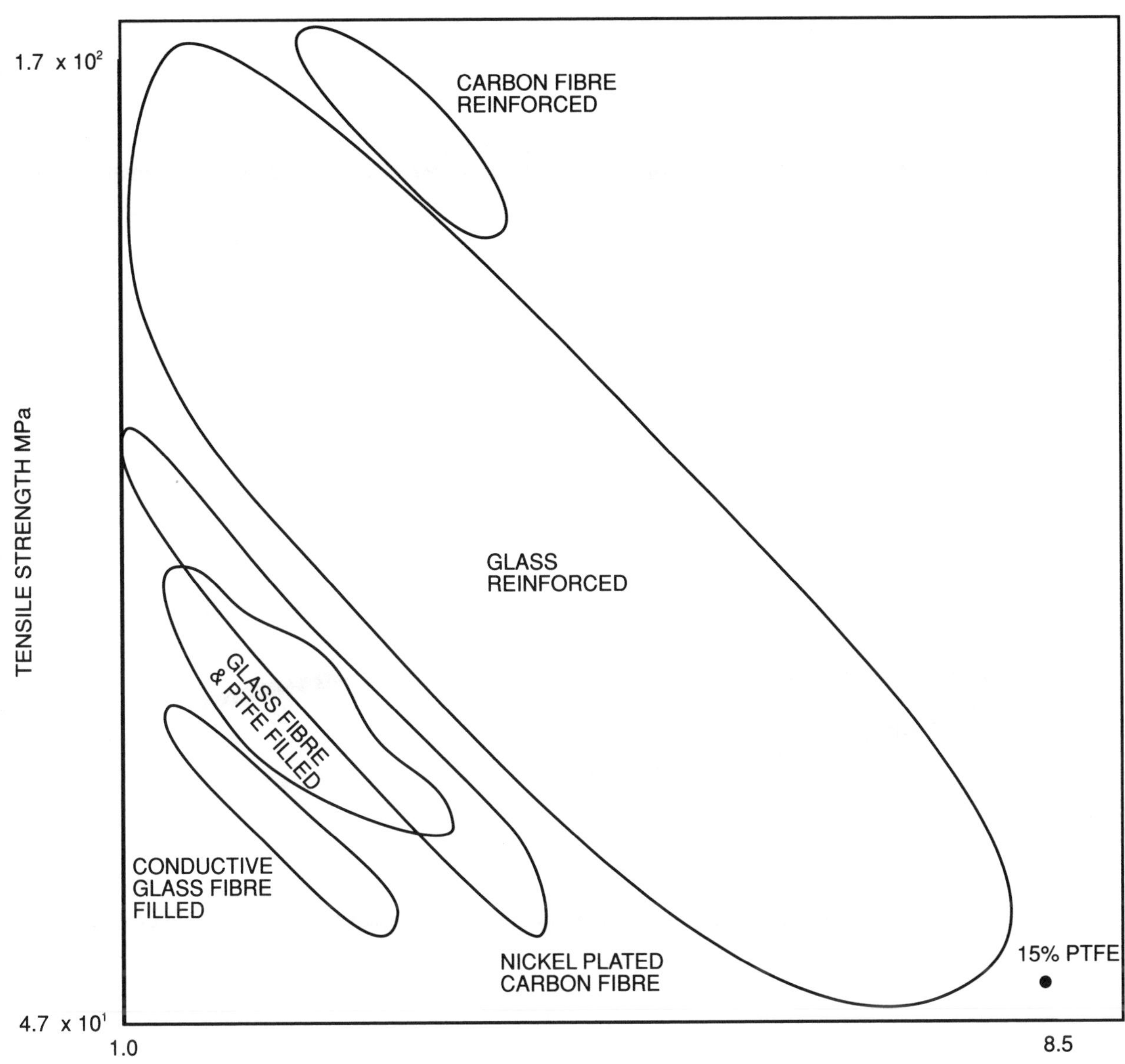

FIG 3.2.42 Tensile strength vs elongation for filled polycarbonates

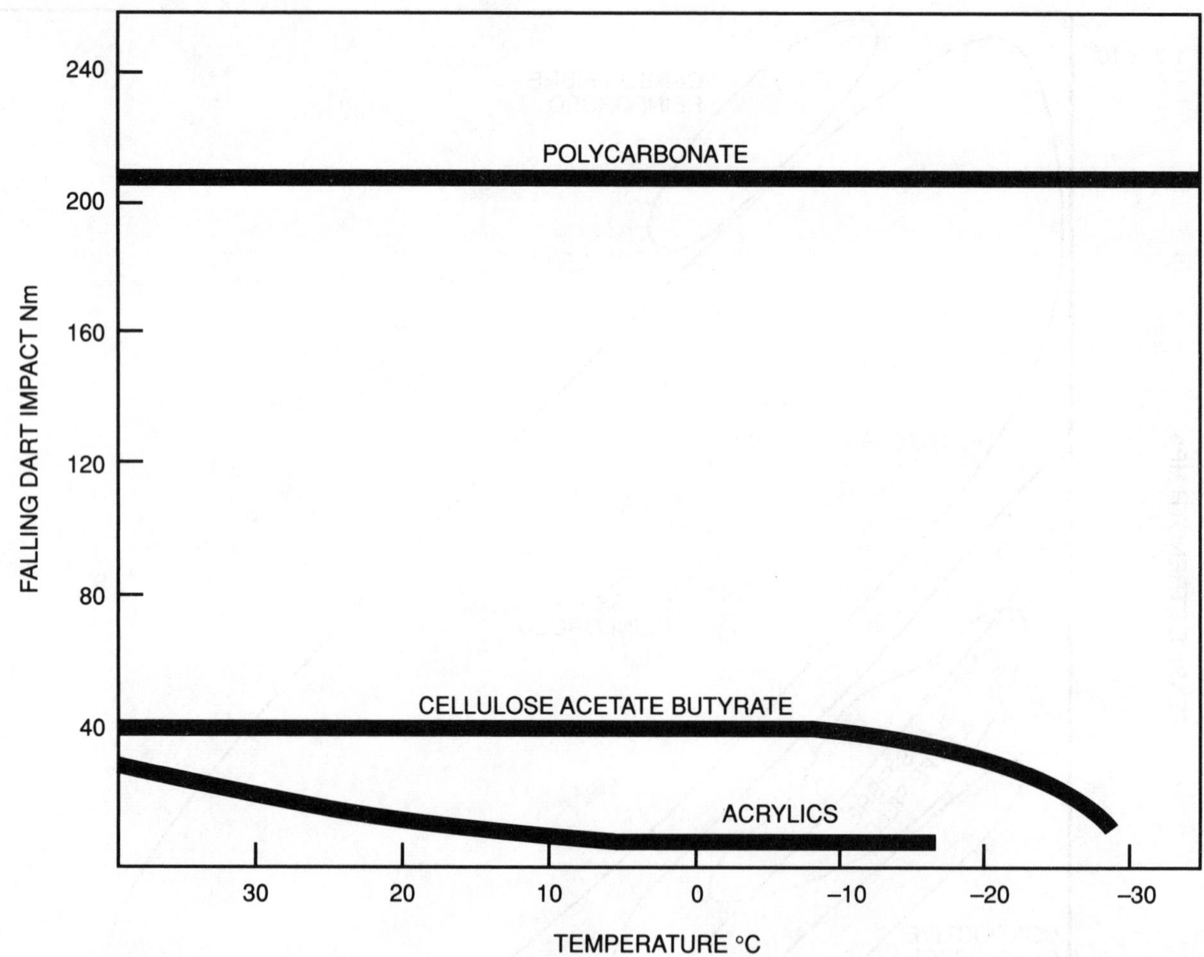

FIG 3.2.43 Comparison of impact strength of polycarbonate with other polymers over a range of temperatures

Fig 3.2.43

3.2.14 Polyphenylene oxide/polystyrene blends (modified PPO, noryl(TM))

3.2.14.1 General characteristics and properties

The advantages and limitations of modified polyphenylene oxide are listed in:
Table 3.2.79 *Characteristics of modified polyphenylene oxide*

AVAILABLE TYPES

Polyphenylene oxide is a linear crystalline polymer of good thermal stability but limited flow properties. To improve its performance so that it can be extruded or injection moulded it is blended with polystyrene.

The available types of modified polyphenylene oxide are listed in:
Table 3.2.80 *Comparison of available grades of modified polyphenylene oxide*

APPLICATIONS

Application	*Reasons for use*
Business machines, appliance and television housings.	Dimensional stability and heat resistance, mouldability.
Pumps, impellers and elevated temperature moist condition applications (e.g. coffee pot parts, washing machines).	Water resistance, strength and dimensional stability at elevated temperatures.
Replacement for zinc die-castings.	Adequate properties. Lower total cost. (No finishing costs).
Replacement for ABS automotive applications (instrument housings, dashboards, grilles, wheelcovers).	Better strength and dimensional stability.

TYPICAL PROPERTIES

The properties of the more important grades of modified polyphenylene oxide are listed in:
Table 3.2.81 *Typical properties of modified polyphenylene oxide*

Table 3.2.81 Includes data obtained by a standard method of measuring wear resistance. PPO is compared to polyphenylene sulphide and ether in Fig 3.2.46 (in Section 3.2.15.1)

3.2.14.2 Mechanical properties (creep)

Beside the standard mechanical property data included in Table 3.2.81 additional information on creep is presented in:
Fig 3.2.44 *Creep of modified polyphenylene oxide*

In addition, Fig 3.2.48, Section 3.2.16 presents the variation of apparent creep modulus of modified polyphenylene oxide with time at 100°C and 20MN/m^2.

3.2.14.3 Assembly techniques

MECHANICAL

Modified PPO mouldings can be joined to other components by self-tapping screws, snap fitting, press fitting or by the use of inserts. Moulded-in inserts are possible, but not recommended. Expansion, heli-coil, ultrasonically or induction heat installed inserts are preferred. Rivetting, the use of boss caps or moulded-in hinges are other mechanical assembly possibilities.

BONDING

Normal ultrasonic bonding techniques may be employed and shear strengths of 20.6 MN/m^2 can be obtained in the bond area. Glass-reinforced resins are less readily ultrasonically welded. If modified PPO is to be ultrasonically welded to other plastics, use of a common solvent is recommended. Heat or ultrasonic sticking can also be used.

Solvent cementing or adhesive bonding can be employed to join modified PPO to other plastics. The choice of bonding agent will depend on the other material and the conditions under which the assembly will have to operate.

3.2.14.4 Trade names and suppliers

The more important suppliers and the trade names of their products are listed in:

Table 3.2.82 *Suppliers and trade names of modified polyphenylene oxide*

TABLE 3.2.79 Characteristics of modified polyphenylene oxide (Noryl™)

	Advantages	***Limitations***
Mechanical properties	Good mechanical properties over broad temperature range from –40°C up to 105°C. High impact strength which is retained at –40°C. Excellent creep resistance.	
Wear resistance	Good resistance to wear.	
Dimensional stability	Outstanding dimensional stability, particularly under stress at elevated temperatures and in humid conditions.	
Thermal properties	High heat deflection temperature. Heat resistance intermediate between ABS and polycarbonate. Low coefficient of thermal expansion. Self-extinguishing grades available.	Upper service temperature limit 105°C.
Environmental	Resistant to strong acids and alkalis and detergents.	Attacked by many organic solvents, particularly aromatic and chlorinated aliphatics. Prone to environmental stress cracking. Prolonged outdoor exposure will cause parts made in standard grade to discolour, craze and lose properties—can be protected by organic or metal coatings.
Electrical properties	High volume resistivity and dielectric strength unaffected by changes in humidity.	
Food and medicine	Some grades suitable for use with foodstuffs.	
Processing	Mould shrinkage low and reproducible. Can be extruded or injection moulded and fabricated by solvent, ultrasonic or heat welding. Grades available which are suitable as substrate for electroplating.	Moulded components are more notch-sensitive than ABS. PPO homopolymers extremely difficult to process; hence all PPO is blended, or modified with styrene-based polymers.

TABLE 3.2.80 Comparison of available grades of modified polyphenylene oxide (Noryl™)

NORYL grade	*General comments*		*Filler content*	*Fire resistance*	*Characteristics*	*Relative cost per unit volume (10 t lots)*
110	*Unfilled grades*	Better impact, strength and lower density than filled grades.		Rated as slow burning by UL.	Cheapest resin but lowest strength and heat resistance.	100
731					General purpose resin. Higher strength and heat resistance than 110.	135
SE100				Rated as self-extinguishing SE 1 by UL	Inferior mechanical properties to 731 but less expensive.	120
SE1					Mechanical properties equivalent to 731 but more expensive.	151
GFN 2	*Glass-filled grades*	Better strength, stiffness and lower thermal expansion than unfilled grades.	20%	Rated as slow burning by UL.	Lower mechanical properties than GFN 3 but also lower cost. More expensive than 731 but better strength and stiffness.	177
GFN 3			30%		With GFN 3-SE 1 best mechanical strength and stiffness but less expensive than self-extinguishing grade.	198
GFN 2 - SE1			20%	Rated as SE1 by UL.	Mechanical properties equal to GFN 2 but with UL self-extinguishing rating at extra cost	192
GFN 3 - SE1			30%		Best combination of strength, stiffness and self-extinguishing qualities. Highest cost.	215

Other grades include (i) NORYL KB250 which is formulated for tracking resistance, self extinguishing qualities, dimensional stability and detergent resistance.
(ii) NORYL PN-235 – a platable resin
(iii) NORYL N190 formulated to provide PPO like properties at minimum cost, but has a significantly lower heat deflection temperature (see Table 3.2.81).
(iv) a structural foam.
(v) GFN1 – 10% glass-filled (cheaper than GFN2, 3).

TABLE 3.2.81 Typical properties of modified polyphenylene oxide (Noryl®)

	Properties		Units	ASTM	731	110	SE1
Mechanical	Tensile strength yield	23°C	MN/m^2	D638	64	49.0	64
	break	23°C	MN/m^2	D638/D651	61	49.0	61
	yield	95°C	MN/m^2	—	—	45	45
	Elongation at break	23°C	%	D638	60	60	60
	yield	23°C	%	D638	7	7	7
	Tensile Modulus	23°C	GN/m^2	D638	2.5	2.5	2.5
		95°C	GN/m^2	—	1.6	—	1.6
	Flexural strength	23°C	MN/m^2	D790	93.1	88.2	93.1
		95°C	MN/m^2	—	51	—	51
	Flexural Modulus	23°C	GN/m^2	D790	2.5	2.5	2.5
		95°C	GN/m^2	—	1.8	—	1.8
	Impact strength (Izod)	23°C	J/cm notch	D256	2.1	2.1	2.1
		–40°C	—	—	1.4	1.4	1.4
	Flexural fatigue endurance limit 2.5 × 10^6 cycles at 23°C		MN/m^2	—	17	—	—
	Hardness — Rockwell		—	D785	R119	R115	R119
Physical	Specific gravity		—	D792	1.06	1.06	1.06
	Thermal conductivity		W/m K	C177	0.22	0.23	0.22
	Coefficient of linear thermal expansion		$K^{-1} \times 10^{-5}$	D696	6	6.7	6
Electrical	Volume resistivity		ohm–m	D257	10^{15}	10^{15}	10^{15}
	Dielectric strength (3mm short time)		10^2 kV/m	D149	220	195	200
	Dielectric constant	60Hz 20°C	—	D150	2.65	2.65	2.69
		10^6 Hz 20°C	—	D150	2.64	2.64	2.68
	Dissipation factor	60 Hz 20°C	—	D150	0.0004	0.0004	0.0007
		10^6 Hz 20°C	—	D150	0.0009	0.0009	0.0024
	Arc resistance (tungsten)		s	D495	75	75	75
Thermal/Environmental	Abrasion resistance taber CS17 1000 cyc.		mg	D1044	20	20	20
	Continuous use temperature (mechanical with impact)		°C	—	90	—	105
	Continuous use temperature (mechanical and electrical without impact)		°C	—	105	—	110
	Heat deflection temperature	0.45 MN/m^2	°C	D648	140	—	140
		1.8 MN/m^2	°C	—	133 (90[b])	110	133
	Inflammability			D635	Self-Ext Non-dripping	Self-Ext Non-dripping	Self-Ext Non-dripping
				UL94	V–0[a]		SE1
	Water absorption	24 h 23°C	%	D570	0.066	0.07	0.08
		Equilibrium 23°C	%	—	0.14	0.15	0.21
		Equilibrium 100°C	%	—	0.3	0.3	0.45
Processing	Mould shrinkage		%	D1299	0.5–0.7	0.5–0.7	0.5–0.7

Data courtesy of General Electric Company (USA). [a] More recent grades. [b] Grade N190.

Table 3.2.81

TABLE 3.2.81 Typical properties of modified polyphenylene oxide (Noryl™)—*continued*

SE100	*GFN 2*	*GFN 3*	*SE1 GFN 2*	*SE1 GFN3*	*Mineral filled*		*PPO fire retardant*	*PPO structural foam*	*PPO/polyamide alloy*
					PX1554	*PX1555*			
54	140	120	—	45			55	21	55
49	100	118	100	118	—	—			
—	74	94	70	78	—	—			
50	4–6	4–6	4–6	4–6	—	—	35	20	100
7	4–6	4–6	4–6	4–6	—	—			
2.5	6.5	8.4	6.5	8.4	—	—			
—	5.25	7.8	5.25	7.8	—	—			
88.2	132	142	132	142	—	—			
—	93	126	92	101	—	—			
2.5	5.3	7.7	5.3	7.7	—	—	2.4	1.8	2
—	4.1	7.0	4.0	6.0	—	—			
2.7	0.9	0.9	0.9	0.9	—	—	160	0.4	2
1.4	0.7	0.7	0.7	0.7	—	—			
12	27	32	—	—	—	—			
R115	L106	L108	L106	L109	—	—	R116	R100	R101
1.10	1.21	1.27	1.23	1.29	—	—	1.09	0.85	1.1
0.23	0.17	0.16	0.17	0.16	—	—			
6.7	3.6	2.5	3.6	2.5	—	—	6	6	9
10^{15}	10^{15}	10^{15}	10^{15}	10^{15}	—	—	10^{15}	10^{14}	10^{13}
160	160	220	240	212	—	—	240	100	210
2.69	2.86	2.93	2.98	3.15	—	—			
2.68	2.85	2.92	2.95	3.11	—	—			
0.0007	0.0008	0.0009	0.0016	0.0020	—	—			
0.0024	0.0014	0.0015	0.0017	0.0021	—	—			
75	110	120	110	120	—	—			
20	35	35	35	35	—	—			
80	90	90	105	105	—	—			
95	90	90	110	110	—	—	80	80	80
110	145	150	140	142	—	—	118	96	150
100	144	150	135	137	110	132	107	82	110
Self-Ext Non-dripping	Self-Ext Non-dripping	Self-Ext Non-dripping	Self-Ext Non-dripping	Self-Ext Non-dripping					
SE1	V–0[a]	V–0[a]	SE1	SE1	V–0/5V	V–0/5V	V0>1.5	V0>3	HB
0.08	0.06	0.06	0.07	0.07	—	—	0.07	0.1	0.4
0.37	0.14	0.12	0.22	0.18	—	—			
0.55	0.32	0.30	0.33	0.33	—	—			
0.5–0.7	0.2–0.4	0.1–0.3	0.2–0.4	0.1–0.3	—	—	0.6	0.8	1.4

Table 3.2.81—*continued*

TABLE 3.2.82 Suppliers and trade names of modified polyphenylene oxide

Supplier	*Trade name*	*Filler*
General Electric	NORYL	U, G, F, S
LNP	Thermocomp	U, G, GB, M, PTFE
Borg Warner Chemicals	Prevex	U, F, G
BASF	Luranyl	U, F, G
	Ultranyl	PPO/Polyamide Alloy
Freeman	Laril	U, F, G

U = Unfilled.
G = Glass-filled.
GB = Glass bead-filled.
M = Mineral-filled.
F = Fire retardant.
S = Structural foam.

Table 3.2.82

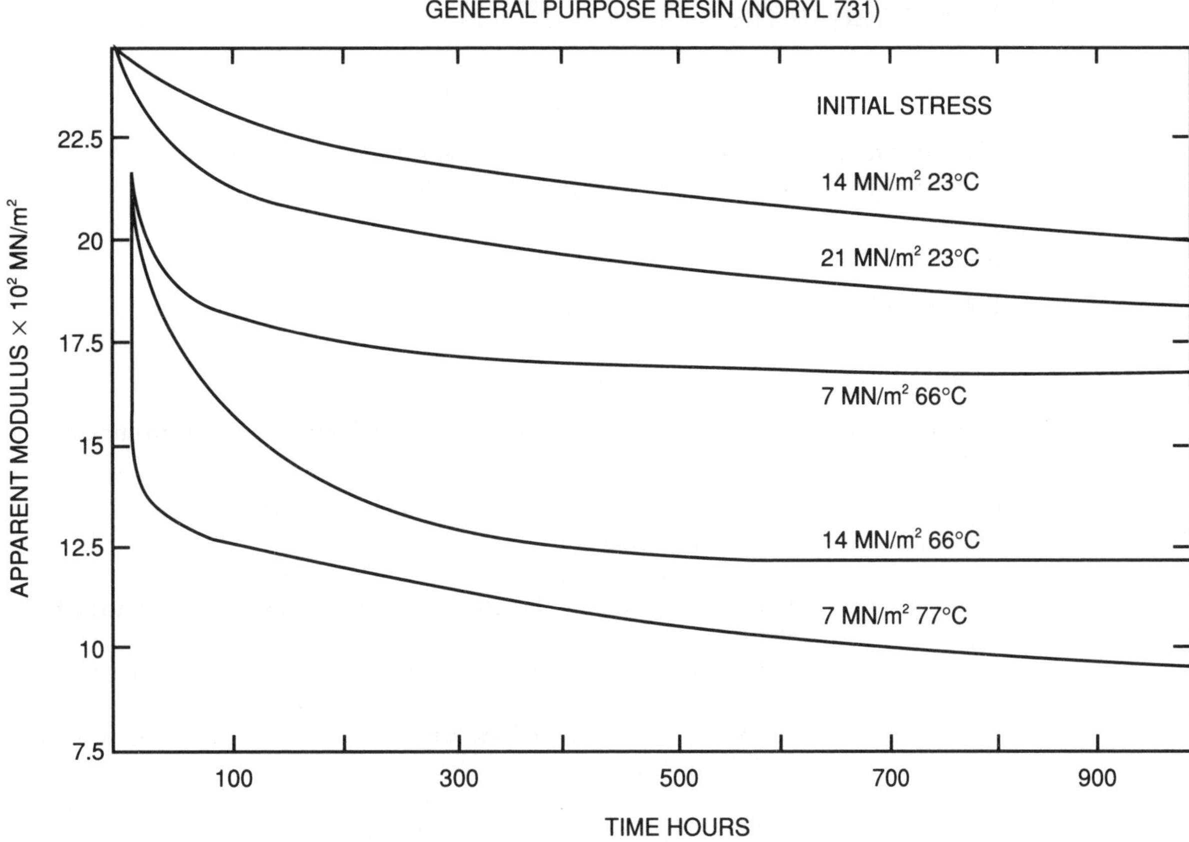

FIG 3.2.44 Creep of modified polyphenylene oxide

3.2.15 Polyphenylene sulphide (PPS)

3.2.15.1 General characteristics and properties

The advantages and limitations of polyphenylene sulphides are listed in:
Table 3.2.83 *Characteristics of polyphenylene sulphide*

AVAILABLE TYPES

Polyphenylene resins are marketed with a variety of additives and fillers.

APPLICATIONS

The major use of polyphenylene sulphide plastics is for electrical and electronic components.

Unfilled	—	Corrosion resistant pump components, valves and pipe.
Ryton R-4	—	Electronics: connectors, sockets, coil-formers, floppy disc heads, cube clamps (used to clamp semiconductor devices). Electrical/appliances: brush-holders, housings and bases, insulating plates. Automotive: under-the-hood components, valves, carburettor parts, lamp sockets. Industrial: pump impellers, cases, metering equipment, medical/dental equipment.
Ryton R-8	—	Components in electrical appliances and automotive industry such as switch components, brush-holders, heating element bases and housings, insulating plates, carriers, conductors, coil formers:
Ryton R-10	—	Electronics: bobbins, coil formers, carriers, connectors, sockets, pulse transformers (replacing transfer-moulded epoxy). Electrical: switch components, brush holders. Appliances: hair-dryer grillers, heating element bases, small housings and bases.
Coatings	—	Chemically resistant coatings and non-stick cookware (also alloyed with PTFE).

TYPICAL PROPERTIES

Because of their major field of application, electrical properties are of major significance for polyphenylene sulphide. These and other properties are listed for typical grades in:
Table 3.2.84 *Typical properties of polyphenylene sulphides as moulded*, and
Fig 3.2.45 *Tensile strength vs elongation for filled and unfilled PPS*

Polyphenylene sulphide is compared to polyphenyl oxide and ether in:
Fig 3.2.46 *Tensile strength vs elongation for PPS, PPO and PPE*

3.2.15.2 Processing

PPS can be extruded, injection moulded and compression moulded and can also be used for coating by certain techniques.

A minimum wall thickness of 1mm for structural strength is recommended, although small sections only 0.25mm thick have been easily moulded. Sections as thick as 13mm with some areas approaching 19mm have been moulded without voids, sink marks or dimensional instability.

PPS can be processed through an injection moulding machine as many as three times without appreciable change in the material. Therefore, regrind can be used with virgin resin with no noticeable change in processability.

3.2.15.3 Trade names and suppliers

The major suppliers of polyphenylene sulphides and the trade names of their products are listed in:

Table 3.2.85 *Suppliers and trade names of polyphenylene sulphides.*

TABLE 3.2.83 Characteristics of polyphenylene sulphide

	Advantages	*Limitations*
Mechanical	Good fatigue and creep resistance. Good dimensional stability.	Low impact strength.
Thermal	Continuous use at 240°C. Non-burning.	Sharp drop in tensile strength at 120°C, and again at 150°C.
Environmental	Very good chemical resistance even up to 190–200°C. Resistant to weak acids, strong alkalis and organic solvents. Low water absorption. Good radiation resistance.	Attacked slowly by strong oxidising acids. Not resistant to halogens or halogenated organics at higher temperatures.
Processing	Glass-filled grades have very low mould shrinkage (0.2%).	High mould temperatures.
Fillers	Filled grades have better impact strength, stiffness and strength. Carbon fibre grades have flexural modulus >16GN/m^2 (30% filled) and improved dimensional stability and mould shrinkage.	Rarely used unfilled.
Miscellaneous		High cost.

TABLE 3.2.84 Typical properties of polyphenylene sulphides as moulded

	Property		*ASTM test*	*Units*	*Grades*			
					Unfilled	*R-4*	*R-8*	*R-10*
						(40% glass)	*(glass, mineral)*	
Mechanical	Tensile yield strength		D638	MN/m^2	75	135	92	70–80
	Elongation at break		D638	%	1.6	1.3	0.7	0.5–0.6
	Tensile Elastic Modulus		D638	GN/m^2	3.3	—	—	—
	Flexural Modulus		D790	GN/m^2	3.8	11.7	13.1	14.0–16.5
	Flexural strength		D790	MN/m^2	140	200	141	105–120
	Compressive strength		D695	MN/m^2	—	145	110	100–128
	Izod impact strength		D256	J/m notch	20	75	31	30–43
	Hardness		D785	Rockwell	R124	R123	R121	—
Physical	Density		D1505	kg/dm^{-3}	1.34	1.6	1.8	1.96–2.07
	Thermal expansion coefficient		—	$10^{-5} K^{-1}$	4.9	2.2	2.8	—
Electrical	Dielectric strength (3.2 mm)		D149	10^2 kV/m	150	177	134	126–157
	Dielectric constant	10^3 Hz	D150	—	3.1	3.9	4.6	5.1–6.6
		10^6 Hz	D150	—	—	3.8	4.3	4.8–6.1
	Power factor	10^3 Hz	D150	—	0.0005	0.0010	0.017	0.01–0.08
		10^6 Hz	D150	—	—	0.0013	0.016	0.01–0.02
	Volume resistivity		D257	ohm cm	10^{16}	4.5×10^{16}	2.0×10^{15}	$1–3 \times 10^{15}$
	Surface resistivity		D257	ohm	—	4.6×10^{16}	0.6×10^{16}	—
	Arc resistance		D495	s	—	34	182	116–211
	Tracking resistance		DIN 53480	—	—	KC 180	KC 235	KC 150–260
Thermal and environmental	Heat distortion temperature 1.8 MN/m^2		D648	°C	137	243 (>260[a])	244 (>260[a])	(>260[a])
	UL long term index	(1)		°C	—	170–200	200–220	220–240
		(2)	—	°C	—	—	—	200–220
		(3)		°C	—	—	—	240
	Flammability		UL94	—	Non-burning Non-drip	V–0/5V	V–0/5V	V–0
	Water absorption (24 h)		D570	%	<0.02	<0.05	0.03	—
Processing	Mould shrinkage		—	%	1.0	0.2–0.6	0.2–0.6	0.2–0.6

[a] Annealed.
(1) Electrical.
(2) Mechanical with impact.
(3) Mechanical without impact.

TABLE 3.2.85 Suppliers and trade names of polyphenylene sulphides

Supplier	*Trade name*	*Filler*
Phillips Petroleum	Ryton	U, GF, G/M, A
Fiberite		G, CF, PTFE
LNP		
Thermofil		

U = Unfilled.
G = Glass.
GF = Glass fibre.
M = Mineral.
A = Asbestos.
CF = Carbon fibre.

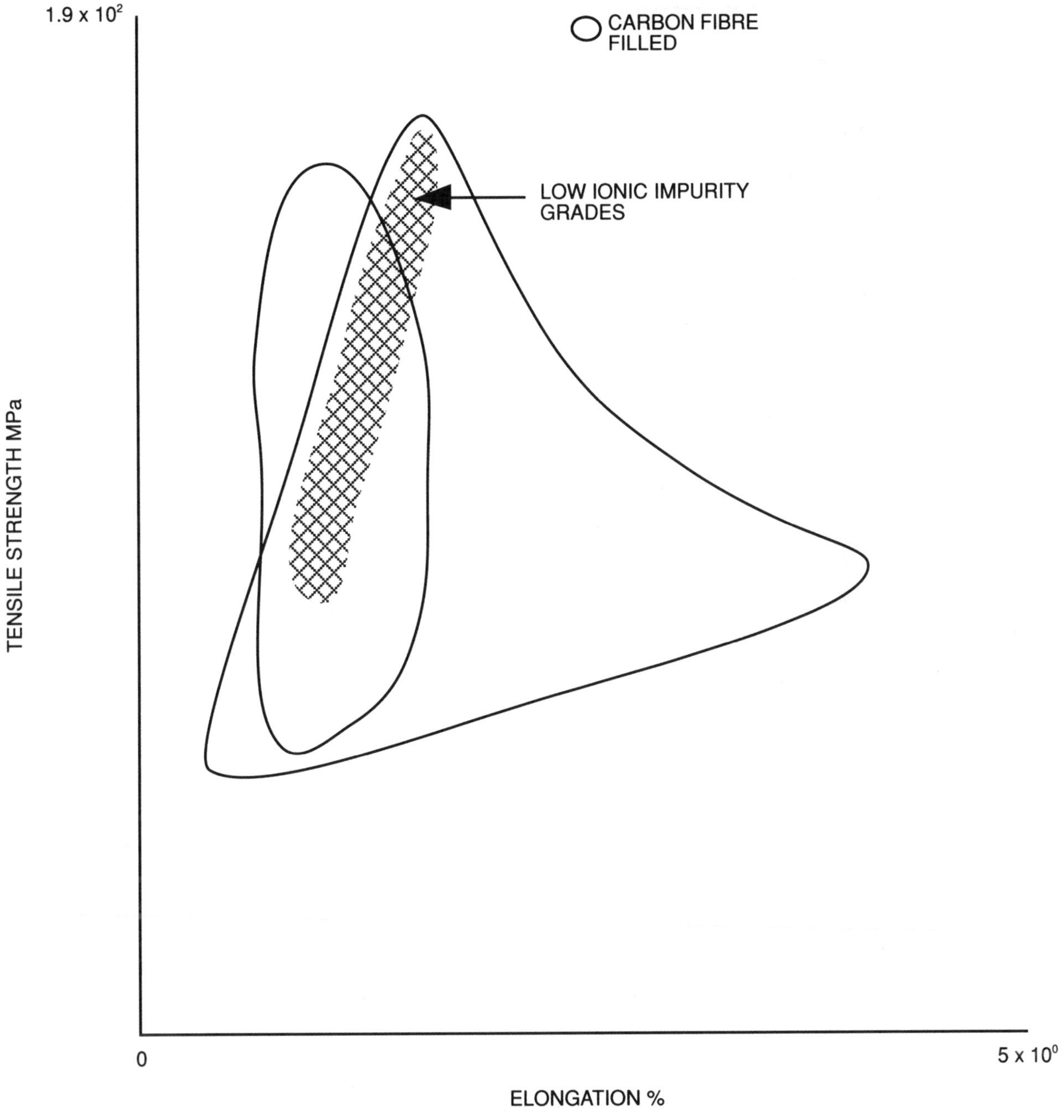
1.9 x 10²
CARBON FIBRE FILLED
LOW IONIC IMPURITY GRADES
TENSILE STRENGTH MPa
0
5 x 10⁰
ELONGATION %

FIG 3.2.45 Tensile strength vs elongation for filled and unfilled PPS

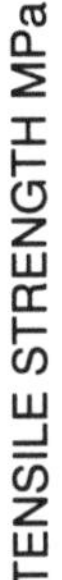

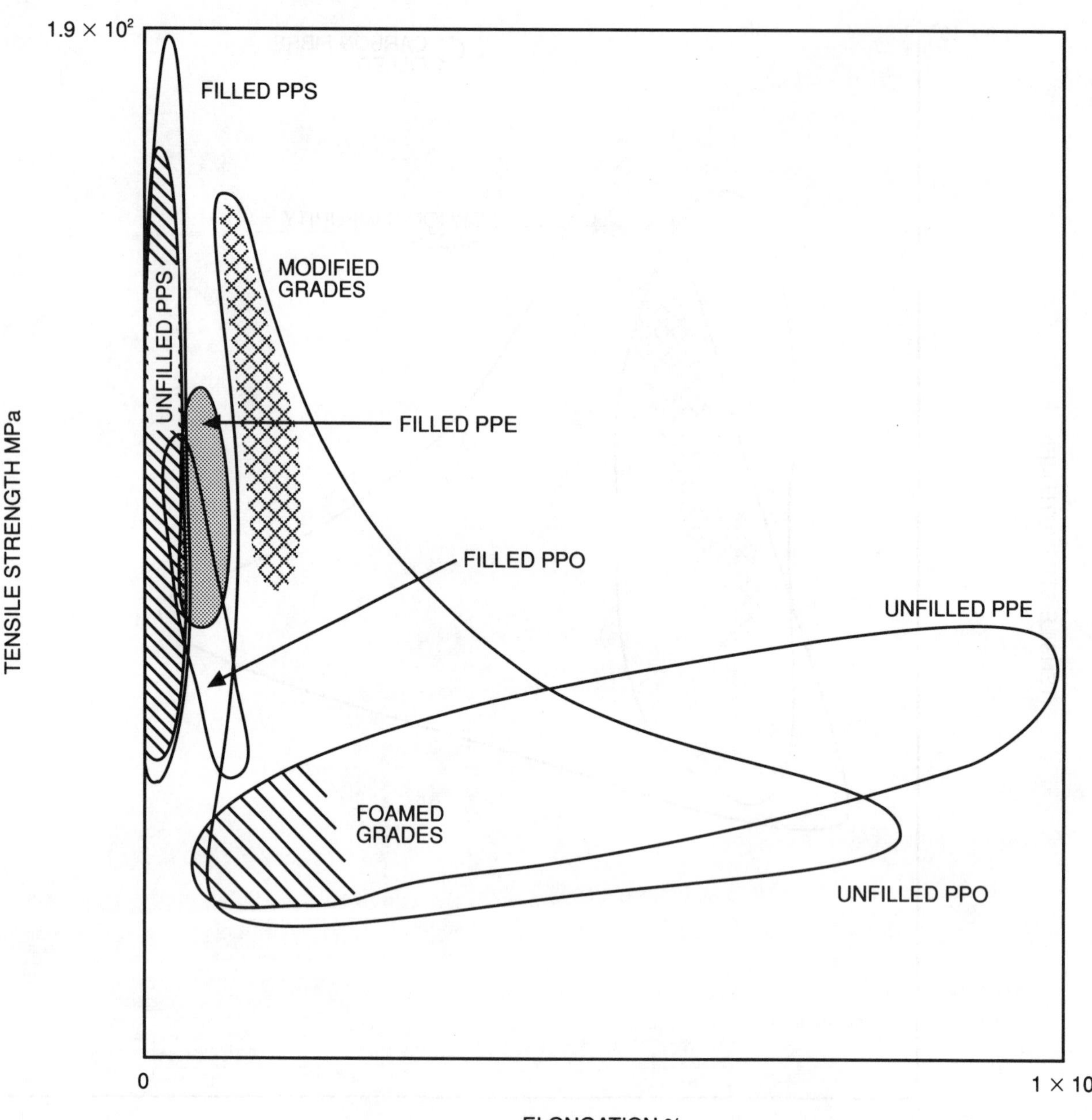

FIG 3.2.46 Tensile strength vs elongation for PPS, PPO and PPE

PPO – Polyphenylene Oxide
PPS – Polyphenylene Sulphide
PPE – Polyphenylene Ether

3.2.16 Polysulphones

(Polysulphone, polyethersulphone, polyphenylsulphone, modified polysulphones, ABS/polysulphone alloy, modified glass-filled polysulphone and SAN/polysulphone alloy.)

3.2.16.1 General characteristics and properties

Polysulphones are one of three groups of thermoplastics which are suitable for operation at high temperatures (for thermoplastics). Because of their elevated temperature properties all these groups require high pressures and temperature for processing.

The advantages and limitations of polysulphones compared with those of the materials described in Section 3.2.17 are listed in:

Table 3.2.86 *Comparative characteristics of polysulphones, -ethers, -imides and -etherether-ketones*

AVAILABLE TYPES

Polysulphones are noted for good oxidative stability at elevated temperatures and compare in high temperature performance with the melt-processable fluorocarbons (see Section 3.2.7), over which they have considerable economic and processing advantages.

A third group of high-temperature thermoplastics, polyamide-imide, polyether-imide and polyetheretherketone are described in Section 3.2.17.

Their stress relaxation characteristics are compared in:

Table 3.2.87 *Ranking of stress relaxation characteristics of high-temperature polymers.*

Polysulphone

The basic repeating structure of polysulphone consists of benzene rings connected by a sulphone group, an ether linkage and an isopropylidene group.

$$\left[-C_6H_4-C(CH_3)_2-C_6H_4-O-C_6H_4-SO_2-C_6H_4-O- \right]_n$$

The sulphone group imparts thermal stability and oxidation resistance; the other two groups make the backbone of the polymer flexible, imparting toughness and improving the processability.

The natural colour of a polysulphone is a transparent amber. However, resins are available in a broad range of transparent or opaque colours. Glass-filled grades are available where higher strength and stiffness and low thermal expansion are required and impact strength is not at a premium.

Polyethersulphone

Polyethersulphone consists essentially of a repeating unit with the following structure.

$$\left[-C_6H_4-SO_2-C_6H_4-O- \right]_n$$

The presence of the sulphone group ensures high temperature performance while the ether group improves processability by allowing mobility of the polymer chain when in the melt phase.

The structure of para-phenylene units linked by alternate sulphone and ether groups gives an amorphous polymer which, since it contains no aliphatic units, possesses only bonds of particularly high thermal and oxidative stability. Further, these linkages are stable in a wide range of chemical environments.

Two unfilled grades, 200P and 300P are availablefrom ICI. The 200P grade is mainly for general purpose injection moulding and has better flow characteristics than the 300P grade which is preferred for precision engineering applications where good creep resistance at high temperatures and/or heat resistance are of prime importance. An additional special grade, 720P has a maximum operating temperature of 175°C, about 25°C higher than 200P and 300P. 420P and 430P are respectively 20 and 30% glass-filled grades.

Modified polysulphones

ABS/polysulphone alloy can be electroplated and will withstand a temperature of 150°C. It has hydrolytic stability and it will withstand paint bake cycles.

Mindel B is a proprietary modification of polysulphone containing glass which exhibits, as an amorphous polymer, lower warpage than PBT.

SAN–polysulphone alloy is another modified polysulphone.

APPLICATIONS

Polysulphones

Injection moulded polysulphones find electrical, domestic, medical and automotive application.

Electrical—electronic connectors, integrated circuit boards, coil bobbins, switchgears, computer parts, desk lamps.

Appliances—coffee pots, humidifiers, TV components, insulated power tool components, microwave oven grills, dishes and trays for food service.

Hygiene and medical—surgical instrument trays, milking machines, inhalators and other chemotherapy devices, electrosurgical and cryogenic surgical tools.

Automotive—under bonnet parts.

Extrusions, shapes, sheet and film for chemical and heat resistant piping, thermoformed access covers, heat resistant film for overhead projectors.

Extruded cable insulation—thin, high temperature, abrasion and flame resistant, tough and with good dielectric properties.

Polyethersulphones

The physical properties of polyethersulphone and its suitability for processing on conventional machinery indicate uses for the material in a wide range of applications throughout industry.

Actual applications include:

Moulded circuit boards, flat or profiled, where design freedom and resistance to high temperatures and aggressive atmospheres are required.
Industrial control relays.
High performance terminal blocks for aircraft
Weapon system components.
Heated hair-styling brushes.
Car heater fans.
Plastic bearing cages.

Hot and pressurised water meter and valve components.
High temperature lamp reflectors.
Aircraft radomes.
Sterilisable medical products.
Spectrophotometer mirror supports.

Polyphenyl sulphone

Potentiometer covers.
High temperature electrical connectors.
Oil drilling electrical cable insulation.
Wire insulation.
Fuel-cell parts.
Aircraft window reveals and passenger service units.
Hospital operating tables.
Nuclear reactor thermocouple 'corks'.

ABS/polysulphone alloy

Food service (particularly hospital trays).
Automotive parts which must withstand paint bake cycles.

Mindel B

Electrical connectors.

TYPICAL PROPERTIES

The properties of polysulphones are listed in:
Table 3.2.88 *Typical properties of polysulphones*

3.2.16.2 Mechanical properties

The basic mechanical properties of polysulphones are listed in Table 3.2.88. The most significant mechanical property is creep. The creep characteristics of polysulphones are compared with those of a number of other thermoplastics in:
Fig 3.2.47 *Time dependence of tensile creep modulus of some unfilled thermoplastics at 20°C.*
Fig 3.2.48 *Time dependence of tensile creep modulus of some unfilled thermoplastics at elevated temperatures.*

3.2.16.3 Electrical properties

Polyethersulphone retains its low loss factor, good permeability and high resistivity at temperatures in excess of 200°C. It is suitable for 'class H' electrical applications.

Polysulphone also has very good electrical properties which remain good under exposure to hot and/or humid environments.

Polyphenylsulphone offers good electrical properties (see applications) and the latest information suggests retention of these after ageing at elevated temperatures.

3.2.16.4 Environmental properties

The environmental properties of polysulphones are included in Table 3.2.86.

STERILISATION OF POLYSULPHONES

Polysulphone moulded and formed parts can be sterilised using a variety of methods, including steam and dry heat without loss of properties. This is of great importance for applications such as food processing and medical equipment.

RADIATION RESISTANCE OF POLYETHERSULPHONE

Polyethersulphone possesses good resistance to X-rays, β-rays and γ-rays in the range 20–200°C.
The influence of water on the permissible stress for polysulphone is listed in:
Table 3.2.89 *Maximum working loads—polysulphone*

3.2.16.5 Processing

POLYSULPHONES

Polysulphones can be processed readily by the standard thermoplastic techniques of injection moulding, blow moulding, extrusion and thermoforming. However, for optimum results some special considerations are required due to the materials high heat resistance. Specific temperatures required should be discussed with suppliers to ensure that these are achievable on in-house equipment.

The material must be thoroughly pre-dried before processing. Only certain types of dry colouring dye may be used.

POLYETHERSULPHONES

Polyethersulphones can be processed by injection moulding, extrusion and blow moulding techniques. Prior to processing the stock should be dried at 150°C for 3 h.

(i) Injection moulding	200P—Barrel temperature: 340–360°C	
	300P	
	Glass-filled grades	Barrel temperature: 360–390°C
	Mould temperature:	150°C
	Mould shrinkage:	0.6% unfilled 0.2–0.5% glass filled
(ii) Extrusion	300P preferred	Barrel temperature: 340–390°C Polyethylene-type screw—L/D:24/1 Compression ratio 2.5/1
(iii) Solution casting	Mainly 100P, also 200P, 300P for glass cloth and fibres (see applications).	
	Curing temperature	360–400°C
(iv) Finishing operations.	The following treatments are suitable:	

machining
adhesive bonding
solvent welding
thermal and ultrasonic welding
vacuum forming
painting and etching
electroplating and metallising (Cu, Au, Al, stainless steel)

3.2.16.6 Trade names and suppliers

Trade names and suppliers are listed in:
Table 3.2.90 *Suppliers and familiar trade names of polysulphones*

TABLE 3.2.86 Comparative characteristics of polysulphones, -ethers, -imides and -etheretherketones

Material	*Advantages*	*Limitations*
1. Poly-sulphones	Creep resistant, dimensionally stable, high service temperature polymer. Good combination of high temperature properties and melt processability. Does not degrade, discolour or cross-link with continuous service at 160°C; higher heat stability than polycarbonate. Resistant to mineral acids, alkalis, salts, detergents, oils and alcohols. Electrical properties retained to 175°C. Self-extinguishing. Approved grades available for use with foodstuffs and medical applications and can be repeatedly sterilised. Injection moulding, extrusion and extrusion cable insulation grades available. Good low temperature toughness down to –100°C. Glass-filled grades have greater strength and rigidity with lower thermal expansion coefficient but lower impact strength. Mineral filled grades reduce cost, enhance stress-crack resistance and are tougher and easier to process than PPS. Parts may be transparent or opaque and are readily coloured. Can be electroplated, and withstand soldering temperatures for short periods (electronic applications).	Upper service temperature limit of 170°C. (Tensile strength drops sharply above 150°C.) Swollen by aromatics and stress-cracked by several organics including acetone, ethyl acetate, trichlorethylene and carbon tetrachloride. Before moulding materials must be dried at 120°C for 5 h. Notch-sensitive. Rigid molecules and high setting temperatures conducive to residual stresses. Not recommended for outdoor service unless painted or plated; lower resistance to UV light than polycarbonate. Inferior fatigue endurance to nylon and acetal, but similar to polycarbonate. High processing temperatures require that equipment must be purged before and after moulding.
2. Polyether-sulphones	Good combination of high temperature resistance and processability. Creep resistant up to 150°C. Load bearing up to 180°C plus retention of electrical properties but considerable tendency to stress relaxation. Dimensionally stable up to 200°C. Can withstand soldering temperatures for short periods. Resists oxidation and degradation at processing temperature. Good resistance to radiation. Resistant to attack by weak acids, alkalis, oils and greases, alcohols and aliphatic hydrocarbons. Sterilizing solutions, anaesthetics and foodstuffs can be used in contact with PES. Injection moulding and extrusion grades available. Low mould shrinkage—close tolerances possible. Parts can be solvent cemented using methylene dichloride or dimethyl formamide.	Resistance to repeated exposure to hot water can cause crazing. Absorbs moisture with consequent slight plasticisation and dimensional changes. Attacked by polar materials such as ketones, esters, chloroform, etc. Classified as non-burning. Unpigmented resins not recommended for outdoor use—additions of carbon black recommended to reduce degradation.
3. Polyaryl–sulphones	Continuous use up to 260°C. Good creep and fatigue resistance and dimensional stability. Excellent compressive strength at 260°C. Resistant to weak acids, alkalis, fuels and	No longer commercially available. Dissolved by highly polar solvents such as *M*-cresol, dimethyl acetamide, dimethyl formamide and *N*-methyl pyrrolidone. Highly swollen by hot (120°C) acetic acid.

Table 3.2.86

TABLE 3.2.86 Comparative characteristics of polysulphones, -ethers, -imides and -etheretherketones—*continued*

Material	*Advantages*	*Limitations*
3. Polyaryl–sulphones *(continued)*	oils and most solvents even at high temperatures. Fluorinated solvents such as freons have little effect. Processable by extrusion, injection or compression moulding. Gamma radiation has little effect on properties. May be welded ultrasonically or solvent welded using *N*-methyl pyrrolidone.	Cracked and delaminated by acetone and hot (108°C) trichloroethylene. Ketones cause stress-cracking. High temperature (>400°C) and high pressure (>30 000 psi, 200 MN/m^2) equipment needed for moulding. High processing temperature requires that equipment must be purged before and after moulding. Moulding material should be dried for 3 h at 260°C or longer at lower temperatures before conversion—the presence of moisture during processing will cause degradation.
4. Polyaryl-ethers	Good combination of heat resistance and high impact strength. Relatively inexpensive. Good creep and fatigue resistance. Dimensional stability withstands prolonged immersion in boiling water. Can be welded with common solvents or ultrasonics. Can be plated without annealing or solvent conditioning. Better processability than polycarbonate or modified PPO.	Upper service temperature limit 150°C. Attacked by chlorinated aromatics, esters and ketones. Not self-extinguishing. Poor weather resistance.
5. Polyphenyl-sulphone	Excellent toughness especially at low temperatures. Exceptional creep resistance. Good stress-crack resistance (better than polysulphone). Thermal stability to 190°C for continuous use. Resistant to steam and boiling water. High heat deflection temperature. Good electrical properties over wide temperature and frequency range. Transparent. Resistant to detergents, hydrocarbon oils, chlorinated solvents. Good at elevated temperatures. Resistant to mineral acids, alkalis and salts.	Stress-crack resistance depends upon stress, temperature and exposure time. Evolves smoke (at low rate) under combustion. Soluble in dimethyl formamide and dimethyl acetamide. Ketones, esters and aromatic solvents attack polyphenylsulphone.
6. Polyamide-imide (see also Section 3.2.17)	Very high strength and impact resistance. Best creep resistance of unreinforced thermoplastic. Can be compounded with PTFE and graphite to produce lubricated or non-lubricated bearings with low friction. [§]Wear resistance superior to polyimide at PV<50 000 ft lb/in^2 per min. Structural integrity maintained up to 260°C under extreme load bearing situations; better screw-torque retention than polyaryl-sulphone, polyimide and poly-p-benzoate which have been replaced by polyamide-imide in connector applications where insert pin retention is critical.	High cost. Attacked by high-temperature caustic systems and some high temperature acid systems. Maximum water pick-up of about 5% after about 3 months which affects dimensions, physical properties and heat distortion temperature. Exposure to humid environments or immersion in water leads to the absorption of small quantities of water, which may need to be driven off to achieve optimum high temperature performance.

Table 3.2.86—*continued*

TABLE 3.2.86 Comparative characteristics of polysulphones, -ethers, -imides and -etheretherketones—*continued*

Material	*Advantages*	*Limitations*
6. Polyamide-imide (see also Section 3.2.17) *(continued)*	Low coefficient of thermal expansion—when glass-filled can be comparable to metals. Non-flammable (to UL94 V-O). Excellent electrical insulator. Resistant to most chemicals and to radiation. Can be injection moulded, extruded or machined.	
7. Polyetheri–mide (see also Section 3.2.17)	Exceptional dielectric properties and flame retardance with good heat resistance, toughness and chemical stability. Transparent. Very good processability. Electrical properties sustained at high temperatures and frequencies.	Fluoropolymers have better flame resistance. Stress-cracks in diesel fuel, concentrated antifreeze and some brake fluids. Moderate temperature capability relative to some others. Moderate impact resistance and notch sensitive (amorphous).
8. Polyether-etherketone (see also Section 3.2.17)	Modification of polyethersulphone, but rystalline. Withstands severe conditions of temperature, burning and chemicals. Can be processed on most extrusion and injection moulding equipment. Very little smoke or gas produced if burnt. Radiation-resistant (better gamma-resistance than polystyrene).	Very high cost.

Table 3.2.86—*continued*

TABLE 3.2.87 Ranking[a] of stress relaxation characteristics of high temperature polymers

Base resin	***% Glass***	***Temperature (°C)***	
		20	***150***
Polyamide-imide	0	1	1
Nylon 6/6	50	2	3
Polyphenylene sulphide	40	3	4
Polyarylsulphone	0	4	2
Polysulphone	40	5	10
Poly-*p*-oxybenzoate	0	6	7
Polyethersulphone	40	7	8
Polyimide	30	8	6
ETFE	20	9	5
Thermoplastic polyester	40	10	9
FEP	20	11	11

[a]The lower the number, the higher the retained stress at the test temperature indicated.
(Source: LNP corporation.)

TABLE 3.2.88 Typical properties of polysulphones

	Property		ASTM Test	Units	Polysulphone	
					Standard	30% glass-filled
Mechanical	Tensile strength	20°C	D638	MN/m^2	60–74 (yld)	117
		150°C	—	MN/m^2	—	—
		180°C	—	MN/m^2	—	—
	Elongation at break		D638	%	50–100	2–3
	Flexural strength		D790	MN/m^2	106	172
	Modulus of Elasticity in Tension at 20°C		D638	MN/m^2	2480	7515
	Flexural Modulus	20°C	D790	MN/m^2	2690	8270
		150°C	—	MN/m^2	—	—
		180°C	—	MN/m^2	—	—
	Impact strength	20°C	D638	J/m notch	70	96
	(Izod notched)	–40°C	—	—	60	—
	Hardness (Rockwell)		D785	—	120R	M84
Physical	Specific gravity		D792	—	1.24–1.25	1.41
	Thermal conductivity		—	W/m per K	0.26	0.32
	Thermal expansion		D696	10^{-6} K^{-1}	56	25
Electrical	Volume resistivity		D257	ohm m	5×10^{14}	10^{15}
	Dielectric strength		D149	10^2 kV/m	150–170	190
	Dielectric constant	60 Hz	D150	—	3.07	3.55
		10^6 Hz	—	—	3.03	—
	Power factor	60 Hz	D150	—	0.008	0.0019
		10^6 Hz	—	—	0.0034	—
	Arc resistance		D495	s	60–122	114
Miscellaneous	Max. continuous service temperature		—	°C	150	177
	Heat distortion temp.	0.45/MN/m^2	D648	°C	187	198
		1.8/MN/m^2	D648	°C	174	185
	Flammability		D635 or UL94	—	Non-burning	—
	Water absorption		D570	% (24 h)	0.3	0.22
	Taber abrasion CS17		D1044	mg/1000 rev.	—	—
	Mould shrinkage		—	%	0.7	—

Table 3.2.88

TABLE 3.2.88 Typical properties of polysulphones—*continued*

Polysulphone		*Polyethersulphone*		*Polyarylsulphone*	*Polyarylether*
Mineral-filled		*Unfilled*	*20–30% glass-filled*		
Mindel M800	*Mindel M825*				
65.5	68	84	124–140	90	55
—	—	55	78–100	—	—
—	—	41	60–76	29	—
2	5	40–80	3	13	25–90
98.6	106	129	172–190	129	119
4480	3790	2440	6900–8620	2550	2206
5170	4140	2600	5900–8400	2780	2720
—	—	2500	5800–8200	—	—
—	—	2300	5600–8000	1800	—
35	50	84	75–80	80	430
—	—	—	—	—	—
M74	M70	M88	M98	M110	R117
1.61	1.48	1.37	1.51–1.60	1.36	1.14
—	—	0.17	0.33	0.19	0.3
—	—	55	23–26	47	65
5×10^{12}	10^{12}	10^{15}–10^{16}	$>10^{14}$	3.2×10^{14}	1.5×10^{14}
177	173	160	160	138	169
3.8	3.7	3.5	3.8	3.9	3.1
3.8	3.7	3.5	3.76	—	—
0.003	0.003	0.001	—	0.003	0.006
0.003	0.006	0.0035	—	—	—
125	125	20–120	—	67	>180
—	—	175–200	—	260	150
—	—	—	210–221	—	160
179	174	203	210–216	270	150
V–0 (1 mm)	V–0 (1.8 mm)	Non-burning	Non-burning	Self-ext. Non dripping	Slow burning
0.46	0.7	0.43	0.28–0.35	1.1	0.25
—	—	6	8	—	—
0.4	0.5	0.60	0.2–0.3	0.7–0.9	0.7

Table 3.2.88—*continued*

TABLE 3.2.88 Typical properties of polysulphones—*continued*

	Property		ASTM Test	Units	Polyphenylsulphone	ABS/ Polysulphone (Mindel A650)
Mechanical	Tensile strength	20°C	D638	MN/m²	72	45
		150°C	—	MN/m²	—	—
		180°C	—	MN/m²	29 at 200°C	—
	Elongation at break		D638	%	60	30
	Flexural strength		D790	MN/m²	85	92
	Modulus of Elasticity in Tension at 20°C		D638	MN/m²	2140	—
	Flexural Modulus	20°C	D790	MN/m²	2280	2.51
		150°C	—	MN/m²	—	—
		180°C	—	MN/m²	—	—
	Impact strength	20°C	D638	J/m notch	640	513
	(Izod notched)	−40°C	—	—	430	—
	Hardness (Rockwell)		D785	—	—	—
Physical	Specific gravity		D792	—	1.29	1.13
	Thermal conductivity		—	W/m per K	—	—
	Thermal expansion		D696	10^{-6} K^{-1}	55	—
Electrical	Volume resistivity		D257	ohm m	9×10^{14}	10^{13}
	Dielectric strength		D149	10^2 kV/m	146	170
	Dielectric constant	60 Hz	D150	—	3.44	3.13
		10^6 Hz	—	—	3.45	3.26
	Power factor	60 Hz	D150	—	0.006	—
		10^6 Hz	—	—	0.007	—
	Arc resistance		D495	s	—	—
Miscellaneous	Max. continuous service temperature		—	°C	>190°C	—
	Heat distortion temp.	0.45/MN/m²	D648	°C	—	—
		1.8/MN/m²	D648	°C	204	150
	Flammability		D635 or UL94	—	Slow burning	—
	Water absorption		D570	% (24 h)	—	—
	Taber abrasion CS17		D1044	mg/1000 rev.	—	—
	Mould shrinkage		—	%	0.8	0.7

Table 3.2.88—*continued*

TABLE 3.2.88 Typical properties of polysulphones—*continued*

Modifed glass-filled (Mindel B-322)	*PES 30% C fibre*	*Polysulphone 30% C fibre*	*Polysulphone 15% PTFE*	*PPS 40% glass-fibre*	*PPS 30% C fibre*	*PPS 20% PTFE*
103	195	159	53	130	156	58
—	—	—	—	—	—	—
—	—	—	—	—	—	—
2.5	1.8	2.5	40	1.2	1.4	0.6
159	—	—	—	—	—	—
—	—	—	—	—	—	—
6.2	15200	14100	2620	11700	17200	3760
—	—	—	—	—	—	—
—	—	—	—	—	—	—
54	53	60	6	60	60	15
—	—	—	—	—	—	—
M80	R123	M90	M69	R123	R123	R110
1.47	1.45	1.37	1.32	1.6	1.46	1.45
—	—	—	—	—	—	—
—	14	10.8	59	30	16	59
—	—	—	10^{14}	$10^{14.6}$	—	10^{13}
160	—	—	160	177	—	134
3.7	—	—	—	—	—	—
3.7	—	—	—	—	—	—
0.003	—	—	—	—	—	—
0.009	—	—	—	—	—	—
125	—	—	—	—	—	—
—	180	150	150	200	200	200
—	218	197	—	—	—	—
166	212	185	177	240	260	140
V–0	V–0	V–0	V–1	V–0	V–0	V–0
0.14	0.29	0.15	0.15	0.05	0.12	0.04
—	—	—	—	—	—	
0.2–0.3	0.2	0.15	0.85	0.3	0.1	0.3

Table 3.2.88—*continued*

TABLE 3.2.89 Maximum working loads—polysulphone

°C	*Steady load air (MN/m²)*	*Steady load water (MN/m²)*	*Intermittent load water (MN/m²)*
20	28	21	25
60	25	11	14
85	—	4	7
100	21	0	4
130	14	—	—
150	7	—	—

TABLE 3.2.90 Suppliers and familiar trade names of polysulphones

Supplier	*Trade name(s)*	*Notes*
Polysulphone		
Amoco	Udel, Mindel M	Also known as 'Polysulphone'
Courtaulds	Grafil	C fibre-filled
Wilson Fibrefil International SA	Sulfil	Glass-filled, PTFE-filled
Vigilant Plastics	RTP	Glass-filled
LNP	Thermocomp	Glass-filled, C fibre-filled, PTFE-filled
Freeman Chemicals Ltd	Lasulf	Unfilled, glass-filled
BASF	Ultrason	Unfilled, glass-filled
Ferro (GB) Ltd	Starglass	Glass-filled
Polyether sulphone		
ICI	Victrex, PES	Unfilled, glass-filled
BASF UK	Ultrason F	Unfilled, glass-filled
LNP UK	Thermocomp	Glass-filled, C fibre-filled
Vigilant Plastics	RTP	Glass-filled
Courtaulds	Grafil	C fibre-filled
Polyphenylsulphone		
Union Carbide	Radel	Not available in Europe
Bayer UK	Tedur	Glass-filled
LNP UK	Thermocomp	Glass-filled, C fibre-filled, PTFE-filled
Phillips Quadrant	Ryton	Glass-filled
Vigilant Plastics	RTP	Glass-filled, C fibre-filled
Freeman Chemicals Ltd	Larton	Glass-filled
Celanese Corp	Fortron	Glass-filled + beads
Ferro (GB) Ltd	Star–C	C fibre-filled
Courtaulds	Grafil	C fibre-filled
ABS/Polysulphone alloy		
Amoco	Mindel A	Previously Uniroyal 'Arylon T'
BASF (UK)	Terluran	
Glass Filled Modified Polysulphone		
Union Carbide	Mindel B	
SAN/Polysulphone Alloy		
Union Carbide	Ucardel	
Polyarylsulphone		
Amoco Chemicals		

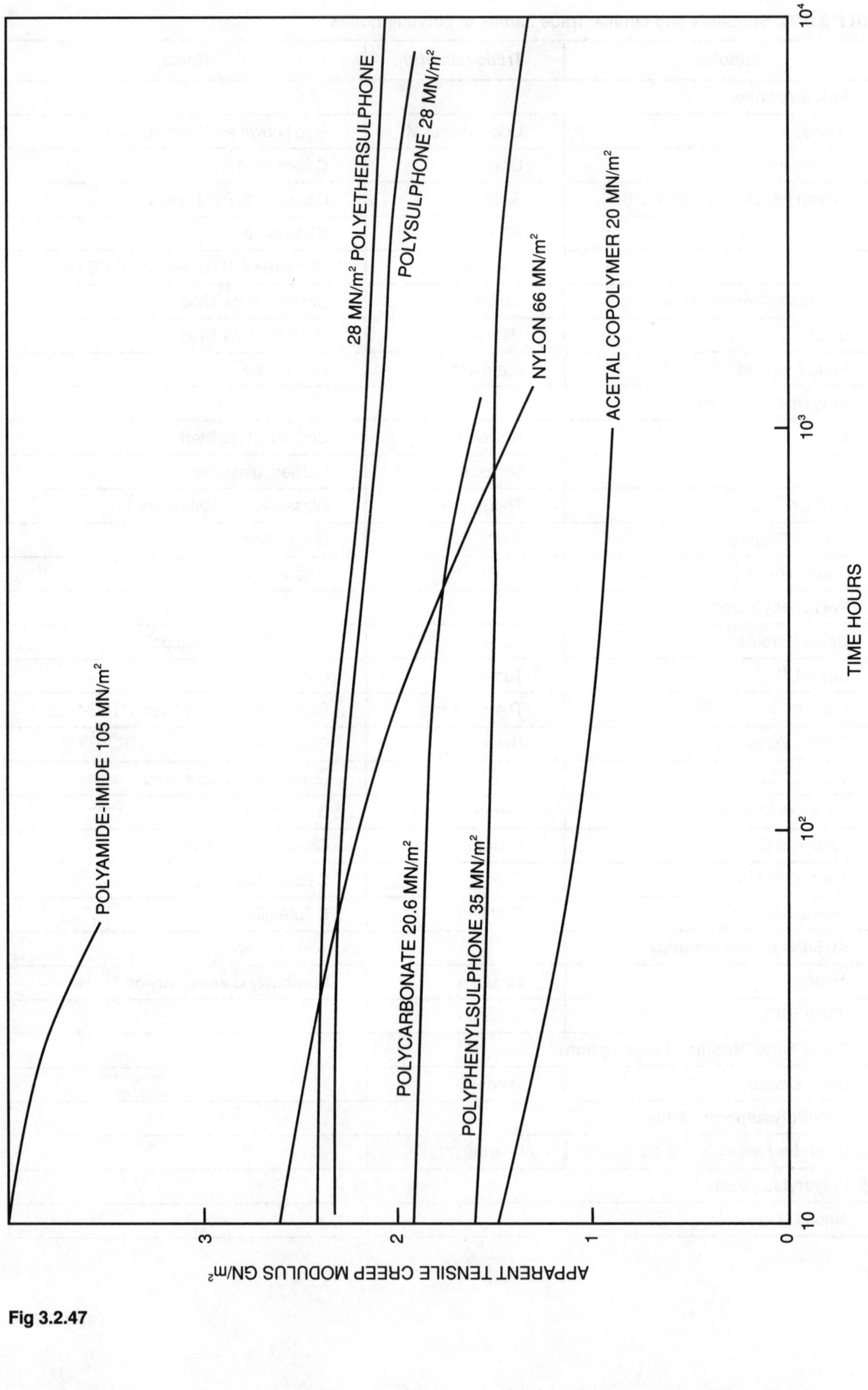

FIG 3.2.47 Time dependence of Tensile Creep Modulus of some unfilled thermoplastics at 20°C

Fig 3.2.47

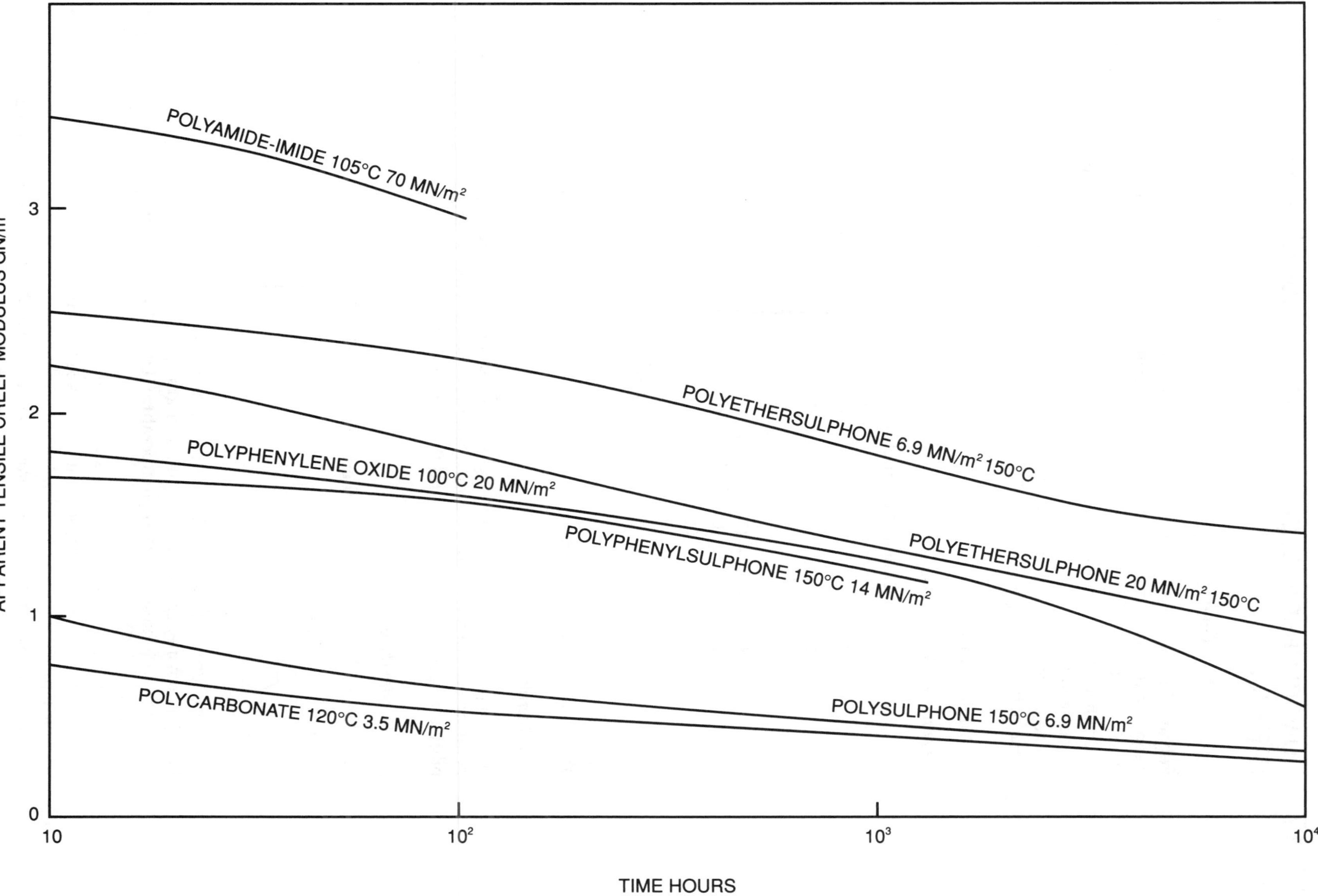

FIG 3.2.48 Time dependence of Tensile Creep Modulus of some unfilled thermoplastics at elevated temperatures

Fig 3.2.48

3.2.17 Polyamide-imide, polyetherimide, polyetheretherketone (PEEK)

3.2.17.1 General characteristics and properties

Polyamide-imide, polyetheretherketone (a crystalline modification of polyethersulphone) and most recently polyetherimide augment the polysulphones, as high-temperature polymers. The highest temperature performance is obtained from polyetheretherketone, whilst polyetherimide offers an alternative profile and very good processability.

The advantages and limitations of the imide ketone polymers are compared with those of the polysulphones in Table 3.2.86, Section 3.2.16.

AVAILABLE TYPES

Polyamide-imide

The general repeating chemical structure of polyamide imide is:

$$\left[-R-\overset{H}{\underset{}{N}}-\underset{\underset{O}{\|}}{C}-C_6H_3\begin{matrix}\overset{\overset{O}{\|}}{C} \\ \underset{\underset{O}{\|}}{C}\end{matrix} N \right]_n$$

The aromatic imide group confers rigidity while the amide link provides some flexibility for processing.
There are a number of grades with different fillings listed in:
Table 3.2.91 *Polyamide-imide, grades, fillings, properties and applications*

Polyetherimide

As with polyamide-imide, high rigidity, creep resistance and high heat deflection temperature are conferred by an aromatic imide group while a long linking confers flexibility for good melt processing and flow behaviour.

Polyetheretherketone

The chemical structure of polyetheretherketone (PEEK) is:

$$\left[-C_6H_4-\overset{\overset{O}{\|}}{C}-C_6H_4-O-C_6H_4-O- \right]_n$$

This structure confers a number of key properties:
Excellent tensile, impact and fatigue strength.
Very high continuous working temperature (UL 250°C), with good retention of properties at over 300°C.
Excellent flammability rating and low emission of smoke and toxic gas during combustion.
Insolubility in all common solvents and excellent resistance to most organic and inorganic liquids.
Resistance to hydrolysis at temperatures in excess of 250°C in steam and high-pressure water.

Exceptional resistance to γ radiation (for a plastic).
Excellent tribological properties.
Excellent electrical properties over a wide range of temperatures and frequencies.
Easy processing both unreinforced and reinforced with glass or carbon fibre.

APPLICATIONS

Polyamide-imide

The applications of polyamide-imide are listed in Table 3.2.91.

Polyetherimide

Polyetherimide is suitable for medical products that come into contact with body fluids and require repeated sterilisation by dry heat, autoclaving, ethylene oxide, gamma radiation, disinfectants and germicides. These include:

Nerve stimulators.
Patient humidifiers.
Surgical lights.
Manifold valves for ventilators.
Endodontic cases.

Other uses include:

Steam iron components.
Microwave ovens, parts and trays.

Polyetheretherketone (PEEK)

PEEK is used in applications where its exceptional high temperature resistance, burning characteristics, and environmental resistance justify its relatively high cost.
These include:

Cable insulation for oil and gas platforms and in warships where high temperature resistance and freedom from smoke generation is essential.
Bearings
Electrical connectors resistant to high temperature and soldering.
Pressurised water pumps and valves.
In film form for high performance filters and belting.
Demonstrator components for small, secondary structure aircraft parts.

TYPICAL PROPERTIES

The properties of the several grades of polyamide-imide, polyetherimide and polyetheretherketone are listed in:
Table 3.2.92 *Typical properties of high temperature polymers (imides and etheretherketone)*

3.2.17.2 Mechanical properties

The mechanical properties of the high-temperature polymers are listed in table 3.2.92 and the creep properties compared with those of other polymers in Figs 3.2.47 and 3.2.48, Section 3.2.16. The influence of temperature on strength is shown in:
Fig 3.2.49 *Tensile and flexural strengths of polyamide-imide as a function of temperature.*

3.2.17.3 Environmental resistance

The more important effects of environment on the high-temperature polymers are listed in Table 3.2.92.

Polyamide swells and its heat distortion temperature is reduced by absorption of moisture. Irradiation stiffens but weakens polyamide-imide.

3.2.17.4 Friction and wear properties

Polyamide-imide is an excellent material for bearings and its properties are enhanced by the addition of friction reducing additives.

Further information can be found in Vol 1, Chapter 1.5.

3.2.17.5 Processing

POLYAMIDE-IMIDE (TORLON®)

Injection moulding

Eight hours drying at 130°C is required to avoid brittle parts and/or melt foaming. Polyamide-imide is best processed at high injection speeds at which complex parts can be easily produced at modest pressure.

Gates should be kept as large as possible. The distance the resin is required to flow should be kept to a minimum. Weld lines should be kept to a minimum. Mould cavities should be well vented to prevent gas burns during rapid fill. Because polyamide-imide resins have a relatively low mould shrinkage (0.6–0.8%) moulds must be designed with large draft angles (c. 1.5°). Undercuts must be avoided and moulds should be draw polished to facilitate ejection. Highly-filled grades may show shrinkage values as low as 0.1%.

Moulds must be designed to be uniformly heated to 200°C or higher with good control. Typical cycle times for a 50 g component are in the region of 15–30s. Melt temperatures of 350–360°C are required.

Extrusion

Polyamide-imide may be extruded into profiles and shapes. Extruders must have heaters capable of operating at 340°C. The high melt viscosity of the resin restricts the complexity of parts able to be extruded.

POLYETHERIMIDE

Injection moulding is possible in sections down to 0.25mm. Melt temperatures are high, but a wide processing window (350—425°C) is permitted. Throat cooling is recommended at the top end of range.

Extrusion and blow moulding is possible at low shear rates (high melt strength). Moisture content must be reduced to 0.05% max. by drying.

PEEK

PEEK is marketed in granular form and can be processed on most forms of extrusion and injection moulding equipment, provided heating capacity is adequate: the processing temperature lies between 350 and 400°C.

3.2.17.6 Trade names and suppliers

Suppliers and familiar trade names are listed in:

Table 3.2.93 *Suppliers and familiar trade names of polyamide-imide, polyetherimide and polyetheretherketone*

TABLE 3.2.91 Polyamide–imide grades, fillings, properties and applications

Grade no.	*Nominal composition*	*Description of properties*	*Applications areas*
4203	3% TiO_2	Best impact resistance. Good electricals. Most elongation.	Electrical, electronic, insulating parts. Structural uses where impact is important. Extrusion.
4203L	3% TiO_2 ½ % PTFE	Same as 4203, but better mould release.	Same as 4203, except designed for injection moulding.
4301	12% Graphite powder, 3% PTFE	Designed for bearing use, good wear resistance, low coefficient of friction. High compressive strength.	Bearings, thrust washers, wear pads, or strips. Piston rings, seals.
4275	20% Graphite powder, 3% PTFE	Similar to 4301 but with better wear resistance.	Bearings, thrust washers, wear pads, or strips. Piston rings, seals.
5030	30% Glass fibre, 1% PTFE	High stiffness, good retention of stiffness at elevated temperature, very low creep, high strength.	Structural, electrical, valve plates, pistons. Metal replacement.
6000	30% Mineral, 1% PTFE	Lower cost, good high temperature properties, good electricals.	Electrical, structural, thermal insulation.
7030 (XG549)	30% Graphite, 1% PTFE	Similar to 5030, but higher stiffness. Best retention of stiffness at high temperature. Best fatigue resistance.	Metal replacement, where ultimate performance is required.

TABLE 3.2.92 Typical properties of high-temperature polymers (imides and etheretherketone)

	Property		ASTM Test	Units	Polyamide-imide			
					4203/ 4203L	4301	4275	5030
Mechanical	Tensile strength	20°C	D638	MN/m^2	185	135	125	195
		150°C	—	MN/m^2	105	73	93	136
		260°C	—	MN/m^2	52	46	44	84
	Elongation at break	20°C	D638	%	12	6	6	5
	Flexural strength	20°C	D790	MN/m^2	212	182	175	318
	Modulus of Elasticity in Tension at 20°C		D638	GN/m^2	—	—	—	—
	Flexural Modulus	20°C	D790	GN/m^2	4.54	6.34	7.17	11.09
		150°C	—	GN/m^2	3.58	5.03	5.93	10.47
		260°C	—	GN/m^2	2.96	4.00	4.00	8.41
	Impact strength	20°C	D638	J/m notch	135	59	65	108
	(Izod notched)	–50°C	—	—	49	—	—	—
	Hardness (Rockwell)		D785	—	E78	M109	—	—
Physical	Specific gravity		D792	—	1.40	1.45	1.46	1.57
	Thermal conductivity		—	W/m per K	0.24	0.36	—	—
	Thermal expansion		D696	10^{-6} K^{-1}	36	27	23	18
Electrical	Volume resistivity		D257	ohm m	1.2×10^{15}	—	—	—
	Dielectric strength		D149	10^2 kV/m	236	—	—	—
	Dielectric constant	10^3 Hz	D150	—	3.5	—	—	—
		10^6 Hz	—	—	4.0	—	—	—
	Power factor	10^3 Hz	D150	—	0.001	—	—	—
		10^6 Hz	—	—	0.009	—	—	—
	Arc resistance		D495	s	125	—	—	—
Miscellaneous	Max. continuous service temperature		—	°C	210	210	210	210
	Hear distortion temp.	0.45/MN/m^2	D648	°C	—	—	—	—
		1.8/MN/m^2	D648	°C	274	274	266	274
	Flammability		UL94	—	V–0	V–0	—	—
	Water absorption		D570	% (24 h)	0.28	0.22	0.19	0.22
	Taber abrasion CS17		D1044	mg/1000 rev.	—	—	—	—
	Mould shrinkage		—	%	0.7	—	—	

[a] O^2 index 47 ‡ 40% glass.

Table 3.2.92

TABLE 3.2.92 Typical properties of high-temperature polymers (imides and etheretherketone)—*continued*

Polyamide-imide		*Polyetherimide*		*Polyetheretherketone*		
6000	*7030 (XG 549)*	*Unfilled*	*30% glass*	*Unfilled 450G*	*30% glass-fibre 450GL3D*	*30% carbon-fibre 450CA3D*
145	205	105	160	62	172	240
107	142	—	—	—	70	120
28	74	—	—	—	40	50
5	6	60	3	4.7	3.6	1.6
212	316	145	230	—	234	—
—	—	3.0	9.0	—	8.65	—
7.85	17.85	3.3	8.3	3.8	7.6	13.7
6.34	15.09	—	—	7	8	12
3.79	14.12	—	—	0.5	1.7	4
76	65	50	100	85	110	98
—	—	—	—	—	—	—
—	—	M109	M125	M90	M98	M107
1.54	1.42	1.27	1.51	1.31	1.49	1.4
—	—	0.22	—	—	—	—
29	11	62	20	48	22	14
—	—	6.7×10^{15}	3×10^{14}	10^{14}	10^{14}	—
—	—	330	300	190	190	—
—	—	3.15	3.5	3.2	3.3	—
—	—	—	—	—	—	—
—	—	0.0013	0.0015	0.0016	0.002	—
—	—	—	—	—	—	—
—	—	128	85	—	—	—
210	210	170	170	250	250	250
—	—	210	212	—	—	—
270	277	200	210	160	288	292
—	—	V–0[a] (0.76 mm)	V–0 (1.6 mm)	V–0	V–0	V–0
0.20	0.22	0.25	0.18	0.15	0.18	0.15
—	—	10	—	—	—	—
—	—	0.5–0.7	0.2	1.1	0.5	0.9

Table 3.2.92—*continued*

TABLE 3.2.93 Suppliers and familiar trade names of polyamide-imide, polyetherimide and polyetheretherketone

Supplier	*Trade name*	
Polyamide-Imide		
Amoco	Torlon	UK distributor Polypenco
Polyetherimide		
General Electric	Ultem	UK distributor Engineering Polymers
Polyetheretherketone		
ICI	Victrex PEEK	

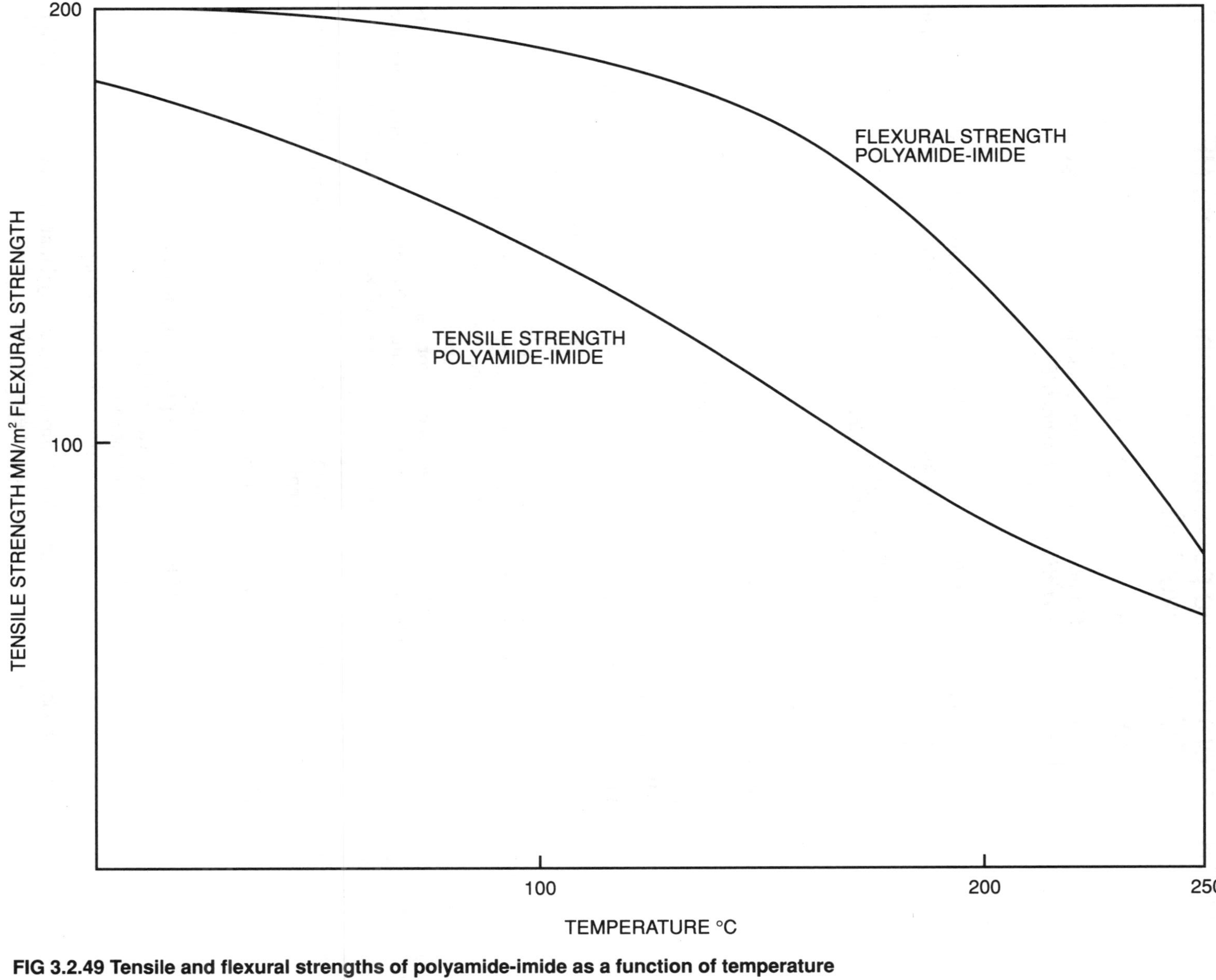

FIG 3.2.49 Tensile and flexural strengths of polyamide-imide as a function of temperature

3.2.18 Thermoplastic polyesters (terephthalates) PETP, PBT, PCTG (copolyester)

3.2.18.1 General characteristics and properties

The advantages and limitations of polyesters (terephthalates) are listed in:
Table 3.2.94 *Characteristics of terephthalates (PBT, PETP and PCTG)*

AVAILABLE TYPES

The thermoplastic polyesters described in this section should not be confused with unsaturated polyesters which are thermosetting materials and are described in Chapter 3.3. Two principal types of thermoplastic polyester resins are available and there are a number of copolyesters. Polybutylene terephthalate (PBT) (also known as polytetramethylene terephthalate PTMT) has the formula:

$$HO — (OC —\langle\bigcirc\rangle— COOCH_2CH_2CH_2CH_2O)_nH$$

and polyethylene terephthalate (PETP or PET) the formula:

$$HO — (OC —\langle\bigcirc\rangle— COOCH_2CH_2O)_nH$$

APPLICATIONS

PBT	Electrical	Coil bobbins, connectors, TV tuners, high voltage components, relays, potentiometers, terminal and motor blocks, fuse cases, edge block connectors for electrical/ electronic interfacing.
	Automotive	Fuel injection controls, ignition coil caps, coil bobbins, distributor caps, housings and rotors, rotor end-caps, speedometer frames, gears.
	Domestic appliances	Base plates for electric coffee makers, toasters, switches, coil cylinders; control programmers in washing machines and dishwashers; control levels, housings and covers.
	Industrial	Pump housings, support brackets, waterheater components.
PETP	Synthetic fibre	Screen and sewing thread.
	Biaxially oriented film	Magnetic tape. X-ray and photographic film. Electrical insulation. Printing sheets. Food packaging (boil-in-bag)
Amorphous PETP		Most important industrial material for blow moulded components (bottles).
	Metal-filled and glass-reinforced	Automotive components. Electrical components Appliance bases Power tool housings.
PCTG		Packaging and bottles

TYPICAL PROPERTIES

Thermoplastic polyesters have the following characteristics:

Unfilled crystalline PETP	Excellent mechanical, electrical and chemical properties. Poor processability, low heat distortion temperature, poor impact resistance.
Unfilled amorphous PETP	Improved processability and transparency at the expense of stiffness and dimensional stability.
Glass fibre-reinforced PETP	Spectacular increase in heat distortion temperature. Significant improvement in impact strength. Processability remains poor.
PBT unfilled	Slightly inferior mechanical properties to unfilled PETP, but equivalent electrical and chemical properties. Superior impact and toughness. Excellent processability. Transparency not possible with unfilled PBT.
Filled PBT	Only slightly lower thermal and mechanical properties to filled PETP. Equivalent chemical and electrical properties. Smooth appearance
PCTG (filled or unfilled)	Where weight or surface finish is important.

Typical properties of thermoplastic polyesters are listed in:
Table 3.2.95 *Typical properties of thermoplastic polyesters (PBT, PETP, PCTG)*
and shown in
Fig 3.2.50 *Tensile strength vs elongation for PBT and PETP*

The water absorption characteristics of reinforced PBT resins are among the lowest of all thermoplastics and lower than most thermosetting materials. Electrical properties (dissipation factor, dielectric constant, dielectric strength and insulation resistance) are hardly affected. However, the resins cannot be used in water above 70°C for long periods.

3.2.18.2 Creep properties

The creep resistance of reinforced PBT rivals thermosets such as general purpose phenolics, alkyds and DAP. Creep strength of reinforced PBT is shown in:
Fig 3.2.51 *Flexural creep of glass-reinforced PBT*

3.2.18.3 Processing

INJECTION MOULDING

Typical mould shrinkage figures are as follows:

		Parallel to direction of flow	Perpendicular to direction of flow
PETP	Unfilled amorphous: mould 20°C	0.2–0.4%	0.2–0.4%
	Unfilled crystalline: mould 130°C	1.6–1.9%	1.6–1.9%
	Glass-reinforced: mould 130°C	~0.3%	0.8–1.4%
PBT	Unfilled: mould 90°C	1.3–1.8%	1.3–1.8%
	Glass reinforced: mould 130°C	~0.3%	0.7–1.7%

PETP

PETP has poor processability due to slow crystallisation rate. Higher mould temperatures (130–150°C) and long cycle times are necessary to overcome differential shrinkage, increased stresses, non-uniform properties and warpage. Amorphous grades have better processability with mould temperature down to 20–40°C.

Modified grades are available which are as easy to process as PBT. These include Rynite 530 (30% glass) and 545 (45% glass) (Du Pont) and a 30% glass-filled PETP–acrylic alloy (Rohm & Haas).

The oriented structures which are essential in fibre or film for subsequent blow moulding are developed from an amorphous preform. This structure results from very rapid chilling of the melt; quenching in air is adequate for fine filaments, a water cooled roll for thin film. Heavier gauge filaments require direct water cooling. The PETP is then heated above its Tg and drawn with a ratio between 4 and 7 to 1 for uniaxial drawing and between 1.5 and 4 to 1 for biaxial drawing.

In bottle-making, PET preforms are injection moulded in a water cooled mould and blow moulded in a second mould above the Tg.

PBT

Excellent processability, very fast cycle times, wide range of suitable mould temperatures (30–90°C, preferably over 60°C).

EXTRUSION

Thermoplastic polyesters can be extruded instead of injection moulded and lend themselves to advanced processing techniques based upon stretching the extruded product. Examples are the production of stretched tapes, monofilament and tubing (e.g. synthetic grass, woven fabric, hinge pins).

FABRICATION

Machining is possible (e.g. for short run injection moulded components with large section changes).
Solvent bonding is not effective; thermosetting adhesives may be used.
Friction and ultrasonic welding give good results. Heated tool welding and thermal impulse welding are less suitable.

3.2.18.4 Trade names and suppliers

Suppliers of terephthalates and familiar trade names are listed in:
Table 3.2.96 *Major suppliers and familiar trade names of thermoplastic polyesters*

TABLE 3.2.94 Characteristics of terephthalates (PBT, PETP and PCTG)

	Advantages	*Limitations*
Mechanical properties	Good creep and fatigue resistance. 45% glass-filled grade of PETP introduced by Flexible copolyesters obtained by the addition of butadiene-containing polymers or thermoplastic urethane elastomers have improved notch toughness; at equivalent stiffness the flexible grades of PBT have better temperature than nylon 11/12 and lower water absorption.	Inferior screw-torque retention to nylon. Notch sensitive.
Specific gravity	For equivalent glass content PCTG is lower.	
Wear	Good wear resistance and low friction.	
Environment	Resistant to dilute acids and alkalis, aliphatic hydrocarbons, carbon tetrachloride, oils, fats, alcohols, ethers, esters, high mol.wt ketones and detergents.	Swollen by ethylene dichloride, dimethyl, formamide, low mol.wt ketones and some aromatics including phenol. Attacked by strong acids and alkalis, formic acid. Prolonged use in water or aqueous solutions above 70°C not recommended. Black pigmented grades necessary for outdoor use.
Electrical properties	Better consistency of electrical properties than nylons. Good under severe moisture and temperature conditions.	
Thermal properties	Good heat resistance. Continuous use 120–200°C depending on grade. Unnotched sample rated as non-breaking down to –40°C. Lower water absorption. Good colour retention during high temperature ageing.	Modulus of unreinforced PBT decreases by a factor of 10 between 20 and 80°C.
Processing	Low mould shrinkage and high dimensional stability (including wet environments). Most types have faster processing cycles than acetals, nylons and polypropylenes. 15–25% CDPE may be added to improve processability and decrease moisture absorption. 10–30% nylon additions assist ability to accept high glass fibre loadings. PCTG is claimed to have better mouldability and better surface finish on moulded items.	Crystalline PETPs have poor processability compared to amorphous PETPs. Expensive on raw material cost basis but can be offset by processing advantages against competitive materials. Painting of PBT more difficult than amorphous plastics. Inorganic pigments can reduce impact strength by 20–25%.
Flammability	Flame retardancy in PBT may be by additives or copolymerisation.	Additives reduce toughness. All flame retardant grades have reduced tracking resistance.

TABLE 3.2.95 Typical properties of thermoplastic polyesters (PBT, PETP, PCTG)

	Property		ASTM Test	Units	PETP amorphous unfilled IM grade	PETP crystalline unfilled IM grade	PETP crystalline 33% glass fibre IM grade
Mechanical	Ultimate tensile strength		D638	MN/m^2	—	—	—
	Tensile yield strength		D638	MN/m^2	55	73–81	165
	Flexural strength		D790	MN/m^2	85	118	250
	Elongation at break		D638	%	300	45–300	2
	Tensile Elastic Modulus		D638	GN/m^2	2.2–2.5	2.8	11.5
	Flexural Modulus		D790	GN/m^2	2.3	3.0–3.3	8.1-12.0
	Impact (Izod)	23°C	D256A	J/cm notch	0.2	0.2–0.5	0.6–1.0
		–40°C			—	—	—
	Hardness		D785	Rockwell M	25	85	102
Physical	Specific gravity		D792	—	1.32–1.34	1.37	1.73
	Thermal conductivity		C177	W/m per K	0.24	0.29	0.33
	Thermal expansion coefficient		D696	10^{-5} K^{-1}	8	7	2–7[a]
Electrical	Volume resistivity		D257	ohm m	2×10^{12}	2×10^{12}	2×10^{12}
	Dielectric strength (short time)		D149	10^2 kV/m	180	200	170–190
	Dielectric constant	60Hz	D150	—	3.6–4.2	3.4–4.3	4.0
		10^6 Hz			3.2–3.4	3.2–3.5	3.8
	Power factor	60 Hz	D150	—	0.0015–0.0020	0.0015–0.0020	0.002
		10^6 Hz			0.020–0.030	0.019–0.021	0.015
	Arc resistance		D495	s	70	50	—
Thermal/Environmental	Maximum continuous service temperature (air)			°C	—	—	>200
	Heat deflection temperature	0.45 MN/m^2	D645	°C	72	98–115	>250
		1.8 MN/m^2			66–70	50–82	205–235
	Flammability		D635	UL94	94HB	94HB	94 V–0
	Water absorption (24 h)		D570	%	0.16	0.10	0.04
Processing	Mould shrinkage		D955	%	0.2	1.2–2.0	0.2–0.9

[a] Higher figure across flow direction grade.
[b] In water.

Table 3.2.95

TABLE 3.2.95 Typical properties of thermoplastic polyesters (PBT, PETP, PCTG)—*continued*

PBT crystalline unfilled IM grade	*PBT crystalline 30% glass fibre IM grade*	*PCTG modified 20% glass 7403*	*PCTG modified 25% glass 7401*	*PCTG modified 30% glass 7402*	*PCTG modified 30% glass 7401*	*PETP UV/ stablilised*	*PETP Fire retardant*
—	—	100	86	93	110	—	—
52–60	124–135	—	—	—	—	50	50
85	130–200	145	124	138	159	—	—
200–250	2–3	<5	<5	<5	<5	200	180
2.6–2.7	6.0–10.2	—	—	—	—	—	—
1.6–2.8	6.5–10.3	6.8	5.4	6.2	7.6	2.3	2.3
0.25–0.6	0.6–1.0	0.64	0.96	0.96	0.96	0.30	0.3
—	—	0.64	0.64	0.96	0.96	—	—
80	85–87	82	—	—	—	R68	R75
1.30–1.31	1.53–1.69	1.39	1.40	1.42	1.45	1.38	1.55
0.21	0.24	—	—	—	—	—	—
7	4–8[a]	—	—	—	—	8	8
5×10^{12}	$7–8 \times 10^{12}$	—	—	—	—	10^{11}	10^{10}
150	170	—	—	—	—	200	150
3.0–3.8	3.3–4.4	—	—	—	—	—	—
2.8–3.2	3.2–3.8	—	—	—	—	—	—
0.001–0.002	0.0025	—	—	—	—	—	—
0.017–0.020	0.012–0.018	—	—	—	—	—	—
150	110–175	—	—	—	—	—	—
120–137 (65[b])	120 (65[b])	—	—	—	—	115	115
160–170	215–220	—	—	—	—	150	150
50–70	200–205	215	218	>220	>220	70	70
94HB/V–0	94HB/V–0	—	—	—	—	HB	V0>1
0.09	0.06	—	—	—	—	0.15	0.15
1.2–2.2	0.2–0.8	—	—	—	—	2	2

Table 3.2.95—*continued*

TABLE 3.2.95 Typical properties of thermoplastic polyesters (PBT, PETP, PCTG)—*continued*

	Property		ASTM Test	Units	PETP Mineral filled	PBT UV/ stabilised	PBT Fire retardant	PBT 30% glass bead
Mechanical	Ultimate tensile strength		D638	MN/m²	—	—	—	—
	Tensile yield strength		D638	MN/m²	100	50	50	80
	Flexural strength		D790	MN/m²	—	—	—	—
	Elongation at break		D638	%	2.2	200	15	3
	Tensile Elastic Modulus		D638	GN/m²	—	—	—	—
	Flexural Modulus		D790	GN/m²	9.5	2.1	2.7	5
	Impact (Izod)	23°C	D256A	J/cm notch	0.6	0.5	0.5	0.3
		–40°C			—	—	—	—
	Hardness		D785	Rockwell M	M100	M70	M80	M90
Physical	Specific gravity		D792	—	1.58	1.30	1.43	1.52
	Thermal conductivity		C177	W/m per K	—	—	—	—
	Thermal expansion coefficient		D696	10^{-5} K^{-1}	2	12	11	6
Electrical	Volume resistivity		D257	ohm m	10^{11}	10^{12}	10^{13}	10^{11}
	Dielectric strength (short time)		D149	10^2 kV/m	200	200	200	180
	Dielectric constant	60Hz	D150	—	—	—	—	—
		10^6 Hz			—	—	—	—
	Power factor	60 Hz	D150	—	—	—	—	—
		10^6 Hz			—	—	—	—
	Arc resistance		D495	s	—	—	—	—
Thermal/Environmental	Maximum continuous service temperature (air)			°C	140	120	120	120
	Heat deflection temperature	0.45 MN/m²	D645	°C	220	150	180	220
		1.8 MN/m²			215	60	60	200
	Flammability		D635	UL94	HB	HB	V0>1	HB
	Water absorption (24 h)		D570	%	0.06	0.13	0.1	0.15
Processing	Mould shrinkage		D955	%	0.5	2	2	0.6

Table 3.2.95—*continued*

TABLE 3.2.95 Typical properties of thermoplastic polyesters (PBT, PETP, PCTG)—*continued*

PBT 30% carbon fibre	*PBT 2% silicone lubricated*	*PBT PTFE lubricated*	*PBT 45% mineral & glass-filled*	*PBT Structural foam*	*PET 55% glass-filled*	*PET 35% Mica/GFR*	*PET 35% gfr super tough*
—	—	—	—	—	—	—	—
152	55	42	85	55	196	94	103
—	—	—	—	—	—	—	—
2	150	15	2	4	2	2	6
—	—	—	—	—	—	—	—
16	2.3	1.7	8	2.2	17.9	9.6	6.9
0.6	0.5	0.5	0.4	0.5	1.2	0.6	2.4
—	—	—	—	—	—	—	—
M90	M70	M70	M85	M50	M100	M85	M62
1.41	1.29	1.42	1.8	1.29	1.8	1.58	1.51
—	—	—	—	—	—	—	—
1	10	10	7	9	1.4	2.7	2
1	10^{14}	10^{14}	10^{14}	10^{13}	10^{11}	10^{11}	10^{11}
—	200	200	170	180	200	230	210
—	—	—	—	—	—	—	—
—	—	—	—	—	—	—	—
—	—	—	—	—	—	—	—
—	—	—	—	—	—	—	—
—	—	—	—	—	—	—	—
115	120	120	120	120	140	140	140
225	180	190	210	140	245	230	235
221	80	80	190	80	229	215	220
HB	HB	HB	V–0	HB	HB	HB	HB
0.04	0.07	0.05	0.07	0.07	0.04	0.05	0.25
0.15	2	2	0.7	1.5	0.45	0.6	0.55

Table 3.2.95—*continued*

TABLE 3.2.96 Major suppliers and familiar trade names of thermoplastic polyesters

Company	*Trade name*	*PBT*	*PETP*
Akzo	Arnite	U, G	U, G, V, F
Amcel[a] (UK)	Kelanex	U, G	—
Allied Chemicals	Versel	U, G	—
(Atochem)	Orgator	U, G, V, F, S	—
BASF	Ultradur	U, G, S, V, F, M	—
Bayer	Pocan	U, G, V, F, S	—
BIP	Beetle	—	U, V, G
Celanese	Celanex	U, G, V, F	—
Ciba Geigy	Crastine *(formerly Tenatine)*	U, G, S, V	—
Du Pont	Dacron	Fibre	—
	Rynite	—	G, GX, M
Dynamit Nobel	PTMT	U, G	U, G
Eastman	Tenite, Kodar	U, G	U, G
	Kodapak	—	Film
EMS Grilon	Grilpet	—	U, V, F, M, G
Ferro (UK) Ltd	Starglas, Star	G, C	—
Freeman Chemicals Ltd	Latex	U, V, F, G, S, PTFE, M	—
GAF	Gafite	U, G,	—
General Electric[b]	Valox	U, G, M, V, F, Struct. foam	—
Goodyear	Videne	—	Film
Hoechst	Hostadur, Polyclear	U, G, V	U, G, V
	Hostaphan	—	Film
Huls	Vestodur	U, V, F, G, S, M	—
ICI	Terylene, Melinar	—	Fibre, U
	Melinex	—	Film
LNP	Thermocomp	G, C, PTFE, Silicone, F	—
3M	Scotchpark	—	Film
Mitsui	PBT	U, G	—
Mobay (US)	Petlon	—	U, G, M
Montecatini	Pibiter	U, G	—
Polyplastics[a] (Japan)	Duranex	U, G	—
Rohm & Haas	Ropet	—	U, Acrylic
Rhone Poulenc	Rhodester, Techster	U, G, V, F, M	G
Vigilant	RTP	G, S, PTFE, F	G
Copolyester			
Eastman	Ektar	Mr	
	Kodar A150	Mr	

[a]Subsidiary of Celanese
[b]UK — Engineering Polymers

U = Unfilled.
G = Glass-filled.
C = Carbon fibre-filled.
S = Glass sphere-filled.
M = Mineral-filled.
GX = Glass–mica-filled.
Mr = Modifier.
V = UV stabilised.
F = Fire retardant.

Table 3.2.96

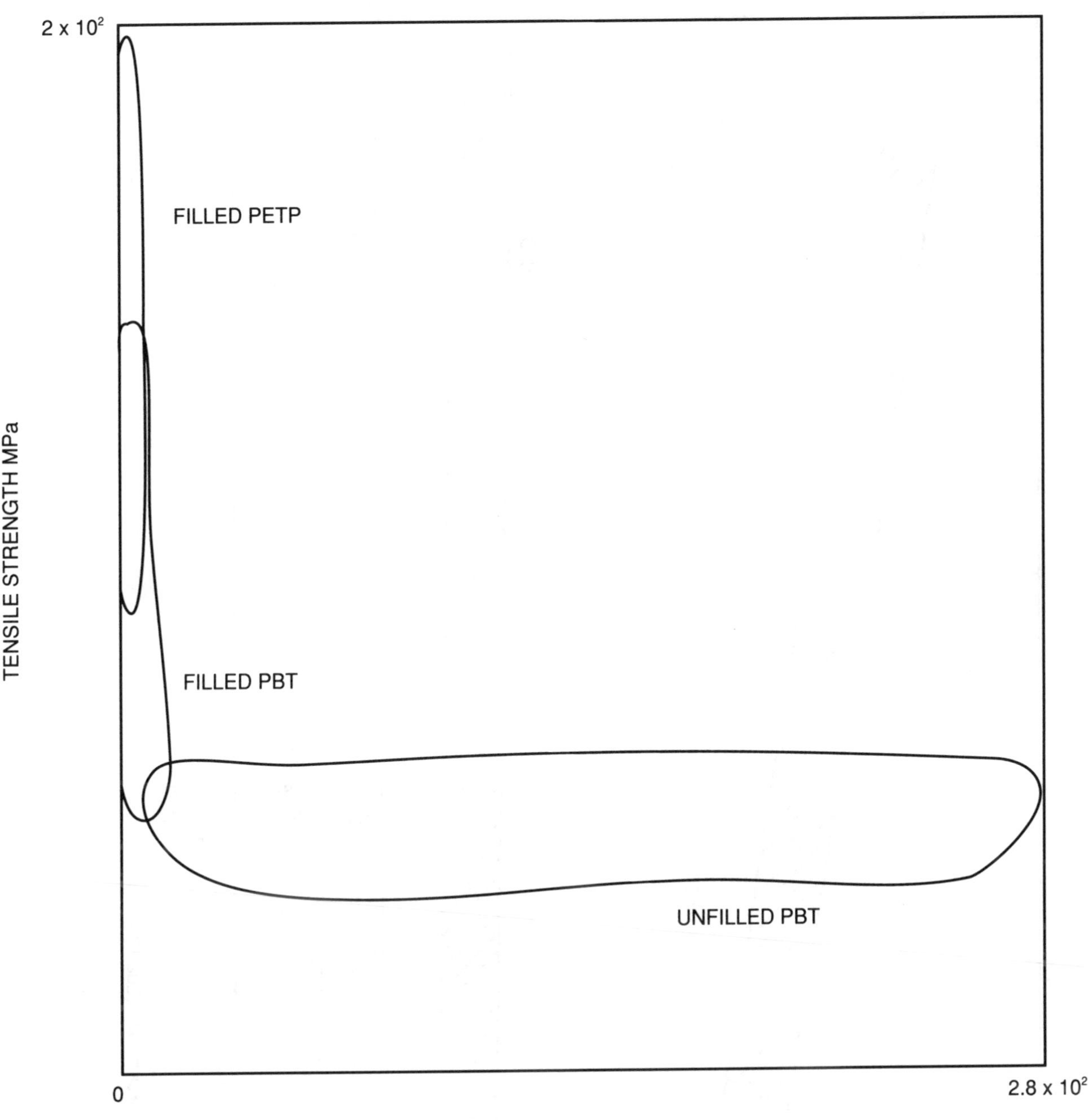

FIG 3.2.50 Tensile strength vs elongation for PBT and PETP

PBT – Polybutylene Terephthalate
PETP – Polyethylene Terephthalate

Fig 3.2.51

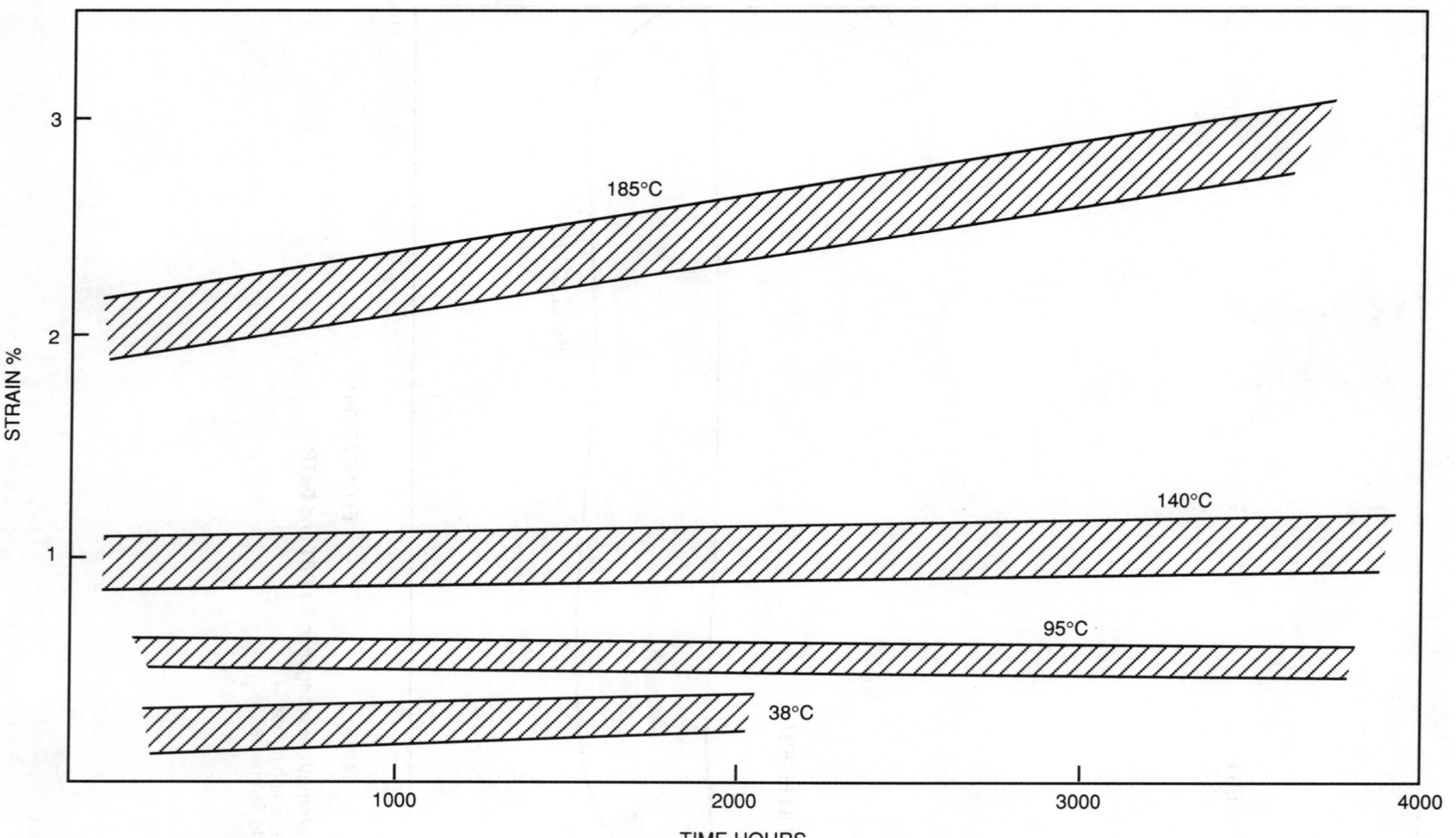

FIG 3.2.51 Flexural creep of glass-reinforced PBT (Load 14 MN/m^2)

3.2.19 Oxybenzoyl polyesters and polyarylates (aromatic polyesters)

3.2.19.1 General characteristics and properties

The advantages and limitations of oxybenzoyl polyesters and polyarylates are listed in:
Table 3.2.97 *Characteristics of aromatic polyesters*

AVAILABLE TYPES

Oxybenzoyl polyesters

Oxybenzoyl homopolyester is a linear polymer of repeating polyoxybenzoyl units with a high degree of crystallinity. It does not melt below its decomposition temperature (550°C) but can be compression sintered in the range 330–360°C.

Plasma-sprayed coatings made from the homopolyester are self-lubricating, wear and corrosion resistant and thermally stable. Spraying techniques are relatively inexpensive. Coating thickness 0.01–3 mm. The material is impervious when polished and has good dielectric properties. The homopolymer can be blended with up to 25% PTFE. Copolyester can be compression sintered or injection moulded. Both grades are available with 30% glass filling.

Polyarylates

Polyarylates are amorphous and naturally transparent (70–75% light transmittance) with a light golden colour. Polyarylate is available as the homopolymer or grades with up to 40% glass fibre.

APPLICATIONS

The major applications for oxybenzoyl polyesters depend on its friction, wear and anti-seizure properties.

Plasma-sprayed coatings and PTFE blended homopolyester: Self-lubricating bearings, bearing races, abradable seals, rotors or vanes in processing pumps.

Compression moulded copolyester: Self-lubricating bearings, bearing cages, valve seats, abradable seals, electrical connectors, packaging equipment, radar windows, heat sealing devices.

Injection moulded copolyester: Electrical connectors, valve seats, high performance aircraft parts and automotive parts.

The high flexural recovery and high modulus of polyarylate makes it a candidate for snap fit applications such as connectors, fasteners, and hinges. Its additional properties of thermal resistance and transparency lead to its use in hot combs, high temperature lenses, solar energy collectors, lighting and tinted glazing. Its properties are favourable for appliances, and electrical/electronic hardware.

TYPICAL PROPERTIES

The properties of grades of aromatic polyesters are listed in:
Table 3.2.98 *Typical properties of aromatic polyesters (oxybenzoyl and polyarylate)*

3.2.19.2 Processing

Benzoyl homopolyester and blends with powdered aluminium, bronze, nickel, chrome and PTFE can be plasma-and flame-sprayed. Compression sintering grades are also available. The compression grade of benzoyl copolyester has no true melting range and is thus compression sintered. Injection moulding is easily achievable on conventional equipment at barrel temperatures between 388 and 416°C, die temperature 93–315°C and pressures of 35–105 MN/m^2 (5000–15000 psi).

3.2.19.3 Trade names and suppliers

The suppliers and familiar trade names of aromatic polyesters are listed in:
Table 3.2.99 *Trade names and suppliers of benzoyl polyesters and polyarylates.*

TABLE 3.2.97 Characteristics of aromatic polyesters

	Oxybenzoyl		*Polyarylates*	
	Advantages	*Limitations*	*Advantages*	*Limitations*
Mechanical properties	Mechanical properties retained at high temperature. Outstanding in range 260–316°C. Compression copolyester has better compressive strength and heat deflection temperature than injection grade. Homopolymer exhibits no flow or creep below 330°C. All grades characterised by high flexural modulus.	Other materials have better mechanical properties in range 150–260°C.	Higher mechanical strength than PPS. High modulus and flexural recovery. Low notch sensitivity. Low creep rate.	
Thermal properties	Continuous use temperature 270–300°C. High temperature mechanical properties are improved by exposure to, and ageing at temperature in the range 260–290°C.	Polyamide-imide, polyimide, PFA, FEP have higher service temperatures.	Better heat stability than modified PPO.	Polysulphone and polyethersulphone have superior heat stability. Opaque grades have lower heat distortion temperature.
Wear products	Free sintered homopolymer–PTFE blends have better wear resistance than pure PTFE and do not wear the mating surfaces. Max. temp. 315°C. Compression copolyester can be filled with PTFE or graphite for longer life and higher temperature capability than PTFE or PTFE graphite.			
Environmental properties	Resistant to most aromatic and chlorinated hydrocarbons. Excellent resistance to radiation (10^{10} rads). Low water absorption. No corrosive outgassing at elevated temperatures.	Range of solvents so far tested not extensive. Cobalt 60 resistance under evaluation.	Self-extinguishing; some grades achieve UL94 V-O. Opaque grades have better chemical resistance. High stability to UV Radiation.	Environmentally induced stress-cracking—similar performance to polycarbonate.
Electrical properties	Potential for miniaturisation in electronics where higher heats are encountered.	Best electrical properties at high temperatures are obtained with polysulphones and fluoropolymers.		
Optical			Naturally transparent—light golden.	Transparency limited to 70–75%.
Processing	Homopolyester by compression sintering and plasma sprayed coatings. Copolyester can be compression sintered or injection moulded. Extremely thin sections, or thick sections without voids or sink marks can be moulded.		Can be injection moulded, extruded or blow moulded. Opaque grades easier to process.	Reinforced (opaque) grades have limited application potential.
Availability and cost		Very high cost. New material.		Much more expensive than PPO, PPS or thermoplastic polyesters, especially transparent grade.

TABLE 3.2.98 Typical properties of aromatic polyesters (oxybenzoyl and polyarylate)

	Property		ASTM Test	Units	Ozybenzoyl	
					Homopolymer	Copolymer compression grade C-1000 unfilled
Mechanical	Tensile strength	20°C	D638	MN/m²	18	69
		260°C		MN/m²	—	—
	Tensile Modulus			GN/m²	4	8
	Elongation			%	6.5	1.3
	Flexural strength		D790	MN/m²	35–40	103.4
	Flexural Modulus	20°C		GN/m²	7	3.1
		260°C		GN/m²	—	—
	Compressive strength		D695	MN/m²	110	138
	Impact toughness Izod	20°C	D256	J/cm notch	—	0.2
		–40°C			—	—
Physical	Specific gravity			—	1.45	1.35
	Thermal conductivity		C177	W/m per K	0.72	0.36
	Linear expansion coefficient		D696	10^{-5} K^{-1}	5.6	5.6
Electrical	Dielectric constant	60 Hz	D150	—	3.22	3.72
		10^6 Hz		—	3.28	3.64
	Power factor	60 Hz		—	—	0.005
		10^6 Hz		—	—	0.0151
	Dielectric strength		D149	10^2 kV/m	—	177
	Arc resistance		D495	s	—	—
Optical	Refractive index		—	—	—	—
Thermal/Envinmetal	Maximum use temperature		—	°C	315	285
	Deflection temperature	1.8/MN/m²	D648	°C	—	295
	Flammability		—	UL94	V–0	V–0
	Water absorption (24h)		D570	%	0.02	0.04
Processing	Mould shrinkage		D955	%	0.7	1.2

[a] High temperature impact strength after ageing 600 h at 360°C rises from 3.25 to 17 J/cm.

Table 3.2.98

TABLE 3.2.98 Typical properties of aromatic polyesters (oxybenzoyl and polyarylate)—*continued*

Oxybenzoyl			*Polyarylate*
Copolymer compression grade C-1006 30% glass bead	***Copolymer injection grade I-2000 unfilled***	***Copolymer injection grade I-2006 30% glass bead***	***Homopolymer***
75	96	76	75–71
—	17	—	—
5.2	2.4	2.4	2
2.8	8	7.5	8
90	117	76	75–103
6.65	4.8	6.2	2.0–2.1
—	1.7	—	—
151	124	62	—
0.3	0.5	1.0[a]	294
—	—	—	214
1.56	1.4	1.57	1.2
0.23	0.98	—	—
6.35	2.9	10.44	6.3
—	—	—	2.73–3.08
3.45	2.9	—	2.62–2.96
—	—	—	0.0001–0.001
0.009	0.025	—	0.020–0.022
>236	138	—	165–170
—	—	—	125
—	—	—	164
—	288	—	—
318	282–293	310	160
—	—	—	V–0 (1.6 mm)
0.05	0.01	0.025	0.2–0.27
—	1.2	0.9	0.7–0.9

Table 3.2.98—*continued*

TABLE 3.2.99 Trade names and suppliers of benzoyl polyesters and polyarylates

Supplier	*Trade name*	*Grade and filling*
Oxybenzoyl polyester		
Carborundum Co.	Ekonel P3000	Plasma-sprayed homopolyester
	Ekonel T4000	Polyester–PTFE homopolymer blend
	Ekonel C1000	Compression sintered
	Ekcel C1006	Compression sintered 30% glass copolymer
	Ekcel I2000	Injection moulded copolymer
	Ekcel I2006	Injection moulded 30% glass copolymer
Polyarylate		
Bayer	APE	
Hooker	Durel	
Solvay	Arylef	Arylef is U-Polymer manufactured under licence
Union Carbide	Ardel	
Unitika (Japan)	U-Polymer	

3.2.20 Chlorinated polyether

3.2.20.1 General characteristics and properties

ADVANTAGES

Greater thermal stability than other chlorinated resins. Outstanding chemical resistance to most reagents except oxidising acids.

LIMITATIONS

Low impact strength.
Service temperature limited to 140°C.
Very limited availability.
High cost.

APPLICATIONS

Mainly for corrosion resistance, for example:

drums for water meters and counting devices;
bearing separators;
pipe fittings;
tank and pipe linings.

TYPICAL PROPERTIES

Typical properties are given in:
Table 3.2.100 *Properties of chlorinated polyether*

3.2.20.2 Environmental resistance

Environment	*Behaviour*
Dilute mineral acids	Excellent
Concentrated mineral acids	Good (oxidising acids poor)
Alkalis	Excellent
Alcohols	Excellent
Ketones	Good
Aromatic hydrocarbons	Good
Chlorinated hydrocarbons	Fair to good
Detergents	Excellent
Greases and oils	Excellent

3.2.20.3 Processing and fabrication

Good processability—thin sections easily filled.
Can be injection moulded, extruded in pipe linings, also available as fluidised bed and dispersion coating material and prefabricated sheet.
Can be jointed with adhesives, and joints sealed by hot gas welding.

3.2.20.4 Supplier

Dow Chemical.

TABLE 3.2.100 Properties of chlorinated polyether

Property		*ASTM test method*	*Unit*	*Value*
Yield strength		D638	MNm^{-2}	41
Elongation		D638	%	130
Flexural strength		D790	MNm^{-2}	35
Impact strength Izod	−40 °C	D256	$J\ cm^{-1}$ notch	0.16
	+20 °C	D256	$J\ cm^{-1}$ notch	0.22
Hardness (Rockwell)		D785	Rockwell R	100
Coefficient of Thermal Expansion		D96	K^{-1}	66×10^{-6}
Thermal conductivity		—	$Wm^{-1}\ K^{-1}$	0.13
Dielectric constant 60–10^6 Hz		D150	—	2.8–3.0
Volume resistivity		D257	ohm–m	10^{15}
Heat deflection temperature (1.84 MN/m^2)		D648	°C	210
Max. continuous service temperature		—	°C	140
Water absorption		D570	% 24 h^{-1}	0.01

3.2.21 Polyvinylcarbazole (PVK)

PVK is a thermoplastic with exceptionally high resistance to heat and very good dielectric properties. It is used for high-grade insulation in high-frequency applications, such as television broadcasting. Its most important property is photoconductivity, which can be reinforced by sensitisers and is exploited in photocopying and dataprocessing.

The more important properties of PVK are listed in:

Table 3.2.101 *Typical properties of PVK*

PVK is not attacked by most liquids, but is swollen by cyclohexanone, dimethylformamide and mixtures of fuels; it is partially or entirely dissolved by aromatic hydrocarbons, chlorinated hydrocarbons and tetrahydrofuran.

Processing is by injection moulding, extrusion and almost all conventional means. There is possible skin irritation and some persons are hypersensitive to contact with PVK.

BASF supplies PVK under the trade name 'Luvican'.

TABLE 3.2.101 Typical properties of PVK

	Property	DIN test method	Unit	Value
Mechanical	Tensile strength (break)	53 455	MN/m^2	10
	Elongation (break)	53 455	%	0.5
	Elastic Modulus	53 457	GN/m^2	3.7
	Impact strength	53 453	kJ/m^2	5–10
	Ball indentation hardness	53 456	—	100 (H358/60) 120 (H358/10)
Physical	Specific gravity	53 479	—	1.19
Electrical	Volume resistivity	53 482	ohm m	10^{14}
	Dielectric constant (10^6 Hz)	53 483	—	3.1
	Loss factor (10^6 Hz)	53 483	—	0.0001
	Dielectric strength	53 481	kV/mm	–50
Thermal and environmental	Vicat softening temperature	53 460	°C	~195 (VST/B/50)
	Water absorption	53 495	%	<0.1

Data supplied by BASF.

3.2.22 Poly-4-methylpentene-1 (TPX)

3.2.22.1 General characteristics and properties

The advantages and limitations of TPX are listed in:
Table 3.2.102 *Characteristics of poly-4-methlypentene-1 (TPX)*

Methylpentene polymers are suitable for applications which require a combination of properties including transparency, lightness, chemical resistance, sterilisability, heat stability, non-toxicity, good impact resistance and good electrical properties.

APPLICATIONS

The good combination of properties of TPX makes it suitable for a range of applications.

In particular, its low density has led to its use for aircraft tableware and textile bobbins.

Its surface properties, which produce a flat meniscus on contained water, have led to its use for laboratory fluid measuring equipment and spirit dispenser optics.

Other applications include:

medical ware
lamp components
microwave applications
food containers

TYPICAL PROPERTIES

Typical properties of TPX are listed in:
Table 3.2.103 *Properties of poly-4-methylpentene-1 (TPX)*

3.2.22.2 Processing

Most of the normal thermoplastic techniques may be used with methylpentene polymers.

INJECTION MOULDING

Methylpentene polymers are supplied as granules which may be natural (transparent, colourless) or white. If coloured mouldings are required these granules may be dry-coloured before moulding—advice will be necessary regarding suitable colourants—or they may be decorated by printing, vacuum metallisation, painting or hot foil stamping after moulding. If decorated by such post-moulding operations, except hot foil stamping, the surface must first be pre-treated by acid etching or corona discharge or a special primer must be used.

Mould design for use with methylpentene polymers requires standards similar to other thermoplastics. For transparent mouldings the surface of the mould must be highly polished and for the best clarity it is preferable that the mould should be polished to lens quality in the line of draw. To facilitate removal of mouldings from the mould a generous draft taper of at least 1° is necessary. Ejector pins may be used but their diameters must be large enough to avoid punching through the moulding or causing white stress marks. Ideally the bearing area for ejection should be large and the thickness of the moulding sufficient to prevent buckling. Stripper plates are preferable. For thin-walled, deep-draught mouldings a poppet valve to relieve vacuum is helpful. Mouldings with small undercuts may be ejected by 'jumping', the depth that can be cleared being dependent on the shapes of the moulding and, because of the rigidity of methylpentene polymers, upon the ability of the walls to flex during ejection. The use of silicone-based mould release agents should be avoided when possible as they tend to mar the surface and may even cause cracking, but if essential they should be used sparingly.

3.2.22.3 Trade name and supplier

Mitsui & Co., Petrochemicals division, trade name TPX.

TABLE 3.2.102 Characteristics of poly-4-methylpentene-1 (TPX)

	Advantages	***Limitations***
Mechanical properties	Impact strength superior to transparent polystyrene.	Low tensile strength. Inferior impact strength to nylon 66 and acetal copolymer.
Optical properties	White light transmittance of 90% slightly inferior to polystyrene and acrylic.	
Environment	Resistant to weak acids, strong alkalis, oil and greases.	Attacked by strong oxidising acids. May show environmental stress cracking with detergents. Degrades by UV light and must be protected by black pigment if used outdoors.
Food and Medicine	Can be used with foodstuffs and sterilised by any of the conventional methods—steam, hot air (up to 1 h at 160°C), ethylene oxide, disinfectants or irradiation.	
Thermal	Excellent stability at high temperatures. Can be used continuously at 134°C.	Not self-extinguishing.
Miscellaneous	Lowest density (0.83 g/cm^3) of all thermoplastics.	Limited availability.

TABLE 3.2.103 Properties of poly–4–methylpentene–1 (TPX)

Property		*ASTM test method*	*Units*	*Unfilled*	*Glass-filled*[a]
Tensile strength (yield)		D638	MN/m^2	22–25	—
Tensile strength (break)			MN/m^2	16–22	56
Elongation			%	10–55	—
Tensile Elastic Modulus			GN/m^2	1.1–1.4	—
Flexural strength		D790	MN/m^2	29–42	96
Flexural Modulus			GN/m^2	0.08–1.8	—
Impact strength Izod		D256	J/cm notch	0.2–0.8	—
Hardness		D785	Rockwell R	70–90	—
Specific gravity		D792	—	0.835–0.845	—
Thermal Expansion Coefficient		D696	10^{-5} K^{-1}	11.7	—
Thermal conductivity		—	W/m per K	0.17	—
Volume resistivity		D257	ohm m	>10^{14}	—
Dielectric constant	60 Hz	D150	—	2.1–2.12	—
	10^6 Hz		—	2.12	—
Dissipation factor	60 Hz		—	0.00007	—
	10^6 Hz		—	0.000025	—
Dielectric strength		D149	10^2 kV/m	276	—
Refractive index		D542	—	1.463	—
Treansmittance		D1003–61	%	>90%	—
Haze		—	%	<5%	—
Maximum continuous service temperature		—	°C	134	—
Heat deflection temperature (0.45 MN/m^2)		D648	°C	100–105	—
Flammability		D635	mm/min	25.4	—
Water absorption		D570	%	0.01	—
Mould shrinkage		D955	%	1.5–3.0	—

[a]% Unspecified.

Table 3.2.103

3.2.23 Polyarylamide

3.2.23.1 Advantages and limitations

The advantages and limitations of polyarylamide are given in:
Table 3.2.104 *Advantages and limitations of polyarylamide*

3.2.23.2 Typical properties

Typical properties are shown in:
Table 3.2.105 *Typical properties of polyarylamide*

3.2.23.3 Applications

Mechanical components
Electrical connectors

3.2.23.4 Trade name and supplier

Solvay (Laporte), 30% Glass fibre-reinforced, trade name IXEF.

TABLE 3.2.104 Advantages and limitations of polyarylamide

Advantages	***Limitations***
Able to accept high filler loadings	Absorbs some water (less than Nylon 6.6)
Reasonable surface finish	
High crystallinity possible	

TABLE 3.2.105 Typical properties of polyarylamide

Property			ASTM tests	Units	Polyarylamide 30% glass fibre	PBT 30% glass fibre (for comparison)
Mechanical	Tensile strength	20 °C	D638	MN/m^2	185	130
		150 °C	—	MN/m^2	—	—
		260 °C	—	MN/m^2	—	—
	Elongation at break	20 °C	D638	%	2.5	2.6
	Flexural strength	20 °C	D790	MN/m^2	—	—
	Modulus of Elasticity in Tension at 20 °C		D638	GN/m^2	—	—
	Flexural Modulus	20 °C	D790	GN/m^2	11.3	8.9
		150 °C	—	GN/m^2	—	—
		260 °C	—	GN/m^2	—	—
	Impact strength	20 °C	D638	J/m notch	75	70
	(Izod notched)	−50 °C	—	—	—	—
	Hardness (Rockwell)		D785	—	M112	R119
Physical	Specific gravity		D792	—	1.43	1.6
	Thermal conductivity		—	W/mK	—	—
	Thermal expansion		D696	10^{-6} K^{-1}	15	5
Electrical	Volume resistivity		D257	ohm m	10^{14}	10^{12}
	Dielectric strength		D149	10^2 kV/m	300	190
	Dielectric constant	10^3 Hz	D150	—	3.9	3.8
		10^6 Hz	—	—	—	—
	Power factor	10^3 Hz	D150	—	0.01	0.002
		10^6 Hz	—	—	—	—
	Arc resistance		D495	s	—	—
Miscellaneous	Maximum continuous service temperature		—	°C	125	120
	Heat distortion temperature	0.45/MN/m^2	D648	°C	240	220
		1.8/MN/m^2	D648	°C	228	204
	Flammability		UL94	—	HB	VO
	Water absorption		D570	% (24 h)	0.2	0.06
	Taber abrasion CS17		D1044	Mg/ 1000 rev.	—	—
	Mould shrinkage		—	%	0.6	0.6

3–3

Thermosets

Contents

List of tables

List of figures

3.3.1 Phenolics, including melamine phenolics

3.3.1.1 General characteristics and properties

The advantages and limitations of phenolic plastics are listed in:
Table 3.3.1 *Comparative characteristics of phenolics*

AVAILABLE TYPES

Phenolic resins are produced by combining phenol and formaldehyde in the presence of a catalyst. The resins are used for bonding (see Vol. 1, Chapter 1.7) and coating (see Vol. 1, Chapter 1.4) and as moulding compounds (this section).

Mouldings invariably contain fillers (50–60%) as the unfilled resin is extremely brittle. A wide range of formulations has been developed, however, and main types are as follows:

General purpose

These are the cheapest grades, normally containing wood flour as a filler and are intended for applications involving low working stresses, e.g. knobs, covers, handles. Variations include fine powders for high gloss finishes, and controlled particle size materials for accurate automatic moulding machines.

High impact

These are higher cost compounds and normally have cotton cloth or glass fibre as a filler. Variations include zero shrinkage materials for accurate moulding. Elastomer-modified types have better heat resistance and impact strength than glass-filled types.

Electrical

These are specially formulated to give higher dielectric strength, lower loss or better arc resistance and mica is often used as a filler.

Odour-free

These are ammonia-free grades which find application as bottle tops, caps etc. for cosmetic or drug containers.

Heat resistant

These normally contain asbestos fibre as a filler and are used for applications requiring higher heat resistance either continuously or for only short times, e.g. pot and pan handles.

Chemical and water resistant

These may contain nylon as a filler and find use as developer tanks, pump housing, working machine parts etc.

Evidently these grades are not exclusive and many applications will require a number of these property improvements, e.g. an electrical component may also require heat and impact resistance.

Melamine-phenolics

Combine the decorative possibilities and tracking resistance of melamine with the easier moulding and cheapness of phenolics. They have better resistance to electrical surface tracking, but are not as resistant as aminos. In contrast to phenolics they are colour stable to light and can be supplied in light colours.

APPLICATIONS

Wood-flour-filled

Fuse boxes, electron meter cases, domestic electrical components, bottle closures, kitchen utensil handles, iron handles, car instrument panels, toilet seats.

Mica-filled

Computer plug board, electronic condensers, volume controls.

Glass-filled

Electric drill handles, fishing reels, selector switch bases.

Asbestos-filled

Heat-resisting lamp holders, terminal blocks, washing machine pulsators, ball valve seats (special grade). Felt-impregnated grades used for heat-shielding in automobiles (aluminium foil-faced).

Mineral-filled

Whilst there is no filler that can match asbestos in all its applications, recent disquiet regarding health hazards has promoted the use of mineral-filled grades. In many cases a suitable replacement is available, often at reduced cost, except at the highest service temperatures.

Cotton-filled

Dynamo fan, washing machine agitators.

Graphite-containing

Moulded bearings, sliding valves.

Metal-containing

Magnetic components, X-ray shields, components for microwave ovens.

Elastomer modified

Components of food processors, cooking and serving utensils, coffee makers, brewing machines, irons, fans, vacuum cleaners.

Candidate automotive applications include servo-pistons, gear shifts, stator retainers, thrust washers, transmission accumulators and front engine covers.

Melamine-phenolics

Domestic kitchenware, electrical sockets and light-switch bases where colour stability is important; contactors, circuit breakers (good track resistance). Melamine-phenolics have failed to make much headway in the UK, where most applications are satisfied by melamine and urea. There is a steady European market.

TYPICAL PROPERTIES

The more important properties of phenolic resins are listed in:
Table 3.3.2 *Typical properties of phenolics*, and
Table 3.3.3 *Typical properties of melamine-phenolics*

3.3.1.2 Environmental resistance

Phenolics have good fire resistance. Their behaviour is compared with other plastics in:
Table 3.3.4 *Fire properties of phenolic/glass composites compared with urethane and polyester*

3.3.1.3 Processing

GENERAL

Phenolic compounds can be formed by compression, transfer or injection moulding. Extrusion is also possible. Generally, injection moulding will give the fastest production rate for most components. Long fibre-filled material should not be injection or transfer moulded if maximum impact strengths are required—however, these methods are recommended for maximum tensile or flexural strength with other grades. Some new low-viscosity grades can be processed on runnerless injection equipment, giving up to 50% increase in manifold life and 5% decrease in cure time.

FINISHING OPERATIONS

Parting line fins must be removed.

Machinability will depend on the filler: good for wood-flour, poor for mineral or glass-filled materials.

REINFORCED REACTION INJECTION MOULDABLE PHENOLICS

Phenolic resins have been developed specifically for RRIM and acid catalysts have been developed to give suitable cure rates. Although few machines are available at present for the handling of phenolic resins in this application, those which are have been used successfully.

Phenolic RRIM, with in-mould reinforcement, is currently being used to produce a variety of heat shields for motor vehicle exhaust systems. Filled resin has also been used and consideration is being given to the injection of the glass with the resin.

For the future. RRIM has been considered for vehicle body panels although vacuum assisted moulding processes (VAMF/VARI/VIM) contend prominently. In railway vehicles which need to satisfy stringent fire retardancy requirements, phenolic RRIM may be attractive.

3.3.1.4 Trade names and suppliers

The suppliers of phenolics and their trade names are listed in:
Table 3.3.5 *Major suppliers and familiar trade names of phenolics*

TABLE 3.3.1 Comparative characteristics of phenolics

	Advantages	*Limitations*
Mechanical properties	High modulus—outstanding resistance to deformation under load. Good compressive strength. Resilient materials obtained by modifying with rubber or nylon.	Forms are hard and brittle. Impact strength not high except with cord, fabric, glass and butadiene-filled grades. Hence high breakage rates during transportation have been reported.
Thermal properties	Good dimensional stability over wide temperature range. Self-extinguishing grades available.	Upper temperature limit 150 °C (continuous) and 220 °C for short-term operation. This can be extended depending upon fillers used.
Electrical properties	Mineral-filled grades used for highest electrical properties, although power factor still variable.	Tendency to form tracking paths at high voltages. Electrical properties adversely affected by hot, wet conditions, especially wood-filled grades.
Environmental properties	Chemical resistance to common solvents, weak acids and most detergents. Very low water absorption.	Not stable in presence of alkalis or strong oxidising acids. Surface finish adversely affected by hot, wet conditions, especially wood-filled grades.
Processing	Injection moulding and transfer moulding grades available. Special fillers incorporated to produce magnetic compounds, provision of X-ray shielding, or for microwave absorption, or for low friction applications. Melamine-phenolics available in light colours.	High-pressure processing equipment needed. Limited to dark colours owing to oxidation discolouration. Fillers required for mouldings. Volatiles released during cure.
Miscellaneous	Low cost. Large amount of application experience (most widely used thermoset).	
	Melamine–phenolics	
General	As phenolics but in addition colour stable so available in light colours.	
Electrical	Better tracking resistance than phenolics.	Not as resistant to tracking as aminos.

TABLE 3.3.2 Typical properties of phenolics

	Property		***Units***	***Asbestos fibre-filled***	***Cold-moulded asbestos-filled 28–30% phenolic resin***
Mechanical	Tensile strength		MN/m^2	30–50	12
	Tensile Modulus		GN/m^2	7–21	—
	Impact strength (Izod)		J/cm	0.1–0.2 (0.75[a])	0.23
	Flexural strength		MN/m^2	49–98	38–42
	Flexural Modulus		GN/m^2	7–15	—
	Compressive strength		MN/m^2	140–250	63–126
Physical	Specific gravity		—	1.4–2.0	1.85–2.0
	Thermal conductivity		W/mK	0.24–0.9	—
	Thermal expansion coefficient		$K^{-1} \times 10^{-5}$	0.8–4.0	—
Electrical	Dielectric constant	60 Hz	—	5–20	—
		10^3 Hz	—	6–16	—
		10^6 Hz	—	5–10	—
	Dielectric strength		10^2 kV/m	80–140	24–40
	Arc resistance		s	10–190	120–240
Thermal/Environmental	Maximum use temperature		°C	180–260	175
	1.8 MN/m^2 deflection temperature		°C	150–260	215
	Flammability		—	Non-burning	Non-burning
	Water absorption		(48 h)%	0.1–0.5 (24 h)	0.7–0.9
Optical	Transparency		—	No	No
Processing	Mould shrinkage		%	0.2–0.9% (<0.01[a])	—
	Remarks			Detergent-resistant grade available with water absorption 0.04% (48 hr). [a] Properties of felt grade used for ball valve seating; it has low friction coefficient and good sealing.	Squeezed into shape by pressure alone and baked to achieve final properties. Enables more rapid production than hot-moulding but lacks surface smoothness. Lower strength. Improved impact resistance. Main applications for insulation, especially in oil.
				Asbestos-filled phenolics are under pressure from asbestos-free grades containing mineral and other fillers. The latter cannot match the heat resistance.	

Table 3.3.2

TABLE 3.3.2 Typical properties of phenolics—*continued*

Wood-flour-filled	*Cotton-flock or fabric-filled*	*Cellulose filled*	*Cellulose mineral-filled*	*Glass-filled*	
				Chopped glass	*Glass fibre*
30–60	20–60	45	38	41	30–120
5–12	6–10	—	—	—	13–23
0.1–0.3	0.4–4.3	2.1–2.5	4.8	5.34	0.2–9.2
62	62–69	62	59	103	62–75
7–9	—	—	—	—	14–23
165	138–172	152–200	172	186	260–280
1.3–1.5	1.37–1.40	1.36–1.38	1.42	1.76–1.80	1.70–2.0
0.16–0.3	—	—	—	—	0.32–0.88
3–4.5	—	—	—	—	0.8–2.0
5–13.0	5.2–21	—	5.7	5.2	5.0–7.1
4–9	—	—	—	—	5.0–6.9
4–6	—	—	—	—	4.5–6.6
100–160	89–98 (25°C)	80–150	118	—	56–160
Tracks	28–49 (100°C)	Tracks	—	160	180 max.
100–150	100–121	100	100	—	175–290
130–190	120–165	160	133	260	150–315
Self-extinguishing	Non-burning	—	Non-burning	UL94 V–0	UL94 V–0
1.0	0.9–1.0	0.8–1.2	0.9	1.0	0.03–1.20
No	No	No	No	No	No
0.4–0.9	0.4–0.9%	—	—	0.08–0.14	0.15–0.30
Lowest cost phenolic plastic used for wall panels, cases, closures. Suitable for mechanical and electrical applications. Smooth glossy finish obtainable. Detergent resistant grade available.	Similar to wood-flour-filled phenolic but with improved impact resistance.	Detergent resistant grade available	Used for air-conditioned housings, solenoid housings and similar parts.	Where high temperature capability not required, cellulose or wood-filled grades can be employed. Asbestos or mineral grades are alternatives at high temperatures.	

Table 3.3.2—*continued*

TABLE 3.3.2 Typical properties of phenolics—*continued*

	Property		Units	Mica-filled	Phosphoric acid/ asbesto-filled	Metal-filled (iron lead)
Mechanical	Tensile strength		MN/m^2	38–50	—	28–35
	Tensile modulus		GN/m^2	17–34	—	—
	Impact strength (Izod)		J/cm	0.14–0.20	0.16	0.14–0.24
	Flexural strength		MN/m^2	63	27.6	50–84
	Flexural Modulus		GN/m^2	—	1.72	12–18
	Compressive strength		MN/m^2	175	152	105-210
Physical	Specific gravity		—	1.7–1.9	2.1	2.0–4.2
	Thermal conductivity		W/mK	—	—	0.56–0.80
	Thermal expansion coefficient		$K^{-1} \times 10^{-5}$	—	0.9	—
Electrical	Dielectric constant	60 Hz	—	4.7–6.0	—	—
		10^3 Hz	—	—	—	—
		10^6 Hz	—	—	—	—
	Dielectric strength		10^2 kV/m	140–160	20	—
	Arc resistance		s	—	500	—
Thermal/Environmental	Maximum use temperature		°C	120–150	400	—
	1.8 MN/m^2 deflection temperature		°C	150–180	>200	175–215
	Flammability		—	SE–0	—	—
	Water absorption		(48 h)%	0.05	6	0.03–0.15 (24 h)
Optical	Transparency		—	No	No	No
Processing	Mould shrinkage		%	0.2–0.6	—	0.2–0.4
	Remarks			Outstanding for low loss factor at low and high temperatures. High stability of electrical properties with temperature and humidity variation.	Recommended for arc chutes and other parts where high power arcs are encountered at high voltage.	

Table 3.3.2—*continued*

TABLE 3.3.2 Typical properties of phenolics—*continued*

Butadiene-modified phenolic (GE-MX582P)	*(Electrical) mica and glass fibre reinforced*	*Phenolic foam*	*Phenolic laminates*		
			Paper	*Cotton fabric*	*Glass fabric*
—	52	0.43	90	70	151
—	—	—	—	—	—
1.62	0.2	0.02	0.24	0.5	2.9
1.03	—	—	—	—	—
—	8.2	0.02	8.2	6.3	10.3
—	—	—	—	—	—
1.54	1.7	0.08	1.32	1.33	1.9
—	—	—	—	—	—
—	2.2	1.1	2	2	1.2
—	—	—	—	—	—
—	5.2	2.8	5.1	4.8	7.5
—	—	—	—	—	—
—	120	70	150	100	80
—	—	—	—	—	—
—	180	120	90	105	150
>290	180	90	160	170	200
UL94 V–0 (0.25)	V–0	V–0	HB	HB	V–0
—	0.2	15	0.25	0.78	0.08
—	—	—	—	—	—
—	0.2	—	—	—	—
Higher impact strength and heat resistance than 20% glass-filled phenolic. Toughness not adversely affected by 5 h ageing 175°C.					

Table 3.3.2—*continued*

TABLE 3.3.3 Typical properties of melamine-phenolics

Property	*Units*	*Wood-flour-filled*	*Cellulose-filled*	*Wood-flour Inorganic*	*Cellulose Inorganic*
Flexural strength	MN/m^2	80	80	70	70
Compressive strength	MN/m^2	6.0	7.0	4.0	5.0
Tensile strength	MN/m^2	25–30	25–30	25–30	25–30
Tensile Modulus	GN/m^2	6–10	6–10	6–10	6–10
Charpy impact strength	k.J/m^2	>1.5	>1.5	>1.2	>1.5
Specific gravity	—	1.50–1.55	1.50–1.65	1.55–1.60	1.55–1.65
Volume resistivity	ohm m	—	10^{10}	10^{10}	10^{10}
Arc resistance	s	—	120–130	120–130	160–180
Dielectric constant	—	5–7	5–8	8–10	6–9
Dissipation factor	—	0.2	0.2	0.2	0.2
Dielectric strength	10^2 kV/m	70–120	80–150	70–120	80–150
Heat stability (Martens)	°C	>120	>120	>120	>120
Heat distortion temp. (A)	°C	—	170–180	160–170	160–170
Maximum cont. use temp.	°C	—	130–140	130–140	130–140
Flammability	UL94	—	V–0	V–0	V–0
Water absorption (24 h)	%	—	0.1–0.2	0.2–0.3	0.1–0.2
Mould shrinkage	%	0.8	0.8	0.8	0.8
Post-mould shrinkage	%	1.0–1.5	0.9–1.3	1.0–1.5	0.9–1.3

TABLE 3.3.4 Fire properties of phenolic/glass composites compared with urethane and polyester

Property	***Phenolic***	***Polyester***		***Rigid polyurethane***
		General purpose	***Highly fire retardant***	
ASTM D229 Oxygen Index	45–55[a]	18	~40	20–25
ASTM D229 Rate of burning	Does not ignite	—	—	—
BS476 Pt. 7 Spread of Flame	Class 1 (Zero spread[b])	Class 4 or unclassified	Class 1	Class 1–[c] Class 4
Toxicity of combustion products	Very low	Low toxicity	Toxic	Toxic
Smoke	Virtually nil	High	Very high	Very high

[a] Catalyst dependent. [b] 10cm permitted. [c] May be optimistic.

TABLE 3.3.5 Major suppliers and familiar trade names of phenolics

Company	*Trade names*	*Fillers*
Allied Chem.	Plaskon	G, W, C, Cl, A
BP Chem.	Cellobond	Foam
Budd Chem.	Polychem	G, NBR, Nylon, Teflon, Mica.
Cdf Chimie	Norsophen, Gedelite	—
Dynamit Nobel	Trolitan	A, W, Stone, Cl, Paper, C, M. Butadiene, G
Fiberite	FM-	Cl, Macerated Fabric, A, Nylon, C, G, Cord.
General Electric	GE-; Genal	W, Cl, M, G, Butadiene
Hooker/Durez	Durez	W, Flock, M, G
Huls (Chemische Werke)	Alberit; Alresin	—
Kooltherm Insulations Prods.	Koolphen	Foam
Monsanto	Resinox	—
Montecatini	Fluosite	—
Perstorp/Ferguson	Nestorite	W, W/Rubber, C, Nylon/W Nylon/Cl, A/Cl, Nylon/C, G, G/M, M, G/W, A, Nylon, Nylon/Mica, W/Mica, M.
Plastics Engineering	Plenco	M, Al, C, W, Mica, Fe, A, G
Reichhold	RCI	A, Carbon Black, Sisal, Rubber
RM Friction Mat.	Pyrotex	A
Rhone Poulenc	Ablaphene, Phenorez, Progilite, Resophene.	—
Rogers	Rogers	Cl, A, M, G, W
SIR	Sirfen	W, A, Mica, C
Sterling	Sternite	W/M, W, W/A, W/Mica, A/C, A, C, C/W
Synthetic Resins	Uravar	—
Toshiba	Tecolite	Filled
TBA	Durestos	A (fibre and felt)
Union Carbide/BXL (Vynckler)	Bakelite	C, W, G, Mica, M, PL, GL
Melamine-Phenolics		
Resart/Mercury (UK)	Resart MP, W, Cl	—
Tufnol	Tufnol Phenolic	PL, CL, GL
Railko	Railko	CL

KEY: A = asbestos; Al = aluminium; C = cotton; Cl = cellulose; Fe = iron powder; G = glass; M = mineral; W = wood-flour; F = foam; PL = paper laminate; CL = cotton fabric laminate; GL = glass fibre laminate.

Table 3.3.5

3.3.2 Phenol aralkyl resins (Xylok®)

3.3.2.1 General characteristics and properties

The characteristics of phenol aralkyl resins, compared to those of phenolics are given in:
Table 3.3.6 *Advantages and limitations of phenol aralkyl resins compared to phenolics*

AVAILABLE TYPES

The structure of Xylok polymers is closely related to that of the phenolics.

Xylok resins can be processed into laminates, tapes, moulding compounds, capping cements and friction material compositions on the same industrial equipment employed for phenolics, epoxies and silicones.

Xylok 225 is specifically developed for use in the formation of high performance moulding compounds.
Principal reasons for selection of Xylok 225 moulding compound are:

(a) Compatible with phenolic resins (for blending), and a wide variety of fillers.
(b) Easy to process using standard phenolic moulding equipment.
(c) Good mechanical strength and dimensional stability.
(d) Suitable for prolonged exposure at temperatures in the range 150–250°C.
(e) Electrical insulation.
(f) Low water absorption.

Although Xylok 225 can be used alone to produce moulding compounds, it is often advisable to use it in conjunction with phenol–formaldehyde resins, preferably of the novolac type. The effect of changes in resin composition on thermal stability are shown in:
Fig 3.3.1 *Comparison of heat stability of asbestos-filled mouldings based on phenolic, Xylok 225 and various blends of the two resins.*

In addition to thermal stability, the solvent resistance, water absorption, dimensional stability and dielectric properties of phenolics are improved by the introduction of Xylok.

Common fillers employed in these moulding compounds include asbestos (flour or fibre), glass (powder or fibre), silica, mica and graphite.

Xylok 210 is a laminating varnish used extensively for the impregnation of a wide variety of reinforcements. The properties of composite materials based on this resin are discussed more extensively in Chapter 3.8. The resin is competitive with epoxy and silicone for lamination and bonding of glass-fabrics.

The epoxy cured Xylok resins (231, 235C, 236, 237) also find application as impregnating resins.

	Applications	*Reasons for use*
ELECTRICAL	Commutator insulation Spacers Coil separators Encapsulation	Electrical insulation suitable for class F, H or C temperatures (BS 2757).
TRIBOLOGY	Dry and boundary lubricated bearings Seals Bushings Rocking and sliding members.	Creep resistance Dimensional stability Operate at higher temperatures than reinforced phenolics. Chemical resistance.
DOMESTIC APPLIANCE	Terminal blocks in cookers, hair dryers, fires, kettles. Oven resistant handles. Impeller blades in washing machines.	Electrical insulation and heat resistance. Chemical resistance.
AEROSPACE	Heat shields for rocket motors.	Ablative property.

TYPICAL PROPERTIES

The more important properties of plastics based on Xylok 225 resin are listed in:
Table 3.3.7 *Typical properties of Xylok 225 based phenol aralkyl plastics*

3.3.2.2 Environmental resistance

Xylok mouldings resist many of the environments encountered in industry, as shown in:
Table 3.3.8 *Chemical resistance of Xylok mouldings*

3.3.2.3 Processing

The processing of phenol–aralkyl resins differs from that of phenols only in requiring slightly higher temperatures (or longer times).

3.3.2.4 Trade name and supplier

Albright & Wilson Ltd, trade name, Xylok.

TABLE 3.3.6 Advantages and limitations of phenol aralkyl resins compared to phenolics

Property	*Advantages and limitations*
Chemical reactivity:	Slower curing than phenolics. In laminating this problem is overcome by increasing moulding temperature.
Thermal stability:	Far more heat stable than phenolics.
Electrical properties:	Xylok resins have lower dielectric properties and are therefore superior insulating materials to phenolics. This property is exploited in glasscloth-reinforced laminates.
Water absorption:	About one third that of phenolics—this in turn leads to better electrical properties and dimensional stability.
Chemical resistance:	Significantly better than phenolics.

TABLE 3.3.7 Typical properties of Xylok 225 based phenol aralkyl plastics

	Property		*Units*	*Asbestos-filled (flour)*	*Mica-filled*	*Silica-filled*	*Chopped glass strand-filled*	*Carbon fibre-filled*
Mechanical	Flexural strength	(RT)	MN/m^2	70	67	85	195	68
		250°C	MN/m^2	58	43	52	124	49
	Tensile strength		MN/m^2	52	30	47	81	42
	Impact toughness Izod notched		J/cm notch	0.15	0.17	0.16	4.80	0.20
Physical	Specific gravity		—	1.68	1.79	1.87	1.62	1.34
Electrical	Volume resistivity		Ω m	10^{12}	10^{10}	10^9	10^{14}	—
	Dielectric strength		10^2 kV/m	10^2	94	109	106	118
	Dielectric constant 10^6 Hz	Dry	—	4.80	4.14	4.94	4.4	—
		24 h, H_2O	—	4.85–4.94	4.30	5.01	4.4	—
	Dissipation Factor 10^6 Hz	Dry	—	0.053–0.076	0.017	0.026	0.014	—
		24 h, H_2O	—	0.062–0.077	0.020	0.028	0.016	—
Environmental	Water absorption		mg	5.5–8.9	6.3	6.5	5.3	9.6
Processing	Linear mould shrinkage		%	0.4	0.3	0.3	0.2	0.4

TABLE 3.3.8 Chemical resistance of Xylok mouldings

Moulding compound	*Environment*	*Test temperature (°C)*	*% Change in*		
			Strength	*Weight*	*Width*
Asbestos-filled	Sulphuric acid 98%	90°C	Destroyed		
	Hydrochloric acid 37%		–27	+3.54	+2.4
	Sodium hydroxide 10%		–12	+1.43	+5.5
	Skydrol 500		–10	+0.42	Nil
	Dimethylformamide		Destroyed		
	Toluene		–11	+1.15	Nil
	Trichloroethylene	20°C	–6	+0.16	Nil
	Car engine oil	150°C	–3	+0.43	+0.1
	Distilled water	100°C	–8	+1.11	Nil
Silica-filled	Sulphuric acid 98%	90°C	Destroyed		
	Hydrochloric acid 37%		–20	+2.11	+2.3
	Sodium hydroxide 10%		–26	–0.55	+1.7
	Skydrol 500		–11	Nil	Nil
	Dimethylformamide		–63	+12.15	+13.0
	Toluene		+6	–0.06	Nil
	Trichloroethylene	20°C	+8	+0.06	Nil
	Car engine oil	150°C	–1.5	–0.23	+0.2
	Distilled water	100°C	–7	+0.41	+0.1
Carbon fibre-filled	Sulphuric acid 98%	90°C	Destroyed		
	Hydrochloric acid 37%		–4	+3.68	+1.1
	Sodium hydroxide 10%		–2	+0.57	Nil
	Skydrol 500		+2	–0.03	Nil
	Dimethylformamide		Destroyed		
	Toluene		–3	+0.39	Nil
	Trichloroethylene	20°C	+1	+0.20	Nil
	Car engine oil	150°C	–4	+0.24	+1.1
	Distilled water	100°C	+2	+1.00	Nil

Conditions: Test pieces pressed for 10 min at 170°C and exposed 100 h.

Table 3.3.8

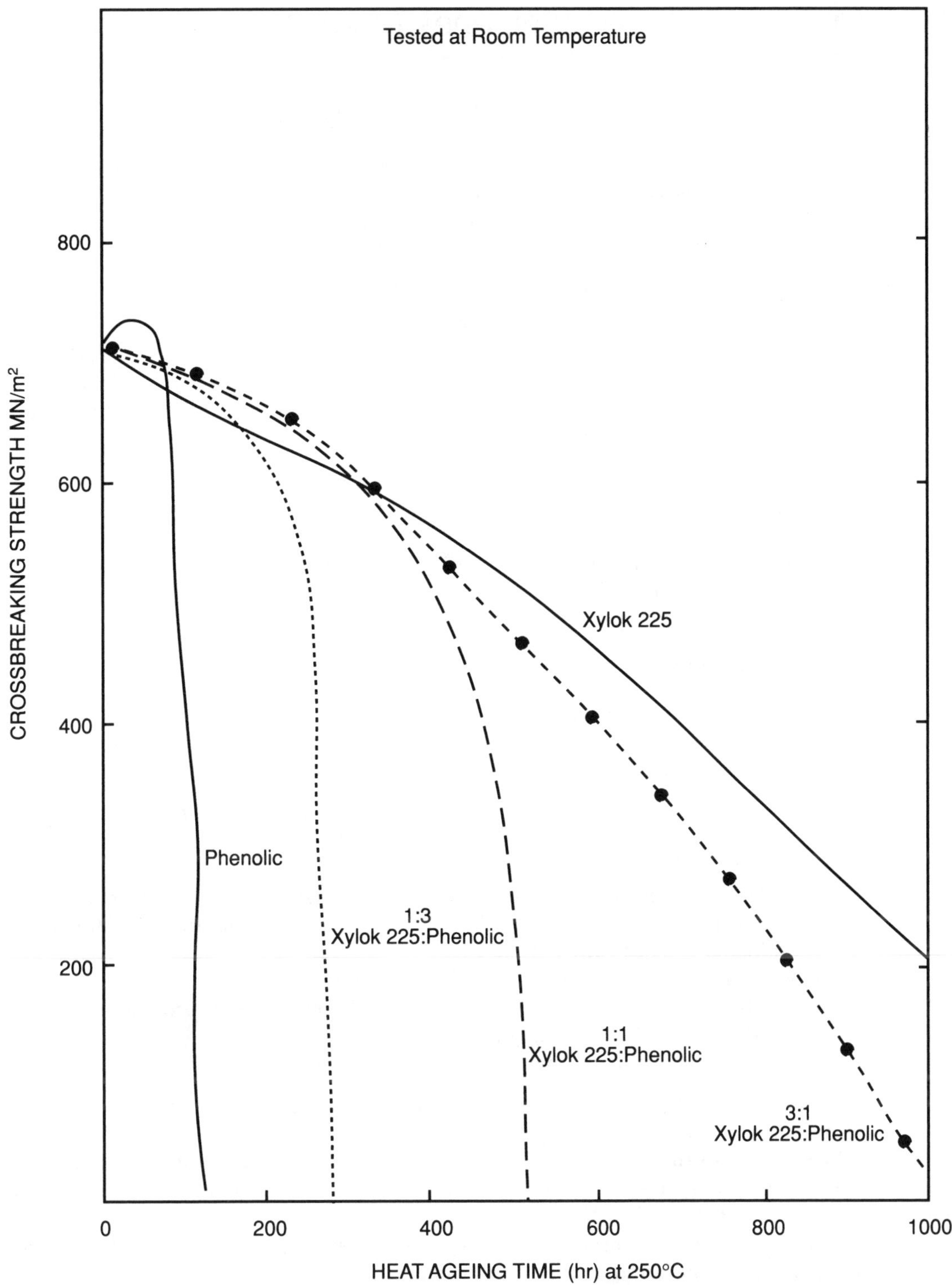

FIG 3.3.1 Comparison of heat stability of asbestos-filled mouldings based on phenolic, Xylok 225, and various blends of the two resins

Fig. 3.3.1

3.3.3 Epoxies and polybutadienes (epoxide resins and epoxidised polybutadiene resins)

3.3.3.1 General characteristics and properties

The advantages and limitations of epoxies and polybutadienes are listed in:
Table 3.3.9 *Characteristics of epoxies compared to polyester*

Epoxies cure much more rapidly than polybutadienes by the addition of a wide variety of hardening agents, usually classified into amine and acid systems:

Amine hardening systems : Primary and secondary amines.
Acid hardening systems : Anhydrides.

The structural variety available, the multitude of possible linking points, the adhesiveness of the oxygen bond and the wide variety of hardeners make it possible to formulate epoxide plastics to meet a large variety of requirements. Epoxide resins are usually characterised by six parameters:

Resin viscosity (of liquid resin).
Epoxide equivalent.
Hydroxyl equivalent.
Average molecular weight, and distribution.
Melting point (of solid resin).
Heat deflection temperature (of solid resin).

AVAILABLE TYPES

Epoxy resins are thermoplastic liquids or solids which may be cured to thermosets by homopolymerising with a catalyst or by copolymerising with a hardener. Characteristics of the resin types included:

Bisphenol A—most widely used resin available in liquid form for processing into castings (rigid or flexible depending on formulation and curing agent), mouldings, laminates or filament composites.
Novolacs—higher heat resistance, chemical resistance, stiffness, improved flow, faster curing and lower cost. Available as rigid castings, moulding and laminates.
Brominated epoxies—fire resistance.
Glycidyls—flexibility, increased elongation, improved flow.
Cycloaliphatics—basic properties of epoxides but with improved mechanical strength, arc resistance and, contrary to other epoxides, good weathering resistance. The rigid structure may however result in brittleness.

Cost, handling characteristics, optimum cure cycles and effect on properties of cured resins vary widely.

In addition to the resin and curing agents, epoxides are usually formulated with fillers to improve specific application properties such as strength, stiffness, impact strength and chemical resistance.

Glass fibre added to improve strength and stiffness may impair chemical resistance; asbestos fibre is preferred in these circumstances.

Some commonly used fillers are listed in:
Table 3.3.10 *Effect of fillers on epoxy resin properties*

Polybutadienes are available as compounds for compression transfer and injection moulding. They have particularly good encapsulating properties. They are available as solutions for liquid potting, impregnating and filament winding applications.

Epoxies are also available in the liquid and powder form for surface coatings (see Vol.1, Chapter 1.4) and in the liquid form as adhesives.

APPLICATIONS

Epoxies

Encapsulation of electrical and electronic devices including pacemakers.
Electrical insulators.
Chemical resistant pumps and pipes.
Impellers.
Pottery moulds.
Press and vacuum forming tools.
Patterns and coreboxes.
Moulds for RIM.
Laminates for aircraft and automobile components.
Fibre wound products.
Laboratory bench tops.
Floorings.
Road surfacing.

Epoxy RIM and RRIM systems have potential applications as automotive radiator supports, transmission supports and other structural components.

Polybutadienes

Electrical components and laminations.

TYPICAL PROPERTIES

The spread of properties available in epoxy resins systems is listed in:
Table 3.3.11 *Typical properties of a range of epoxy casting resins*
Table 3.3.12 *Typical properties of epoxy laminates*

Typical properties of components produced for epoxy moulding compounds are listed in:
Table 3.3.13 *Typical properties of epoxy mouldings*

Typical properties of components for epoxy RIM/RRIM systems are listed in:
Table 3.3.14 *Typical properties of epoxy RIM/RRIM systems*

The more important properties of polybutadiene are listed in:
Table 3.3.15 *Typical properties of polybutadiene*

3.3.3.2 Processing

Epoxies can be processed by all the methods normally used for thermosets. They are normally supplied in two or three component systems: resin, catalyst and for some systems, accelerator. The components must be mixed at a temperature varying, according to the system, between 23 and 100°C and processed not later than a time varying, according to the system and temperature, between half an hour and 6 months.

3.3.3.3 Trade names and suppliers

The suppliers of epoxy and polybutadiene resins and the commonly used trade names of their products are listed in:
Table 3.3.16 *Trade names and suppliers of epoxide resins*

TABLE 3.3.9 Characteristics of epoxies compared to polyester

	Advantages	*Limitations*
Mechanical properties	High strength and modulus. Filament-wound pipes have a high strength/weight ratio. Good wear resistance. High resistance to cracking when used to encapsulate metal components.	
Environmental properties	Good chemical resistance. Better resistance to alkalis and solvents than polyester. High dimensional stability.	Poor oxidative stability, some moisture sensitivity. Susceptible to UV degradation and not used in external applications where appearance is of prime importance (some special grades available). Weathering resistance poor, except cycloaliphatics.
Electrical properties	High volume resistivity. Good dielectric strength. Special grades offer good tracking resistance (better than phenolics). Better electrical insulation and dimensional stability than polyesters under high humidity conditions.	
Thermal properties	Good thermal stability. Better thermal ageing than polyesters.	Limited to 180–240 °C. High thermal expansion and conductivity when unfilled.
Processing and fabrication	Can be processed by all thermosetting methods. One or two part systems available. Convenient range of cure conditions. No volatiles formed during cure. Low curing shrinkage (filled). Excellent adhesion, especially to metal, wood and other polar substrates.	Many grades have a short shelf life, therefore cool storage is necessary. High shrinkage (unfilled). High exotherm.
	Good flow properties permit processing at low pressures and consequent extension of mould life.	
Miscellaneous	Exceptionally suited for encapsulation purposes because of low viscosity, good adhesion. Some grades have a low level of impurity and a high glass transition temperature.	High cost, in particular speciality grades (except Novolacs).

Table 3.3.9

TABLE 3.3.10 Effect of fillers on epoxy resin properties

Filler/reinforcement	*Property improved*
Silica, calcium carbonate	Cost
Glass microballoons	Density
Fibrous glass, chopped nylon	Impact strength Tensile strength Flexural strength
Aluminium powder, calcium carbonate	Machinability
Carbon black, asbestos	Heat resistance
Silica, metal powders	Dimensional stability Thermal conductivity
Lithium aluminium silicate	Thermal expansion coefficient
Inorganic fillers	Fire retardance
Asbestos, powdered coal	Chemical resistance
Mica, silica, powdered coal	Moisture resistance
Mica, silica, asbestos	Electrical insulation
Silver or aluminium powder, carbon, graphite	Electrical conductivity
Graphite, mica, molybdenum disulphide	Lubricating.

TABLE 3.3.11 Typical properties of a range of epoxy casting resins

Cured property		***Units***	***Low-cost versatile***	***Low cost good flow***	***Low viscosity***	***Vacuum grade***
Mix preparation		—	Add Cat.	Add Cat.	Add Cat.	Add Cat.
Mixed viscosity		cps	25,000	6,000	4,000	3,500
Room temperature cure		—	Yes	Yes	Yes	Yes
Cure shrinkage		%	0.2	0.2	0.2	0.2
Specific gravity		kg/dm^3	1.55	1.55	1.45	1.55
Hardness		Shore D	88	89	87	88
Compressive strength		MN/m^2	110	107	107	107
Compressive Elastic Modulus		GN/m^2	5.5	4.8	5.5	4.8
Izod Impact		J/m	16	13.5	16	13.5
Water absorption (24 h)		%	0.1	0.1	0.1	0.1
Service temperature		°C	177	177	177	177
Thermal conductivity		W/m per K	8.0	7.3	6.6	4.8
Thermal expansion coefficient.		$10^{-6}\ K^{-1}$	40	40	45	38
Volume resistivity		ohm m	10^{12}	10^{12}	10^{12}	10^{13}
Dielectric constant	60 Hz	—	4.8	4.6	4.7	4.4
	10^3 Hz	—	4.6	4.4	4.5	4.1
	10^6 Hz	—	3.9	3.7	3.8	3.8
Dissipation factor	60 Hz	—	0.02	0.02	0.02	0.02
	10^3 Hz	—	0.01	0.01	0.01	0.01
	10^6 Hz	—	0.02	0.02	0.02	0.01
Dielectric strength		10^2 kV/m	180	180	180	170
Relative cost		—	Low	Low	Medium	Low

Data supplied by Emerson & Cuming.

Table 3.3.11

TABLE 3.3.11 Typical properties of a range of epoxy casting resins—*continued*

Lowest cost Easy 1 : 1 mix	*Low cost electrical grade*	*Low density*	*Lowest density*	*Low cost Lowest viscosity*	*High thermal conductivity*	*Good thermal shock*
MIX A & B	MIX A & B	Add Cat.	Add Cat.	Add Cat.	Add Cat.	Add Cat.
15,000	21,000	30,000	1,800	500	90,000	100,000
Yes	Yes	Yes	Yes	Yes	Yes	Yes
0.08	0.3	0.3	0.3	0.4	0.1	0.1
1.40	1.65	0.88	0.78	1.55	2.30	2.30
88	96	82	78	88	94	95
55	110	103	69	83	114	124
2.1	4.8	2.8	1.4	4.1	7.6	9.0
16	16	16	14.5	16	16	22
0.1	0.1	0.4	0.4	0.2	0.15	0.15
125	125	177	107	125	205	205
4.8	9.0	2.2	2.1	4.8	17.3	18.3
56	38	40	54	40	25	25
10^{10}	10^{14}	10^{10}	10^{11}	10^{12}	10^{14}	10^{14}
4.5	4.6	3.9	3.7	4.4	6.5	6.5
4.2	4.4	3.3	3.1	4.2	6.3	6.3
3.7	3.7	3.1	2.9	3.9	5.9	5.9
0.08	0.20	0.02	0.02	0.01	0.02	0.02
0.06	0.01	0.01	0.01	0.02	0.008	0.008
0.07	0.02	0.01	0.01	0.04	0.02	0.02
172	160	150	150	160	220	152
Lowest	Low	Medium	High	Low	Low	Low

Table 3.3.11—*continued*

TABLE 3.3.11 Typical properties of a range of epoxy casting resins—*continued*

Cured property		*Units*	*Highest thermal conductivity*	*High temperature thermal shock*	*High temperature low expansion*	*High temperature easy to handle*
Mix preparation		—	Add Cat.	MIX A & B	Add Cat.	Add Cat.
Mixed viscosity		cps	170,000	10,000	50,000	60,000
Room temperature cure		—	Yes	No	No	No
Cure shrinkage		%	0.1	0.5	0.2	0.2
Specific gravity		kg/dm^3	2.80	1.70	2.30	2.30
Hardness		Shore D	92	80	96	96
Compressive strength		MN/m^2	131	17	124	117
Compressive Elastic Modulus		GN/m^2	9.6	1.4	8.3	8.3
Izod Impact		J/m	13.5	32	13.5	13.5
Water absorption (24 h)		%	<0.1	0.08	0.02	0.2
Service temperature		°C	205	177	260	260
Thermal conductivity		W/m per K	52	4.8	16.6	16.1
Thermal expansion coefficient.		10^{-6} K^{-1}	29	120	27	39
Volume resistivity		ohm m	10^{13}	10^{12}	10^{14}	10^{14}
Dielectric constant	60 Hz	—	—	5.5	4.3	6.3
	10^3 Hz	—	7.0	5.2	3.9	—
	10^6 Hz	—	—	4.5	3.3	5.8
Dissipation factor	60 Hz	—	—	0.035	0.007	0.007
	10^3 Hz	—	0.01	0.009	0.009	0.009
	10^6 Hz	—	—	0.036	0.012	0.012
Dielectric strength		10^2 kV/m	164	184	164	184
Relative cost		—	Medium	Medium	High	High

Data supplied by Emerson & Cuming.
[a] Contains glass microballoons.

Table 3.3.11—*continued*

TABLE 3.3.11 Typical properties of a range of epoxy casting resins—*continued*

Fire retardant	*Adjustable flexibility*	*Transparent grades*				*Repairable expoxy gel*
		Crystal clear and tough	*High temperature*	*High impact*	*RT cure epoxy*	
Add Cat.	Add Cat.	MIX A & B	Add Cat.	MIX A & B	MIX A & B	MIX A & B
14,000	40,000	4,000	3,000	900	1,000	600
Yes	Yes	No	No	No	Yes	Yes
0.4	0.2	0.6	0.5	0.5	0.6	0.8
1.65	1.40	1.20	1.20	1.10	1.18	1.2
87	Variable	85	82	78	75	25A
96	Up to 55	62	107	76	69	—
2.8	Variable	3.4	3.4	0.7	1.4	—
16	21–49	27	17	113	108	Very high
0.4	0.5	0.1	0.05	0.3	0.2	0.4
125	110	125	150	110	120	110
3.8	4.2	3.3	3.1	3.1	3.1	3.1
48	50	75	59	125	80	59
10^{12}	10^{11}	10^{13}	10^{11}	10^{13}	10^{12}	10^{10}
4.6	4.4	—	4.2	3.7	—	3.8
4.8	4.1	—	4.2	—	—	3.6
4.4	3.4	3.8	3.6	3.3	3.5	3.4
0.007	0.04	—	0.01	0.008	—	0.03
0.024	0.04	—	0.02	—	—	0.02
0.02	0.03	<0.005	0.03	0.03	0.02	0.02
148	160	172	172	160	160	152
Medium	Medium	Medium	Medium	Medium	Medium	Medium

Table 3.3.11—*continued*

TABLE 3.3.11 Typical properties of a range of epoxy casting resins—*continued*

Cured property		*Units*	*Glass-filled*	*Mineral-filled*	*Aluminium-filled*	*Silicone-filled*
Mix preparation		—	—	—	—	—
Mixed viscosity		cps	—	—	—	—
Room temperature cure		—	—	—	—	—
Cure shrinkage		%	—	—	—	—
Specific gravity		kg/dm^3	2	1.92	1.6	1.95
Hardness		Shore D	M115	M106	D89	D82
Compressive strength		Mn/m^2	—	—	—	—
Compressive Elastic Modulus		GN/m^2	—	—	—	—
Izod Impact		J/m	20	20	40	20
Water absorption (24 h)		%	0.1	0.1	0.2	0.1
Service temperature		°C	130	130	130	130
Thermal conductivity		W/m per K	—	—	—	—
Thermal expansion coefficient.		10^{-6} K^{-1}	23	27	72	40
Volume resistivity		ohm m	10^{13}	10^{13}	—	10^{13}
Dielectric constant	60 Hz	—				
	10^3 Hz	—	4.5	4	—	3.8
	10^6 Hz	—				
Dissipation factor	60 Hz	—				
	10^3 Hz	—	0.01	0.01	—	0.025
	10^6 Hz	—				
Dielectric strength		10^2 kV/m	160	160	—	220
Relative cost		—	—	—	—	—

Data supplied by Emerson & Cuming.
[a] Contains glass microballoons.

Table 3.3.11—*continued*

TABLE 3.3.12 Typical properties of epoxy laminates

	Property		ASTM Test	Units	Epoxy laminate			
					Glass fibre	Kevlar prepreg	C-fibre prepreg	Glass prepreg
Mechanical	Tensile yield strength		D638	MN/m^{-2}	240	500	600	285
	Elongation at break		D638	%	1.7	1.7	1.3	1.4
	Tensile Elastic Modulus		D638	GN/m^2	—	—	—	—
	Flexural Modulus		D790	GN/m^2	17	25	80	16
	Flexural strength		D790	MN/m^2	—	—	—	—
	Compressive strength		D695	MN/m^2	—	—	—	—
	IZOD impact strength		D256	J/m notch	800	1000	500	600
	Hardness (Rockwell)		D785	Rockwell	M115	M112	M113	M110
Physical	Specific gravity		—	—	1.7	1.33	1.5	1.7
	Thermal expansion Coeff.		—	$10^{-5} K^{-1}$	1.5	1	0.05	1.8
Electrical	Dielectric strength (3.2mm)		D149	10^2 kV/m	200	180	—	180
	Dielectric constant	10^3 Hz	D150	—	4.5	3.5	—	4.6
		10^6 Hz	D150	—	—	—	—	—
	Power factor	10^3 Hz	D150	—	0.01	0.1	—	0.01
		10^6 Hz	D150	—	—	—	—	—
	Volume resistivity		D257	ohm m	10^{15}	10^{15}	10^2	10^{15}
	Surface resistivity		D257	ohm m	—	—	—	—
	Arc resistance		D495	s	—	—	—	—
	Tracking resistance		DIN 53480	ohm-m	—	—	—	—
Thermal and Environmental	Heat distortion temp. 1.8 MN/m^2		D648	°C	220	210	230	210
	UL Long term index	(1)	—	°C	—	—	—	—
		(2)	—	°C	—	—	—	—
		(3)	—	°C	—	—	—	—
	Flammability		UL94	—	V0>5	V0>5	V0>5	V0
	Water absorption (24h)		D570	%	0.1	0.2	0.2	0.1
Processing	Mould shrinkage		—	%	—	—	—	—

(1) Electrical. (2) Mechanical with impact. (3) Mechanical without impact.

Table 3.3.12

TABLE 3.3.13 Typical properties of epoxy mouldings

	Property		Units	*General purpose granular*	*High heat resistance granular*	
Mechanical	Filler		—	Short glass fibre		
	Flexural strength		MN/m^2	90–110	90–110	90–110
	Flexural Modulus		MN/m^2	12-14	12–14	12–14
	Impact strength	Izod notch	J/m	100–1500[a]		
		ISO-R-179 notch	kJ/m^2	1.5–2.0	1.5–2.0	2.0–2.5
Physical	Specific gravity		kg/dm^3	1.90–1.95	1.83–1.87	1.92–1.96
	Thermal expansion coefficient		$10^{-6}\ K^{-1}$	30–35	25–30	20–25
Electrical	Dielectric constant	60 Hz	—	—	—	—
		10^3 Hz	—	4.4–4.6	5.0–5.2	5.0–5.2
		10^6 Hz	—	4.3–4.5	4.9–5.1	4.8–5.0
	Dissipation factor	60 Hz	—	—	—	—
		10^3 Hz	—	0.030–0.031	0.024–0.025	0.014–0.015
		10^6 Hz	—	0.033–0.034	0.028–0.029	0.017–0.018
	Arc resistance		s	>180	>180	>180
	Track resistance		V	200	275	300
	Dielectric strength		10^2 kV/m	170–190	160–180	130–150
Optical	Refractive index		—	—	—	—
Thermal and Environmental	Flammability		UL94	—	—	—
	Heat resistance	continuous	°C	200	220	220
		short-term	°C	300–400	350–450	400–500
	Heat distortion temperature		°C	170–180	170–180	170–180
	Water absorption 24h boiling		%	0.4–0.6	0.4–0.6	0.3–0.5
Processing	Post-mould shrinkage		%	<0.05	<0.05	<0.05

Data supplied by Allied Chemicals (except [a]).
Properties of long glass and directional glass reinforced epoxies are detailed in Chapter 3.8.
[a] Silica, carbonates, oxides, silicates to reduce cure shrinkage or raise thermal conductivity or other properties.

Table 3.3.13

TABLE 3.3.13 Typical properties of epoxy mouldings—*continued*

Electrical/electronic/encapsulation						
Standard granular	***Standard granular***		***Flame resistant granular***		***Standard granular***	***Flame resistant granular***
Unfilled	Silica-filled				Short glass fibre and mineral	
40–50	100–120	100–130	100–130	90–110	70–120	70-100
3–5	11-14	11–14	12–16	11-12	10–12	9-12
—	—	—	—	—	—	—
1.0–1.5	2.5–3.0	2.5–3.0	2.5–3.0	1.70–2.0	2.5–3.0	2.0–2.5
1.18–1.22	1.93–1.97	1.93–1.97	1.90–1.94	1.90–2.10	1.70–1.80	1.70–1.80
60–65	25-30	30	25-30	30–32	30–50	40–50
—	—	—	4.2	—	—	—
4.2–4.3	4.2–4.3	4.2	—	3.6–3.8	4.6–4.9	4.6–4.9
4.0–4.1	4.1–4.2	—	—	3.6–3.8	4.5–4.8	4.5–4.8
—	—	—	0.004	—	—	—
0.02–0.03	0.007–0.009	0.004	—	0.008–0.010	0.008–0.010	0.008–0.010
0.04–0.045	0.01–0.015	—	—	0.020–0.025	0.020–0.025	0.020–0.025
—	>180	>180	>180	>180	175	>180
—	450	450	450	—	250	200
—	110–130	110–130	120–140	110–140	120–140	120-140
1.50–1.52	—	—	—	—	—	—
—	—	—	V–0	V–0	—	—
80-100	195	195	195	195	195	195
200–400	300–400	300–400	300–400	300–400	300–400	300–400
110–120	160–180	160–180	160–180	>180	160–180	140–160
3.0–6.0	0.3–0.4	0.2–0.4	0.3–0.6	0.5–0.7	0.5–0.8	0.5–0.8
0.02–0.08	<0.05	<0.05	<0.05	<0.05	<0.05	<0.05

Table 3.3.13—*continued*

TABLE 3.3.13 Typical properties of epoxy mouldings—*continued*

	Property		Units	*Electrical/electronic/encapsulation*		
				Standard granular		
Mechanical	Filler		—	Short glass fibre and mineral		Mineral[a]
	Flexural strength		MN/m²	95–130	100–130	55–105
	Flexural Modulus		MN/m²	13-17	13–17	10–16
	Impact strength	Izod notch	J/m	—	—	7.5–12.5
		ISO-R-179 notch	kJ/m²	2.0–2.5	2.0–2.5	—
Physical	Specific gravity		kg/dm³	1.80–1.90	1.80–1.90	1.6–2.0
	Thermal expansion coefficient		10^{-6} K^{-1}	30–35	30–35	20–50
Electrical	Dielectric constant	60 Hz	—	—	—	3.5–5.0
		10^3 Hz	—	3.6–3.8	3.6–3.8	—
		10^6 Hz	—	3.6–3.8	3.6–3.8	3.5–5.0
	Dissipation factor	60 Hz	—	—	—	0.01
		10^3 Hz	—	0.007–0.009	0.007–0.009	—
		10^6 Hz	—	0.01–0.015	0.01–0.015	0.01
	Arc resistance		s	>180	>180	150–190
	Track resistance		V	400	300	—
	Dielectric strength		10^2 kV/m	160–180	160–180	120–160
Optical	Refractive index		—	—	—	—
Thermal and Environmental	Flammability		UL94	—	—	—
	Heat resistance	continuous	°C	195	195	—
		short-term	°C	300–400	300–400	—
	Heat distortion temperature		°C	160–180	160–180	95–120
	Water absorption 24h boiling		%	0.3–0.7	0.3–0.5	0.04 (RT)
Processing	Post-mould shrinkage		%	<0.05	<0.05	0.2–0.8

Data supplied by Allied Chemicals (except [a]).
Properties of long glass and directional glass reinforced epoxies are detailed in Chapter 3.8.
[a] Silica, carbonates, oxides, silicates to reduce cure shrinkage or raise thermal conductivity or other properties.

Table 3.3.13—*continued*

TABLE 3.3.14 Typical properties of a range of epoxy RIM/RRIM systems

Properties		*Units*	*Supplier designation*		
			DRH504 unreinforced	*DRH506 unreinforced*	*DRH506 reinforced*
Glass content		%	—	—	45–52
Flexural Modulus	–30°C	GN/m^2	2.93	3.10	20.7–27.6
	20°C	GN/m^2	2.74	2.86	17.9–25.5
	70°C	GN/m^2	2.51	2.62	17.9–23.4
Tensile strength		MN/m^2	69	68	193–303
Tensile Modulus		GN/m^2	2.96	2.68	15.2–17.9
Elongation		%	6	9	—
Heat sag	120°C	mm/h	0.127	5.84	—
Heat deflection temperature 1.8 MN/m^2		°C	118	102	—
Coefficient of thermal expansion		10^{-6} K^{-1}	56	41	—
Impact strength Izod notched		J/m	24	28	1782–2052

Data supplied by Shell.

TABLE 3.3.15 Typical properties of polybutadiene

	Property		*Units*	*Silica-filled*	*Fibrous-reinforced*
Mechanical	Tensile strength		MN/m^2	40–50	
	Tensile Modulus		GN/m^2	4–5.1	
	Flexural strength		MN/m^2	75–100	
	Flexural Modulus		GN/m^2	9–11	
	Compressive strength		MN/m^2	125–175	
	Impact strength		J/cm	0.15–0.25	Up to 6.0
	Hardness		Rockwell E	70–95	
Physical	Density		—	0.89–1.00	
Electrical	Dielectric strength		10^2 kV/m	200–400	
	Dielectric constant	60 Hz	—	3.7	
		10^3 Hz	—	3.7	
		10^6 Hz	—	3.7	
	Dissipation factor	60 Hz	—	0.007	
		10^3 Hz	—	0.004	
		10^6 Hz	—	0.002	
	Volume resistivity		ohm m	$1–7 \times 10^{13}$	
	Arc resistance		s	252	
Thermal and Environmental	Maximum service temperature		°C	180	
	1.8 MN/m^2 Distortion temperature		°C	>285	
	Flammability		—	NB	
	Water absorption (24h)		%	0–1%	
Processing	Mould shrinkage		%	1.1–1.5%	0.3–0.8%

Table 3.3.15

TABLE 3.3.16 Trade names and suppliers of epoxide resins

Company	*Trade names*	*Fillers*
Allied Chem.	Epiall; Plaskon	G, G/M
BXL/Union Carbide	Bakelite	Includes cycloaliphatic grade
BTR Permali	Permaglass	G
Budd	Polychem	—
CdF Chimie	Lopos	—
Ciba-Geigy	Araldite, Fibredux	G, M, K, C, CM, Al, Silica
Conap	Conap	—
Dow Chem.	DEN; DER	Bromine
Emerson-Cuming	Eccogel; Eccomold; Eccoseal	M, GS, GM, G/M
EMS/Grilon UK	Grilonit	C, CM, Al, Silica
Epoxylite	Epoxylite	T
Epoxy Prod.	E-Form	G,M
Fiberite (Vigilant)	Fiberite	G,M, Carbon/Graphite
General Electric	Arnox	M, G
Hysol/Dexter	—	U, M, S, Silica, G
Isochem	Isochemrex, Isochemduct	Bromine, Ag
Lavorazione	EMC	G, M
Morton	Polyset	M, G, G/M
Rhone Poulenc	Ervamix; Stratyl; Stratyrex	—
SIR	Eposir	G
Shell	Epikote; Epikure	C, CM, Al, Silica
Synres Almoco	Synres	G, Silica, M, G/M
Bayer UK	Lekutherm	Al, C, U, CM, Silica
3M (UK)	Scotchply	K, Glass fabric laminate
Tufnol	Tufnol	Glass fabric laminate
BTR Permali	Permaglas	Glass fabric laminate
Cyanamid Aerospace Products	Carboform	K
Epoxy RIM/RRIM systems		
Shell		
Polybutadiene		
Firestone Synthetic Rubber	FCR 1261	
ISR	Intene	

Key: Al = aluminium; Ag = silver; C = casting—glass filled; CM = casting—mineral filled; G = glass; GS = glass sphere; GM = glass microballoon; K = Kevlar Prepreg laminate; M = mineral; S = silicon; T = talc; U = unfilled.

3.3.4 Unsaturated polyesters, including alkyds

3.3.4.1 General characteristics and properties

Unsaturated polyesters only are dealt with in this section. Saturated (thermoplastic) polyesters and aromatic polyesters and copolyesters (*p*-oxybenzoyl) are dealt with in Section 3.2.19. Allylics, which are polyesters of allyl carbonate, are dealt with separately (Section 3.3.5). Laminated and highly reinforced polyesters are considered in Chapter 3.8.

The advantages and limitations of filled and unfilled polyesters and alkyds are listed in:

Table 3.3.17 *Comparative characteristics of unsaturated polyesters*

Polyester resins are supplied as liquids which are converted to the solid state by adding a catalyst or as moulding compounds, the most familiar of these being SMC, sheet moulding compound, and DMC, dough moulding compound.

Liquid resins can be reacted at room temperature (i.e. without heat or pressure) with simple equipment by hand lay-up or casting. Reinforcing with glass fibres is particularly easy and effective, e.g. by impregnation of a mat of glass fibres with activated resin whilst pressing it into the contours of a mould. When the resin sets or hardens the GRP (glass reinforced polyester) can be removed.

Sheet moulding compound (SMC) is long glass fibre-filled sheet of chemically thickened polyester resin. It is dry enough to handle and has a viscosity at moulding temperature low enough to allow flow at moderate pressures and high enough to cause the fibre to flow with the resin and uniformly fill the mould. Properties of parts thus manufactured are similar to those made by preform wet lay-up. SMC has the advantages of smoother surfaces and little or no fibre pattern. Typical glass fibre loadings are in the range 20–35%.

More recently the terms SMC-C, SMC-D and SMC-R have been adopted to signify continuous, directional and random fibre orientation. Combinations of letters such as C/R can be used to describe composites containing different kinds of reinforcements.

Dough moulding compound (DMC) can be moulded by compression, transfer or injection and due to the pre-mix of constituents will produce a more homogeneous product. Many formulations are commercially available as blended mixtures of resin, filler, pigments, catalyst and fibrous reinforcement, having a dough-like consistency. Even minor changes in formulation can have a major effect on properties (e.g. pigments can affect mechanical and electrical properties and storage life). Special formulations can accentuate particular characteristics. DMC is known in the USA as bulk moulding compound (BMC). Glass loading is 10–30%.

In the USA many proprietary trade marks offer similar abbreviations. To avoid confusion these are summarised below:

Abbreviation	*Meaning*	*Description*	*Holding company*
HMC	High-strength moulding compound	SMC containing 65% chopped glass fibre	PPG
SPMC	Solid polyester moulding compound	DMC for use on automatic equipment	Koppers
TMC	Thick moulding compound	SMC-type material for mouldings up to 50mm thickness; SMC suitable only to 5mm.	USS
XMC	Directionally reinforced moulding compound	75% glass reinforced polyester for filament winding	PPG
UMC	Unidirectional moulding compound	Mixed chopped and continuous fibres, containing in some cases, fibres other than glass.	Armco

In addition, the term LMC (low pressure moulding compound) designates SMC made with polyester resins formulated to allow moulding at greatly reduced pressure. These allow the use of less expensive presses and moulds and the manufacture of larger parts.

Dry granulated polyesters are an alternative to DMC for moulding, claiming advantages in ease of use and storage life. May be known as PMC (pelletised moulding compound). PMC is seen as intermediate between DMC and phenolic, having twice the arc-resistance and five times the impact strength of phenolics.

The essential differences between 'dry granulated' and 'wet dough' polyesters are listed in:

Table 3.3.18 *Comparison of dry granulated and wet dough polyesters*

Besides the classifications depending on filler and form, polyesters can be designated according to the characteristics produced by formulation. The relative advantages and drawbacks of the types of resins are listed in:

Table 3.3.19 *Characteristics of polyester types*

Alkyds are available in a number of grades differentiated by filling (mineral or fibre) and by degree of fire resistance.

APPLICATIONS

Unreinforced Polyester

Buttons, surface coating, imitation marble, embedding and potting, flooring, nut locking, pipe joints, mortars.

Reinforced Polyester

Chemical processing
Tanks, pipes, ducting.
Filament winding is commonly used to achieve great tensile strengths by high degrees of reinforcement orientation.

Transport Industry
Boats and ships by hand lay-up and spray-up.
Hoods for heavy-duty trucks, recreational vehicles.
Automotive components: front end panels, deck lids, air conditioner housings.
Miscellaneous
Appliance housings
Business machines
Furniture
Electrical and electronic components
Bathroom furniture
Luggage
Corrugated sheeting

Alkyds

Automotive distributor caps and rotors, switch gear, circuit breakers, coloured appliance housings and similar components requiring high arc and track resistance and good dielectric strength at high temperatures.

TYPICAL PROPERTIES

The more important properties of unsaturated polyesters and alkyds are listed in:
Table 3.3.20 *Typical properties of unsaturated polyesters*
Table 3.3.21 *Typical properties of alkyd mouldings (AMC)*

3.3.4.2 Mechanical properties

The mechanical properties of unsaturated polyesters and alkyds are listed in Tables 3.3.20 and 3.3.21

These fall off with increase in temperature as shown in:
Fig 3.3.2 *Typical variation of properties of unsaturated polyesters with temperature.*

3.3.4.3 Environmental resistance

Polyesters can be compounded to optimise resistance to most environments.

3.3.4.4 Processing

POLYESTERS

Polyesters lend themselves to a very great variety of processing methods most of which are listed in Section 3.3.4.1.

The major advantage of the injection moulding process for thermosets is the faster rate of cure that is obtained compared with compression moulding. The higher cost of injection machines is said to be a disadvantage, but this is usually more than offset by the gain in production rate.

The greatest potential disadvantage is the reduction in strength due to the breakdown in glass fibres as DMC passes through the machine and the mould. The magnitude of this effect varies and although reductions in strength of 50% can occur in extreme situations, they are usually much less and may be negligible in certain circumstances.

ALKYDS

If the proper temperature is applied to the resin-catalyst system during moulding a quick reaction occurs; however with insufficient heat there is a danger of the system remaining thermoplastic. Hence a pre-heat temperature of 105°C should be maintained and the injection time short. Die temperatures are relatively critical (not less than 165°C). Moulds should be well polished and chromed and entrapped air must be avoided (i.e. good venting required with avoidance of dead space).

Alkyds are very easy to mould using the transfer and reciprocating-screw injection methods, but considerable care is required with compression moulding because of low melt viscosity, fast setting and tendency to entrap air.

3.3.4.5 Trade names and suppliers

The suppliers of unsaturated polyesters and alkyds, and the trade names of their products, are listed in:

Table 3.3.22 *Trade names and suppliers of unsaturated polyester and alkyds*

TABLE 3.3.17 Comparative characteristics of unsaturated polyesters

	Filled and unfilled SMC and DMC/BMC		*Alkyds*	
	Advantages	***Limitations***	***Advantages***	***Limitations***
Mechanical	High strength when filled. Grades range from flexible and resilient to hard and brittle. Polyester moulding compounds exhibit lower creep than filled or unfilled thermoplastics.	DMC mouldings do not develop their full strength for some 24 h after removal from the mould. Tensile strength much lower in DMC than flexural strength.	More rigid than phenolics.	More brittle than phenolics. Weaker than allylics and lower resistance cracking around inserts.
Environmental	Grades range from water sensitive to chemical and weather resistant.	High strength laminates (>70% glass fibre) have poor chemical resistance. Generally poor solvent resistance.	Fungus resistant	Attacked by alkalis and strong acids.
Thermal	Non-burning grades available, without incorporating additives to the extent necessary with some thermo-plastics. Better thermal properties than ABS.	Upper service temperature inferior to allylics.	Excellent long term dimensional stability. No evaporation of volatiles and/or gases. High heat distortion temperature.	
Electrical	Non-tracking. Good impact at subzero temperatures, demanded in new legislation covering electrical equipment, relative to glass-filled thermoplastics.		Very good dielectric properties justify higher cost for some electrical and electronic applications. Non tracking.	Applications confined to electrical or decorative products. Electrical properties not maintained under severely humid conditions.
Processing	Less costly processing equipment than phenolics and widest range of processing conditions of any resin; includes methods requiring neither heat nor pressure and consequent opportunities for cheap moulding equipment and short economical production runs. Injection moulding economical for high volume production. Accepts high filler contents.	Curing reaction is exothermic and unless this affect be reduced by fillers, etc., then internal stresses and cracking may occur, limiting the maximum section thickness of components. This is not normally a limitation because resins are rarely used unfilled. Emit styrene by evaporation during processing (except LSE types). Injection moulding may result in strength reductions of 50% in extreme cases and impact strength reductions of 30–60% although with good design may be far less.	Fast use grades available suitable for injection moulding and rapid cycle compression. Low moulding pressures.	
Optical/ appearance	Mouldings can be transparent, coloured, gloss or matt. Better surface finish than phenolics.	High temperature storing may cause surface quality problems, reduced with low profile types.	Good colour stability.	Restricted colour possibilities.
Miscellaneous	Low cost raw material and easily available. When used as housing, for example, property profile allows parts consolidation.	More expensive raw material than phenolics.	Lower cost than allylics (DAP, DAIP).	More expensive than phenolics or melamines.

TABLE 3.3.18 Comparison of dry granulated and wet dough polyesters

	Dry type polyester moulding compound	*Conventional wet dough polyester moulding compound*
Form	Pellet shaped, unique among thermosetting plastics. Dust free.	Bulk form, extruded rope, cut slugs, logs
Use in injection moulding	This pellet form enables use of standard thermosetting injection machines because of easy feeding into the hopper and screw. Automatic feeding into the hopper is also possible. Compression and transfer moulding are also available.	Dough forms require special equipment for automatic feeding due to their bulk form and stickiness.
Mould filling	Internal voids, cracks and surface porosity at the dead end of the mould cavity are eliminated. The unique dry form allows complete filling.	Internal voids and surface porosity at the dead end of the mould cavity often result because the material flows too easily even under low pressure, thus flashing out at mould gaps and entrapping air. This is especially true for large moulds.
Storage life	Storage life is longer because of the special composition. The shelf life is approximately 6 months at 30 °C. This longer life results in stable moulding quality and allows sufficient time for production preparation.	Storage life is 3 months at 20 °C. This short life can give too short time allowance in use and for production arrangements.
Handling and weighing	The dust-free pellets make handling clean and weighing simple. Automatic weighing is very easy.	The stickiness of the wet dough form makes handling messy and weighing difficult, as this is mostly done by tearing or cutting off the material. Automatic weighing is very difficult.

TABLE 3.3.19 Characteristics of polyester types

Polyester type	*Advantages*	*Limitations*	*Typical applications*
General purpose (G.P.)	Low cost Rigid. Good tensile and flexural strengths. Low water absorption.	Low abrasion resistance. Maximum service temperature 150°C.	Trays, boats, sinks, bathroom fittings.
Resilient	Improved impact resistance. High flexural strength.	Increased water absorption. Some temperature limitations. Lower stiffness and tensile strength.	Housings, machine guards and covers, safety helmets, bowling balls, gelcoats.
Heat resistant	Maximum service temperature 220°C. Much easier to process than polyimides or phenolics.	High costs. Not as heat resistant as polyimides or special phenolics.	Aircraft and electronic components.
Chemical resistant	High chemical resistance. Selective resins can be produced (e.g. for resistance to water and automotive fuels in a low profile form).	High cost. Not resistant to strong acids and alkalis, solvents. Same limitations as G.P.	Pipes, tanks, ducts, fume stacks.
Water resistant	High resistance to weathering and UV degradation.	Slightly higher cost than G.P. Same limitations as G.P.	Structural panels, glazing, skylighting, gelcoats.
Flame retardant	Self extinguishing	High cost. Higher density (inorganic additives) Lower chemical and weather resistance.	Building panels. Ships. Ducts.
Low shrink (low profile)	Good surface finish, and distortion control.		Automotive and aesthetic parts.

TABLE 3.3.20 Typical properties of unsaturated polyesters

	Properties		Units	Sheet moulding compounds (SMC)[a]							
				General purpose		Low profile flame retardant		Resilient		Low profile[b]	
Mechanical	Flexural strength		MN/m²	170	200	190	235	190	235	170	200
	Flexural Modulus		GN/m²	11	14	12.5	15	9	13	11	14
	Tensile strength		MN/m²	70	85	70	85	80	110	70	85
	Tensile Modulus		GN/m²	10	13	13	15	10	13	12	14.5
	Izod impact strength		J/m notch	880	1040	880	1040	1280	1840	800	960
Physical	Specific gravity		—	1.81	1.87	1.81	1.87	1.80	1.86	1.78	1.84
	Thermal Expansion Coefficient		10^{-6} K^{-1}	20	—	20	—	20	—	20	—
Electrical	Arc resistance		s	120	—	140	—	120	—	120	—
	Dielectric strength		10^2 kV/m	100	—	100	—	100	—	120	—
	Dielectric constant	10^3 Hz	—	—	—	—	—	—	—	—	—
		10^6 Hz	—	4.4	—	4.4	—	4.2	—	4.4	—
	Power factor	10^3 Hz	—	—	—	—	—	—	—	—	—
		10^6 Hz	—	0.015	—	0.015	—	0.015	—	0.015	—
Thermal and Environmental	Heat distortion temp. (1.8 MN/m²)		°C	>200	—	>200	—	>200	—	>200	—
	UL thermal index	Electrical	°C	—	—	—	—	—	—	—	—
		Mech W/O Imp	°C	—	—	—	—	—	—	—	—
		Mech W Imp	°C	—	—	—	—	—	—	—	—
	UL flammability	0.06 in	UL94	—	—	—	—	—	—	—	—
		0.12 in		—	—	—	—	—	—	—	—
	Water absorption (24 h)		%	—	—	—	—	—	—	—	—
Processing	Glass fibre	(6mm)	%	25	35	25	35	25	35	25	35
		(25mm)	%	—	—	—	—	—	—	—	—
	Mould shrinkage		%	0.2	—	Zero	—	0.2	—	0.12	—

[a] Data supplied by Freeman Chemicals Ltd., Rostone Inc., Takeda (Japan), and Ciba Geigy Ltd.

[b] These grades may contain elastomer modifiers, such as CDB (conjugated diene butyl) for enhanced impact strength in a low profile grade.

Table 3.3.20

TABLE 3.3.20 Typical properties of unsaturated polyesters—*continued*

Sheet moulding compounds (SMC) [a]						*SMC*	*Dough moulding compounds (DMC/BMC)* [a]		
Low density		*Chemical resistant*		*Automative painting grade* [b]		*High impact*	*Low shrink electrical grades*		
140	165	170	200	190	235	—	48–70	110–138	124–152
8	10	9	13	9.5	12	9.5	10.3–13.8	10.3–13.8	10.3–13.8
60	70	70	85	70	85	80	13–27	20–35	20–35
9	11	12	14.5	13	15	—	—	—	—
760	920	960	1120	880	1040	1000	108–216	320–540	320–540
1.30	1.41	1.70	1.77	1.81	1.87	1.9	2.08	1.94	1.88
20	—	20	—	20	—	30	15	15	15
140	—	140	—	140	—	—	>240	180–225	180–225
120	—	120	—	100	—	120	120	120	120
—	—	—	—	—	—	5	6–8	6–8	6–8
4.6	—	4.4	—	4.4	—	—	—	—	—
—	—	—	—	—	—	0.01	0.017–0.019	0.017–0.019	0.017–0.019
0.015	—	0.015	—	0.015	—	—	—	—	—
>200	—	>200	—	>200	—	180	>200	>200	>200
—	—	—	—	—	—	—	130	130	130
—	—	—	—	—	—	—	180	180	180
—	—	—	—	—	—	—	150	150	150
—	—	—	—	—	—	V1>3	V–0	V–0	V–0
—	—	—	—	—	—	—	V–0	V–0	V–0
—	—	—	—	—	—	0.2	0.2	0.2	0.2
25	35	25	35	25	35	—	5	15	20
—	—	—	—	—	—	—	—	—	—
Zero	—	0.2	—	Zero	—	0.2	0.05	0.05	0.05

Table 3.3.20—*continued*

TABLE 3.3.20 Typical properties of unsaturated polyesters—*continued*

	Properties		Units	*Dough moulding compounds (DMC/MBC)* [a]				*DMC*	
				General purpose electrical	*Corrosion resistant*		*General purpose low profile*	*Fire retardant*	*High heat*
Mechanical	Flexural strength		MN/m^2	140–160	48–70	100–140	40–70	—	—
	Flexural Modulus		GN/m^2	—	—	—	—	9	9.5
	Tensile strength		MN/m^2	27–41	13–27	20–35	20–35	40	45
	Tensile Modulus		GN/m^2	—	—	—	—	—	—
	Izod impact strength		J/m notch	370–595	100–200	160–260	100–200	400	300
Physical	Specific gravity		—	1.94	2.08	1.94	1.94	1.9	1.8
	Thermal Expansion Coeff.		$10^{-6}\ K^{-1}$	—	—	—	—	25	25
Electrical	Arc resistance		s	180–240	>240	189	183	—	—
	Dielectric strength		10^2 kV/m	120	120	120	120	130	150
	Dielectric constant	10^3 Hz	—	4–6	—	—	—	5	5
		10^6 Hz	—	—	—	—	—		
	Power factor	10^3 Hz	—	0.041	—	—	—	0.01	0.01
		10^6 Hz	—	—	—	—	—	—	—
Thermal and Environmental	Heat distortion temp.(1.8 MN/m^2)		°C	>200	>200	>200	>200	210	260
	UL thermal index	Electrical	°C	130	130	130	—	—	—
		Mech W/O Imp	°C	180	130	130	—	—	—
		Mech W Imp	°C	150	130	130	—	—	—
	UL flammability	0.06 in	UL94	V–0	HB	V–0	HB	V0>1	HB
		0.12 in		V–0	V–0	V–0	HB	—	—
	Water absorption (24 h)		%	0.2	0.1	0.1	0.4	0.25	0.25
Processing	Glass fibre	(6mm)	%	15	5	15	15	—	—
		(25mm)	%	—	—	—	—	—	—
	Mould shrinkage		%	0.2–0.4	0.2–0.4	0.2–0.4	0.05	0.25	0.25

Table 3.3.20—*continued*

TABLE 3.3.20 Typical properties of unsaturated polyesters—*continued*

Cellulose-filled polyester granulate	*Thick moulding compound (TMC)*			*Cast polyester*		*GMC*		
				Rigid	*Flexible*	*General*	*Fire retardant*	*High impact*
60–70	185	107	139	60–160	—	—	—	—
—	10.0	10.0	10.8	—	—	7	6.5	6
—	98	47	56	40–90	4-20	32	27	25
—	11.1	10.6	11.6	2–4	—	—	—	—
—	—	—	—	10–20	>350	100	100	200
1.75	—	—	—	1.1–1.5	1.01–1.20	1.95	2.2	2.0
—	—	—	—	—	—	20	20	20
130–140	—	—	—	125	135	—	—	—
110-120	—	—	—	150–200	100-160	120	120	120
7–8	—	—	—	2.8–5.2	4.5–7.1	5.5	5.5	5.5
—	—	—	—	2.8–4.1	4.1–5.9	—	—	—
0.03	—	—	—	0.005–0.025	0.016–0.050	0.01	0.01	0.01
—	—	—	—	0.006–0.026	0.023–0.060	—	—	—
—	—	—	—	110–200	—	210	210	180
—	—	—	—	—	—	—	—	—
—	—	—	—	—	—	—	—	—
—	—	—	—	—	—	—	—	—
HB	—	—	—	—	—	V1>3	V0>1	V1>3
—	—	—	—	—	—	—	—	—
—	—	—	—	0.15–0.60	0.5–2.5	0.2	0.2	0.2
—	28	15	20	—	—	—	—	—
—	—	—	—	—	—	—	—	—
0.9–1.0	—	—	—	—	—	0.3	0.4	0.4

Table 3.3.20—*continued*

TABLE 3.3.20 Typical properties of unsaturated polyesters—*continued*

	Properties		Units	*Polyester laminates*		*Bisphenol polyester laminate (glass-filled)*
				Chopped glass-filled	*Woven glass roving*	
Mechanical	Flexural strength		MN/m^2	—	—	—
	Flexural Modulus		GN/m^2	7	16	16
	Tensile strength		MN/m^2	100	280	280
	Tensile Modulus		GN/m^2	—	—	—
	Izod impact strength		J/m notch	700	1060+	1060+
Physical	Specific gravity		—	1.4	1.7	1.7
	Thermal Expansion Coeff.		10^{-6} K^{-1}	30	20	20
Electrical	Arc resistance		s	—	—	—
	Dielectric strength		10^2 kV/m	130	130	130
	Dielectric constant	10^3 Hz	—	5	5	5
		10^6 Hz	—	—	—	—
	Power factor	10^3 Hz	—	0.2	0.02	0.02
		10^6 Hz	—	—	—	—
Thermal and Environmental	Heat distortion temp.(1.8 MN/m^2)		°C	215	230	230
	UL thermal index	Electrical	°C	—	—	—
		Mech W/O Imp	°C	—	—	—
		Mech W Imp	°C	—	—	—
	UL flammability	0.06 in	UL94	HB	HB	V2>3
		0.12 in		—	—	—
	Water absorption (24 h)		%	0.3	0.3	0.25
Processing	Glass fibre	(6mm)	%	—	—	—
		(25mm)	%	—	—	—
	Mould shrinkage		%	—	—	—

Table 3.3.20—*continued*

TABLE 3.3.21 Typical properties of alkyd mouldings (AMC)

Property		*Units*	*Mineral*			*Short glass fibre*		*Long glass*	*Mineral and glass*
			Standard	*Flame retardant*		*Standard*	*Flame resistant*	*Fibre, flame retardant*	
Form		—	Granular	Putty	Granular	Granular	Granular	Flake	
Flexural strength		MN/m^2	50–60	60–85	50–60	55–75	50–85	70–100	
Flexural Modulus		GN/m^2	—	—	—	7–12	8–14	9–12	8.6
Tensile strength		MN/m^2	20–60*	—	—	25–65*	—	—	72
Tensile Modulus		GN/m^2	3–21*	—	—	14–20*	—	—	—
Specific gravity		—	2.1–2.2	2.1–2.2	2.2–2.3	2.02–2.20	2.04–2.20	2.0–2.2	2.2
Thermal Expansion Coefficient		10^{-6} K^{-1}	—	—	—	40–60	40–50	50–60	37.5
Thermal conductivity		W/m per K	—	—	—	0.6–0.8	0.6–0.8	0.4–0.6	
Dielectric constant	10^3 Hz	—	—	—	—	5.9–6.6	4.9–5.9	4.6–4.8	6.1
	10^6 Hz	—	3.5–4.2	3.8–4.2	4.0–4.3	5.7–6.4	4.7–5.7	4.4–4.6	
Power factor	10^3 Hz	—	—	—	—	0.007–0.020	0.006–0.016	0.010–0.012	0.012
	10^6 Hz	—	0.018–0.022	0.011–0.023	0.014–0.023	0.010–0.024	0.009–0.020	0.017–0.019	
Arc resistance		s	>180	>180	>180	>180	>180	>180	
Tracking resistance		V	>600	—	—	>575	>550	>550	
Dielectric strength		10^2 kV/m	100–120	100–120	100–120	120–150	120–150	80–140	130
Heat distortion temperature		°C	>180	>180	>180	>180–>200	>220	>200	>200
Heat resistance	continuous	°C	160	160	160	160	160	160	
	short-term		200–300	200–300	200–300	200–300	200–350	200–300	
Flammability	1.47 mm	UL94	—	—	—	—	V–1/V–0	—	
	3.05 mm		—	—	—	—	V–1/V–0	—	V2
	6.10 mm		—	—	—	—	V–0	—	
Water absorption, boiling, 24 h		%	—	—	—	0.4–0.8	0.4–1.4	1.0–1.3	
Post-mould shrinkage		%	<0.05	<0.05	<0.05	<0.05	<0.05	<0.05	0.4

Data (except *) supplied by Synves Almoco

TABLE 3.3.22 Trade names and suppliers of unsaturated polyesters and alkyds

Company	*Trade names*	*Type*	*Fillers*
Unsaturated polyesters			
Altulor	Altuglas, Altulite	—	—
American Cyanamid	Cyglas	—	G
Armco	Armco	SMC	G
Ashland	Aropol, Hetron	—	—
BASF	Palatal	—	—
BIP	Beetle	DMC	G
Bayer	Leguval	—	—
BTR	Permachem, Permaglass, Pemiglas	DMC, SMC	—
CdF Chimie	Norsomix	DMC	G
	Norsodyne	—	—
	Norsopreg	SMC	G
Chemische Werke	Vestopal	—	—
Ciba Geigy	Ampal	Granulated	G, Cl
	Melopas	Melamine modified	—
Dynamit Nobel UK	Polydur	DMC	—
Eagle/Picher	—	DMC	G
		SMC	G
Emerson–Cuming	Stycast	Cast	G
Fiberite	RTP	—	G
Freeman Chemicals	Freemix	DMC, PMC	G
	Freflow	GMC	(Fire retardant)
	Flomat	SMC	G
GAF	Gafite	—	G
Goodyear	Vituf	—	G
Haysite	Haysite	DMC	G
		SMC	G
ICI	Melinex, Atlac	—	—

G = Glass
Cl = Cellulose.
M = Mineral.

DMC (dough moulding compound) is known as BMC (bulk moulding compound) in the USA.

Table 3.3.22

TABLE 3.3.22 Trade names and suppliers of unsaturated polyesters and alkyds—*continued*

Company	***Trade names***	***Type***	***Fillers***
Unsaturated polyesters—*continued*			
International Paint	Interes	—	—
Koppers	Koppers	Cast	—
Perstorp Ferguson	—	Granulated DMC	G
Plumb	Fibercore	DMC	G
Premix	Premi-Glas	DMC	G
		SMC	G
Rohm & Haas	Paraplex	SMC	G
Rostone	Rosite	DMC	G
		SMC	G
SIR	Sirester, Sirmasse	Unfilled, SMC, DMC	G
Scott Bader	Crystic	SMC, GMC	G
Shell	Epocryl	—	—
Synthetic Resins	Filobond, Uralam	—	—
Ugine Kuhlmann	Ukadiol; Ukapreg;	—	—
	Ukapon	—	—
Alkyds			
Allied Chem.	Plaskon	—	G
Amer. Cyanamid	Glaskyd	—	G
BIP	Injak	—	G
Hooker/Durez	Durez	—	M
Occidental Chemical	Durez	—	G, M
Perstorp Ferguson	Peropal	—	G/M
Plastics Eng.	Plenco	—	M, Cl, G
SIR	Siralkyd	—	—
Synres Amoco	AMC	—	G, M

Table 3.3.22—*continued*

TABLE 3.3.22 Trade names and suppliers of unsaturated polyesters and alkyds—*continued*

Company	***Trade names***	***Type***
Polyester Laminate		
BTR Permali RP Ltd	Permaglas Polyester	C, W
Alan Butcher Associates	Butch Polyester Laminates	C, W
Atlas Chemicals	Atlac	Bisphenol Polyester Laminate (glass filled)

C = Chopped glass-filled.
W = Woven glass roving-filled.

Table 3.3.22—*continued*

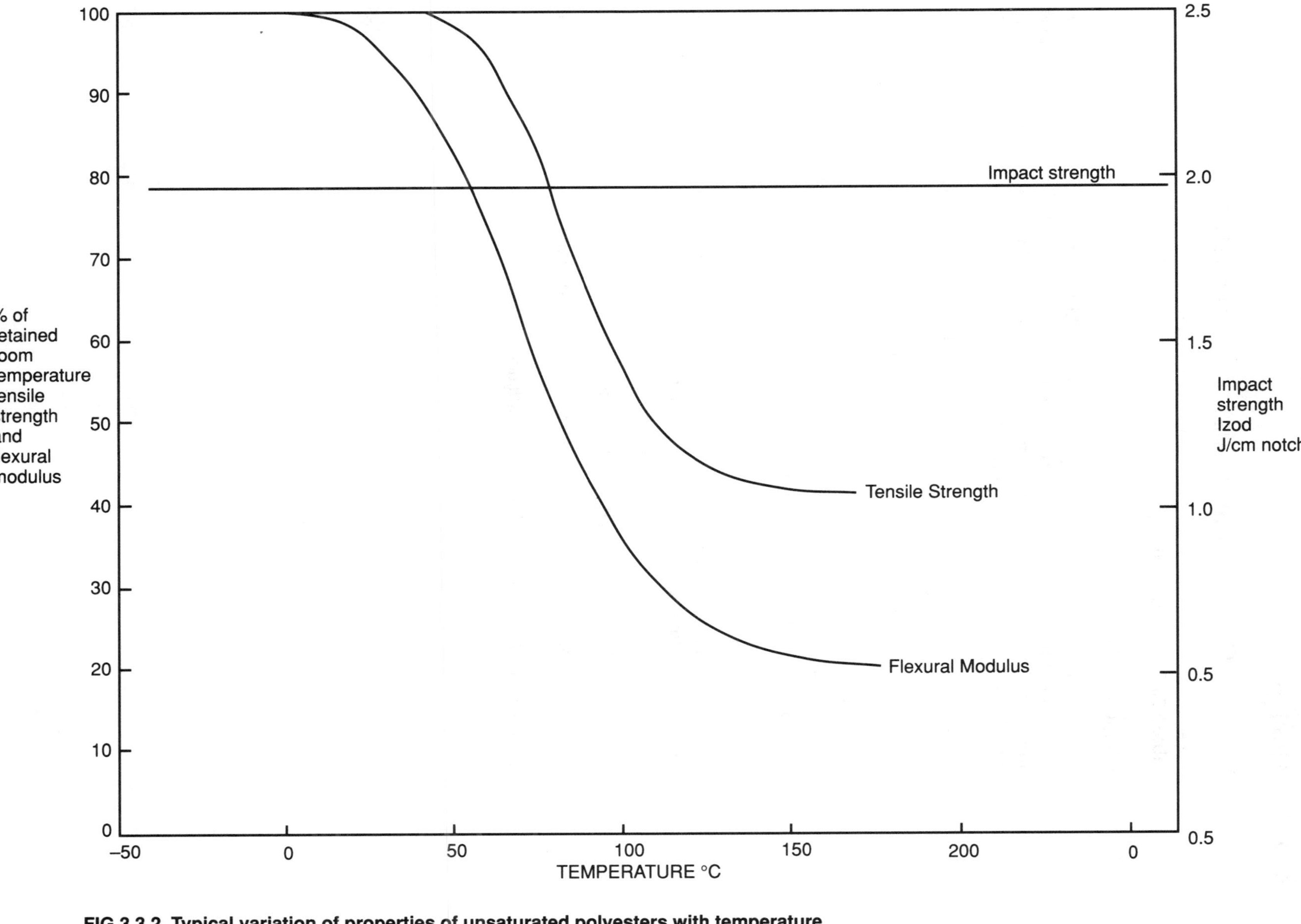

FIG 3.3.2 Typical variation of properties of unsaturated polyesters with temperature

Fig. 3.3.2

3.3.5 Allylics

3.3.5.1 General characteristics and properties

The advantages and limitations of allylics are listed in:
Table 3.3.23 *Characteristics of allylics*

APPLICATIONS

Diallyl phthalate (DAP) and diallyl isophthalate (DAIP) are readily converted into thermoset moulding compounds and preimpregnated glass cloths and papers, and are also employed as sealants for metal castings and surface coatings. Most applications are critical electrical applications requiring high reliability under long-term adverse conditions.

Diethylene glycol bis (allyl carbonate), CR-39, is employed as plastic lenses for eyeglasses because of its impact resistance, light weight, and scratch and abrasion resistance. Applications such as face shields and transparent panels are typical.

TYPICAL PROPERTIES

The more important properties of allylics are listed in:
Table 3.3.24 *Typical properties of allylics*

3.3.5.2 Processing

Shrinkage of allylic monomers during cure is 12% by volume. Compounds for in-line injection moulding equipment have been developed for production of quality parts on short moulding cycles with no volatile by-product.

3.3.5.3 Trade names and suppliers

The more important suppliers of allylics and the trade names of their products are listed in:
Table 3.3.25 *Trade names and suppliers of allylics*

TABLE 3.3.23 Characteristics of allylics

	Advantages	*Limitations*
Electrical properties	Exceptional electrical insulation characteristics under dry and wet heat conditions (better than phenolics or alkyds). Optimum combination of flame retardance and contaminated arc track resistance. No 'brown powder' effect on silver relay contacts.	Applications (of DAP and DAIP) confined to electrical components where service conditions justify cost.
Environmental properties	Combination of dimensional stability, chemical resistance and mechanical strength is outstanding. Resistant to almost all chemicals (except to strong oxidising acids), moisture and weathering. Fungus-resistant.	
Thermal properties	Short-term use to 260 °C. Flame-retardant grades available. DAIP has higher heat resistance than DAP.	
Optical properties	Diglycol carbonate (CR39) is transparent and maintains stable optical properties over wide range of conditions, in combination with hardness and scratch resistance. Can be dyed.	
Processing	Can be processed by modern thermoset techniques. DAIP has improved hot strength c.f. DAP and can be moulded faster.	Higher shrinkage on cure than phenolics or epoxies.
Cost		Higher than alkyds or phenolics.

Table 3.3.24

TABLE 3.3.24 Typical properties of allylics

	Property		Units	DAP					DAIP		Allyl Diglycol Carbonate (CR39)
				Glass filled	Mineral filled	Asbestos filled	Fire retardant	Mineral and synthetic fibre-filled	Glass filled	Mineral filled	cast
Mechanical	Tensile strength		MN/m^2	40–80	35–60	25–55	55–66	45	25–35	65	35–40
	Tensile Modulus		GN/m^2	10–15	8–15	8–75			—		2
	Elongation		%	—	—	—	0.8	1.1	—	0.9	—
	Flexural strength		MN/m^2	55–175	60–80	50			60–65		40–100
	Flexural Modulus		GN/m^2	8–11	11.7	8.2–9.6	8.5–9.6	6.8	—	9.5	1.8–2.4
	Compressive strength		MN/m^2	150–250	140–230	130			110		140–160
	Impact strength Izod		J/cm	0.16–7.5	0.15–0.23	0.2–0.3	3–3.8	3.8	0.15	3.2	0.1–0.2
Physical	Specific gravity			1.5–2.08	1.2–2.3	1.5–1.75	1.8–1.82	1.6	1.61–1.79	1.65	1.3–1.4
	Thermal conductivity		W/mK	0.2–0.6	0.3–1.0	—			—		0.19–0.20
	Linear thermal expansion		10^{-5} K^{-1}	1.0–3.6	1.0–4.2	—	3.1–3.3	3.4	—	3.2	8.1–14.3
Electrical	Volume resistivity		ohm m	10^{11}–$>10^{14}$	$>10^{11}$	—	10^{11}–10^{12}	10^{13}	—	10^{12}	$>4 \times 10^{12}$
	Surface resistance		MΩ	—	—	5.0			10.0	—	—
	Dielectric strength dry		10^2 kV/m	130–220	140–210	130–180	150	160	130	150	150
	Dielectric strength (wet 48 h)		10^2 kV/m	—	—	56			130		—
	Dielectric constant	60 Hz	—	4.3–4.6	5.2	4.6–6.2	4.6–4.8	5	—		4.4
		10^3 Hz	—	4.1–4.5	4.8–5.3	—			—	5.2	4.2–4.4
		10^6 Hz	—	3.5–3.9	3.9–4.0	6.0			4.7		3.5–3.9
	Dissipation factor	60 Hz	—	0.006–0.019	0.03–0.06	—	0.003–0.004	0.035	—		0.006–0.019
		10^3 Hz	—	0.1	0.03–0.10	—			—	0.03	0.01
		10^6 Hz	—	0.04–0.06	0.02–0.04	0.14			0.17		0.04–0.06
	Arc resistance		—	115–250	140–190	115			115		120–250
Optical	Clarity		—	Opaque	Opaque	Opaque			Opaque		Transparent
Thermal and environmental	Maximum cont. service temp.		°C	150–230	150–205	150–200	160	160	230–260	180	100
	1.8 MN/m2 Defl^n temp.		°C	150–260	160–>300	150–260	195–205	170	160		60–90
	Flammability		—	SE–NB*	SE–NB	SE–NB	V0	HB	SE	V2	—
	Water absorption (24 h)		%	0.12–0.60	0.2–0.5	0.70 (max.)	0.22	0.3	0.50 (max.)	0.3	0.2
Processing	Mould shrinkage		%	<0.1–0.5	0.5–0.7	0.6–0.8	0.3–0.6	0.6	0.3–0.5	0.3	—
	Remarks			Also contain mineral fibres in some cases. Glass fibres improve electrical properties strength and toughness.		Preheat at 65°C. Some grades also contain alpha-flock filler. Pre-heat at 65°C for best electrical properties.			Higher heat resistance, strength and stiffness at higher temperatures than DAP.		

TABLE 3.3.25 Trade names and suppliers of allylics

Company	*Trade names*	*Fillers*
DAP/DAIP		
Acme	Dapex	M, Orlon, G
Allied Chem.	Diall	G, A, Dacron, Orlon, M
Budd	Polychem	—
Cosmic	CP	G, Asbestos yarn, A, Dacron, Orlon
FMC	Dapon	G, M, U, Acrylic Fibre, Polyester Fibre
Hooker Durez	Durez	—
Oxidental Chemical	Durez	G, M
Rogers	Rogers	Synthetic Fibre, G
Synres Almoco	Synres, DAP	M
US Polymeric	Parr; Poly-Dap	Orlon, Dacron, G, M
Allyl Diglycol Carbonate		
PPG Ind.	CR 39	Registered Trade Mark

A = asbestos; G = glass; M = Mineral; U = unfilled.

3.3.6 Polyimide and aramid resins

3.3.6.1 General characteristics and properties

The advantages and limitations of polyimide and aramid resins are listed in:
Table 3.3.26 *Characteristics of polyimide and aramid resins*

Polyimides are a family of highly heat resistant polymers ranging from thermoplastics to thermosets.

They are available in the forms of moulding compounds, film, coatings or laminates, or fabrics (fibres). Films are generally employed unfilled, whilst laminates and moulded compounds rely respectively on continuous or discrete fibres or fillers. Moulding fillers include chopped glass fibres, graphite powder, molybdenum disulphide, PTFE and asbestos fibres.

Recommended fillers for various service conditions are:

Unfilled	High temperature mechanical and electrical parts.
15% graphite (by weight)	Non-lubricated bearings and seals.
40% graphite (by weight)	As above requiring low thermal expansion.
15% graphite plus 10% PTFE (by weight)	As above requiring low initial friction.
15% MoS_2 (by weight)	As above for vacuum or dry environments.
30% sheet glass fibres (by volume)	High temperature mechanical parts requiring low thermal expansion.

Polyimide film has good mechanical properties from liquid helium temperature to 600°C. It has high tensile and impact strength and high resistance to tear initiation. Room temperature properties are comparable to polyester film.

APPLICATIONS

The high price of polyimides has restricted applications to quality and performance-centred industries. More recently price reductions have resulted in polyimides finding wider applications, particularly as replacements for epoxy and phenolic.

Thermoplastic polyimide

Moulding compounds filled with graphite powder or fibre, molybdenum disulphide or PTFE are used for product parts with self-lubricating wear surfaces: piston rings, valve seats, bearing, seals.

Filled with glass, boron or graphite fibre, thermoplastic polyimide is moulded into high-strength structural components (see below).

Thermoset polyimide mouldings

Applications include:
- Jet engine parts.
- Automobile turbine engines.
- Automobile wheels.
- Electrical connector housings on aircraft.
- High speed/load bearings.
- Precision bearing cogs.
- Insulating bushings for arc-welding torch.
- Self-lubricating wear surfaces (as thermoplastic polyimide).
- Soldering fixtures.

Film

Wire and cable wrap, motor slot liners, flexible printed circuits, tubing (can be combined with FEP in insulation for cables).

Coatings

In semi-conductor devices and electrical components where low thermal expansion coefficient and high heat resistance are required.

Laminated and prepregs

Structural shapes, honeycomb structures, high temperature adhesives and coatings.

TYPICAL PROPERTIES

The more important properties of polyimides and aramids are listed in:
Table 3.3.27 *Typical properties of polyimide and aramid mouldings*

3.3.6.2 Mechanical properties

The most significant characteristic of polyimide is the way it retains its strength and other properties on heating, and that of aramid, its room temperature strength.

These characteristics are illustrated in:
Fig 3.3.3 *Comparative tensile strengths of unfilled resins vs temperature*, and
Fig 3.3.4 *Modulus of elasticity of unfilled resins vs temperature*

3.3.6.3 Electrical properties

The electrical applications of polyimides normally arise in combination with severe thermal or radiation environments.

Their electrical properties at room temperature (which are detailed in Table 3.3.27 are however comparable with these of the better insulating plastics, as shown in:
Table 3.3.28 *Comparative electrical data for typical insulating plastics.*

3.3.6.4 Wear properties

The wear properties of polyimides are outstanding with characteristics comparable with those of polyamide-imide, as follows:

Pressure P (psi)	*Velocity V (fpm)*	*Wear factor K (10^{-10} in^3 min/ft lb h)*	
		Polyimide (high performance)	*Polyamide-Imide*
50	200	6	8
1000	50	35	30
50	900	43	32

Wear = $KPVt$ (in) where t = time (h)
10^{-10} in^3 min/ft lb h ≡ 2×10^{-17} m^3/J

At best the bearing performance of the unlubricated polyimide plastics approaches that obtained in lubricated wear, as shown in:
Fig 3.3.5 *Comparative wear rate of bearing quality plastics against mild steel at room temperature*

3.3.6.5 Processing

Polyimide parts are fabricated by techniques ranging from powder metallurgical methods to more conventional injection, transfer, extrusion and compression moulding.

Currently, there are a number of different, commercialised polyimide resins available. Whilst all have good thermal properties, the physical properties cover a fairly broad range. In general, the resins modified for easier processing have the lower properties.

Polyimide parts are manufactured by hot compression moulding, machining from hot compression moulded or extruded stock shapes, direct forming (a process similar to the powder metallurgy process) and injection or transfer moulding. Because of the inherent lack of melt flow of polyimide resins, injection or transfer mouldability can only be achieved at a sacrifice of physical properties. The most common methods of manufacturing parts having the highest strength and temperature properties are hot compression moulding, machining from stock shapes and direct forming.

Stock shapes for machining are available in a wide range of shapes and sizes. The machinability of polyimides depends upon the resin type. Those polyimides readily available as stock shapes can be easily machined using conventional metalworking equipment and tooling. Normal precautions must be taken to prevent distortion in thin-wall parts and to prevent chipping on tool breakthrough, as in cast iron.

The direct forming of polyimides is done using procedures quite similar to those commonly used in powder metallurgy. One must work with the same design considerations and limitations. The most suitable parts for direct forming are those having a uniform cross-section in the pressing direction.

The obvious advantage of direct forming over machining is the considerable reduction of material waste. On the other hand, the direct forming process requires tooling and is therefore limited to quantity production.

3.3.6.6 Trade names and suppliers

The suppliers of polyimide and aramid moulding compounds and the trade names of their products are listed in:

Table 3.3.29 *Trade names and suppliers of polyimides and aramids*

TABLE 3.3.26 Characteristics of polyimide and aramid resins

	Advantages	*Limitations*
Mechanical properties	Retained at high temperatures. Aramid has best strength and toughness	Resins modified for easier processing have inferior properties.
Wear properties	Excellent dry bearing properties. PTFE-filled grades exhibit better wear friction characteristics than any filled PTFE compound.	Aramid resins have inferior wear properties to polyimides.
Thermal properties	Continuous service temperatures up to 250 °C, intermittent to 480 °C. Exceptional combination of thermal and electrical insulation. Dimensional stability, strength maintained at high temperatures. Creep resistant. Non-flammable.	Compounds having the highest heat resistance are generally most difficult to fabricate. Aramid resins have inferior mechanical properties at high temperature to polyimides.
Environmental	Unaffected by exposure to dilute acids, aromatic and aliphatic hydro-carbons, esters, ethers, alcohols, fluorinated refrigerants, aeration spirit and kerosene. Resistant to superheated steam. Excellent resistance to ionising radiation.	Attacked by dilute alkalis and concentrated inorganic acids.
Electrical	Exceptional combination of thermal and electrical insulation. Very good electrical properties over wide range of temperature and humidity conditions.	
Processing	Some grades suitable for hot compression moulding, injection or transfer moulding.	Some grades can only be sinter-processed—known as 'direct forming' and similar to powder metallurgy. Other products have to be machined from hot compression-moulded or extruded stock shapes. Because of poor inherent melt flow, injection or transfer mouldability can be achieved only by a sacrifice in physical properties.
Cost		High

TABLE 3.3.27 Typical properties of polyimide and aramid mouldings

	Property		Units	Polyimide[a]					
				Unfilled		15% graphite		40% graphite	
				M	DF	M	DF	M	DF
Mechanical	Tensile strength	20°C	MN/m²	86	72	65	62	52	48
		250°C		41	37	38	30	23	26
		290°C		—	—	—	—	—	—
	Tensile Modulus	20°C	GN/m²	3.2	—	—	—	—	—
		290°C		1.8	—	—	—	—	—
	Elongation (break)	20°C	%	7.5	8.0	4.5	6.0	3.0	2.5
		250°C		7.0	7.0	2.5	5.2	2.5	2.0
	Flexural strength	20°C	MN/m²	117	—	103	—	90	—
		250°C		62	—	>55	—	>48	—
		290°C		—	—	—	—	—	—
	Flexural Modulus	20°C	GN/m²	3.1	—	3.7	—	5.2	—
		250°C		2.0	—	2.6	—	3.7	—
		290°C		—	—	—	—	—	—
	Compressive strength	20°C	MN/m²	>275	—	220	—	124	—
		250°C		>138	—	90	—	83	—
	Impact strength, Izod notch		J/m	80	—	43	—	—	—
Physical	Specific gravity		kg/dm³	1.43	1.36	1.51	1.43	1.65	1.58
	Thermal Expansion Coefficient	20–250°C	10^{-6} K^{-1}	54	50	49	41	38	27
		–40–140°C		—	—	—	—	—	—

[a] Data supplied by Du Pont and Upjohn, filled grades also available.

TABLE 3.3.27 Typical properties of polyimide and aramid mouldings—*continued*

Polyimide[a]						*Thermoplastic polyimide*	*Aramid resin[a]*		
15% graphite 10% PTFE		*Porous*		*15% MoS_2*	*40% glass fibre*		*Unfilled*	*12% graphite*	*12% PTFE*
		Unfilled	*15% graphite 10% PTFE*						
M	*DF*	*DF*	*DF*	*M*			*DF*	*DF*	*DF*
45	52	31	24	56	79	118	121	65	64
24	24	19	14	—	—	—	37	27	24
—	—	—	—	—	—	30	4.4	—	—
—	—	—	—	—	—	1.3	—	—	—
—	—	—	—	—	—	0.67	—	—	—
3.5	5.5	4.0	3.0	4.0	1.2	10	4.8	2.9	2.8
3.0	5.3	4.5	2.5	—	—	—	5.1	2.9	4.1
70	—	—	—	131	—	198	—	—	—
—	—	—	—	—	—	—	—	—	—
—	—	—	—	—	—	34	—	—	—
3.2	—	—	—	3.44	13.2	3.3	—	—	—
—	—	—	—	—	—	—	—	—	—
—	—	—	—	—	—	1.1	—	—	—
125	—	—	—	—	—	206	—	—	—
—	—	—	—	—	—	—	—	—	—
—	—	—	—	—	240	38	—	—	—
1.55	1.46	1.08	1.19	1.60	1.65	1.40	1.30	1.37	1.36
54	41	56	52	—	24	28	—	—	—
—	—	—	—	—	—	—	40	29	36

Table 3.3.27—*continued*

TABLE 3.3.27 Typical properties of polyimide and aramid mouldings—*continued*

	Property			Units	Polyimide[a]					
					Unfilled		15% graphite		40% graphite	
					M	DF	M	DF	M	DF
Electrical	Dielectric constant 20°C, 100 Hz			—	3.62	—	13.53	—	—	—
	Dissipation factor 20°C, 100 Hz			—	0.0018	—	0.0053	—	—	—
	Dielectric strength			10^2 kV/m	220	—	98	—	—	—
Thermal and environmental	Heat deflection temperature 1.8 MN/m²			°C	360	—	360	—	—	—
	Water absorption (24 h)			%	0.24	—	0.19	—	0.14	—
Wear	Wear rate (unlubricated)	P = 50 psi V = 500 fpm	N_2	in/1000 h	0.01–0.015	—	0.004	—	—	—
			Air		0.25–1.2	—	0.09	—	0.06	—
	Coefficient of Friction (unlubricated)		Vacuo	—	—	—	—	—	—	—
			N_2		0.04–0.09	—	0.06–0.08	—	—	—
			Air		0.29	—	0.24	—	0.30	—
		PV = 100 000	Air		—	—	0.12	—	0.09	—
Processing	Mould shrinkage			%	—	—	—	0.6	—	—

[a] Data supplied by Du Pont and Upjohn, filled grades also available.

Table 3.3.27—*continued*

TABLE 3.3.27 Typical properties of polyimide and aramid mouldings—*continued*

Polyimide[a]						*Thermos plastic polymide*	*Aramid resin*[a]		
15% graphite 10% PTFE		*Porous*		*15% MoS_2*	*40% glass fibre*		*Unfilled*	*12% graphite*	*12% PTFE*
		Unfilled	*15% graphite 10% PTFE*						
M	*DF*	*DF*	*DF*	*M*			*DF*	*DF*	*DF*
—	—	—	—	—	—	3.43 (60 Hz)	5	—	
—	—	—	—	—	—	0.0055 (60 Hz)	0.008	—	—
—	—	—	—	—	20	—	315	—	—
—	—	—	—	—	380	270–280	260	—	—
0.21	—	—	—	0.23	0.7	—	0.6	6.0	0.55
—	—	—	—	—	—	—	—	—	—
0.07	—	0.002	0.001	0.25–0.33	—	—	1.0	0.075	0.125
—	—	—	—	0.03	—	—	—	—	—
—	—	—	—	0.25	—	—	—	—	—
0.12	—	0.05	0.05	0.25	—	—	0.25	0.20	0.15
—	—	—	—	0.17	—	—	—	—	—
—	—	—	—	—	—	0.12	—	—	—

Table 3.3.27—*continued*

TABLE 3.3.28 Comparative electrical data for typical insulating plastics

Property	*Polyimide*	*DAP*	*Silicone*
Dielectric strength (10^2 kV/m)	160	180	140–160
Dielectric constant (10^3 Hz)	3.5	3.6	4.0
Dissipation factor (10^3 Hz)	0.002	0.009	0.009
Volume resistivity (ohm–m)	10^{14}–10^{15}	10^{14}	>10^{11}
Arc resistance (s)	230	118	300–420

TABLE 3.3.29 Trade names and suppliers of polyimides and aramids

Company	*Trade names*	*Fillers*
Polyimide		
ASEA Isolation AB	Polyimider	—
Ciba Geigy	Ciba–Geigy	—
Du Pont	Vespel SP	U, Graphite, M, P
	Kapton (Film)	—
Fiberite (Vigilant)		G, Graphite, M
Fluorocarbon	Tribolon	PTFE
Monsanto	Skybond	—
Rhodia/Rhone Poulenc	Kinel	U, Graphite, G, PTFE, M, P
Rohm & Haas	Kamax	—
Upjohn	Polyimide	—
Aramid		
Du Pont	Vespel KS	U Graphite, PTFE
Mektron	Envex	U, G, Graphite, M, P

U = unfilled. G = Glass. M = MoS_2. P = PTFE.

Table 3.3.28 and Table 3.3.29

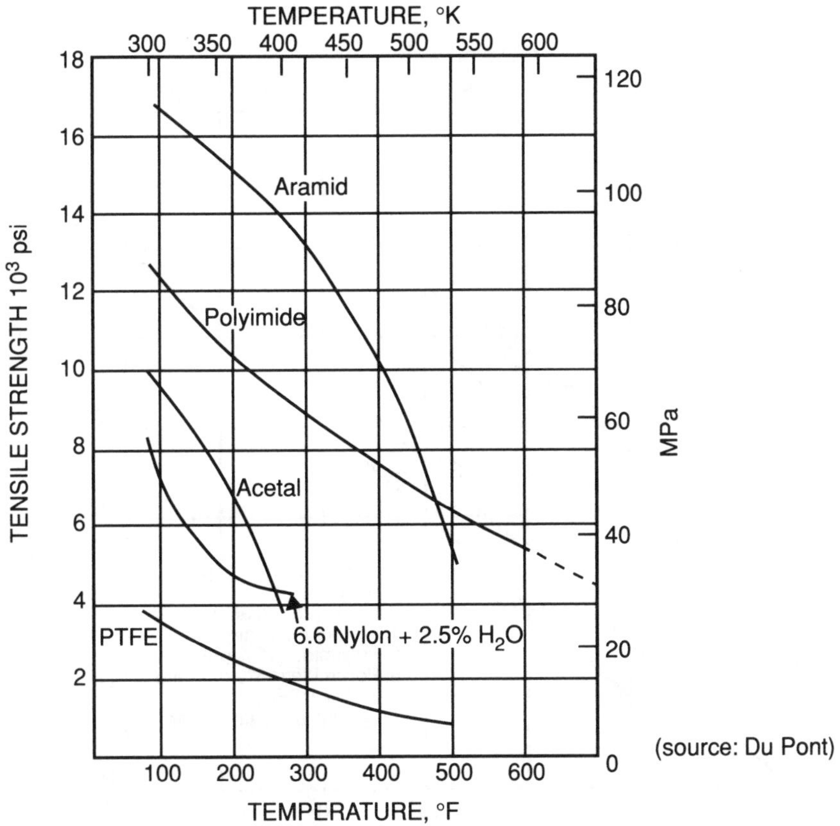

FIG 3.3.3 Comparative tensile strengths of unfilled resins vs temperature

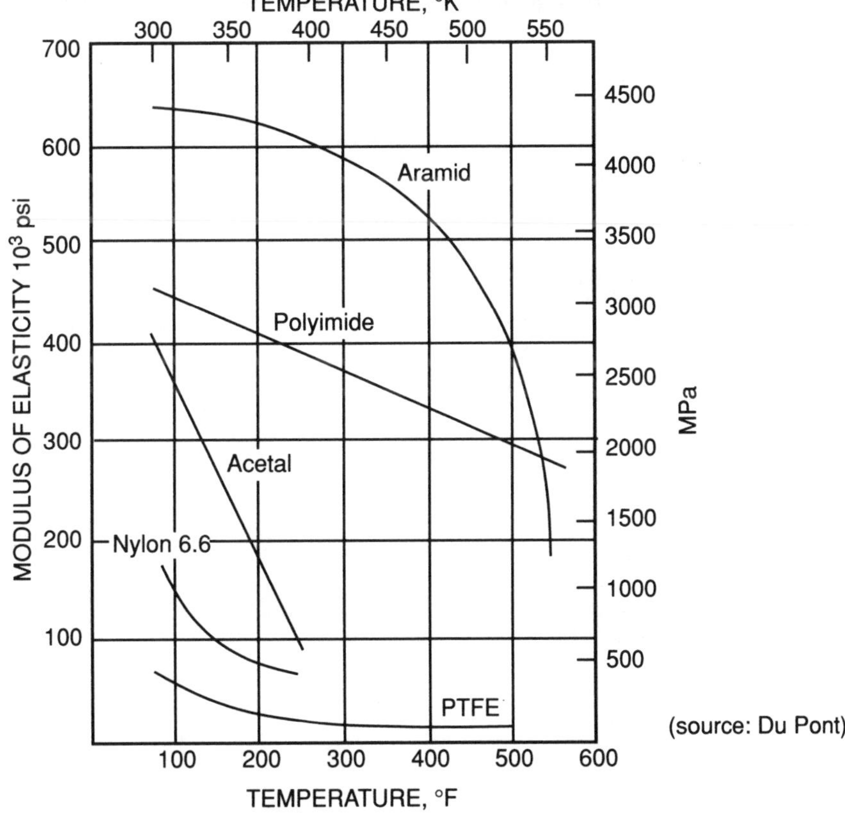

FIG 3.3.4 Modulus of elasticity of unfilled resins vs temperature

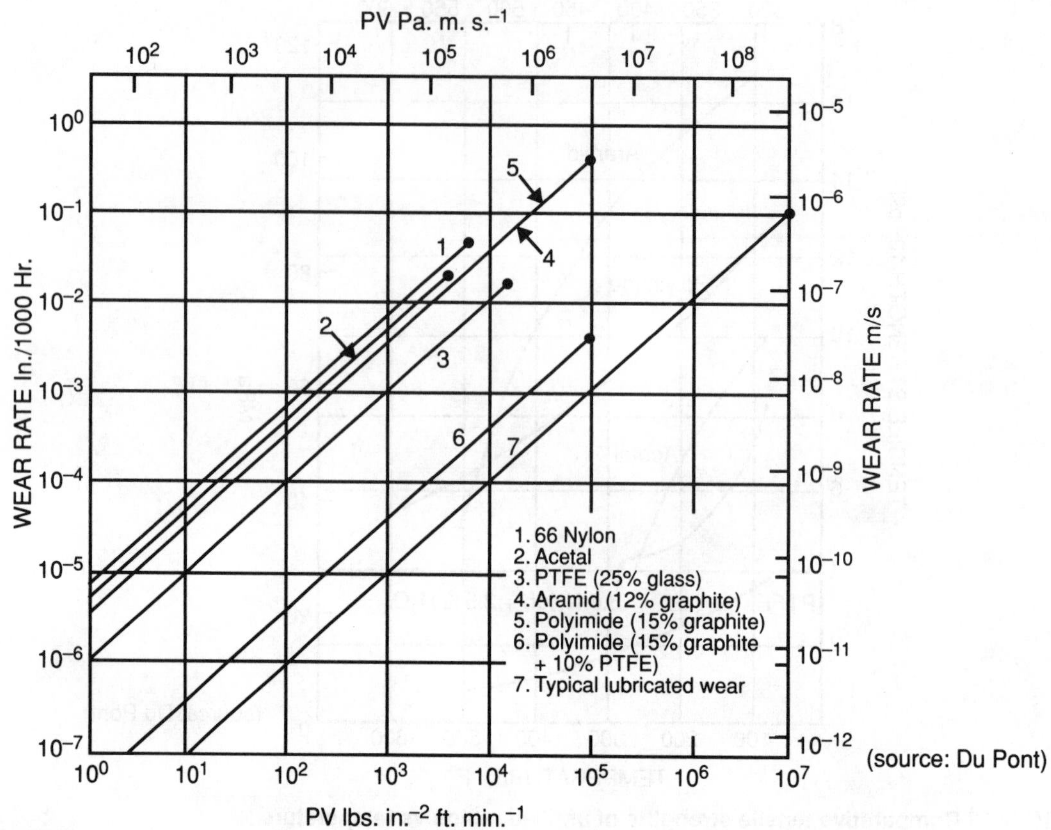

FIG 3.3.5 Comparative wear rate of bearing quality plastics against mild steel at room temperature

Fig. 3.3.5

3.3.7 Silicones

3.3.7.1 General characteristics and properties

Because of the silicon oxide linkage the properties of silicones have wider temperature capability, improved water resistance and better oxidative stability than polymers which depend on carbon. The advantages and limitations of silicone resins are listed in:
Table 3.3.30 *Characteristics of silicone mouldings*

AVAILABLE TYPES

Silicones are significantly different from other resins and plastics because the polymer's backbone contains no carbon; it is exclusively silicon and oxygen.

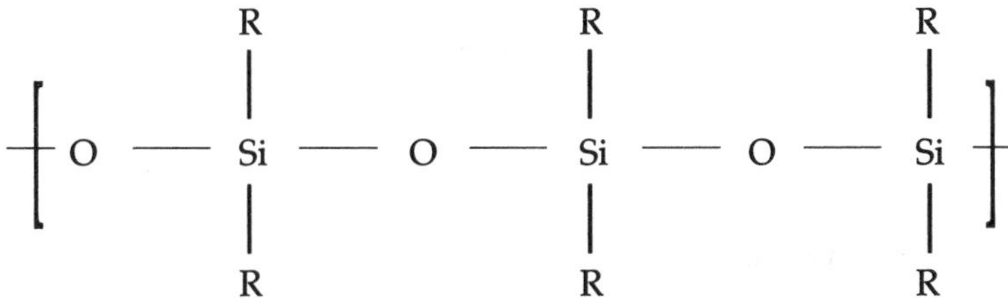

APPLICATIONS

Glass fibre-filled moulding compound

High temperature structural applications.
Encapsulation of electronic devices and other electronic applications.
Heat barriers in welding equipment, fuse holders and fire wall connectors including coil forms, technical boards, connector plugs and other Class H insulating components.

Asbestos–silicone moulding compound

Moulded boards for service in temperature extremes requiring high strength to weight ratios, stiffness, chemical resistance and electrical properties.

Other applications of resins

Formation of glass, asbestos and microlaminates for electrical applications.
Scratch resistant coatings for acrylic and polycarbonate sheets, lenses windshields and windows.
Coatings, adhesives, sealants, paints.
Medical implants.
Epoxy silicone moulding compounds.
Encapsulants for semi-conductor devices.

TYPICAL PROPERTIES

Silicone casting resins can be compounded to produce a wide range of properties, ranging for example in tensile strength from 0.9 to 5.5 MN/m^2.

Properties are given in:
Table 3.3.31 *Typical properties of silicone casting resins (room temperature cure, or RTV)*

Reinforced moulding compounds are available with tensile strengths higher by an order of magnitude. Properties are given in:
Table 3.3.32 *Typical properties of silicone compounds*

3.3.7.2 Resistance to temperature and environment

Silicones have better resistance to heating and aqueous environments (or a combinations of both) than the majority of plastics. This characteristic is illustrated in:
Fig 3.3.6 *Weight gain in boiling water of a silicone compared with other polymers*

3.3.7.3 Processing

Moulding compounds are available with long glass fibres or refractory fillers. The former are suitable for high temperature structural applications, the latter can be fused to produce ceramic articles.

Asbestos silicone moulding compound is a dry 30% silicone powder mixture which can be moulded and cured under heat and pressure to produce shapes and panels. Oven post-curing enhances the properties.

3.3.7.4 Suppliers of silicones

BTR Permali
Dow Corning
Emerson Cuming
General Electric (silica-filled)
RM Friction Materials (asbestos-filled)
Rhone Poulenc
Wacker
Synres Almoco

TABLE 3.3.30 Characteristics of silicone mouldings

	Advantages	*Limitations*
Mechanical properties	Wide range of mechanical properties, range from rubbers (see also Chapter 3.4) to hard resins.	Low strength.
Thermal properties	Relatively uniform properties over a wide temperature range. Flame retardant grades available. High temperatures can be sustained.	
Environmental properties	Very low water absorption. Good chemical resistance. Weather resistant. High humidities sustainable. Inertness: compatible physiologically and in electronic applications.	Not resistant to strong alkalis. Attacked by halogenated solvents.
Electrical properties	Good electrical properties (Class H) maintained over a wide range of temperature and frequency.	
Processing	Excellent release properties.	Mouldings require high pressure equipment.
Miscellaneous		High cost.

Table 3.3.31

TABLE 3.3.31 Typical properties of silicone casting resins (room temperature cure, or RTV)

	Property		Units	General purpose encapsulant	Versatile low cost	Low viscosity	High thermal conductivity	Fibre reinforced	Low density	Foam-in-place closed cell	High temperature	Clear resin	Thick section cure	Glass fibre and mineral
Mechanical	Tensile strength		MN/m^2	5.5	4.8	2.7	4.5	3.8	5.2	0.9	5.9	low	0.9–2.0	35
	Elongation		%	140	>100	180	>100	>100	135	>300	130	60	30–60	1.5
	Hardness		Shore A	60	60	45	70	72	55	Variable	60	22	55–70	M90
Physical	Specific gravity			1.70	1.65	1.15	2.28	1.60	0.75	0.33–0.5	1.50	0.99	0.75–2.3	1.85
	Thermal conductivity		W/m per K	0.46	0.40	0.33	1.08	0.36	0.14	0.06–0.10	0.35	0.17	0.15–1.79	
Electrical	Dielectric strength		10^2 kV/m	270	210	200	210	210	100	40	210	180	100–270	15
	Volume resistivity		ohm m	$>10^{12}$	$>10^{12}$	$>10^{12}$	$>10^{12}$	$>10^{12}$	$>10^{12}$	$>10^{12}$	$>10^{12}$	$>10^{12}$	$>10^{12}$	10^{12}
	Dielectric constant 10^6 Hz		—	4.0	4.0	3.1	5.2	3.9	2.0	1.3–1.4	3.6	3.0	2.0–4.0	
	Dissipation factor 10^6 Hz			0.007	0.005	0.004	<0.01	0.005	0.004	<0.01	0.015	0.001	0.004–0.007	
Optical	Colour		—	White	White	White	Red	White	White	White	Red	Trans-parent	White	
Thermal and Environmental	Service Temp.	continuous	°C	204	204	204	260	204	204	204	274	204	219–260	240
		short-term	°C	260	260	260	316	260	260	260	329	232	274–316	

Data supplied by Emerson and Cuming.

TABLE 3.3.32 Typical properties of silicone compounds

	Property	*Units*	*Glass fibre filled*	*Asbestos–silicone*	*Glass fibre and mineral filled*
Mechanical	Tensile strength	MN/m^2	28–45	21	240
	Tensile Modulus	GN/m^2	—	8.3	
	Elongation	%	—	0.24	1.5
	Flexural strength	MN/m^2	70–100	—	
	Flexural Modulus	GN/m^2	7–18	—	14
	Compressive strength	MN/m^2	70–105	70–200	
	Impact strength (Izod)	J/cm	0.15–5	—	0.5
	Hardness	—	M80–90	61–67 Barcol	M90
Physical	Specific gravity	—	1.8–1.9	1.8–1.95	1.85
	Thermal conductivity	W/mK	0.28–0.36	0.36	
	Linear Thermal Exp. Coeff.	$K^{-1} \times 10^{-5}$	2.0–5.0	—	4
Electrical	Volume resistivity	ohm m	10^{12}	—	10^{12}
	Dielectric strength, air	10^2 kV/m	80–160	—	150
	Dielectric strength, oil	10^2 kV/m	100	—	
	Dielectric constant 60 Hz	—	3.3–5.0	—	
	10^3 Hz	—	3.2–4.5	—	3.9
	10^6 Hz	—	3.2–4.3	—	
	wet 10^6 Hz	—	4.4	—	
	Dissipation factor 60 Hz	—	0.004–0.030	—	
	10^3 Hz	—	0.0035–0.020	—	0.002
	10^6 Hz	—	0.002–0.02	—	
	wet 10^6 Hz	—	0.16	—	
	Arc resistance	s	175–250	—	
Optical	Clarity	—	Opaque	— —	
Thermal and Environmental	Max. service temperature	°C	>335	—	240
	1.8 MN/m^2 Deflection temperature	°C	>500	—	260
	Flammability	—	—	—	VO >3
	Water absorption (24 h)	%	0.2	—	0.1
Processing	Mould shrinkage	%	0–0.5	—	0.3
	Mould pressure: transfer	psi	5000–15000	—	
	compression	psi	1000–5000	—	

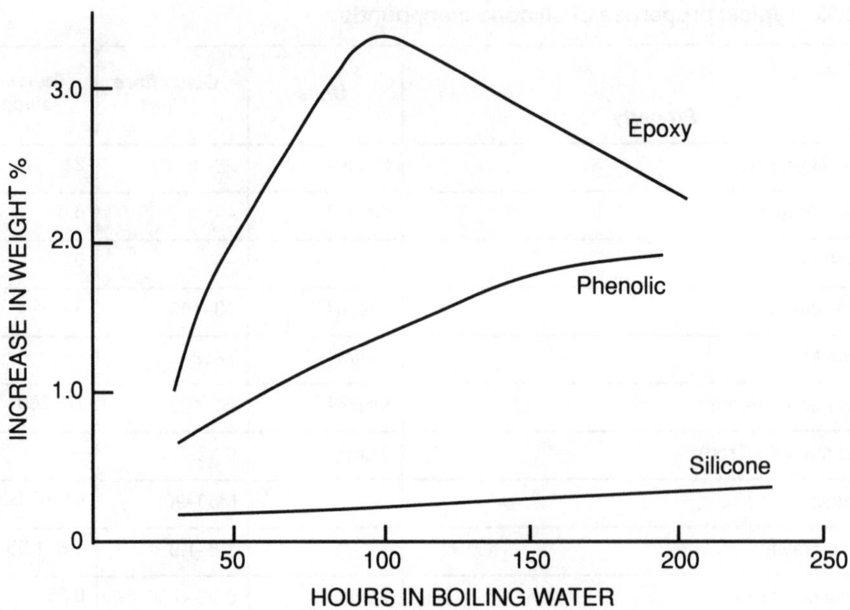

FIG 3.3.6 Weight gain in boiling water of a silicone compared with other polymers

Fig. 3.3.6

3.3.8 Amino resins (melamines and ureas)

3.3.8.1 General characteristics and properties

The advantages and limitations of aminos are listed in:
Table 3.3.33 *Characteristics of melamines and ureas*

AVAILABLE TYPES

Aminos are polymers formed by the reaction of formaldehyde with compounds containing the amino group. A wide variety of resins can be obtained by careful selection of pH, reaction temperature, reactant ratio, amino monomer and degree of polymerisation.

Although usually supplied with a cellulose filler, melamine plastics may be compounded with a variety of fillers. The effects of these are compared in:
Table 3.3.34 *Modifications of melamine moulding compounds*

Because it is primarily used in electrical components the range of fillers for urea is smaller.

APPLICATIONS

Urea–formaldehyde moulding compound

Electrical wiring devices and closures.

Melamine–formaldehyde moulding compound

Greatest volume for dinnerware.

Mineral-filled melamine–formaldehyde is used in heavy-duty industrial electrical equipment requiring high arc resistance and track resistance, coupled with high heat resistance.

Other applications for both types include housings for electric domestic equipment, soap dispensers, toilet seats, utensil handles.

Resins are employed as adhesives and bonding resins, laminating resins (melamine), paper resins and surface coatings.

TYPICAL PROPERTIES

The more important properties of melamine and urea–formaldehyde mouldings are listed in:
Table 3.3.35 *Typical properties of amino mouldings*

3.3.8.2 Processing

Amino resins are processed as aqueous alcoholic solutions or as spray-dried powders. Cure is by condensation under the influence of heat, catalyst or both. Urea–formaldehyde resins require the presence of acid to convert them to an infusible state and, properly catalyzed, can be cured in the absence of heat. Melamine resins are cured by heating, in the presence of an acid catalyst. No system is available for the low-temperature cure of melamine resins.

Moulding compounds in the main employ alpha-cellulose filler, although various other fillers are used. Both melamine–formaldehyde and urea–formaldehyde exhibit good preforming qualities; melamines may be pre-heated to 100–130°C and ureas 60–85°C. Moulding temperature for ureas vary from 125 to 170°C and slightly higher temperatures for melamines.

Compression pressures of 2000–8000 psi are used. In transfer and injection moulding process, pressures of up to 140 MN/m^2 are required. With melamine, moulded parts may be easily decorated at the moulding stage.

3.3.8.3 Trade names and suppliers

The suppliers of amino resins and the trade names of their products are listed in:
Table 3.3.36 *Trade names and suppliers of amino moulding compounds*

TABLE 3.3.33 Characteristics of melamines and ureas

Advantages	*Limitations*
1. Melamines Very hard and scratch resistant. Resistant to staining, water and detergents. Self-extinguishing. Good electrical insulator (non-tracking). Wide range of colours including white and pastel. Opaque to transparent. Easy to decorate mouldings. No tendency to yellowing.	Must be filled for successful moulding. Long-term oxidation resistance is poor. Attacked by strong acids and bases. Poor dimensional stability, particularly over 80 °C.
2. Ureas Low cost. High mechanical strength and wear resistance possible. Freedom from taste and odours (compatible with foodstuffs). Self-extinguishing. Good electrical properties—high tracking resistance. Unaffected by organic solvents, oils, greases and fluorinated hydrocarbons. Range of colours including pastels and opal white. Opaque to transparent. No tendency to yellowing.	As melamine plus: Not recommended for use outside range –55–77 °C. Low impact strength at room temperature. High moulding and post moulding shrinkage. Lower humidity resistance than melamine.

TABLE 3.3.34 Modifications of melamine moulding compounds

Effect on[a]	*Glass fibre*	*Wood-flour*	*Asbestos*	*Cotton flock*	*Chopped fabric*	*Phenol modified*
Specific gravity	↓	(↓)	↑			
Flexural strength			↓			
Impact strength	↑↑			↑	↑	
Resistance to cracking						↑
Heat resistance	↑		↑			
Heat discolouration						↓
Dielectric strength		↑	↑↑			
Arc resistance	↑	↓	↑			
Dimensional stability	↑			↑		(↑)
Light stability						↓
Colour range		↓	↓			↓
Processability	↓			↓	↓	↑
Cost		↓				

[a]Relative to cellulose-filled.

TABLE 3.3.35 Typical properties of amino mouldings

	Properties		Units	Melamine–formaldehyde		
				Cellulose-filled	Wood-flour filled	Asbestos-filled
Mechanical	Tensile strength		MN/m²	41–70	35–65	30–50
	Tensile Modulus		GN/m²	7–11	8	13
	Elongation		%	0.6–0.9	0.6	0.3–0.45
	Flexural strength		MN/m²	70–115	60–85	50–80
	Flexural Modulus		GN/m²	6.9–12.4	—	—
	Compressive strength		MN/m²	280–330	230	175–210
	Impact strength (Izod)		J/m notch	0.1–0.3	0.13–0.20	0.14–0.20
	Hardness		Rockwell	M115–M125	M120	M110
Physical	Specific gravity		—	1.47–1.55	1.45–1.55	1.7–2.0
	Thermal conductivity		W/mK	0.29–0.42	0.26–0.42	0.52–0.68
	Expansion Coefficient		$K^{-1} \times 10^{-6}$	2–5	3–5	2–4.5
Electrical	Dielectric constant	60 Hz	—	6.0–9.0	6.0–7.6	6.4–10.2
		10^3 Hz	—	8.0–10.0	6.0–7.5	9.0
		10^6 Hz	—	7.0–9.0	4.7–7.0	6.1–6.7
	Power factor	60 Hz	—	0.04–0.08	0.02–0.04	0.07–0.017
		10^3 Hz	—	0.02–0.05	0.01–0.035	0.07
		10^6 Hz	—	0.025–0.045	0.025–0.06	0.041–0.05
	Volume resistivity		ohm m	10^9–10^{11}	10^9–10^{11}	10^9
	Dielectric strength		10^2 kV/m	110–120	140–160	140–170
	Arc resistance[a]		s	110–140	70–135	120–180
Optical	Clarity			Translucent	Opaque	Opaque
Thermal & Environmental	Maximum service temperature		°C	100	100–120	115–205
	1.81 MN/m² Deflection temp.		°C	180	130	125
	Flammability		—	S.E.	S.E.	S.E.
	Water absorption		% (24 h)	0.1–0.6	0.3–0.80	0.08–0.14
Processing	Mould shrinkage		%	0.4–1.5	0.6–0.8	0.5–0.7
				Higher arc resistance grades have lower mechanical properties than shown.		

[a] Tracking resistance >600 V in all cases.

Table 3.3.35

TABLE 3.3.35 Typical properties of amino mouldings—*continued*

Melamine–formaldehyde			*Urea–formaldehyde*		
Cotton flock	*Phenol modified*	*Glass-filled*	*Alpha cellulose-filled*	*Wood-flour filed*	*Foam*
50–103	40–55	35–138	40–90	40–55	0.1
—	5.5–12	17	7–10	7–10	—
—	0.4–0.8	—	0.5–1.0	—	0.2
90	55–70	105–160	70–120	75–115	—
12.4	7–8.5	13.8–18.6	9–11	—	0.02
210–250	180–210	140–245	175–315	—	—
0.2–0.23	—	0.3–9.0	0.12–0.28	0.12–0.15	0.2
M115	E95–100	—	M115	M115	D50
1.5–1.6	1.5-1.7	1.8–2.0	1.47–1.55	1.4–1.5	0.01
1.15	0.16–0.28	0.46–1.66	0.28–0.40	0.25–0.38	—
2.97	1.0–4.0	1.5–1.7	2.2–3.6	3–6	3
—	7.0–7.7	9.7–11.1	7.0–9.5	7.0–9.0	—
—	—	—	6.5–7.5	6.5–8.5	1.7
—	5.2–6.0	6.6–7.5	6.5–7.0	6.0–8.5	—
—	0.02–0.04	0.14-0.23	0.025-0.043	0.025-0.045	—
—	—	—	0.020–0.035	0.020–0.040	0.05
—	0.04–0.06	0.013–0.015	0.020–0.035	0.025-0.040	—
—	—	2×10^9	$10^{10} - 10^{14}$	$10^9 - 10^{12}$	10^3
120-138	85–130	70–120	120–160	120-180	100
110–150	130–180	180	80–150	80–130	—
Opaque	Opaque	Opaque	Transparent-Opaque	Opaque	—
120	135–165	150–205	75	75	90
150	140-155	205	130–145	—	75
N.B.	S.E.	S.E.	S.E.	S.E.	V0
0.16-0.30	0.30–0.65	0.09–0.21	0.4–0.8	0.5–1.2	20
0.6–1.5	0.4–1.0	0.1–0.4	0.6–1.4	0.6–1.0	3.5

Table 3.3.35—*continued*

TABLE 3.3.36 Trade names and suppliers of amino moulding compounds

Company	*Trade names*	*Fillers*
Melamine formaldehyde		
Allied Chemicals	Plaskon	Cl
American Cyanamid	Cymel	—
BIP	Melmex	Cl
Budd	Polychem.	C, G, Nylon
Dynamit Nobel	Ultrapas	Cl, W, C, A, A/W
Fiberite	M–	G, Cl, M
Hoechst	Hostaset	—
Monsanto	Resimene	—
Perstop Ferguson	Isomin	Cl
	Perstop Melamine	Cl
Plastics Eng.	Plenco	W, M, Cl
SIR	Melsir	—
Urea formaldehyde		
Allied Chem.	Plaskon	Cl, U
BASF	Schaumharz	F
	Basopor	—
BIP	Beetle	Cl
	Mouldrite	—
	Scarab	W
Dynamit Nobel	Pollopas	Cl
Monsanto	Resimene	—
Perstorp-Ferguson	Skanopal	Cl
	Perstorp Urea	Cl
SIR	Sirit, Siritle, Celsir	W
Sterling	Sternite	W

KEY: A = asbestos; C = cotton; Cl = cellulose; G = glass; M = mineral; U = unfilled; W = wood-flour; F = foam.

Table 3.3.36

3.3.9 Furans

3.3.9.1 General characteristics and properties

The advantages and limitations of furans are listed in:
Table 3.3.37 *Characteristics of furans*

APPLICATIONS

In the form of glass fibre-reinforced plastics for chemical processing towers, tanks and pipes for use in corrosive environments—competitive with stainless steel, titanium and high nickel alloys.

As a hot or cold cured binder for moulding sand.

Acid-proof flooring and laboratory table-tops.

Insulating foam for areas requiring low flame spread and smoke development (alcohol-modified).

TYPICAL PROPERTIES

The more important properties of furans are listed in:
Table 3.3.38 *Properties of furans*

3.3.9.2 Environmental resistance

Furans may be used (in the form of glass fibre-reinforced or mineral-filled composites) as building or flooring materials for plant handling many industrial chemicals and solvents.

A list of the reagents for which furan is or is not suitable is provided in:
Table 3.3.39 *Influence of process fluids on furan*

The resistance of furan is compared with that of other process plant materials in:
Table 3.3.40 *Comparative corrosion resistance of furan resin and other corrosion resistant materials*

3.3.9.3 Processing

The resin systems are designed for hand lay-up, spray-up and filament winding fabrication techniques.

3.3.9.4 Trade names and suppliers

Supplier	*Trade Name*
Quaker Oats	Qua Corr Resin
Ashland	Hetron

TABLE 3.3.37 Characteristics of furans

Advantages	*Limitations*
Produced from non-petrochemical sources.	Unpleasant odour.
Excellent chemical resistance to acids, bases, most solvents—comparable to best corrosion resistant polyesters.	Difficult to process: limited to hand lay-up, spray-up or filament winding.
Only **resin** resistant to carbon disulphide.	Longer cure time and more controlled cure than isophthalate polyesters.
Strength retention at elevated temperature.	Opaque, more difficult to quality control than polyester.
Good flame resistance and low smoke emission.	Attacked by halogens.
Moulding resin under evaluation.	

TABLE 3.3.38 Properties of furans

Property		*Units*	*Asbestos-filled*	*32% glass-filled laminate*
Mechanical	Tensile strength	MN/m^2	20–32	82
	Tensile Modulus	GN/m^2	10–12	5.5
	Flexural strength	MN/m^2	4–65	138
	Compressive strength	MN/m^2	70–100	80
	Hardness	Rockwell	R110	—
Physical	Specific gravity	—	1.75	1.4–1.6
Optical	Clarity	—	Opaque	—
Thermal and environmental	Max. Service Temp.	°C	125–165	148
	Water absorption	%	0.01–0.20	—

TABLE 3.3.39 Influence of process fluids on furan

Suitable for:	*Not suitable for:*
Acids (strong and weak) Bases (strong and weak) **Aggressive solvents:** Ketones Chlorinated types Esters Low boiling esters Acrylates Acetates Aromatics Styrene Benzene Toluene Xylene Carbon disulphide	Aqua regia Free bromine Calcium hypochlorite Chlorosulphonic acid Hydrogen peroxide (>5%) Hypochlorous acid Nitric acid (>5%) Oleum Piperidene Phenol (hot or conc.) Phosphorus bromide Potassium peroxide Pyridine Sodium hypochlorite Sulphur trioxide Sulphuric acid (>60%) Thionyl chloride

Table 3.3.37, Table 3.3.38 and Table 3.3.39

TABLE 3.3.40 Comparative corrosion resistance of furan resin and other corrosion resistant materials

Material \ Process fluid	Acid				Base		Salt	Solvent				
	H_2SO_4		HCl		NaOH		CaCl sat.	CS_2	Toluene	MEK	Chlorobenzene	Ethylene dichloride
	dil.	conc.	dil.	conc.	dil.	conc.						
Carbon steel (1020)	A	R Above 85%	A	A	R	R	A	R	R	R	R	R
Stainless steel (316)	R Below 10%	R Above 90%	A	A	R	A	A	R	R	R	R	R
Hastelloy C®	R	R	R	R	R	R	R	R	R	R	R	R
Aluminium	A	A	A	A	A	A	A	R	R	R	R	R
Premium polyester laminates	R	R To 60%	R	R	R	A	R	A	R	A	A	A
Furan laminates	R	R To 60%	R	R	R	R To 50%	R	R	R	R	R	R

(Data supplied by Quaker Oats.) R = resistant, A = attacked.

3.3.10 Vinyl esters

3.3.10.1 General characteristics and properties

The advantages and limitations (compared with isophthalic polyester resins, see Section 3.3.4) of vinyl esters are:

Advantages	*Limitations*
1. Superior to other polyesters and - bisphenol α fumarate for some corrosive environments 2. Low viscosity which facilitates a high degree of glass reinforcement.	1. Unsuitable for: Some acids including chromic and fuming sulphuric. Some halogen substituted compounds including methylene and ethylene chloride, and chloroform. Benzene. Ketones including acetone. Ethers, phenol.

APPLICATIONS

Reinforced structures exposed to corrosive environments. Tanks, ducts, piping and other process equipment but note limitations (above).

TYPICAL PROPERTIES

The more important properties of vinyl esters are:

Property	*Unit*	
Tensile strength	MN/m^2	79.3
Tensile Modulus	GN/m^2	3.1
Elongation	%	5–6
Flexural strength	MN/m^2	134.5
Flexural Modulus	GN/m^2	3.2
Compressive strength	MN/m^2	107
Compressive Modulus	GN/m^2	2.6
Compressive deformation (yield)	%	6.6–6.8
Hardness (Barcol)		33–37
Heat distortion temperature	°C	98–102[a]

[a] Load not quoted: therefore rough guide only.
Data supplied by Synthetic Resins Ltd.

3.3.10.2 Trade names and suppliers

The suppliers of vinyl esters and the commonly used trade names of their products are:

Supplier	*Trade Names*
Synthetic Resins Ltd	Uralam 31-345
Dow Chemicals/Freeman Chemicals	Derakane

3–4

Elastomers

Contents

List of tables

List of figures

3.4.1 Introduction to elastomers

3.4.1.1 Structure

Elastomers are polymers with long flexible chains. These chains are independent in the raw material or `gum', which can be processed by the same methods as are employed for thermosetting plastics. Added `vulcanising' agents introduce cross-linking on heating after, or in the later stages of, processing.

Thermoplastic elastomers (see Chapter 1.5) become fluid at elevated temperatures and can be processed by most of the same methods as thermoplastics.

In the unstressed state the long polymer molecules take up irregular statistically random configurations in thermal motion. On deformation the molecular chains are straightened causing a decrease in entropy and the exertion of a retractive force on the ends of the chains. Chain straightening permits very large increases in length before failure.

Vulcanising is progressive. Increased proportions of vulcanising agent, higher temperatures, longer times of vulcanising or a combination of these parameters increase the amount of cross-linking and shorten average chain lengths between links. This increases elastic modulus and decreases extension before fracture and allows the properties of an elastomer to be tailored to design requirements.

3.4.1.2 Classification

ASTM D1418 recognises four classes of elastomer:

R Class — elastomers with unsaturated carbon chain.
M Class — elastomers having a saturated chain of the polymethylene type.
Q Class — substituted silicone elastomers.
U Class — elastomers containing carbon, nitrogen and oxygen in the polymer chain

The elastomers dealt with in this section are listed according to class together with their ASTM D1418 designations in the table of contents.

There is considerable variation and overlapping of characteristics from one type to another depending upon the compounding and processing specification adopted.

A further method of classification depends on resistance to heating and to oil, see:
Fig 3.4.1 *Approximate heat and oil resistance capability of elastomers*
and
Table 3.4.1 *ASTM D-2000/SAE J2000 polymer classifications*

3.4.1.3 Typical properties

The more important properties of elastomers, including resistance to most industrial environments, from which can be selected the grade most appropriate for specific application, are compared in:
Table 3.4.2 *Comparative properties of elastomers*

3.4.1.4 Fillers

Rubbers containing only processing aids and chemicals for protection, colouring and effecting vulcanisation are known as gum, or unfilled rubbers.

The majority of rubbers used for engineering applications contain a filler, generally one of the many kinds of carbon black, which may comprise up to one-third of the total volume of the vulcanisate. These black fillers fall into two groups:

(i) Reinforcing blacks which improve the tear and abrasion properties of the unfilled gum rubber as well as increasing the Young's Modulus, hysteresis and creep.

(ii) Non-reinforcing blacks which have little effect on tear and abrasion resistance and give only moderate increases in modulus, hysteresis and creep. They can, however, be added in greater volumes than reinforcing blacks.

The major drawback with the addition of Carbon Black is that the resulting compounds are inevitably black. However, inorganic fillers are available which allow non-black synthetic rubbers to be formulated having a level of reinforcement in vulcanisates comparable to 'semi-reinforcing' Carbon Black. Typical applications include washing machine gaskets (essentially white!), cables and radiator hose.

TABLE 3.4.1 ASTM D–2000/SAE J2000 polymer classifications. According to 70-h air ageing tests and volume changes after heating for 70 h in aromatic, high swell ASTM No. 3 oil.

Class	*Type*	*Heat ageing temperature (°C) in air*	*Maximum % swell at 150°C in ASTM No 3 oil*
AA	NR, SBR, IIR, EPM, EPDM, BR, isoprene	70	
AK	PIR		
BA	EPM, EPDM, SBR (high temperature) butyl compounds	100	
BC	CR		
BE	CR		
BF	NBR		60
BG	NBR, urethanes		40
BK	PIR, NBR		10
CA	EPM, EPDM	125	
CE	CSM		
CH	NBR		
DF	Polyacrylate (butyl-acrylate)	150	60
DH	Polyacrylate		30
FC	Silicones (high strength)	200	120
FE	Silicones		80
GE	Silicones	225	40
FK	Fluorinated silicones	200	10
FFKM	Perfluoroelastomers		—
HK	Fluorinated elastomers (FKM)	250	10

TABLE 3.4.2 Comparative properties of elastomers

Material	ASTM D1418 designation	ASTM D2000 designation	Section	Permissible working temperature range (°C)	Tensile strength (MPa)		Hardness range (Shore A)	Specific gravity	Maximum elongation (%) reinforced	Tear strength	Abrasion resistance	Compression set	Rebound	
					Gum	Black							Cold	Hot
Natural rubber	NR	AA	2	–50– +100	>20	>20	30–90	0.93	700	4	4	3–4	4	4
Isoprene	—	AA	2	As natural rubber except lower mechanical properties, easier to process and more consistent										
Styrene butadiene	SBR	AA	4	–60– +90	<7	>14	40–90	0.94	600	2	4	3	3	3
Butyl (isobutene isoprene)	IIR	AA	5	–50– +120	>10	>14	40–80	0.92	800	3	3	2–3	1	4
Butadiene	BR	AA	4	–60– +100	>10	>14	40–80	—	600	2	4	3	—	—
Chloroprene	CR	BC, BE	3	–20– +95	>20	>20	40–95	1.23	600	4	3	2–3	3	3
Nitrile (butadiene acrylonitrile)	NBR	BF, BG, BK, CH	4	–20– +115	<7	>14	40–95	1.00	600	2–3	4	3–4	3	3
Ethylene propylene	EPM, EPDM	CA	6	–40– +150	<7	>20	40–90	0.86	600	2	3	3	3	3
Polyacrylate	ACM,ANM	DF[a], DH	7	–40– +175	<7	>14	40–85	—	400	1–2	2	3	3	—
Ethylene acrylic	—	—	7	–46– +165	—	>17	40–95	1.08–1.12	—	3	3	3	1	2
Chlorosulphonated polyethylene	CSM	CE	8	–40– +135	>17	>20	40–95	1.12–1.28	500	1–2	3–4	1–2	2–3	3
Fluoroelastomer	FKM	HK	9	–40– +205	>12	>14	50–95	1.85	300	2	3	** 2–3	2–3	3–4
Perfluoroelastomer	FFKM	—	9	+290	—	>14	65–95	2.01	—	3	3	2	—	—
Silicone	MQ, PMQ, VMQ, PVMQ	FE, GE	10	‡ –60– +235	<10	>10	40–85	1.14–2.05	700	1	1	2	4	4
Fluorosilicone	FVMQ	FK	10	Improved solvent resistance compared to other silicones; better low temperature capability than other fluoroelastomers										
Urethane – Polyether	EU	BG	14	–50– +85	>27	—	60A–80D	1.06	—	4	4	2	3*	3
Urethane – Polyester	AU	BG	14	More rigid, more abrasion resistant, tougher and more solvent resistant than polyethers, but hydrolysis prone										
Polysulphide (polyethylene sulphide)	PIR	BK	12	–50– +90	—	<9	40–85	—	450	1	1	0	0	0
Epichlorohydrin	CO, ECO	CH	13	–40– +160	<7	17	40–90	1.27–1.36	400	2–3	2–3	1	3	3
Polynorbornene	PNR	—	11	+70	—	—	15–80	—	—	2	—	*	—	—
Chlorinated polyethylene	—	BC, BE, CE	—	+120	>10	>17	60–90	1.16–1.32	—	2	3	3	2	3
Polyester elastomer	—	—	†	+100	>40	—	92A–72D	1.17–1.25	—	4	4	1	2–4	3–4

Notes: a Butyl-acrylate type
† Section 5 (thermoplastic elastomer)
Code: 4 – Excellent; 3 – Good; 2 – Fair; 1 – Poor; 0 – Not recommended.

Notes: § Will melt
** Exc. at high temps.
* Poor at very low temps.
‡ Special grades, to –100 °C.

TABLE 3.4.2 Comparative properties of elastomers—*continued*

Cut growth resistance	*Creep resistance*	*Electrical insulation*	*Dielectric strength*	*Low temperature flexibility*	*Adhesion*		***Resistance to***																	
					Metals	*Fabrics*	*Gas permeability*	*Flame*	*Weathering*	*Oxidation*	*Ozone*	*Water*	*Steam*	*Radiation*	*Dilute acids*	*Conc. acids*	*Dilute alkalis*	*Conc. alkalis*	*Lubrication oils*	*Animal and veg. oils*	*Aliphatic HCs*	*Aromatic HCs*	*Halogenated HCs*	*Oxygenated HCs*
4	4	4	4	4	4	4	3	1	1	3	1	4	2	3	2–3	2–3	4	2–3	0	1–3	0	0	0	3
As natural rubber except lower mechanical properties, easier to process and more consistent																								
3	4	4	4	3	4	3	3	1	1	2	0	3–4	2	3	2–3	2–3	2–3	2–3	0	1–3	0	0	0	3
4	2	4	4	3	3	3	4	1	4	4	4	4	2–4	0	4	3	4	4	0	3–4	0	0	0	3–
—	—	—	—	—	—	—	2	1	1	2	0	3	2	—	2–3	2–3	2–3	2–3	0	1–3	0	0	—	—
2	2–3	2	3	2	4	4	3	3–4	4	4	4	3	2	2	4	3	4	4	3	3	2–3	2	0	0
3	3	1	3	2	4	3	3–4	1	1	3	2	3–4	2–3	2	3	3	3	3	3–4	4	4	4	0	0
—	—	4	4	4	3–4	3	2	1	4	4	4	4	3–4	—	4	4	4	4	0	3–4	0	0	—	1–
3	2	2	—	0–2	—	—	3	1	4	3	4	1	0	2	1–2	1–2	1–2	1–2	1–4	3	0	0	0	0
—	—	3	2–3	3	4	3	4	1	4	4	4	3	0	—	3	0	—	—	3	3	0	3	—	2
2	2	3–4	3	2–3	4	3	3–4	3–4	4	4	4	3–4	3	2	4	4	4	4	3	3	2	1–2	0	0
2	3	3	2–3	3	2–3	4	4	4	4	4	4	4	3	2	4	4	4	4	4	4	4	4	3	4
—	—	4	4	3	2	3	2	4	4	4	4	4	—	—	4	4	4	4	4	4	4	4	—	4
2	3	3	4	4	4	4	2	2–3	4	4	4	4	2–3	2	3	0–3	4	4	2	3–4	0	0	2	2–
Improved solvent resistance compared to other silicones; better low temperature capability than other fluoroelastomers																								
—	—	4	2–3	4	4	4	2	2§	3	4	4	3	0	—	2	0	—	—	4	4	3–4	2–3	—	0
More rigid, more abrasion resistant, tougher and more solvent resistant than polyethers, but hydrolysis prone																								
0	0	2	—	3	—	—	4	1	3	3	4	4	—	0	2	2	3	3	4	4	4	4	3	3
—	—	3	3	3–4	2–3	2–3	3–4	1–3	3	3	3	3	2–3	—	3	2	3	1	4	4	4	4	—	0
—	—	—	—	4	3	—	3	—	—	0	0	3	0	—	3	0–2	—	—	—	—	0	1	0	0
—	—	3	4	4	2–3	2	4	3	4	4	4	4	—	—	4	3	—	—	3	3	3	2	—	2
—	—	2–3	2–3	4	3	3	2–3	3§	4	4	4	4	0	—	2	0	—	—	4	4	4	3	—	2

Table 3.4.2—*continued*

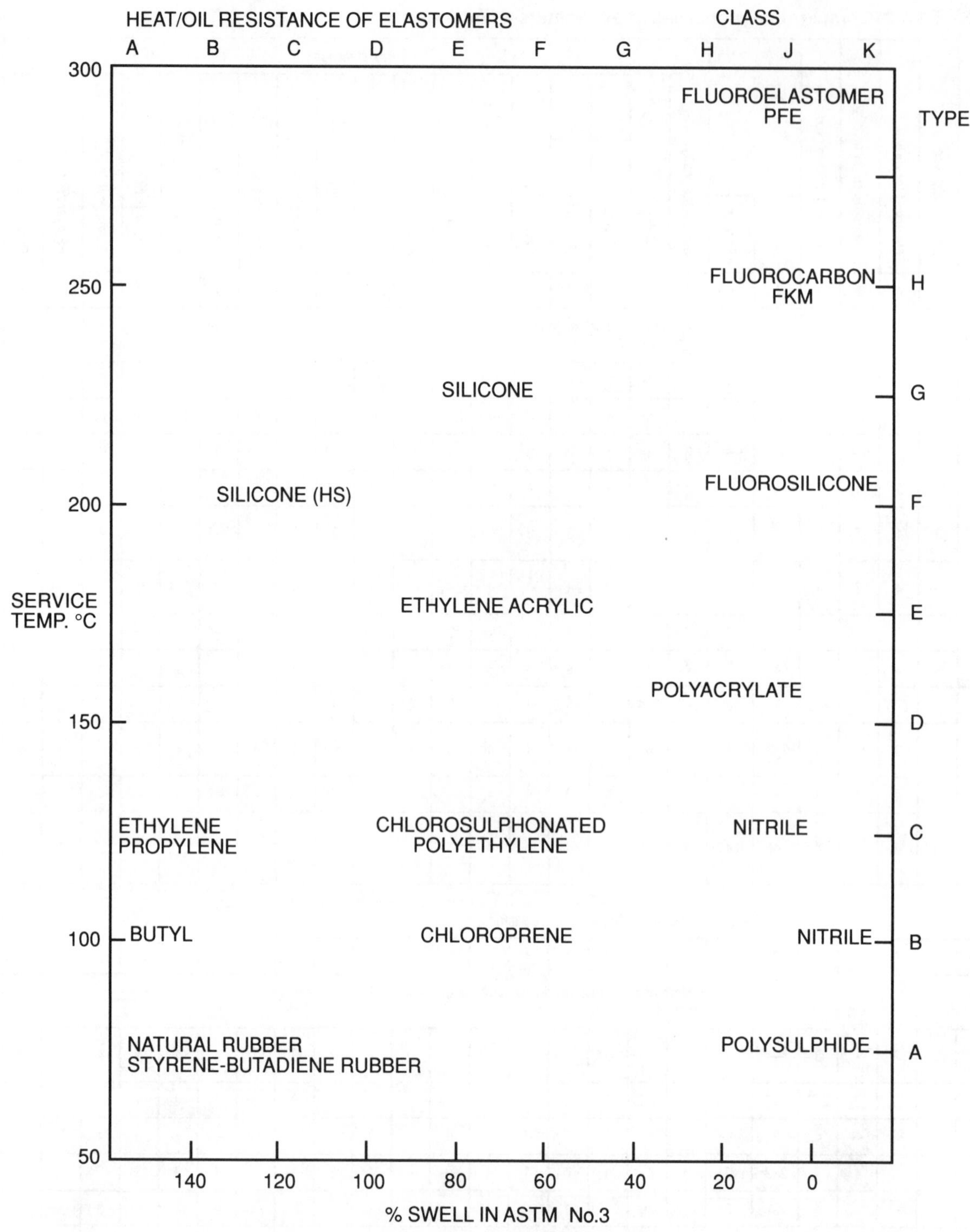

FIG 3.4.1 Approximate heat and oil resistance capability of elastomers

Fig. 3.4.1

3.4.2 Natural rubber and polyisoprene (*cis* 1,4) NR

3.4.2.1 General characteristics and properties

Rubber and polyisoprene (*cis* 1,4) have the repeating chemical formula:

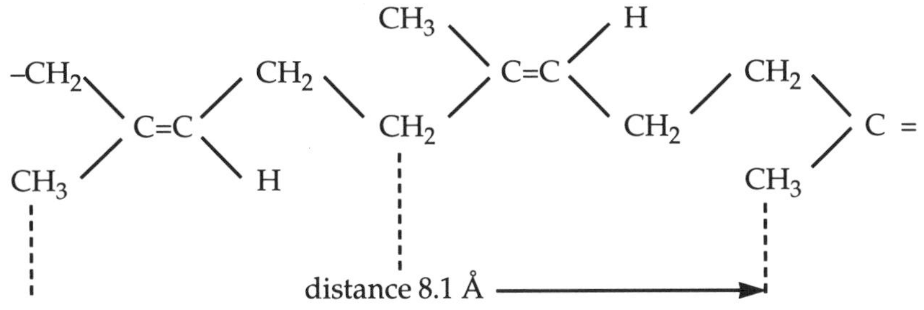

$$-CH_2\diagdown C{=}C \diagup CH_2 \diagdown CH_2 \diagup C{=}C \diagdown CH_2 \diagup CH_2 \diagdown C =$$

$$\text{(substituents: } CH_3,\ H,\ CH_3,\ H,\ CH_3 \text{); distance } 8.1\ \text{Å between } CH_3 \text{ groups)}$$

There exist two tranisomers of polyisoprene which differ only in the spatial arrangement of the atoms in the molecule. These isomers constitute Gutta Percha which has a higher specific gravity, higher melting points and a higher degree of crystallinity than rubber.

The advantages and limitations of natural and synthetic rubber are listed in:
Table 3.4.3 *Characteristics of natural rubber (NR) and polyisoprene (cis 1,4)*

AVAILABLE TYPES

Natural rubber

Rubber is produced at the plantation as 'latex', a colloidal solution.
Rubber is exported as either liquid latex or as the milled and dried coagulant from the latex which must be compounded.

Polyisoprene

After the basic structure of natural rubber was identified it was imitated through a synthetic approach. Polyisoprene is a result of this effort. Polyisoprene is more reproducible in processing and properties. Other advantages are less mastication and lower processing temperatures and less odour than the contaminated natural rubber. The tensile strength is not as high as that of natural rubber, other mechanical properties are similar.

APPLICATIONS

Tyres.
Resilient mountings in automobiles and machines.
Insulated cables.
Flexible hoses for air and water.

PROPERTIES

The more important properties of rubber are listed in Section 3.4.1, Table 3.4.1.

The mechanical properties of some natural rubber formulations, including the effects of ageing for certain times and temperatures, are listed in:
Table 3.4.4 *Typical properties of common natural rubber formulations*

3.4.2.2 Processing

LATEX DIPPING

Rubber gloves and other types of thin rubber products are often produced by dipping a former of the shape required in the latex and then coagulating the film of wet latex by an acid dip.

VULCANISED RUBBER

The basic material rubber is masticated to produce a workable mass to which is added the vulcanising agent, some form of sulphur or sulphur compound, reinforcing filler, carbon blacks or other pigments and small quantities of processing acids and antioxidants and antiozonants. The compound so produced can be calendered to give sheets of unvulcanised compound which can be used in the moulding operation, or the compound can be extruded through dies of the required shape to give continuous lengths which may be subsequently vulcanised by heating in autoclaves or other medium.

With the advent of synthetic materials, users of elastomeric materials began to receive gum stocks with more reproducible properties and behaviour than could be obtained with natural rubber. The rubber industry took steps to develop quality specifications and to define natural rubber types and grades. Natural rubber is produced in an environment where it is not possible to completely control all factors. As a result many of the grades (SMR—Standard Malaysia Rubber) deal with factors such as dirt, mould, bark, colour contamination and with gross physical parameters of packaging.

CV AND LV RUBBER

Natural rubber gum stock can undergo changes in storage and shipment which bring about a hardening of the rubber. Such undesirable effects can be avoided by stabilisation (CV rubbers) which results in retention of the original molecular weight and viscosity for a much longer period of time. Lower viscosity (LV) rubber is attained by also adding a naphthenic process oil as a plasticiser.

Fabricators have found that viscosity stability in the range of 55–65 (Mooney $M_L 1+4$ (100°C)) is reproducible, which means that no mastication is necessary before processing.

3.4.2.3 Trade names and suppliers

The suppliers of polyisoprene and the commonly used trade names of their products are listed in:

Table 3.4.5 *Representative trade names and suppliers of polyisoprene gum stocks*

TABLE 3.4.3 Characteristics of natural rubber (NR) and polyisoprene (*cis* 1,4)

Advantages	*Limitations*
Lowest cost.	Fair weather resistance (shield from sunlight).
Excellent strength, resilience, low hysteresis, tearing resistance and abrasion resistance.	Fair heat resistance.
Low compression set.	Attacked by all hydrocarbons.
Very good electrical resistance.	Inferior oil and ozone resistance to synthetics without compounding.
Very good resistance to cold weather.	Low temperature of –50°C without compounding.
Carbon black reinforcement greatly improves properties.	Attacked by conc. sulphuric acid, halogens and strong oxidising agents.
Resistant to most inorganic acids, salts and alkalis.	
Low swelling in water, acetone, alcohol and vegetable oils.	

TABLE 3.4.4 Typical properties of common natural rubber formulations

State	*Property*	*Unit*	*Standard Malaysian rubber SMR 5–CV*			
As delivered	Hardness classification	IRHD	35–40	45–50	55–60	65–70
	SRF Black	%	10	25	50	80
	Process oil	%	9	—	—	—
	Cure	—	35 min at 140 °C			
	Hardness	IRHD	37	48	56	67
	Tensile strength	MN/m^2	24	29	26	21
	Elongation at break	%	745	660	545	420
	Modulus (MR100)	MN/m^2	0.6	1.0	1.6	2.8
	Compression set 1 day, 70 °C	[a]	26	31	33	34
	Resilience, 50 °C	%	90	88	82	73
	Stress relaxation rate	[b]	2.1	2.7	5.1	6.7
	Young's Modulus	MN/m^2	1.5	2.2	3.0	4.1
Following ageing	**Aged 7 days at 70 °C**[c]					
	Change in hardness	IRHD	+3	+3	+4	+3
	Change in TS	%	+1	+3	−3	−3
	Change in EB	%	−8	−8	−12	−18
	Aged 28 days at 70 °C					
	Change in hardness	IRHD	+5	+8	+7	+8
	Change in TS	%	+3	−15	−12	−13
	Change in EB	%	−13	−19	−23	−37
	Aged 3 days at 100 °C					
	Change in hardness	IRHD	+2	+2	+2	+2
	Change in TS	%	−41	−45	−38	−38
	Change in EB	%	−24	−27	−36	−49

[a] 25% strain, 60 minutes recovery, % of imposed strain (BS AU 106).
[b] 100% extension at 30 °C and 30% RH, % per decade.
SOURCE: Malaysian Rubber Producers' Association.
[c] It is important to note that these general purpose formulations do not contain an antiozonant, but either a wax or chemical antiozonant should be added if necessary. Somewhat greater resistance to ageing can be obtained by replacing some Carbon Black with precipitated silica. However, with the finer grades of silica, the rubber is so hysteresial that in dynamic applications there could be a danger of excessive heat buildup. These effects are lessened when larger particle silica sizes together with coupling agents are used.

TABLE 3.4.4 Typical properties of common natural rubber formulations—*continued*

Conventional deproteinised natural rubber (DPNR)				*Soluble 'efficient vulcanising' (EV) SMR 5–CV*				*Soluble EV DPNR*			
35–40	45–50	55–60	65–70	35–40	45–50	55–60	65–70	35–40	45–50	55–60	65–70
10	25	50	80	10	25	50	80	10	25	50	80
9	—	—	—	9	—	—	—	9	—	—	—
35 min at 140 °C				35 min at 150 °C							
35	45	55	66	36	45	54	65	36	45	54	65
26	30	26	22	23	29	25	20	23	29	26	21
800	690	555	435	710	625	525	385	685	610	515	395
0.6	0.9	1.5	2.6	0.6	0.9	1.4	2.6	0.6	0.9	1.4	2.6
25	26	27	30	11	11	11	11	6	7	8	9
91	89	83	74	92	88	82	71	95	93	88	80
1.6	2.4	3.7	6.0	1.2	2.4	4.3	6.1	0.8	1.7	3.5	6.0
1.3	1.7	2.5	3.9	1.3	1.8	2.5	3.7	1.3	1.9	2.4	3.9
+3 −2 −12	+3 −3 −10	+4 −3 −11	+3 0 −15	+3 +3 −9	+2 −3 −9	+3 −6 −14	+1 −2 −14	+3 0 −9	+1 −4 −6	+1 −5 −10	+2 −5 −10
+4 −8 −16	+6 −10 −16	+6 −11 −21	+8 −12 −34	+2 +4 −10	+3 −2 −9	+4 −8 −15	+5 −7 −24	+2 +4 −11	+2 −9 −12	+3 −5 −13	+5 −10 −20
−1 −42 −25	−1 −48 −20	−1 −36 −29	0 −37 −41	+4 −3 −13	+3 −14 −16	+4 −16 −21	+3 −9 −24	+4 −2 −13	+2 −7 −12	+4 −13 −16	+3 −15 −23

Table 3.4.4—*continued*

TABLE 3.4.5 Representative trade names and suppliers of polyisoprene gum stocks.

Supplier	*Trade name*
Anic	Europrene
B.F. Goodrich	Ameripol
Goodyear	Natsyn
JSR	JSR
Polysar	TRANS-PIP
Hardman	DPR
Arto	—

Table 3.4.5

3.4.3 Chloroprene CR

Chloroprene is the product of chlorination of isoprene. It has superior resistance to heat and lubricating oil but at a higher cost than natural rubber. Its advantages and limitations are listed in:
Table 3.4.6 *Characteristics of chloroprene (CR)*

APPLICATIONS

Car radiator hoses.
Gaskets and seals.
Conveyor belts.

TRADE NAMES AND SUPPLIERS

Trade names and suppliers are listed in:
Table 3.4.7 *Representative trade names and suppliers of chloroprene (neoprene) gum stocks*

TABLE 3.4.6 Characteristics of chloroprene (CR)

Advantages	*Limitations*
Good mechanical properties.	More expensive per unit volume than natural rubber.
Good electrical properties.	Low resilience.
Good resistance to heat (100°C cont., 120°C intermittent).	Inferior oil and aromatic hydrocarbon resistance to nitrile (NBR).
Exceptionally good flexing characteristics down to –35°C (exceptionally –55°C).	Permeability inferior to NR.
Heavy filling results in hard rubbers, at expense of other mechanical properties.	Limited to use above –10°C but can be extended by compounding.
Outdoor exposure does not impair elastomeric qualities.	Not recommended for use with chlorinated or highly aromatic solvents, or oxidising acids.
Good resistance to lubricating oils, animal and vegetable oils and aliphatic hydrocarbons except chlorinated.	
Resistant to most inorganic chemicals.	
Similar abrasion resistance and resilience to NR.	
Self-extinguishing.	

TABLE 3.4.7 Representative trade names and suppliers of chloroprene (neoprene) gum stocks

Manufacturer	*Trade name*	*Notes*
Bayer	Baypren, Levapren	
Denki	Denka	
Du Pont	Neoprene	
Petro-Tex	Neoprene	
Rhone Poulenc/Rhodia/BP	Butaclor	(manufactured by BP/Rhone Poulenc company)
Rhone Poulenc/Rhodia/BP	Distugil SA	
ICI	Alloprene	
Union Carbide/BXL	Rubazote	

3.4.4 Butadiene based rubbers: butadiene polymer (BR); styrene butadiene copolymer (SBR); butadiene acrylonitrile copolymer (nitrile) (NBR)

3.4.4.1 General characteristics and properties

The advantages and limitations of butadiene based rubbers are listed in:
Table 3.4.8 *Characteristics of butadiene based rubbers*
Typical properties are given in:
Table 3.4.9 *Typical properties of butadiene based rubbers*

AVAILABLE TYPES

Butadiene is produced by catalytic dehydrogenation of butane which is a product of the catalytic cracking of petroleum.

It will readily polymerise to long chain polymers, which can be cross-linked by a vulcanising process to produce a rubber-like elastomer, butadiene copolymer (BR).

A large amount of plant suitable for producing butadiene and converting it to synthetic rubber is kept in strategic reserve against a shortage of natural rubber.

Butadiene may be copolymerised with styrene and with acrylonitrile.

Styrene butadiene copolymer SBR is a low-cost elastomer which is the most widely used synthetic rubber and is often employed as a blend in natural rubber compounds.

Depending on the temperature at which the initial emulsion copolymerisation is carried out styrene butadiene can be produced as `hot' or `cold' copolymer. `Cold' copolymers frequently contain staining antioxidants and their use is usually restricted to black products.

The characteristics of styrene butadiene acrylonitrile copolymer vary directly according to the proportions of each constituent.

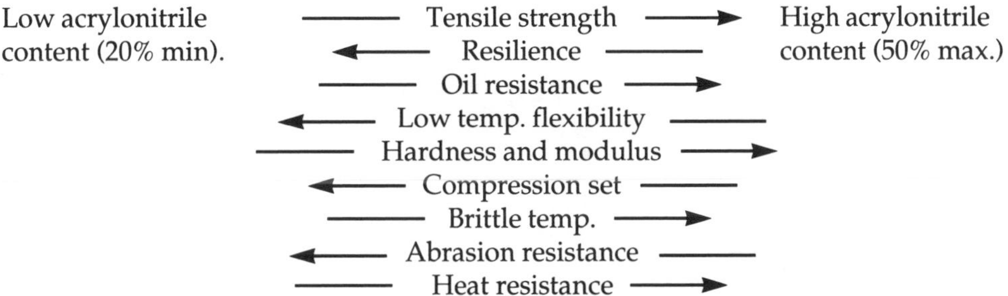

Butadiene acrylonitrile copolymer (nitrile rubber NBR) has a wider useful temperature range than most elastomers and better resistance to oil than styrene butadiene but it is difficult to combine these advantages with good weathering characteristics.

Nitrile rubber can be compounded with PVC to product a product, nitrile-PVC, with a restricted hardness range but improved resistance to oil and excellent weather ageing characteristics.

APPLICATIONS

Butadiene polymer

Used in large quantities as a general purpose rubber: hose, tubing, soles, equipment mounts, tyre tread.

Used for golf balls because of abrasion resistance and resilience.

Liquid forms used for castings and adhesives.

Styrene butadiene copolymer

Tyre treads (passenger vehicles)—often in direct competition with natural rubber; while many properties are not as good as those of natural rubber, its use results in a good overall product often at a lower cost than if natural rubber were used exclusively.

Transmission belting.
Electrical products.
Rollers.
Moulded and extruded mechanical items.

'Hot' polymer is used in white sidewall tyres, light coloured household goods, sporting goods, other products where colour is important.

Butadiene acrylonitrile copolymer (nitrile)

Components needing good oil or petroleum resistance such as hose, processing rolls, liners, seals, hydraulic components.
Refrigerant liners.

PVC - Nitrile compound

Similar applications to NBR but used when the component is exposed to ozone and/or sunlight.

3.4.4.2 Trade names and suppliers

The suppliers of butadiene rubbers and their commonly used trade names are listed in:
Table 3.4.10 *Trade names and suppliers of butadiene rubbers (gum stocks)*

TABLE 3.4.8 Characteristics of butadiene based rubbers

Advantages	*Limitations*
Butadiene polymer (BR)	
1. Good low temperature properties. 2. High resilience. 3. Improved abrasion resistance, flex-cracking, heat build-up and low temperature properties compared with SBR and NR. 4. Frequently blended with other elastomers to improve properties, especially abrasion resistance (e.g. tyre treads).	1. Attacked by all hydrocarbons. 2. Mechanical properties mostly inferior. 3. Much lower tear strength than NR, SBR. 4. Upper service temperature limited to 120°C. 5. Difficult to process unless blended.
Styrene butadiene copolymer (SBR)	
1. Low cost. 2. Similar mechanical properties to natural rubber. 3. Superior ageing characteristics to natural rubber. 4. 'Cold' polymerised SBR has superior strength, modulus and abrasion resistance to 'hot'*. 5. Carbon Black reinforcement is employed to improve tensile strength.	1. Poor mechanical properties without reinforcing fillers. 2. Less abrasion resistant and resilient than natural rubber. 3. Slightly inferior tensile and tear strength to natural rubber. 4. Attacked by all hydrocarbons; inferior chemical resistance to NR. 5. Inferior ageing and weathering characteristics to NR.
Butadiene acrylonitrile copolymer (nitrile NBR)	
1. Fairly good mechanical properties. 2. Good resistance to oils. 3. Good resistance to heat—can be compounded for use at –40–120°C. 4. Low compression set. 5. Can be PVC-modified to improve weathering and ozone resistance. 6. Special grades available with higher temperature capability (over 130°C) and improved abrasion resistance.	1. Types with best oil and heat resistance (high acrylonitrile content) have poor low temperature performance, poor weathering and ozone resistance. 3. High hysteresis and low resilience. 4. More expensive than NR and SBR. 5. Inferior to CR (neoprene) in weathering resistance. 6. Poor resistance to chlorinated hydrocarbons. 7. Inferior elasticity to NR.
PVC– nitrile compound	
1. Better resistance to oil than nitrile. 2. Better resistance to ozone than nitrile. 3. Better resistance to sunlight (UV radiation) than nitrile. 4. Flame resistant grades available.	1. Restricted range of hardness.

TABLE 3.4.9 Typical properties of butadiene based rubbers

	Property		ASTM test	Units	Styrene butadiene styrene based			Styrene ethylene butylene styrene based		
					35A	35D	80A	45A	55A	95A
Mechanical	Tensile yield strength		D638	MN/m^2	10	25	20	6	8	17
	Flexural strength		D790	MN/m^2						
	Elongation		—	%	1000	700	700	800	700	600
	Modulus of Elasticity in Tension		D638	MN/m^2						
	Flexural Modulus		D790	MN/m^2	0.007	0.12	0.05	0.02	0.08	0.21
	Impact strength		D638 (Izod)	J/cm of notch	10.6 +	10.6 +	10.6 +	10.6 +	10.6 +	10.6 +
	Hardness		—	Shore	A35	D35	A80	A45	A55	A95
Thermal	Max. continuous service temperature (no load)		—	°C	50	50	50	50	50	50
	Deflection temperature	0.45 MN/m^2	D648	°C	—					40
		1.81 MN/m^2	D648	°C	—					
	Thermal conductivity		—	W/m per K						
	Thermal expansion		D696	$10^{-6} K^{-1}$	130	130	130	160	150	140
Electrical	Volume resistivity		D257	ohm m	10^{12}	10^{12}	10^{12}	10^{14}	10^{14}	10^{14}
	Dielectric strength		D149	10^2 kV/m	200	200	200	250	250	250
	Dielectric constant	60 Hz	D150	—						
		10^6 Hz								
	Power factor	60 Hz	D150	—						
		10^6 Hz								
	Arc resistance		D495							
Miscellaneous	Specific gravity		D792	—	0.92	0.95	0.94	0.94	0.9	1.14
	Flammability		D635	UL94	HB	HB	HB	HB	HB	HB
	Water absorption		D570	% in 24 h	0.3	0.3	0.3	0.3	0.3	0.3
	Mould shrinkage		—	%	1	1	1	1.5	1.0	1

TABLE 3.4.10 Trade names and suppliers of butadiene rubbers (gum stocks)

Supplier	*Trade names*				
	Butadiene polymer	***Styrene butadiene copolymer***	***Butadiene acrylonyitrile copolymer***	***PVC - nitrile compound***	***Styrene ethylene butylene styrene based***
American Synthetic	Cisdene	ASRC	—	—	
Anic	Europrene	Europrene	Europrene	—	
Arto	—	—	—	—	
Ashland	—	Baytown	—	—	
BASF (UK)	—	—	—	—	
Bayer	—	—	Perbunan	Perbunan	
BP	—	—	—	Breon	
Bunawerke Huls	Buna	Buna huls	—	—	
Chemapol	—	Kralex	—	—	
Chemische Werke	—	Duranit		—	
Copolymer Rubber	—	Carbomix, Copo	Nysin	—	
Firestone	Diene	FR-S, Stereon	—	—	
General Tire	Duragen	Gentro			
B.F. Goodrich	Ameripol, Microblack	Ameripol, Microblack	Hycar, Nitrile	Hycar	
Goodyear	Budane Plioflex	Plioflex	Chemigum	Chemivic	
ICI	—	Butakon	—	—	
ISR	Incarb, Intene	Incarb, Intol, Unidene	—	—	
JSR	JSR	JSR	JSR	—	
LNP	—	PDX	—	—	
Montedison	—	—	Elaprim	—	
Northern Rubber	—	—	Nordoil	Nordoil	
Phillips (Fina Chemicals Ltd)	Solprene	Solprene	—	—	
Polysar	Taktene	Krylene, Krymix, Krynol, Polysars	Krynac	Krynac	
SIR	—	Sirel	—		
Shell	Cariflex	Kraton, Cariflex	—		Kraton
Sumitomo	—	Sumitomo	—		
Texas/Us Chem	Synpol	Synpol	—		
Ugine Kuhlmann	—	—	Butacril	Butacril	
Uniroyal	—	—	Paracril	Paracril	

3.4.5 Butyl rubbers: isobutene isoprene copolymer (butyl rubber)(IIR); chlorinated isobutylene isoprene (CIIR); bromobutyl; conjugated diene butyl (CDB)

3.4.5.1 General characteristics and properties

The advantages and limitations of butyl rubber are listed in:
Table 3.4.11 *Characteristics of isobutene isoprene or butyl rubber (IIR)*
Butyl rubbers are formed by the copolymerisation of isoprene and isobutene

$$\begin{array}{c} \quad CH_3 \\ \quad | \\ CH_2 = C \\ \quad | \\ \quad CH_3 \end{array}$$

They have fewer olefinic links than isoprene homopolymer and consequently better resistance to heat and steam. They are relatively low cost but require more active vulcanising agents than polyisoprene.

Modifications to butyl rubber include halobutyls (chlorinated butylene isoprene and bromobutyl) and conjugated diene butyl.

Chlorinated isobutylene isoprene (CIIR) has resistance to ozone which is normally good, but can be outstanding if the optimum curing system is used. It can withstand exposure to high temperatures and be made to adhere tenaciously.

Bromobutyl exhibits even better adhesion than chlorobutyl.

Cross linked butyl rubbers have superior thermal stability and are resistant to degradation by steam. They have better compression set than chlorobutyl formulations, noted for good set characteristics. They are said to be 'quite resistant' to ozone attack.

APPLICATIONS

Butyl rubber

Inner tubes.
Cable insulation.
Electrical applications.
Linings for chemical applications.

Chlorinated isobutylene isoprene

Inner liners of tubeless tyres (good adhesion to carcass).
Load bearing pads.
Mounts.
Hose.

Bromobutyl

Inner tube liners.

Conjugated diene butyl

Impact additive for polyester resins and plastic blends, where some cross-linking is desirable to achieve a balance between thermoset and thermoplastic properties not previously achievable with regular butyls (e.g. HDPE). 50/50 blends of CDB/HDPE and regular butyl/HDPE show, after cross-linking by radiation, that the former exhibit improved stress and elongation values, especially at elevated temperature.

Other uses include pharmaceutical caps and tubing, where the ability to cure in the absence of sulphur and heavy metal oxides is attractive.

TYPICAL PROPERTIES

The typical properties of Butyl Rubbers are listed in Table 3.4.1, Section 3.4.1.

3.4.5.2 Environmental resistance

The influence of heat on butyl rubber is compared in:
Table 3.4.12 *Comparative dry heat ageing of butyl rubbers*
and of compounds of butyl rubbers in:
Table 3.4.13 *Heat ageing resistance of butyl rubber compounds*

The resistance of butyl rubbers to steam ageing is listed in:
Table 3.4.14 *Comparative steam ageing of butyl rubbers*

3.4.5.3 Trade names and suppliers

The suppliers of butyl rubbers and their commonly used trade names are listed in:
Table 3.4.15 *Trade names and suppliers of butyl rubbers*

TABLE 3.4.11 Characteristics of isobutene isoprene or butyl rubber (IIR)

Advantages	*Limitations*
Fairly good mechanical properties (about the same tensile strength as SBR). Low resilience gives good vibration damping. Excellent electrical insulation. Low gas permeability. Good ozone and weathering resistance. Carbon black reinforcement improves tear resistance. Suitable for use with inorganic chemicals such as metal salts, acids and alkalis to 100°C. Good resistance to vegetable oils and ester-based lubricants. Low cost.	Low resilience at room temperature. Inferior abrasion and tear resistance. Attacked by all hydrocarbons, fuels and oils. No improvement on tensile strength by carbon black reinforcement. Blends with natural rubber not recommended.

TABLE 3.4.12 Comparative dry heat ageing of butyl rubbers

Property	*Units*	*Oven aged at 170°C for:*	*Standard butyl*		*Conjugated diene butyl*	*EPDM*
			1	***2***	***2***	***3***
Tensile strength	MN/m^2	0	9.9	10.8	11.5	16.2
		7 days	4.9	Melted	535	7.0
		14 days	Degraded	—	52	Brittle
Elongation	%	0	680	455	6.8	375
		7 days	50	Melted	210	115
		14 days	Degraded	—	65	Brittle
Hardness	Shore A	0	73	59	3.9	65
		7 days	56	Melted	210	79
		14 days	Degraded	—	60	Brittle

1 Cured 60 min. at 160°C.
2 Cured 20 min. at 160°C.
3 Cured 30 min. at 160°C.
(Data supplied by Exxon Chem. Co.)

Table 3.4.11 and Table 3.4.12

TABLE 3.4.13 Heat ageing resistance of butyl rubber compounds

Elastomer	*Cure*	*Filler*	*Protection*	*Useful life at 170°C (h)*
EP copolymer	Peroxide EDMA	70 black/30 oil	Cadmium oxide	500
CDB	Dienophile	100 black/20 oil	Age Rite Resin D	400+
EP copolymer	Peroxide EDMA	40 black/5 oil	None	400
Butyl (035/218)	Resin	50 black/5 oil	None	280
Chlorobutyl (1066)	Resin	50 black/5 oil	None	230
EP terpolymer	System E	180 black/100 oil	NBC	220
EP terpolymer	System E	180 black/100 oil	None	120
Butyl/chlorobutyl	Sulphur Donor	40 black/20 oil	None	60

Data supplied by Exxon Chemical Co.

TABLE 3.4.14 Comparative steam ageing of butyl rubbers

	Units	*Time (days) in steam at 150°C*	*Butyl*	*Chlorobutyl*	*Conjugated diene butyl*
Cure time at 160°C	min.	—	90	30	30
Tensile strength	MN/m^2	0	9.8	14.6	12.5
		50	4.0	6.2	11.4
Elongation	%	0	540	290	330
		50	500	250	150
Hardness	Shore A	0	75	62	58
		50	70	60	65
Weight change after steam ageing	%	0	—	—	—
		50	0.68	8.44	–1.13

TABLE 3.4.15 Trade names and suppliers of butyl rubbers

Supplier	*Trade name*
Isobutene isoprene copolymer (butyl rubber IIR)	
Cities Service Co.	Bucar
JSR	JSR
Polysar	Polysar
Dunlop Nordac	—
Eriks-Allied Polymer	—
Northern Rubber	Nordoil
Esso/Exxon	Vistane X
Chlorinated isobutylene isoprene (CIIR)	
Northern Rubber	Nordoil
Esso/Exxon	Chlorobutyl
Polysar	Polysar
Bromobutyl	
Esso/Exxon	
Polysar	
Conjugated diene butyl (CDB)	
Esso/Exxon	

3.4.6 Ethylene propylene EPM, EPDM

3.4.6.1 General characteristics and properties

The advantages and limitations of ethylene propylene are listed in:
Table 3.4.16 *Characteristics of ethylene propylene*

AVAILABLE TYPES

The two forms of ethylene propylene are EPM and EPDM. The copolymer EPM has the structure:

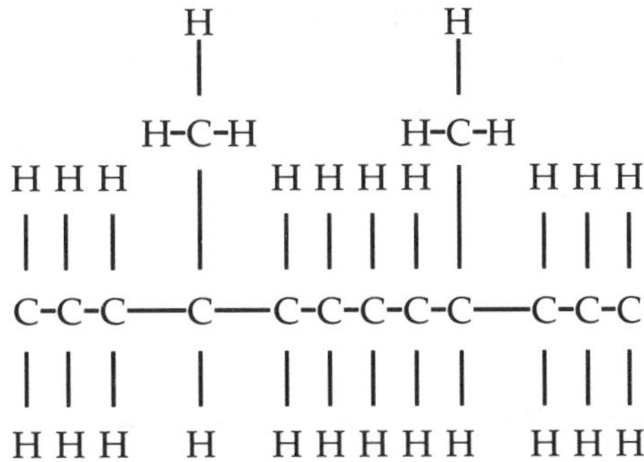

EPDM is a terpolymer which includes as its third constituent dicyclopentadiene or 1, 4 hexadiene which introduces degrees of unsaturation which can be varied between 3 and 8% (mol per cent).

APPLICATIONS

Ethylene propylene elastomers are used in:

Automobiles: for weather strips, radiator hoses, braking circuit gaskets and sound deadening sheets.
Electric cable sheathing.
Domestic gaskets and pipes for washing machines.
Buildings for door and window seals and waterproofing.
Polyolefin modifiers for sports goods and car bumpers and spoilers.

TYPICAL PROPERTIES

The more important properties of EPM and EPDM polymers and unaged vulcanisates, and compounds available from one supplier, are listed in:
Table 3.4.17 *Basic polymer and vulcanisate properties and compounds of ethylene propylene elastomers*

The EPDM properties listed are typical of materials cured with sulphur and accelerators.

Temperature has a relatively minor effect on elasticity modulus of ethylene propylene elastomers as shown in:
Fig 3.4.2 *Variation of Flexural Modulus of EPDM with temperature*

3.4.6.2 Environmental resistance

EPM copolymer vulcanisates offer excellent resistance to organic and inorganic acids, bases, polar-type solvents, antifreeze fluids, bleaching agents, detergents and brake fluids.

They offer poor resistance (on a par with natural rubber and SBR) to aliphatic and aromatic solvents, ethers and mineral oils.

EPDM terpolymer vulcanisates withstand many inorganic chemicals (such as water, acids and alkalis) and many polar organic compounds with a limited number of carbon atoms (such as acetic acid, methyl alcohol, butyl alcohol, acetone, ethyl acetate, ethylene glycol, formaldehyde and nitrobenzene). They swell considerably in aromatic, aliphatic and cycloaliphatic solvents, chlorocarbons, mineral oils and greases.

3.4.6.3 Processing

—Vulcanisation of mouldings can be carried out very rapidly (approx. 60s at 200°C).
—Demoisturising substances (CaO) may be used in extrusions to prevent porosity.
—Extrudates may be cured in steam, by ultra high frequency, in hot air, in molten salts, in a fluid bed of glass balls.
—EPM copolymers must be cured by organic peroxides combined with a co-agent, such as sulphur or an unsaturated organic compound, to improve the cross-linking efficiency.
—EPDM terpolymers may be cured by: (i) organic peroxides; (ii) sulphur and accelerators.

3.4.6.4 Trade names and suppliers

The more important suppliers of ethylene propylene elastomers and their commonly used trade names are listed in:

Table 3.4.18 *Trade names and suppliers of EPM and EPDM gum stocks*

TABLE 3.4.16 Characteristics of ethylene propylene

Advantages	***Limitations***
Good mechanical properties, flexing and resilience characteristics. Outstanding resistance to ozone and weathering. Wide range of operating temperatures: continuous 125°C (specials 175°C) intermittent exposure to 200°C, low temperature flexibility is good at –70°C; retains 60% of tensile strength at 100°C. Resistant to superheated steam and hot water. Good resistance to acids and bases. Accepts paint. Wide range of formulations with compounds exhibiting high and low hysteresis for energy absorption and resilience respectively. Some formulations suitable for injection moulding. Excellent abrasion resistance.	Moderate adhesion to fabrics and metals. Must be reinforced with Carbon Black or special white fillers. Very poor oil and hydrocarbon resistance. Tear resistance not as good as NR but similar to SBR. Relatively new elastomer class.

TABLE 3.4.17 Basic polymer and vulcanisate properties and compounds of ethylene propylene elastomers

Property / Montedison grade		EPM saturated copolymer DUTRAL-CO				
		054	034[b,c]	038/EP[a,b]	059	554/P
Properties of the product						
Mooney viscosity [ML (1 + 4) at 100°C]		43	42	—	—	—
Mooney viscosity [ML (1 + 4) at 121°C]		—	—	65	80	—
Oil content	% wt	—	—	—	—	17
Curing rate		—	—	—	—	50
Stabiliser		non-staining				
Volatiles	% wt. max	0.5	0.5	0.7	0.5	0.5
Specific gravity	g/cm³	0.865	0.865	0.865	0.865	0.890
Typical compounds						
DUTRAL-CO		100	100	100	100	100
FEF Carbon Black		55	55	80	80	150
Zinc oxide		5	5	5	5	5
Stearic acid		—	—	—	—	—
Polyalkyl benzene, naphthenic or paraffinic oil		30	30	55	55	—
TMTMS		—	—	—	—	—
MBT		—	—	—	—	—
Sulphur		0.97	0.37	0.45	0.45	0.79
Peroximon F-40		5	5	6	6	10
Compounds characteristics						
Mooney viscosity [ML (1 + 4) at 100°C]		—	—	—	—	—
Specific gravity	g/cm³	45	40	60	70	75
Press curing		40 min at 165°C				
Properties of the unaged vulcanisate						
Tensile strength	MPa	13	15	17.5	15	13
Elongation at break	%	500	540	550	610	450
300% modulus	MPa	9	8	10	7	10
200% tension set	%	6	10	13	5	8
Hardness, Shore A	points	60	60	65	50	62

[a] = Available also in free flowing form as DUTRAL-CO 038/FF.
[b] = Available also in pellets form.
[c] = Available also in powder form as DUTRAL-CO 034/MAC.
[d] = Available in soft bales as well as in free flowing form as DUTRAL-TER 038/FF.
Information by courtesy of Montedison UK Ltd.
Table 3.4.17

TABLE 3.4.17 Basic polymer and vulcanisate properties and compounds of ethylene propylene elastomers —*continued*

EPDM Unsaturated terpolymer									
DUTRAL-TER									
054/E	***044/E***	***038/EP[d]***	***048/E***	***334/E***	***436/E***	***535/E***	***235/E2***	***046/E3***	***537/E2***
52	42	—	—	—	—	—	—	65	—
—	—	65	65	30	55	32	35	—	50
0	0	0	0	30	40	50	22	0	50
—	—	—	high	—	—	—	quite high	very high	quite high
non-staining									
0.5	0.5	0.7	0.5	0.5	0.5	0.5	0.5	0.5	0.5
0.865	0.865	0.865	0.865	0.880	0.890	0.890	0.890	0.865	0.890
100	100	100	100	140	170	200	125	100	200
55	55	80	80	80	150	150	80	55	150
5	5	5	5	5	5	5	5	5	5
1	1	1	1	1	1	1	1	1	1
30	30	55	55	15	30	—	30	30	—
1.5	1.5	1.5	1.5	1.5	1.5	1.5	1.5	1.5	1.2
0.75	0.75	0.75	0.75	0.75	0.75	0.75	0.75	0.75	0.60
1.5	1.5	1.5	1.5	1.5	1.5	1.5	1.5	1.5	1.20
—	—	—	—	—	—	—	—	—	—
35	30	60	60	66	85	85	54	50	90
1.05	1.05	1.05	1.05	—	—	—	—	1.05	—
30 min at 160°C							15 min at 160°C	10 min at 160°C	15 min at 160°C
12	13	16	13.5	15	17	15	13	13	16
300	310	400	340	350	270	300	290	300	250
8.5	8	8.5	8	9	—	11.5	5	8	—
5	6	7	4	5	6	11	4	4	7
65	66	70	64	68	66	70	66	65	68

Table 3.4.17—*continued*

TABLE 3.4.18 Trade names and suppliers of EPM and EPDM gum stocks

Supplier	***EDPM***	***EPM***
Bunawerke Huls	Buna AP	Buna AP
Chemische Werke	Bura AP	—
Copolymer Rubber	Epsyn	Epsyn
DSM	Keltan	—
Du Pont	Nordel	—
Esso	Vistalon	—
B.F. Goodrich	Epcar	Epcar
ISR	Intolan	Intolan
Mitsui	EPT	—
Montedison	Dutral-TER	Dutral CO
Northern Rubber	Nordoil	—
Sumitomo	Esprene	—
Uniroyal	Royalene	—

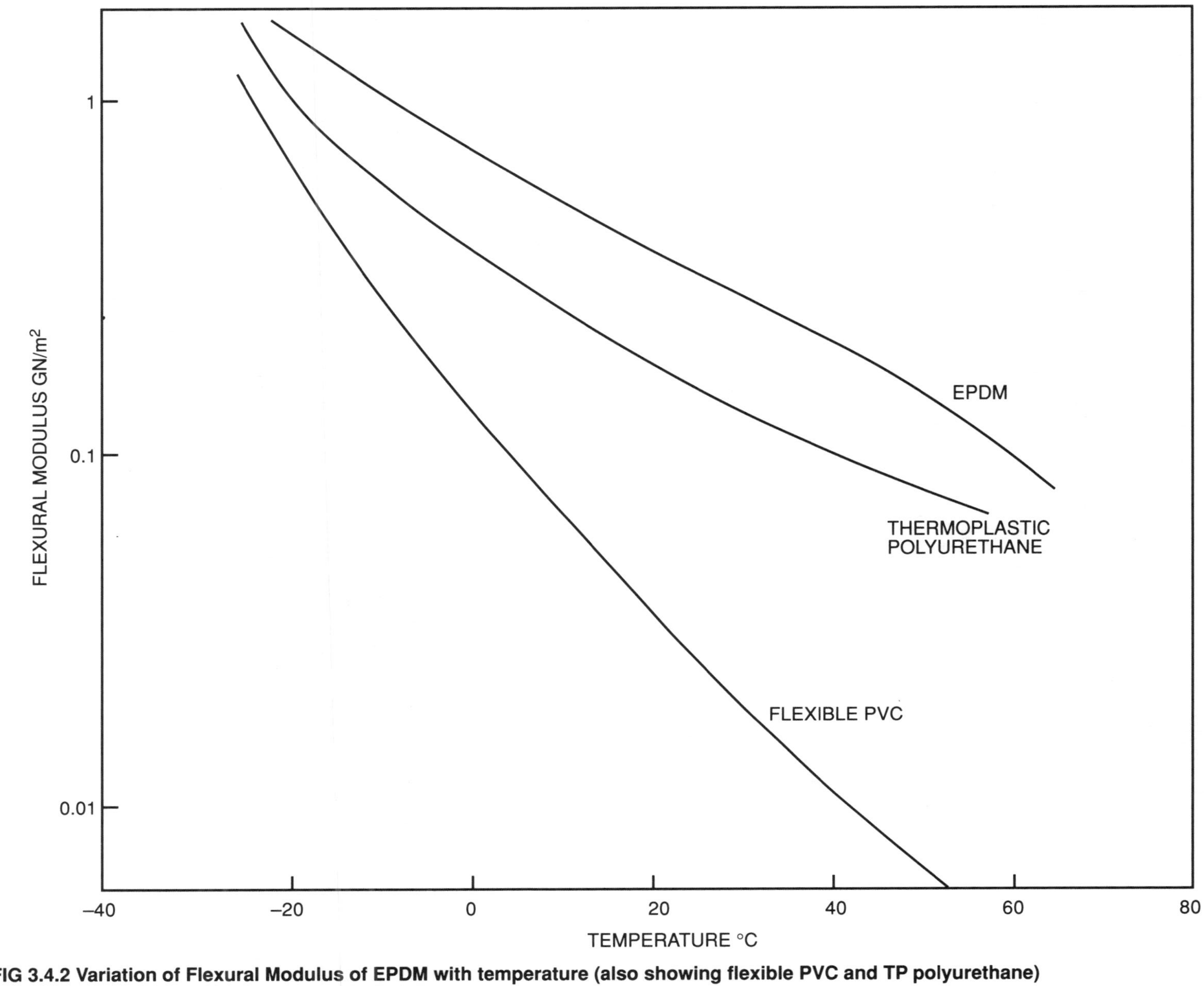

FIG 3.4.2 Variation of Flexural Modulus of EPDM with temperature (also showing flexible PVC and TP polyurethane)

Fig. 3.4.2

3.4.7 Acrylic rubber or polyacrylates (ACM, AR) including ethylene acrylic

3.4.7.1 General characteristics and properties

Acrylic rubbers are copolymers of acrylic esters and olefins, whose performance is distinctly different from those elastomers containing acrylonitrile (see Section 3.4.4). In common usage there is some confusion in terminology and some acrylonitrile-containing elastomers have been called acrylics. However, pure acrylics resist sulphur-containing oils and fluids over the temperature range ±40–150°C, whereas acrylonitrile rubbers become hard on extended exposure.

One specific acrylic rubber is ethylene acrylic elastomer which is a copolymer of ethylene and methyl acrylate.

The advantages and limitations of acrylic rubber are listed in:
Table 3.4.19 *Characteristics of acrylic rubbers*

APPLICATIONS

Polyacrylates

(High temperature) gaskets and oil seals, O-rings, packings, coating cements and adhesives.

Ethylene acrylics

Applications are in areas calling for a tough rubber with good low temperature properties and resistance to the combined deteriorating influences of heat, oil and weather; these include ignition harnesses, coolant hose, power steering hose, hydraulic hose, motor mounts, timing belts, transmission seals, dust boots, and steering and suspension grease seals.

TYPICAL PROPERTIES

The more important properties of polyacrylates and ethylene acrylic are listed in Table 3.4.1, Section 3.4.1, and
Table 3.4.20 *Physical properties and service temperatures of acrylic rubber grades*

3.4.7.2 Trade names and suppliers

The suppliers of acrylic rubbers and the commonly used trade names of their products are listed in:
Table 3.4.21 *Trade names and suppliers of acrylic rubbers*

TABLE 3.4.19 Characteristics of acrylic rubbers

Advantages	*Limitations*
Polyacrylates	
Outstanding resistance to hot oils and automotive transmission fluids, especially sulphur containing oils and fluids. Very good weather resistance. Very good heat resistance.	Only moderate mechanical properties. Tear resistance poor. Fairly expensive. Poor low temperature properties.
Ethylene acrylic	
As polyacrylates. Tough. Better low temperature properties than some other polyacrylates, KFM, chlorosulphonated polyethylene and epichlorohydrin.	Only moderate mechanical properties. Fairly expensive. Tough.

TABLE 3.4.20 Physical properties and service temperatures of acrylic rubber grades

	Specific gravity	*Mooney viscosity ML4–100°C*	*Service temperature (°C)*	*Remarks*
HYCAR® grades	1.1	55	–15–175	Standard
	1.1	45	As above	Soluble version
	1.1	42	–15–175	Easier processing
	1.1	35	As above	Cement grade
	1.1	42	–30–175	Improved low temperature resistance
	1.1	32	–40–175	Exc. low temperature resistance
	1.1	50	–15–175	Low compression set
	1.1	40	–30–175	As above but low temperature grade
	1.1	30	–40–175	Mod. oil res., low temperature grade
VAMAC® ethylene acrylic	1.12	20	–29 (unplast.) Compounded with ester plasticisers to –46 °C.	B–124[a] Black
	1.12	26		VMR[a] 5160 Black
	1.08	30		N–123[a] Non Black
	Useful life		*Exposure (dry heat) to temperature*	
	>15 months		120 °C	
	6 months		150 °C	
	4 weeks		177 °C	
	10 days		191 °C	
	5 days		204 °C	

HYCAR — B.F. Goodrich
VAMAC — Du Pont
[a]Masterbatch identification.

TABLE 3.4.21 Trade names and suppliers of acrylic rubbers

Supplier	*Trade names*
Polyacrylates	
Amer Cyanamid	Cyanacryl
Anchor Chem.	Crestapol
Eriks-Allied Polymer	
B.F. Goodrich	Hycar
Montedison	Elaprim
Ethylene acrylic	
Du Pont	Vamac

3.4.8 Chlorosulphonated polyethylene or 'hypalon' (CSM)

3.4.8.1 General characteristics and properties

The advantages and limitations of chlorosulphonated polyethylene are listed in **Table 3.4.22** *Characteristics of chlorosulphonated polyethylene*

APPLICATIONS

Acid hose
Protective coatings, e.g. auto ignition wire, tanks.
Gasketing.
Ideal for use outdoors or near sparking electrical equipment.

3.4.8.2 Trade name and supplier

Du Pont, trade name Hypalon.

TABLE 3.4.22 Characteristics of chlorosulphonated polyethylene

Advantages	*Limitations*
Excellent resistance to ozone, weathering and chemicals. Moderate mechanical properties. Some formulations resistant to 160°C with some loss of other properties. Very much better abrasion resistance than NR and many synthetics. Good low temperature flexibility. Self-extinguishing. Resistant to oxidising agents and acids.	Poor compression set. Expensive. Not suitable for use with aromatic hydrocarbons.

3.4.9 Fluorine-containing elastomers: fluorocarbon (FKM); perfluoroelastomer (PFE); phosphonitrilic fluoroelastomer (fluoroalkoxyphosphazene PNF)

3.4.9.1 General characteristics and properties

The advantages and limitations of three types of fluorine-containing elastomer are listed and compared in:
Table 3.4.23 *Characteristics of fluorine-containing elastomers*

General purpose fluorocarbon rubbers combine outstanding resistance to heat, fuels, oils and weathering with good mechanical properties and dielectric characteristics. In addition to the basic FKM grades, based upon copolymers of vinylidene fluoride, hexafluoropropylene and tetrafluoroethylene, modified grades are available for improved low temperature characteristics (e.g. Viton GLT), improved heat resistance and grades with improved resistance to specific chemicals (e.g. viton GF, for use with newer types of fuels, KEL-F for high resistance to strong oxidising acids and viton VT-R).

Fluorosilicone elastomers have markedly enhanced resistance to fuels and petroleum products compared to silicone. Improved low temperature performance is obtainable compared to the fluorocarbons, but because the fluorine content is lower the corrosion resistance properties are generally inferior.

For aggressive environments where FKM is unacceptable, the more expensive perfluoroelastomer (PFE), a tetrafluoroethylene, perfluoromethyl vinyl ether is available from Du Pont under the trade name Kalrez.

Phosphonitrilic fluorelastomer (PNF) available from Firestone is less heat resistant than silicone or fluorocarbon elastomers, but offers unique advantages over both, by virtue of a combination of mechanical properties and low temperature capability. It has a back bone of alternating phosphorus and nitrogen atoms with highly fluorinated organic pendant groups.

The relative costs of these materials are listed and compared in:
Table 3.4.24 *Approximate relative costs of fluorine-containing elastomers*

APPLICATIONS

FKM

Up to 75% of fluorocarbon rubbers are used in sealing devices such as O-rings, packings and gaskets. In the automotive field, applications include valve stem seals, heavy duty automatic transmission pinion and crankshaft seals, and diesel engine cylinder liner O-rings. Marine stern tube seals, especially in bulk carriers, form a large and growing market for FKM (see Fig 3.4.3 for the comparative performance of FKM seals).
Fig 3.4.3 *Operating temperature ranges for dynamic lip seals*
Other applications include foam, fabric coatings, process roll coatings. KEL-F is used for O-rings in carbonated beverage dispensers. In aggressive environments or when the upper service temperature range exceeds that which is tolerable by FKM, other flourine-containing elastomers may be more suitable.

PFE

The following is a list of media for which PFE O-rings have been successfully employed. There is either no alternative elastomer or frequent replacement of another elastomer has cause expensive downtime.

Media	*Temperature °C*	*Days Service*
Acrylonitrile	Room	>14
Methyl Methacrylate	Room	>90
Hydrazine + Ammonia	66-93	>84
M–Phenylene Diamine	230	>60
Oil well 'Sour Gas'	260	Satisfactory
Process Steam	265	>365
30% $H_2 SO_4$	50	>270
Xylene	121	>90

Other applications include insulation sleeves for electrical connectors, chevron packings, shaft seals, packings, insulated wire for oil field downhole use, aerospace applications (developmental), injection port septa for liquid chromatographs, valve seals in nuclear power plants.

PNF

Lip seals, O-ring rod seals, air plenum seal in XM-1 Abrahams Army tank, high pressure T-seals in fighter aircraft, O-ring valve seals in petrochemical pipeline.

TYPICAL PROPERTIES

Some of the more important properties of PFE and two grades of FKM are listed in:
Table 3.4.25 *Comparative properties of PFE and FKM fluoroelastomers*
Comparative resistance to compression is shown in:
Table 3.4.26 *Comparative compression set of PFE and FKM fluoroelastomers*

3.4.9.2 Resistance of FKM to heat and environment

The elastic life of FKM heated in air is:

Temperature °C	*Service Time* *Hours*
230	>3000
260	1000
290	240
315	50

The relative resistance of PFE and FKM in air and a selection of solvents is:

Property	*PFE*	*FKM*
Continuous service temperature °C	260+	200
Life at 200°C	several months	7 days
Volume Swell, 7 days at 24°C (%)		
Ketones	<1	>200
Nitrobenzene	<1	24
Ammonia, anhydrous	Satisfactory	Brittle
Nitrogen Tetroxide	Satisfactory	280

The resistance to swelling of grades of FKM in a number of solvents is listed in:
Table 3.4.27 *Volume swell of grades of FKM in a number of solvents*

RADIATION RESISTANCE

FKM ranks about midway among commonly available elastomers with respect to radiation resistance at ambient temperature. At high temperatures, such as are commonly encountered in such applications, many other elastomers and plastics will be ruled out.

For dynamic applications maximum dosage is 10^7 roentgens.

Higher dosages are permissible for static applications.

ELECTRICAL PROPERTIES

Suitable as wire insulation for low voltage, low frequency applications requiring unusual heat and fluid resistance.

3.4.9.3 Trade names and suppliers

The suppliers of fluoroelastomers and the more commonly used trade names of their products are listed in:

Table 3.4.28 *Trade names and suppliers of fluoroelastomers*

TABLE 3.4.23 Characteristics of fluorine-containing elastomers

Advantages	*Limitations*
Fluorocarbon (FKM)	
Continuous resistance to dry heat to temperatures in excess of 200°C. Outstanding resistance to swelling in mineral oils. Resistant to swelling in many solvent/fluids. Resistant to chemicals, including most oil/grease additives. Ozone resistant. Outstanding resistance to weathering. Self-extinguishing/flame retardant. 'Low voltage' electrical goods. Outstanding hot set/stress relaxation resistance. Good/excellent steam, hot water and acid resistance. Fair low temperature flexibility. Low-set, low temperatures, steam and fluid resistant grades available.	Very expensive and cannot extend with fillers and plasticisers to reduce costs. Parts tend to be 3–10 times more expensive than NBR. Extensive swelling in ketones and esters. Eroded by concentrated alkalis. Hardened by ammonia and amines. Attacked by oil well 'sour gas'. Questionable steam resistance above 200°C. Physical properties inferior to hydrocarbon rubbers and decrease with temperature. However, unsuitability due to inadequate strength or elastic properties is rare. Only fair gas permeability. Relatively high brittle point (–40°C for general purpose grades; –50°C for special grades). Dynamic applications limited to –25°C. High density. Speciality grades cost more.
Polyfluoroelastomer (PFE)	
Similar mechanical properties to FKM but greater heat resistance and chemical inertness. Similar thermal, chemical and electrical properties to FKM but excellent resistance to creep and set. Continuous service temperature of 260°C+. Outstanding chemical resistance to almost all chemicals except fluorinated solvents and excellent resistance to permeation by solvents. In particular, PFE is superior to FKM in respect of hot jet fuels, hydraulic fluids, organic fluids, amines, oil well 'sour gas'. Radiation resistance. Resistant to vacuum outgassing.	Cost (roughly 30 times the cost of FKM).
Phosphonitrilic fluoroelastomer (PNF)	
Maximum continuous service temperature 175°C. Can resist intermittent exposure to at least 200°C. Better mechanical properties than silicones. Better low temperature properties than fluorocarbons (down to –65°), but not fluorosilicones. Exceptional flexural fatigue life. Highly resistant to petrol, jet and diesel engine fuels, silicone-based brake fluids, alkyl/aryl phosphate esters and oxygenated solvents. Good vibration damping. Good weathering resistance. Highly flame retardant.	Inferior heat resistance to silicones and fluoroelastomers. Not especially resistant to acid chlorinated hydrocarbons, such as carbon tetrachloride, or animal or vegetable oils. High cost material.

TABLE 3.4.24 Approximate relative costs of fluorine-containing elastomers

Type	*Remarks*	*Relative cost*
FKM	General purpose	100–140
FKM	Low temperature grade	900
FKM	Petrol-resistant grade	n.a.
KEL–F	Acid resistant type	300
PFE	Kalrez, Du Pont	3000
PNF	PNF, Firestone	350
FVMQ	Fluorosilicone	250 (1000[a])
VMQ	Silicone	30–50 (60–65[a])

[a]Speciality grades.

TABLE 3.4.25 Comparative properties of PFE and FKM fluoroelastomers

Property	*Units*	*PFE*				*FKM*			
		24°C	*100°C*	*204°C*	*260°C*	*Low set 24°C*	*LT 24°C*	*150°C*	*200°C*
Hardness	Duromet A	70–95	—	—	—	72	67	—	—
100/ Modulus	MN/m^2	6–13	—	—	—	—	—	—	—
Tensile strength	MN/m^2	14–18	5	2	1.6	14	19	—	—
Tensile set	%	5	2	1	—	—	—	—	—
Elongation	%	60–170	80	45	45	200	185	—	—
Compression set (70 h 204°C)	—	36–50	—	—	—	—	—	—	—
Compression set (70 h 288°C)	—	47–53	—	—	—	—	—	—	—
25% compression Modulus	—	3.5	2.3	2	—	—	—	—	—
Brittle point	°C	−29– −41	—	—	—	—	—	—	—
Specific gravity	—	2.02	—	—	—	—	—	—	—
Linear expansion	$10^{-5}\ K^{-1}$	← 32 →				—	—	—	—
Dielectric constant (10^3 Hz)	—	4.9	—	—	—	10.5	—	7.1	9.1
Dissipation factor (10^3 Hz)	—	5×10^{-3}	—	—	—	34×10^{-3}	—	273×10^{-3}	390–1190 $\times 10^{-3}$
Resistivity	ohm m	5×10^{-15}	—	—	—	2×10^{11}	—	—	—
Dielectric strength	$10^2\ kV\ m^{-1}$	>180	—	—	—	200	—	—	—

Table 3.4.24 and Table 3.4.25

TABLE 3.4.26 Comparative compression set of PFE and FKM fluoroelastomers

Time/temperature (air)	*O Ring set %*				
	Standard FKM	*Low set PKM*	*FKM/LT*	*PFE Acid resistant*	*PFE Heat resistant*
70 h, –30 °C	—	42	73	—	—
70 h, –10 °C	65	45	—	—	—
70 h, 24 °C	9–20	10	—	20	40
70 h, 200 °C	65	15	—	—	—
1000 h, 200 °C	100	50	—	—	—
70 h, 232 °C	32	30	53[a]	70	65
70 h, 288 °C	100	—	—	75	75

[a]Can be improved by compounding.

TABLE 3.4.27 Volume swell of grades of FKM in a number of solvents

Fluid	*Exposure*		*Volume swell (%)*			
			Viton GF	*Viton B*	*Low set*	*LT*
ASTM Ref. Fuel C	7 days	24 °C	3	6		
ASTM Ref. Fuel C	Equilibrium	54 °C	12	14		
Fuel C/Methanol 85/15	Equilibrium	54 °C	20	30		
Methanol	20 days	21 °C	3	27		
Toluene	7 days	24 °C	4	13		
Skydrol 500B	7 days	121 °C	28	127		
Stauffer Blend 7700	7 days	200 °C	9.5	13.0		
Trichloroethylene	7 days	24 °C	6.8	8.0		
Methylene chloride	7 days	24 °C	16	20		
Hydraulic fluid	72 hours	163 °C			<5	<5

Source: Du Pont.
Viton GF surpasses standard (e.g. Viton B) FKM types in resistance to steam, inorganic acids, phosphate esters and polar solvents.

TABLE 3.4.28 Trade names and suppliers of fluoroelastomer

Fluorocarbon FKM	
Du Pont	Viton
Montedison	Technoflon
3M	Fluorol, Kel F
Polyfluoroelastomer	
Du Pont	Kalrez
Phosphonitrilic Fluoroelastomer	
Firestone	PNF

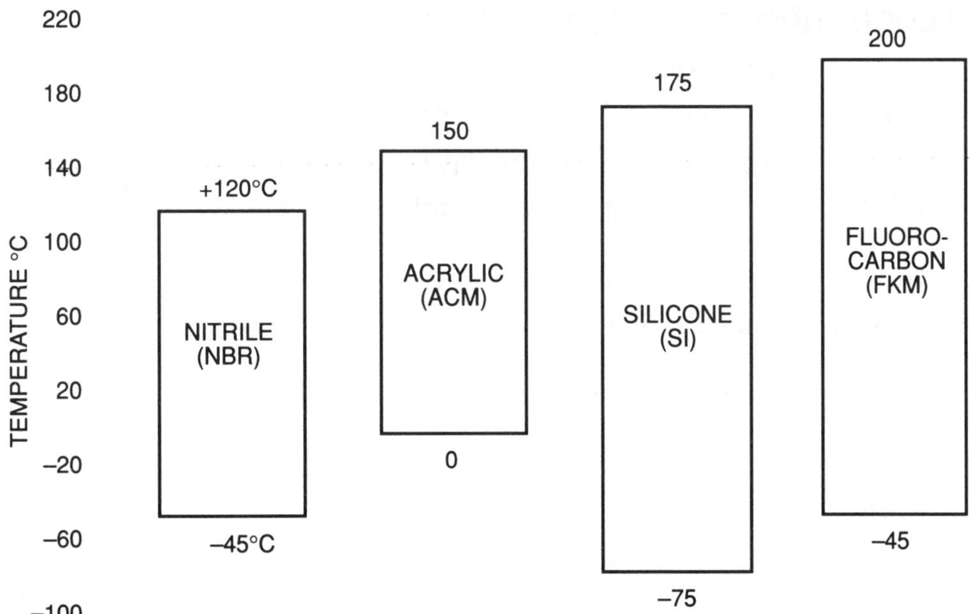

FIG 3.4.3 Operating temperature ranges in °C for dynamic lip seals

3.4.10 Silicone rubber polydimethyl (siloxanes) including fluorosilicones and silicone-based developments: methyl substituted (MQ); methyl and phenyl (PMQ); vinyl and methyl (VMQ); methyl, phenyl and vinyl (PVMQ); fluorine, vinyl and methyl (FVMQ)

3.4.10.1 General characteristics and properties

The advantages and limitations of silicone rubbers are listed in:
Table 3.4.29 *Characteristics of silicone rubbers*

VMQ, which has better tensile strength, tear resistance and resiliency than MQ, is the major general purpose grade. Phenol formulations are noted for superior low temperature resistance. Silicone rubbers are available as gums or as one or two pack liquid systems.

APPLICATIONS

Silicones: electrical connectors, cable shielding, automotive crankshaft seals, check valves, radiator and heater hose, spark plug boots, gaskets, pads, mounts, seals. Frequently chosen to extend the life of products alternatively made in lower cost elastomers. Hot stamp tooling.

Encapsulation and sealing of electrical and electronic components and modules. Solar panel seals.

Fluorosilicones: check valves for automotive accelerator pumps and diaphragms, valves for exhaust gas recirculation. In the aircraft industry, control cable attachment clips, insert grommets for signal circuit connectors and O-rings and seals.

TYPICAL PROPERTIES

The more important properties of grades of silicone rubber are listed in:
Table 3.4.30 *Typical properties of liquid silicone elastomers.*

3.4.10.2 Resistance to heat and environment

A guide to the permissible exposure times and temperatures for silicone rubbers is provided in:
Table 3.4.31 *Useful life of silicone rubber at elevated temperatures*

Silicones have excellent water resistance as shown in:
Table 3.4.32 *Electrical properties of silicone rubber*

Silicones have fair resistance to many commercial solvents, but fluorosilicones are much superior as shown in:
Table 3.4.33 *Environmental properties (volume change) of silicone and fluorosilicone*

The smoke emitted by burning silicones contains fewer really damaging constituents than that emitted by some other polymers as shown in:
Table 3.4.34 *Smoke test values of silicones compared with other polymers*

3.4.10.3 Processing

Silicones can be processed by normal rubber techniques or cast or injection moulded in the liquid form. The economics of these two processes are compared in:
Fig 3.4.4 *Comparative production costs of silicone mouldings by the liquid polymer system compared with the conventional process*

The liquid polymer system (LPS) is a high speed, high volume technique for injection moulding silicone parts and producing supported silicone extrusions from liquid silicone rubber (LSR).

Costs are reported to be about 60% of those experienced in producing the parts from conventional silicone rubbers.

Moulded parts and extruded coatings produced from the liquid materials have slightly inferior toughness and tear strengths compared to those made from conventional silicone rubber gums, but demonstrate superior compression set and inherent flame resistance.

The technology is applicable to the high volume production of small, moulded parts such as connector inserts, spark plug boots, and O-ring sealing devices. It is also suitable for insert moulding, moulding onto a substrate and multimaterial coextrusion and co-moulding. Inserts stay in place without distortion because of the lower moulding pressures involved.

The materials are blends of 100% vinyl-functional dimethylsiloxane polymers with additives. To simplify mixing of the A and B components of the liquid silicone elastomer products, the ratios have been set at 1 : 1 by weight or volume. They are relatively insensitive to mixing errors, tolerating up to 10% variation in the ratio.

Moulding cycle and extrusion speeds can be fast because of the materials' rapid curing characteristics.

Temp. (°C)	*Cure time*
5	3 months
50	2 h
75	15 min
110	110 s
120	95 s
150	50 s
177	20 s
200	5 s

(Data supplied by Dow Corning)

Cure is up to 10 times as fast as older formulations on conventional equipment; a small part can cure in under 30 s. Extrusion on wire, for example, can be done at rates of 150ft/min without the need for multiple passes.

Moulds for producing parts are different from those typical for thermoplastic resins; they must be heated from 150 to 205°C; which requires the use of 10mm thick thermal insulation material between mould and platen. Draft angles should be double those common to conventional injection moulding. Part ejection is the biggest single problem in moulding any kind of elastomer and pulsating hydraulic ejection is recommended.

For extrusion, the rubber should be applied to the substrate by a crosshead die. The substrate is centred by a guider tip. Application of vacuum to the crosshead will improve the penetration of rubber into absorbent substrates.

Suitable materials are available from Dow Corning (Silastic), General Electric and Emerson Cuming.

3.4.10.4 Conducting elastomers

A number of electrically conductive silicone plastics are available, but one has been commercialised which has the unusual ability to change its electrical resistance with applied pressure. In the relaxed state, its resistance is very high and it exhibits good insulation properties. But when this castable silicone is squeezed between two opposing points, its resistance drops evenly from an initial value of 10Ωcm to less than 1Ωcm. Various types of switches, sensors, alarms and transducers were potential targets for the material when launched by American company Dynacon over a decade ago. Current availability status is not known.

Conductive elastomers and metal-on-elastomer connectors provide methods of achieving reliable connection in electronic circuits without many of the limitations of soldering or socketing. The STAX system of connectors from Symot Ltd comprises alternating layers of conductive and non-conductive silicone rubber. The conductive layers are either carbon - or silver filled (the silver is used when very low resistance is specified). The objective is to provide at least one conducting path between contacts and at least one insulating path between adjacent circuits. Since the layers are nominally 0.05mm thick, successful connection can be made in circuits with adjacent contacts as close as 0.25mm. Clearly, greater pad misalignment can be tolerated than with thicker-layered connectors. STAX connectors are usually clamped in place between two circuits and are useful for connecting components such as displays, IC chip carriers, printed circuit boards, leadless hybrid circuits and flat cables.

3.4.10.5 Trade names and suppliers

The suppliers of silicone rubbers and the commonly used trade names of their products are listed in:

Table 3.4.35 *Trade names and suppliers of silicone rubbers*

TABLE 3.4.29 Characteristics of silicone rubbers

Advantages	*Limitations*
Wide working temperature range –60– +230°C (special silicones to –100°C).	Generally moderate to poor mechanical properties but relatively unaffected by temperatures.
Resistant to weathering, ozone and UV light.	Expensive.
Improved water resistance cf. carbon-based rubbers.	Care required to avoid contamination by other elastomers.
Resistant to a wide range of reagents; not usually attacked but may be swollen.	Not resistant to oil, petrol, aviation fuels or aromatic-based lubricating and hydraulic oils.
Excellent resistance to compression set.	Evolve CO on burning, but not HCl, F_2, SO_2, NO_2.
Good electrical properties—best resistance to electrical stress, arcing, ozone and corona.	FLUOROSILICONES limited to 200°C.
Highly flexible.	
Flame retardant grades available.	
Lightest of the high-temperature elastomers.	
FLUOROSILICONES (FVMQ) have a wide temperature capability and improved solvent resistance. Behaviour at cryogenic temperatures is superior to other fluorinated elastomers.	

Table 3.4.30

TABLE 3.4.30 Typical properties of liquid silicone elastomers

Property			Silastic grades[a]				Eccosil grades				
		Units	**590**	**591**	**Q3-6593**	**X3-9595**	**4954**	**4995**	**4122**	**4553**	**2CN**
Durometer hardness		Shore A	32	55	40	40	78	75	45	72	22
Tensile strength		MPa	5.5	3.8	2.1	7.6	3.5	3.5	2.5	5	Low
Elongation		%	520	250	170	450	85	85	180	100	60
Tear strength		kN/m	14	14	5	35	—	—	—	—	—
Specific gravity			—	—	—	—	2.33	1.87	1.15	1.60	0.99
Thermal conductivity		W/m per K	—	—	—	—	2.5	0.8	0.4	0.5	0.2
Service temperature	continuous	°C	—	—	—	—	260	205	205	205	205
	intermittent	°C	—	—	—	—	300	260	260	260	230
Viscosity of solution		CPS	—	—	—	—	40 000	25 000	10 000	36 000	1700
Essential requirement							High thermal conductivity		Low viscosity, flexibility.		Clarity
Application							Encapsulation of heat generating circuitry.		Filling thin sections. Making flexible moulds.		Water clear potting.

[a] Based on press cured samples.
Q3-6593 is an electrically conductive product with a volume resistivity of 0.1–0.2 ohm-m.
X3-9595 is translucent.

TABLE 3.4.31 Useful life of silicone rubber at elevated temperatures

Temperature (°C)	*Useful life[a] (50% elongation retained)*
120	10–20 years
150	5–10 years
200	2–5 years
260	3 months–2 years
260–320	1 week–2 months
320–370	6 h–1 week
370–430	10 min–2 h
430–480	2–10 min

[a] Life in non-flexing applications would be longer.

TABLE 3.4.32 Electrical properties of silicone rubber

	Dry	*Water immersion, 23°C*	
		Original	*10 years*
Volume resistivity	10^{12}–10^{14}	10^{12}	5×10^{13}
Electrical strength ($10^2 \times$ kV/m)	160–280	—	—
Dielectric constant, 60 Hz	2.95–4.0	3.0	3.2
Power factor, 60 Hz	0.001–0.010	0.0011	0.0011

TABLE 3.4.33 Environmental properties (volume change) of silicone and fluorosilicone

Reagent	*Silicone*	*Fluorosilicone*	*Reagent*	*Silicone*	*Fluorosilicone*
Acid solutions[a]			**Solvents and fuels**[a]		
10% hydrochloric	0–2	Nil	Acetone	15–25	180
Conc hydrochloric	0–15	10	Carbon tetrachloride	>150	20
10% nitric	1–10	Nil	Ethyl alcohol	0–20	5
Conc nitric	10–5	5	Iso-octane (gasoline)	>150	20
10% sulphuric	1–5	Nil	Xylene	>150	20
Conc sulphuric	Dec[h]	Dec[h]	Reference Fuel B	>150	20
Conc acetic	5–18	25	JP–4 jet fuel	>150	10
Alkali solutions[a]			**Oils**[b]		
10% ammonium hydroxide	Nil	Nil	ASTM No. 1	5–10	Nil
			ASTM No. 3	35–60	5
Conc ammonium hydroxide	0–7	5	MIL–O–7808 (PQ 8365)[c]	10–30	8
10% sodium hydroxide	0–3	Nil	MIL–O–5606 (PQ 4226)[d]	>100	6
50% sodium hydroxide	0–9	Nil	Silicate ester[e]	>150	5
			Silicone fluid[f]	28–35	Nil
			Phosphate ester[g]	10–20	25

Source: Dow Corning. [a]After seven days at room temperature. [b]After 70 h at 302°F. [c]Oil. [d]Hydraulic fluid. [e]Oronite 8200. [f]Dow Corning 200 fluid, 100 centistokes. [g]Skydrol 500 (after 70 h at 212°F). [h]Decomposed.

TABLE 3.4.34 Smoke test values of silicones compared with other polymers

	Evolution of:			
	CO	*HCl*	*HCN*	*SO_2*
Silicone rubber	Yes	No	No	No
PVC	Yes	Yes	Trace	No
Neoprene	Yes	Yes	Trace	No
Chlorosulphonated polyethylene	Yes	Yes	No	Yes

TABLE 3.4.35 Trade names and suppliers of silicone rubbers

Supplier	*Trade name*
Silicone rubbers	
Dow Corning	Silastic, Sylgard
Rhodia	Rhodorsil
SWS Silicones	SWS (not fluorosilicone)
Emerson Cuming	Eccosil
General Electric	Blensil, Silmate
ICI	Silcopac, Silcoset
Isochem	Isochemsilrub, Epoxsilrub
Northern Rubber	Nordoil
Silicone Fabrications	Silfab
Conducting elastomers	
Emerson Cuming	Eccoshield
Pressure-sensitive resistors/conductors	
Dynacon	Dynacon
Sandwich Conductors	
Symot Ltd	Stax

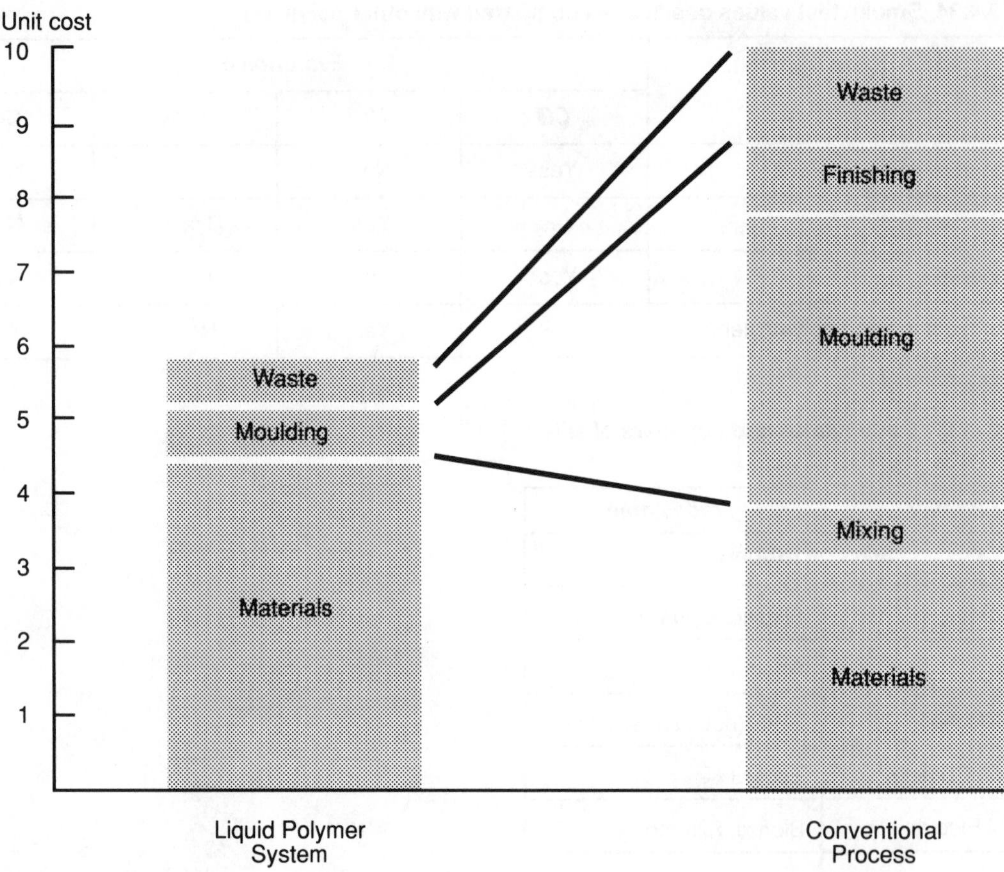

FIG 3.4.4 Comparative production costs of silicone mouldings by the liquid polymer system compared with the conventional process

Fig. 3.4.4

3.4.11 Polynorbornene (poly(1,3 cyclopentylene vinylene))

3.4.11.1 General characteristics and properties

The advantages and limitations of polynorbornene are listed in:
Table 3.4.36 *Characteristics of polynorbornene*

APPLICATIONS

Automotive applications include extruded tail-light assembly seals and window seals; others include turntables, handle grips for tennis racquets and golf clubs, diaphragms, armrests, and shoulder harnesses and driving rollers for paper handling (where low hardness assures positive gripping for paper transfer). Higher hardness, tougher formulations are used in seals, grommets, diaphragms and tubing. The good green strength of formulations, even when highly extended, leads to applications, essentially as a binder:

— viscoelastic aggregates containing highly structured Carbon Blacks having a very high damping factor (cement, antivibrating sheets, flooring tiles and sandwiched metallic structure insertion layers;
— viscoelastic aggregates containing very dense fillers (metal and metal salts) which offer in sheet form, a high insulation to airborne noise.

TYPICAL PROPERTIES

The properties of polynorbornenes depend on the nature and proportion of additives. Typical compounds and properties are listed in:
Table 3.4.37 *Effect of plasticiser concentration and filler loading on mechanical properties of polynorbornenes*

3.4.11.2 Electrical properties

Polynorbornene can be compounded to have high electrical resistance. Lower electrical resistance can also be achieved by changing both the filler and cure system in the compound, thereby (in the limit) obtaining an elastomer with semi-conductive electrical properties.

Some indication of the effect of changing the cure system on both mechanical and electrical properties is given in:
Table 3.4.38 *Electrical and mechanical properties of polynorbornenes cured in different ways*

3.4.11.3 Processing

Compounds based on polynorbornene may be processed on conventional equipment by techniques familiar to the rubber industry. Polynorbornene readily accepts oils, fillers and other compounding ingredients. The processing of stocks proceeds quite easily provided that two guidelines be observed:

(i) The polymer and oil must be presented to the shearing conditions at the same time. This can be achieved for small-scale mill mixing by pre-blending powder and oil. On large-scale mills, it may be desirable to start with masterbatch, which may be purchased or produced in internal mixers.
(ii) Minimum stock temperatures should be 80°C for smooth processing. Maximum stock temperature is not critical as long as an oil of low volatility and a cure system of adequate scorch protection are used.

Mixes prepared by any procedure can be shaped on conventional extruders or calenders. Best processing at this stage is obtained on adjusting conditions to maintain the stock at 80°C. Higher temperatures can be used if a softer consistency is desired, but lower temperatures must be avoided.

Production of vulcanised polynorbornene compounds can be achieved by all current techniques, including compression or transfer moulding, injection moulding, autoclave, microwave and fluidised bed.

3.4.11.4 Trade names and suppliers

Polynorbornenes are supplied under the trade name Norsorex by Atochem. American Cyanamid and BTR act as agents for Atochem, BTR using the trade name Suppelseal.

TABLE 3.4.36 Characteristics of polynorbornene

Advantages	*Limitations*
Available in range of hardnesses from 15 to 80 Shore A.	Not suitable for continuous exposure above 70°C.
High and low damping compounds obtainable.	Ozone resistance poor, unless alloyed with EPDM.
Low hardness formulations offer improved strength and moisture resistance compared to equivalent-hardness foamed or sponge rubber. Dense compounds exhibit superior compression set resistance compared to sponge rubber.	Gas permeability, at best, is slightly inferior to that of butyl rubber.
	Unsatisfactory resistance to aromatic solvents.
	Extreme care required if the elastomer is to be exposed to aliphatic solvents or oxidising agents.
Good resistance to tear initiation at room and elevated temperatures	however, the tear propagation properties are not as good as crystalline elastomers.
Good resistance to hot and cold, fresh and salt water and detergents.	
Resistant to dilute acids	but, as concentration increases, the selection of filler may determine performance.
Good low temperature elastomer.	
Fully compounded dry blends may be directly extruded or injection moulded.	
Can be bonded to metals, or brominated for non-adherence.	

TABLE 3.4.37 Effect of plasticiser concentration and filler loading on mechanical properties of polynorbornenes

Plasticiser	*Filler*	*Property*			
Aromatic oil	***SAF-black***	***Tensile strength (MN/m²)***	***Elongation (%)***	***200% Modulus (MN/m²)***	***Hardness Shore A***
50	50	36	355	9.5	59
100	100	31	406	7.0	55
150	150	26	384	7.0	55
200	200	20	370	6.2	53
250	250	16	380	5.0	52
300	300	12	370	1.5	51
200	100	22	550	1.5	30
200	150	24	464	3.2	40
200	200	20	370	6.5	53
200	250	17	340	6.6	70
200	300	9	200	9.0	78

TABLE 3.4.38 Electrical and mechanical properties of polynorbornenes cured in different ways

Properties	*Units*	*Sulphide cure*	*Peroxide cure*
Hardness	Shore A	43	42
100% Modulus	MNm^{-2}	2.0	1.2
300% Modulus	MNm^{-2}	7.6	4.8
Tensile strength	MNm^{-2}	11.0	4.8
Elongation	%	425	535
Dielectric strength (1.75mm)	10^2 kVm^{-1}	270	260
Dissipation factor	—	0.004	0.006
Dielectric constant	—	2.90	2.91
Volume resistivity	ohm-m	8×10^{13}	3×10^{13}

Table 3.4.38

3.4.12 Polysulphide rubbers or polyethylene sulphides

3.4.12.1 General characteristics and properties

The advantages and limitations of polyethylene sulphides are shown in:
Table 3.4.39 *Characteristics of polyethylene sulphides*

APPLICATIONS

Oil seals. Caulking and sealant applications.

3.4.12.2 Trade names and suppliers

Trade names and suppliers of polysulphide rubbers are:

Supplier	*Trade Name*
Thiokol	ST Polysulphide
Berger	PRC

TABLE 3.4.39 Characteristics of polyethylene sulphides

Advantages	*Limitations*
1. Excellent resistance to oils, and degreasing solvents. Good ozone and weathering resistance. 2. Outstanding resistance to volume swell in organic solvents. 3. Low permeability. 4. Good low temperature flexibility.	1. Poor mechanical properties—reinforcing fillers often required. 2. High compression set. 3. Poor creep performance. 4. Low resilience. 5. Sulphurous odour emitted. 6. Maximum use temperature 90°C.

Table 3.4.39

3.4.13 Epichlorohydrin (CO, ECO)

3.4.13.1 General characteristics and properties

The advantages and limitations of epichlorohydrin are shown in:
Table 3.4.40 *Characteristics of epichlorohydrin*

APPLICATIONS

Liners for gas containers and refrigerant hoses.

3.4.13.2 Trade names and suppliers

Trade names and suppliers of epichlorohydrin are:

Supplier	*Trade Name*
Nippon Zeon	Hydrin Herclor

TABLE 3.4.40 Characteristics of epichlorohydrin

Advantages	*Limitations*
1. Good resistance to solvents, fuels, oils and ozone. 2. Outstandingly low permeability to gases (for air, less than half that of butyl rubber). 3. ECO copolymer especially suitable for temperatures down to –40°C. 4. Upper service temperature limit 160°C.	1. ECO has reduced flame and oil resistance and permeability (× 5 cf. CO). 2. Not readily processable. 3. Can cause corrosion in contact with metals due to release of chlorine.

3.4.14 Thermosetting polyurethane elastomers (AU, EU); cast polyurethane elastomers (solid polyurethane elastomers (SUP)); polyurethane millable gums; reaction injection moulded urethanes (RIM/PU and R-RIM/PU); (see also Section 3.5.2 thermoplastic polyurethanes and Section 3.6.2.2 polyurethane foams)

3.4.14.1 Introduction

The term polyurethane has been derived from the word urea. It is applied to a group of plastics in whose macromolecular structure, urea or urea-like molecules intermittently occur. Fundamentally, the molecules of these plastics consist of two contiguous blocks, the ***isocyanate*** and ***polyol*** blocks.

Linear polyurethane consists of long, mutually entangled molecules which become mobile at elevated temperatures. In this respect, they are similar to other thermoplastics consisting of long molecular chains such as polyethylene or polyamides. Thus they can be moulded, or even remoulded if desired by the application of heat.

Cross-linked products have the widest field of application. The degree of cross-linking may vary from slight to very high. Cross-linked compounds are no longer thermoplastic; their properties are governed by the density of cross-linkage and the length of the links between cross-linkage points (see Fig 3.4.5).

Fig 3.4.5 *Cross-linking of plastics*

Polyurethane elastomers are manufactured in several steps. A basic intermediate is first prepared in the form of a low molecular weight polymer with hydroxyl end groups. This may be either:

(i) a polyester would be derived from ethylene glycol and adipic acid;
(ii) a polyether;
(iii) a mixed polyester amide.

This basic intermediate is then reacted with an aromatic diisocyanate to give a prepolymer. The elastomer is then vulcanised through the isocyanate groups by reaction with glycols, diamines, diacids or amino alcohols. Water can also be used, but then carbon dioxide is eliminated during the cross-linking, as in the production of urethane foams.

The influence of linking on the type and properties of elastomers is illustrated in:

Fig 3.4.6 *Polyurethanes: structure/property relationship*

Another important factor that affects the properties of polyurethanes is their morphological structure, which can range from amorphous to partially crystalline. The flexible types, thermoplastic polyurethanes, elastomers and soft foams, have both amorphous and crystalline structural elements. In rigid types an amorphous character predominates.

Elastomers are available essentially in four types:

thermoplastic polyurethanes (see Section 3.5.2)
cast elastomers
millable gums
reaction injection mouldings

Urethane rubbers are also available as cellular materials and foams (Chapter 3.6) and surface coatings (Vol. 1, Chapter 1.4).

Polyurethane elastomers are noted for extremely good abrasion resistance and hardness combined with good elasticity and resistance to greases, oil and solvents.

TYPICAL PROPERTIES

With polyurethane elastomers it is possible to obtain a wide range of physical properties by merely changing the formulation or raw materials of the system. This type of freedom in compounding does not exist for conventional elastomers.

Physical properties of finished products

Hardness
Materials with a very wide range of hardness can be produced from Shore 10A to Shore 75D. The majority of commercial applications require a range from Shore 60A to 60D.

Elongation at break
The high elongation at break is found with all hardnesses.

Elasticity and low temperature behaviour
Elastomers have relatively high rebound resilience and articles produced have good low temperature behaviour. The dynamic brittle point is about 25 to –30°C.

Abrasion
Polyurethane elastomers have good resistance to abrasion which is superior to that in all other elastomers.

Tensile strength
The tensile strength of vulcanisates is generally between 17 and 40MN/m2.

Compression set
At elevated temperatures articles produced which are cross-linked with isocyanate have a relatively high compression set. Articles cured with peroxides have higher degrees of cross-linkage and relatively favourable compression set.

Imperviousness to gases
The articles are almost as impervious to air and nitrogen as butyl rubber.

Chemical properties (see Section 3.5.1 for distinction between polyester and polyether types)

Resistance to swelling
Little swelling takes place in aliphatics and many other solvents. More swelling takes place in highly polar chlorinated hydrocarbons and in aromatics, esters and ketones. The swelling effect of any solvent on a polyurethane elastomer is more or less inversely proportional to the elastic modulus.

Ageing properties
Components are not attacked by oxygen and ozone and suitable formulations can ensure relatively good resistance to the prolonged action of light and weather.

For exposure to air at high temperatures articles can be used continuously under conditions of peak temperatures of 70–80°C. They can be exposed for limited periods to temperatures up to 100°C and for short periods to temperatures up to about 130°C. However, the extent to which the parts are subjected to mechanical stress or chemical attack should be taken into consideration.

Hydrolysis
Because of their chemical structure polyurethane elastomers are subject to hydrolysis. Hydrolysis may occur in hot water or steam, acids and bases and some technical lubricants especially when high temperatures and prolonged use in a tropical climate are involved.

The properties, processing techniques and applications of the several polyurethane systems are tabulated in:
Table 3.4.41 *Typical comparative properties of solid polyurethanes*

3.4.14.2 Cast polyurethane elastomers (solid urethane elastomers, SUP)

GENERAL CHARACTERISTICS AND PROPERTIES

The cast polyurethanes were the first to be used to any significant degree and can be subdivided into three main groups:

(i) unstable prepolymers
(ii) stable prepolymers
(iii) one-shot systems

In the unstable pre-polymer group the polyurethane is manufactured by a pre-polymer route, but this constituent cannot be stored as undesirable side reactions may occur. The pre-polymer is formed by reaction of the hydroxyl terminated polyester and the diisocyanate at 130°C and the chain extender is added fairly soon after manufacture, whereupon the chain extension and cross-linking are brought about by having an excess of diisocyanate present in the prepolymer. The curing continues rapidly with the final cure of the elastomer being brought about in hot air.

In the stable prepolymer type, the pre-polymer is supplied as a viscous liquid or low melting solid which has to be heated before use and the chain extender added to give chain extension and cross-linking.

The one-shot system disposes of the problems of dispensing and mixing. Both polyether and polyester systems are available and polyol is mixed with the chain extender yielding a stable mixture. The diisocyanate is added later, and catalysts are often used to balance the reactivities of the components to give satisfactory network structure. After mixing the system can be cast or injected at low pressures into moulds.

Solid urethane plastics offer high productivity due to short mould residence times and are ideally suited for the production of large or small mouldings and for the preparation of mouldings required in low production runs, e.g. 30–30 000 off.

Castables comprise about two-thirds of total urethane consumption. They can be tailored to Shore hardnesses ranging from 10A to 90D. Products with hardnesses lower than 50A can compete against other elastomers only when abrasion resistance is crucial as in paint, ink and coating rollers. Principal applications for cast urethanes include the following:

Printing and industrial rolls.
Die pads and flexible tooling moulds.
Industrial solid tyres and wheels.
Automotive, shoe and electrical products.
Furniture.

PROPERTIES

The more important properties of cast polyurethane elastomers are listed in:
Table 3.4.42 *Typical properties of cast polyurethane elastomers*

ECONOMICS

Processing by low pressure casting techniques offers considerable economic advantages to a plastics moulder. The main advantage is the low capital expenditure required for dispensing machines and moulds. Because the only pressure on the mould is the hydrostatic pressure of the mixture, the moulds can be fabricated from relatively cheap materials. This makes the material ideal for prototype and development work where simple matched die moulds can be used for glass fibre-reinforced polyester or epoxide resin which give rise to moulds that are quick to produce and subsequently alter.

Small production runs can be undertaken using cast material tools, such as aluminium, zinc or copper/beryllium alloys, which can be fabricated at considerably lower costs than machined steel tools normally used for injection moulding.

Although the materials are ideal for short production runs, where the low cost of the mould required counteracts to some extent the high price of the material, this price factor is a major factor limiting use.

PROCESSING OF CAST POLYURETHANE ELASTOMERS

Moulding process

The mixing and casting of the polyol and isocyanate may be accomplished by a batch technique or by continuous mixing or dosing machines. The speed of reaction can be controlled by the use of catalysts.

Because the flow of the urethane chemicals into the mould causes little pressure to be exerted on the mould, moulds can be prefabricated from inexpensive materials, e.g. glass fibre-reinforced polyester resins or cast metals, etc. The glass-reinforced moulds are generally used at 40–50°C whilst metal moulds are heated to 70–110°C prior to casting. During the curing reaction a volumetric contraction of 4–6% occurs, but this can be allowed for by using a reservoir during the injection process.

After 3–7 min from casting, the moulding may be removed, but at this stage is still elastomeric and precautions against distortion must be made. This elastomeric state is useful in aiding removal of the casting from the mould, particularly if there are undercuts in the mould.

Post-Curing

Post-curing is required to develop the optimum level of physical properties. A typical cycle is 45 min at 110°C. If the articles are not given sufficient cure, reduced heat distortion temperatures, flexural and impact strength may result.

Reinforcement of SUPs

Where reinforcement may be required, the reinforcing material may be placed in the mould cavity and the urethane system is then injected into the base of the mould so that air is displaced through the reinforcing media. For glass fibre reinforcement it has been found that excellent wetting of the fibre occurs, yielding a high-quality void free casting.

3.4.14.3 Special cast polyurethanes

SORBOTHANE

Sorbothane is a modified low molecular weight polyurethane originally developed to mimic the energy absorbent qualities of human flesh. It was invented by Maurice Hiles of Concept Infinitive and patented by the British National Research and Development Corporation. Bostik Laboratories in Leicester helped to develop the two base materials and BTR Industries of London was granted the exclusive licence for product development and manufacture.

Processing is either by gravity casting into an open mould, or by reaction injection moulding (see Section 3.4.14.5) into closed moulds.

'Sorbothane' exhibits good dimensional stability, very low permanent deformation and extremely high energy absorption. Although it is a solid, Sorbothane behaves much like a liquid. Even after 1 million cycles of 60% compression it is claimed to show negligible distortion.

The material is supplied in various grades with different hysteresis characteristics (i.e. rates of distortion and recovery) and mechanical properties. Shore D hardness values range from 15 to 70. In all but the hardest grades, recovery takes longer than the duration of the impact.

Typical applications include heels for orthopaedic and sports shoes, industrial seals and gaskets and sound damping, the latter in office machinery and sound reproduction equipment, such as turntable pick-up arms.

The more important properties of sorbothane are listed in:
Table 3.4.43 *Typical properties of 'Sorbothane' (special cast polyurethane)*

THORDON

Thordon is a thermosetting elastomer/plastic blend 'based on urethane' and readily cast in formulations giving a low coefficient of friction and improved thermal conductivity relative to other non-metallics. As a bearing material under arduous conditions of dirty environments, shock loading and lubrication abuse, Thordon is claimed to outperform bronze, nylon, PTFE, laminated phenolics, acetal copolymers and ultra-high molecular weight polyethylene (UHMWPE).

3.4.14.4 Polyurethane millable gums

The castable polyurethanes are usually formulated with a slight excess of diisocyanate which causes the cross-linking. However, if a deficiency of diisocyanate is formulated, a hydroxyl terminated polymer results that is relatively stable and non-cross-linked. Such a product, correctly formulated, yields a plastic gum that can be handled on a rubber mill in a similar manner to other elastomers. Cross-linking of the product can be effected by the addition on the mill of more diisocyanate, peroxide or sulphur. Fillers such as Carbon Black can also be added on the mill and the whole process is analogous to the manufacture of conventional rubber product.

Millable gums are made to replace gum rubber when the abrasion, oxygen, ozone, oil, chemical and solvent resistance of polyurethane is needed. Applications include speciality mechanical goods like diaphragms, seals, gaskets, conveyor belts. Millable gums have less mechanical strength than castables or thermoplastics at identical hardness but can withstand temperatures of 120–170°C.

TYPICAL PROPERTIES

The more important properties of millable gums are listed in:
Table 3.4.44 *Typical properties of polyurethane millable gums*

PROCESSING

There are a number of ways in which the cure of millable gums can be brought about. Since the prepared gum is isocyanate deficient, the addition of more isocyanate to provide an excess will promote the cross-linking and cure. If diisocyanate and a short chain polyol are added in equivalent amounts, a harder product results.

The millable gums, however, can be cured by the use or peroxide and sulphur, but such a cured polymer is weaker than the isocyanate cross-linked polymer and usually requires reinforcement by conventional filler such as Carbon Black.

When using conventional peroxides for cure the amount of peroxide has an important effect on the properties of the final product. Too little peroxide results in high compression set and heat buildup on flexing, whilst too much peroxide provides a highly cross-linked structure giving high hardness values and modulus, but poor flex cracking resistance and lower tear properties. The cure times will be dependent on the temperature of cure which can vary from 150°C to 200°C.

Conventional polyurethane compositions are not usually suitable for sulphur cure, but systems can be prepared with suitable structure for the sulphur to combine and form a cure. The sulphur-cured materials are similar to peroxide cured products in that more stable chemical bonds are formed which reduce the strength of the polymers and therefore reinforcing fillers are required to obtain optimum properties. However, the sulphur-cured products generally have similar properties to peroxide cured gums. Typical cure times are:

45 min at 142°C
20 min at 151°C
12 min at 160°C

The disadvantage of the sulphur-cured products is that highly active accelerator systems are required and this can be detrimental to stability of other groups, leading to poor hydrolysis resistance.

A comparison of properties of isocyanate and peroxide cures of millable gum appears in:
Table 3.4.45 *Comparison of isocyanate and peroxide cures for millable gums*

3.4.14.5 Reaction injection moulding (RIM) polyurethanes

GENERAL

Reaction injection moulding (RIM) is a specialised form of casting involving the high-pressure mixing of two or more liquid components and the injection of this mixture into the cavity of a mould. The process is illustrated in:
Fig 3.4.7 *The RIM process*

Although the term is most commonly associated with polyurethanes, RIM is a process, not a material, and suitable formulations for the reaction injection moulding of other polymer systems are available. These include nylons and epoxides.

The term 'moulding' applied to a polymer manufacturing process displaces 'casting' when the liquid material is injected into the mould cavity rather than poured and when a greater degree of automation is introduced. The fact that a chemical reaction takes place means that the process is not reversible and the material is a thermoset.

In the case of urethanes the two liquid components used are polyol and isocyanate. Within seconds the liquids begin a chemical reaction, releasing gases which expand and fill the mould. The liquid solidifies and begins the cooling cycle, which results in a part with a microcellular core, sandwiched between tough, smooth, integral kinds of solid polymer at the surface (below about 3–4mm thickness, significant foaming is difficult to achieve).

APPLICATIONS

Polyurethane RIM is still restricted to vertical components in its automotive uses, whereas SMC, because of its considerably higher flexural modulus, is preferred among plastic materials for horizontal surfaces, where truly self-supporting structures are necessary. Furthermore, among the critical needs of the RIM urethanes is better paint-oven resistance, particulary as the material gains ever-widening use in the automotive field. This problem can be tackled along two fronts; the urethane system itself, or the paint system. One approach on the urethane side is through new extenders, which react with the isocyanate to form the hard block or segment. The extender and the isocyanate can be combined in the proper formulation to improve heat-sag resistance.

TYPICAL PROPERTIES

The more important mechanical and physical properties of grades of polyurethane RIM are listed in:
Table 3.4.46 *Typical properties of polyurethane RIM systems (unreinforced)*

PROCESSING OF RIM POLYURETHANE

Reaction rates should be high enough to give mould cycle times of typically less than 30s, without the evolution of undesirable products of low molecular mass.

The moulded part is then removed in 1–10min, depending upon the reactivity of the chemicals, the part thickness and the processing equipment design. It may be post-cured in an oven to enhance mechanical properties.

The viscosities of the reactant streams must be kept to a minimum to aid pumping into the mixing chamber. Reaction starts to occur in this chamber, giving a solid material just prior to demoulding.

Since the process is a chemical reaction, rather than simple cooling, as in thermoplastic moulding, the material is actually created as the part is moulded; this leads to a relatively stress-free part without sink marks, regardless of configuration.

Low moulding pressures, typically less than 100 psi, are inherent in this process and allow two very important advantages. First, large parts requiring huge platens and clamp tonnage in other processes can be made in low-tonnage RIM presses, saving on production costs. Secondly, these lower pressures allow lower cost tooling in both prototype and production stages. Short runs can therefore still be economical with RIM, whilst full-scale production remains competitive with other processes, although production time cycles tend to be longer.

Tooling

The relatively gentle nature of the RIM process allows a wide range of tooling options. Moulds may be built from epoxy and sprayed metal (for short runs) as well as from cast aluminium; machined tools also perform to rigorous standards. Tooling costs for RIM parts are considerably lower than for most moulding processes. Lead times are also greatly reduced as moulds can be constructed relatively quickly using easily machined and less expensive materials.

3.4.14.6 Reinforced reaction injection moulding (RRIM) polyurethanes

The problems with RIM mentioned above have hastened the development of RRIM systems which differ from RIM in incorporating fillers.

When the high grade microcellular elastomers are stiffened by the incorporation of fillers, new property combinations arise. This development was stimulated by the requirements expressed by the automotive industry for plastics intended for use in exterior body parts:

(a) the thermal coefficient of expansion must be reduced to correspond more closely to sheet metal;
(b) the stiffness must be increased to permit self-supporting parts;
(c) the high temperature stability must be improved to permit painting procedures similar to those for steel—this means in practice, resistance to 120°C for 30–60 min.

The basic requirements of chemical systems for RRIM is essentially the same as that for RIM. Conventionally, inorganic filler, normally glass, is added to the polyol stream, but may be added to both (or either of) the constituents. This imposes stricter constraints on viscosity than for RIM. Also, the handling of reactants containing fillers dictates the redesign of conventional RIM machines.

TAILORING POLYURETHANES

Polyurethane polymers are not defined by a unique set of properties, but comprise a whole family of materials covering a spectrum of physical and property forms. The ability to formulate the material 'to taste' allows immense versatility.

For example, a given stiffness could be obtained by incorporating a large amount of glass into a low modulus system, a moderate amount of glass into a medium modulus system or a small amount of reinforcement into a high modulus system. The use of glass to increase stiffness particulary aids dimensional stability, sag and low coefficient of expansion; reinforced high modulus urethanes have superior elongation and toughness. Processing becomes progressively more difficult for unreinforced systems, however, as the modulus is increased, so the incorporation of glass is desirable for this reason alone.

CHOICE OF GLASS REINFORCEMENT

From the viewpoint of reinforcing any polymeric matrix, long fibres are attractive but limit the processing techniques which may be employed. Short fibres still show reinforcing capability whilst having the advantage of being processable by RIM methods.

RRIM urethane composites are commonly reinforced with hammer milled fibre with a weight average length of about 0.25mm. Milled fibres with a weight average length of 0.5mm have also been used, but are more difficult to process than the shorter fibres. Precision cut 1.5mm strands with a narrow distribution of length are also commercially available, produced by Fibreglass Ltd, under an ICI Patent.

Other prospective fillers are mica, wollastonite, carbon and nylon or polypropylene fibres.

MAXIMUM GLASS EFFICIENCY

The quantity of reinforcing glass which can be added is limited by the resulting increase in viscosity. This in turn is dependent primarily, but not exclusively, on fibre length. Milled glass provides easier processing of typical polyol slurries than the same weight percentage of chopped strand glass. The maximum levels of milled fibres which can be added to a 1 : 1 Polyol/isocyanate mix is generally in the range 25–35%, whilst for 1.5mm chopped strand the limit is about 6–8%.

Properties are shown in:

Fig 3.4.8 *Variation of impact strength of polyurethane RRIM as a function of tensile modulus for two primary forms of glass,* and

Table 3.4.47 *Typical properties of polyurethane RRIM mouldings comparing hammer milled and chopped glass fibre reinforcement*

It is desirable to have a very narrow fibre length distribution with a mean value several times the critical length. The theory predicts that a 1.5mm strand would have a reinforcement efficiency of 87%. If the initial length were halved, as workers using some processing machinery report, the fibres would still retain 73% efficiency. A milled fibre with a length of 0.25mm would have a reinforcement efficiency of 31% and such milled fibres would therefore have to be incorporated at concentrations 2–3 times greater than those used for chopped strands to achieve comparable enhancement of properties.

Composites of high modulus containing chopped strands have superior impact strength; a low concentration of chopped strands would thus seem to be the better method to produce combined stiffness and impact strength than a larger loading of milled fibres, subject to satisfactory processing performance.

3.4.14.7 Trade names and suppliers of thermosetting polyurethane elastomers

The suppliers of thermosetting polyurethanes and the familiar trade names of their products are listed in:

Table 3.4.48 *Trade names and suppliers of polyurethanes*

TABLE 3.4.41 Typical comparative properties of solid polyurethanes

System		***Properties***							***Processing techniques***	***Typical uses***
		Shore hardness Din 53 505	***Rebound resilience DIN 53 512 (%)***	***Stress at break DIN 53 504 (MPa)***	***Elongation at break DIN 53 504 (%)***	***Tear resistance DIN 53 515 (kN/m)***	***Abrasion loss DIN 53 516 (mm^3)***	***Elastic Modulus DIN 53 457 (MPa)***		
Casting systems	Elevated temperature curing	68-99 A to 70 D	35-50	24.5-39	280-800	15-100	45-60	9-620	Casting by specialised firms. Hardness variable by choice of recipe.	Rolls and roll covers for heavy loads; drive and coupling components, linings for prevention of noise and wear.
Casting systems	Elevated temperature curing	55-98 A to 65 D	35-45	10-50	300-500	6-68	30-80	5-50	2-part casting (prepolymer) by hand or with machinery. Reaction mixture is processed in the hot state.	Roll covers for medium loads, sieves, pulley wheel linings. Many industrial applications, including wear resistant coverings and self-supporting parts.
Casting systems	Cold curing	25-70 A	25-40	1.5-5	120-450	3-10			2-part casting by hand or machine. Cold-curing.	Formwork mats for relief pattern concrete. Stoneware pipe seals. Binder for elastic polishing disks.
Thermoplastic polyurethane		80-98 A to 70 D	35-55	25-55	250-600	60-150	20-80	10-650	Injection moulding, extrusion, calendering. Supplied as granules in many different hardness grades.	Drive components, seals, roll covers, damping components, wear-resistant parts, coverings, ski-boots, sports footwear soles, hoses.
Vulcanised urethane rubber		40-97 A to 70 D	24-50	16.5-39	300-800	to 93	35-90	4-260	Techniques used in the rubber industry. Four basic grades available. Hardness and properties variable through choice of grade and filler content.	Dampers, transmission components, seals and wear resistant coverings.

Data supplied by Bayer AG, Leverkusen, FRG

Table 3.4.41

TABLE 3.4.42 Typical properties of cast polyurethane elastomers

Property	*Units*	*Polyether*	*Polyether*	*Polyether*	*Polyether*	*Polyester*	*Polyester*	*Polyester*	*Polyester*
Hardness	***Shore***	***A10–A40***	***A45–A70***	***A88–A98***	***D68–D75***	***A60–A73***	***A78–A83***	***A88–A92***	***A93–A98***
Tensile strength	MN/m^2	1.7–2.9	4.1–31	26–42	27–55	27–42	31–55	27–44	27–38
Elongation	%	425–1000	430–700	200–480	120–270	550–650	475–700	475–600	450–550
100% Modulus	MN/m^2	—	—	5–14	20–26	1–4	3–6	7–12	8–11
300% Modulus	MN/m^2	0.2–1.5	1–8	14–31	—	4–12	9–18	11–22	14–21
Compression set	22 h, %	2–34	7–40	—	—	10–25	20–25	20–25	20–25
Tear strength	Graves lb/in	20	50–225	475–750	—	280–320	450–525	550–600	550–700
Resilience	%	—	—	—	—	45–50	50–60	45–50	45–50

For additional properties the reader is referred to thermoplastic polyurethanes for guidance (Section 3.5.1).

TABLE 3.4.43 Typical properties of 'Sorbothane' (special cast polyurethane)

Property	*Unit*	*Value*
Specific gravity	—	1.34
Elastic Modulus	MN/m^2	0.9 (Shore 20D) over frequency 5.2 (Shore 55D) 10^{-4} to 10^2 Hz
Compression set		0.13 after 0.25 constant strain for 22 h (Shore 20D)
Tensile strength	MN/m^2	0.12 (Shore 20D) 0.36 with 0.5% silica fibres
Elongation (break)	%	610 (Shore 30D) 300 (Shore 55D)
Tear strength	kN/m	0.96 (Shore 20–33D)

(Sorbothane is produced by BTR)

TABLE 3.4.44 Typical properties of polyurethane millable gums

Property	*Units*	*Test*	*Gums*	
Hardness	Shore	ASTM D412	A62–66	A70–95
Tensile strength	MN/m^2	D412	21–38	24–31
Elongation	%	D412	450–500	315–450
300% Modulus	MN/m^2	D412	11–17	19–23
Compression set 22 h, 70°C	%	D395	18–21	14–47
Tear strength (Graves)	lb/in	D624	260–381	—

TABLE 3.4.45 Comparison of isocyanate and peroxide cures for millable gums

Property	*Cross-linking*	
	With isocyanate	*With peroxide*
Store stability of compounds at room temperature.	Strictly limited	Good
Tendency of compounds to scorch under heat.	Pronounced	Practically none
Opportunities for reducing the curing time with accelerator curing temperature	Good in certain cases 130–135°C	None
		150–160°C in special cases up to 190°C.
Surface after hot air curing	Satisfactory	Under cross-linked
Hardness of finished products	High (about 80–99 Shore A)	Moderate (about 60–80 Shore A)
Tear resistance	Usually relatively high	Considerably lower
Compression set at elevated temperature	Relatively high	Often very low

Table 3.4.46

TABLE 3.4.46 Typical properties of polyurethane RIM systems (unreinforced)

		Units	*Very low modulus*		*Low modulus*		*Medium modulus*		*High modulus*
Identification:	Polyol blend		PBA 1467	PBA 1478	PBA 1475	PBA 1534	PBA 1501	PBA 1539	PBA 1548
	Isocyanate		VM10	VM10	VM10	VM10	VM10	VM10	VM10
Hardness		Shore D	52	48	58	60	74	65	85
Tensile strength		MN/m^2	18	17	22	27	30	27	41
Elongation at break		%	150	300	160	220	80	170	67
Angle tear		kN/m	59	64	106	94	131	144	164
Flexural Modulus:	room temp.	MN/m^2	160	120	260	330	600	750	1280
	–30°C		320	330	540	610	1400	1540	2280
	+70°C		80	81	190	250	420	360	750
	RATIO +70/–30		4.0	4.1	2.8	2.4	3.3	4.3	3.0
SAG	30 min, 120°C	mm	5	4	5	0	4	2	3
	60 min, 120°C	mm	8	7	7	1	8	3	5
Specific gravity			1.015	1.100	1.100	1.080	1.100	1.100	1.220
Linear Coefficient of Thermal Expansion		10^{-6} K^{-1}	140	136	130	143	92	96	94

Data supplied by ICI

TABLE 3.4.47 Typical properties of polyurethane RRIM mouldings comparing hammer milled and chopped glass fibre reinforcement

System identification *Polyol blend* *Isocyanate*		*PBA 1478* *VM10*					*PBA 1501* *VM10*						*PBA 1546* *VM10*
Filler content (wt %)		*0*	*8% 1.5mm chopped strand WX6012 glass*		*25% hammer milled glass*		*0*	*5% 1.5mm chopped strand WX6012 glass*		*20% hammer-milled glass*		*0*	*10% 1.5mm chopped strand WX6450 glass*
Property	***Units***												
Direction of test [a]		*b & c*	*b*	*c*	*b*	*c*	*b & c*	*b*	*c*	*b*	*c*	*b*	*b*
Hardness	Shore D	65	64	64	69	60	74	78	78	74	74	66	70
Tensile strength	MPa	23	23	19	22	19	30	32	28	28	24	26	32
Elongation at break	%	130	40	50	40	60	80	25	40	40	70	180	15
Angle tear	kN/m	106	87	85	78	74	131	130	123	107	96	125	140
Flexural Modulus	MPa												
Room temperature		330	817	340	800	460	600	927	660	1300	710	430	1600
−30°C		592	1160	850	1500	990	1400	1600	1200	2300	1800	910	2200
+70°C		160	460	170	550	320	420	831	450	658	320	390	1410
Ratio +70/−30°C		1:3.7	1:2.5	1:5.0	1:2.7	1:3.1	1:3.3	1:1.9	1:2.7	1:3.5	1:5.6	1:2.3	1:1.6
Sag (mm): 30 min at 120°C		5	0	0	1	3	4	1	1	2	3	4	1
60 min at 120°C		7	1	1	1	4	8	2	4	2	4	4	2
Specific gravity		1.200	1.200	1.200	1.200	1.200	1.100	1.200	1.200	1.100	1.100	1.150	1.150
Linear Coefficient of Expansion	$°K^{-1}\times10^{-6}$	130	42	94	45	104	92	66	87	89	88	135	50

[a] Direction of test
b = parallel to flow
c = perpendicular to flow
Source: ICI

TABLE 3.4.48 Trade names and suppliers of polyurethanes

Supplier	*Trade name*	*Notes*
	Casting resins	
Allied Resin	Arcon	
American Cyanamid	Cyanaprene	
Anderson	Andur	
Arnco	Catapol, Fastcast	
BASF	Lupranol, Lupranat, Lupraphen	
Bayer	Vulkollan, Baytec, Desmoflex	
Cal	Calthane	
Conap	Conap, Conathane	
Du Pont	Adiprene	Polyether
Hallam Polymers	V-Thane	
Hexcel	Uralite	
Irathane International	Irathane	
Ren	Rent C; O-Thane	
Thiokol	Solithane	
Uniroyal	Vibrathane	
Upjohn	Castethane	
Witco	Castomer, Formrez	
Emerson Cuming	Stycast	Transparent Grade
	Gum Stocks	
Bayer	Urepan	
Disogrin	Disogrin	
Du Pont	Adiprene	
Phillips	LR-	
Tech. Sales & Eng.	Millathane	
Thiokol	Elastothane	
Uniroyal	Vibrathane	
	RIM/RRIM Systems	
BASF	—	
Bayer/Mobay	Bayflex, Baydur	
Dow	—	
ICI	—	
Texaco	Rimtex	
Tufnol	Tufset (formerly Polyset)	
Union Carbide	RIM/RRIM	
Upjohn	CPR	

Table 3.4.48

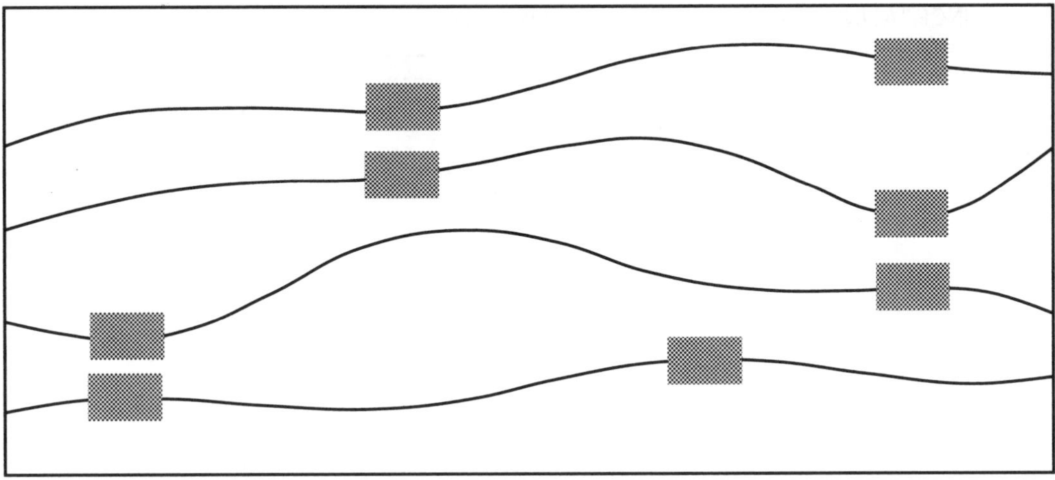

Schematic diagram illustrating the configuration of a thermoplastic polyurethane.

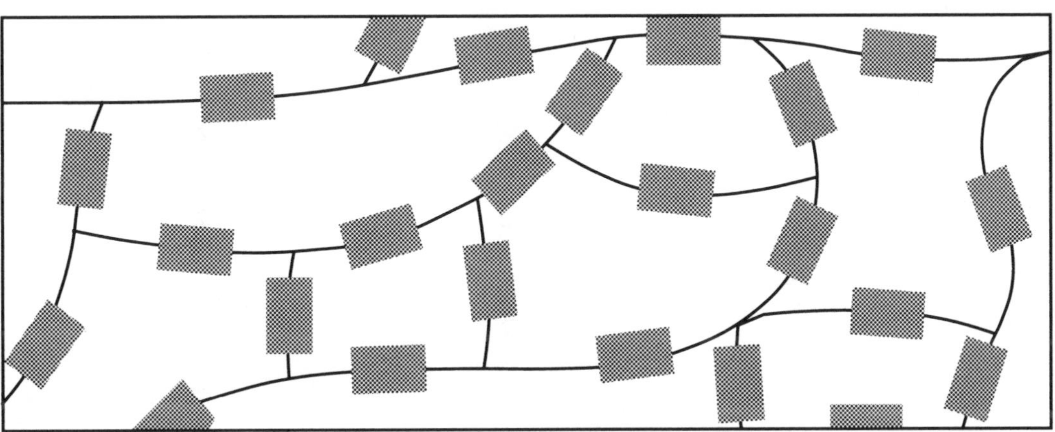

Narrow-meshed crosslinked thermosetting plastic.

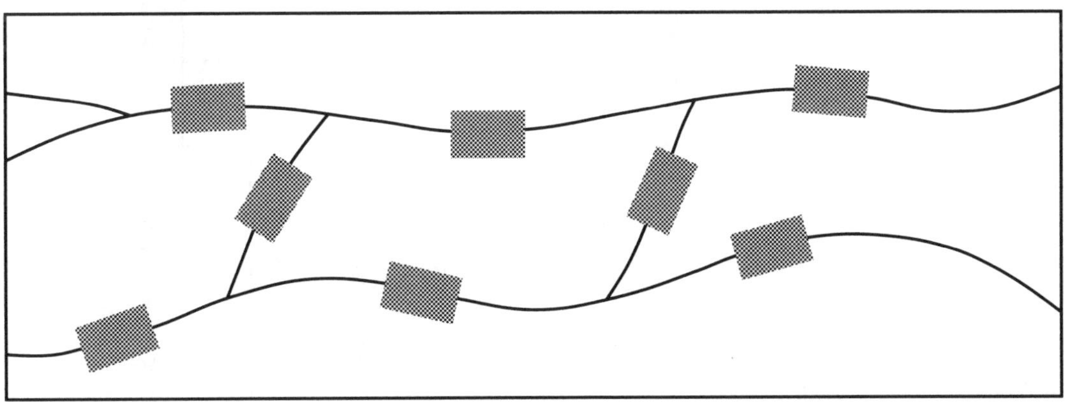

Wide-meshed crosslinked elastomer, e.g. flexible foam.

FIG 3.4.5 Cross-linking of plastics

Fig. 3.4.5

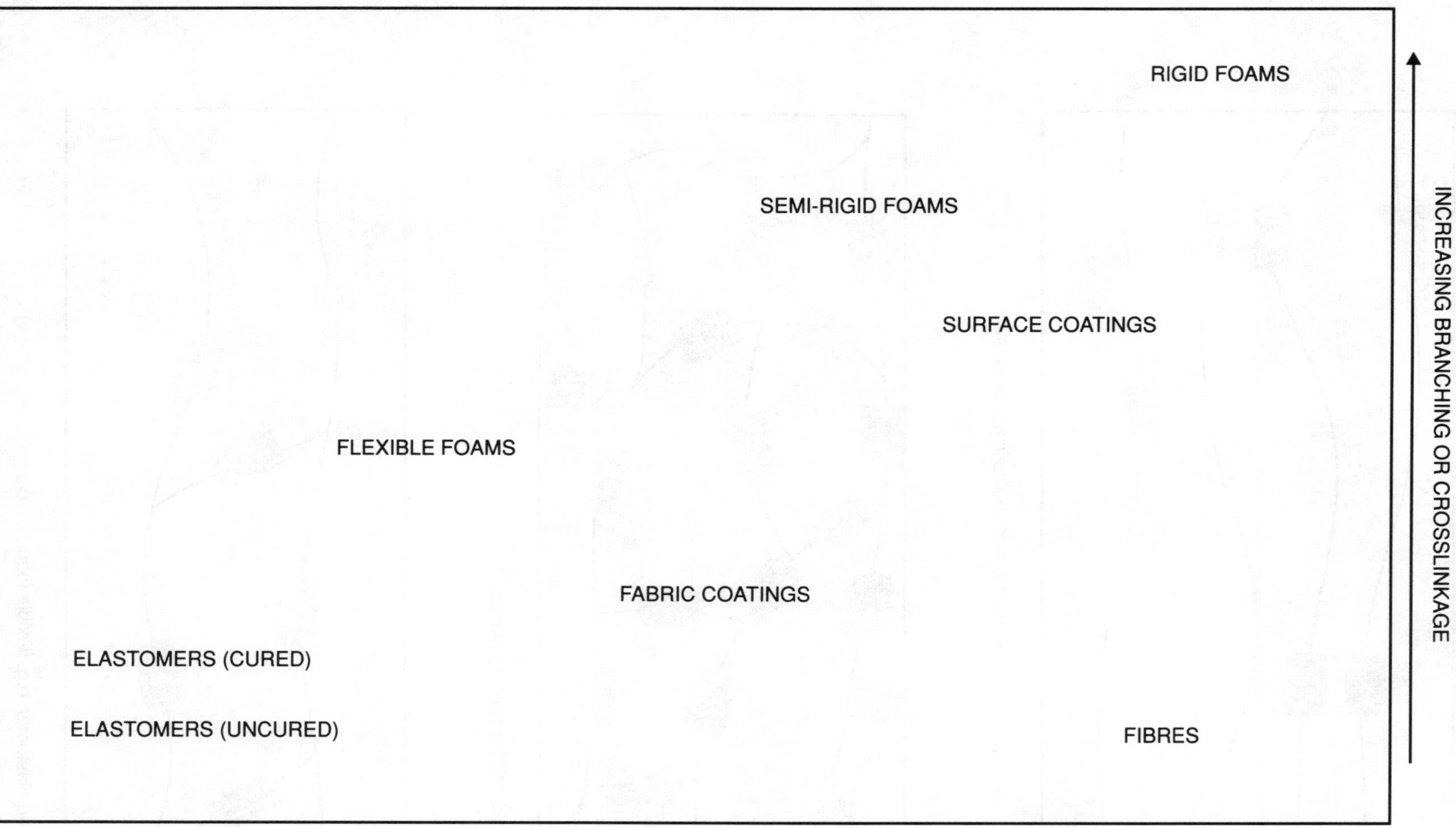

FIG 3.4.6 Polyurethanes: structure/property relationship

Fig. 3.4.6

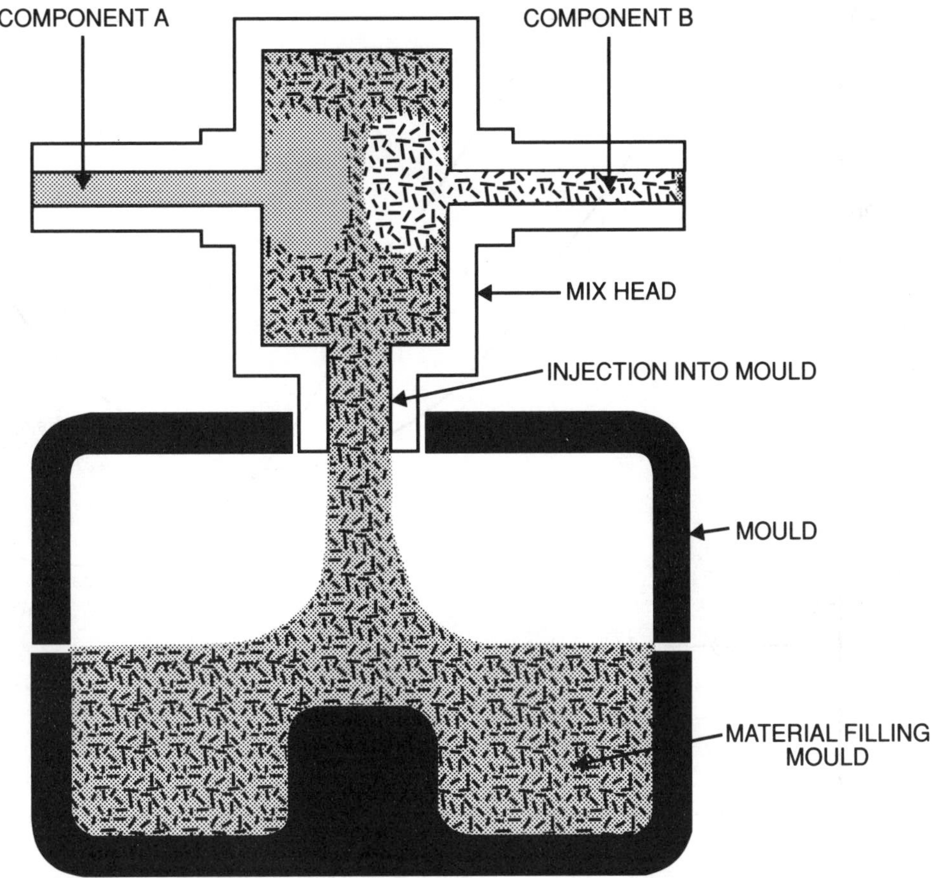

FIG 3.4.7 The RIM process

Fig. 3.4.7

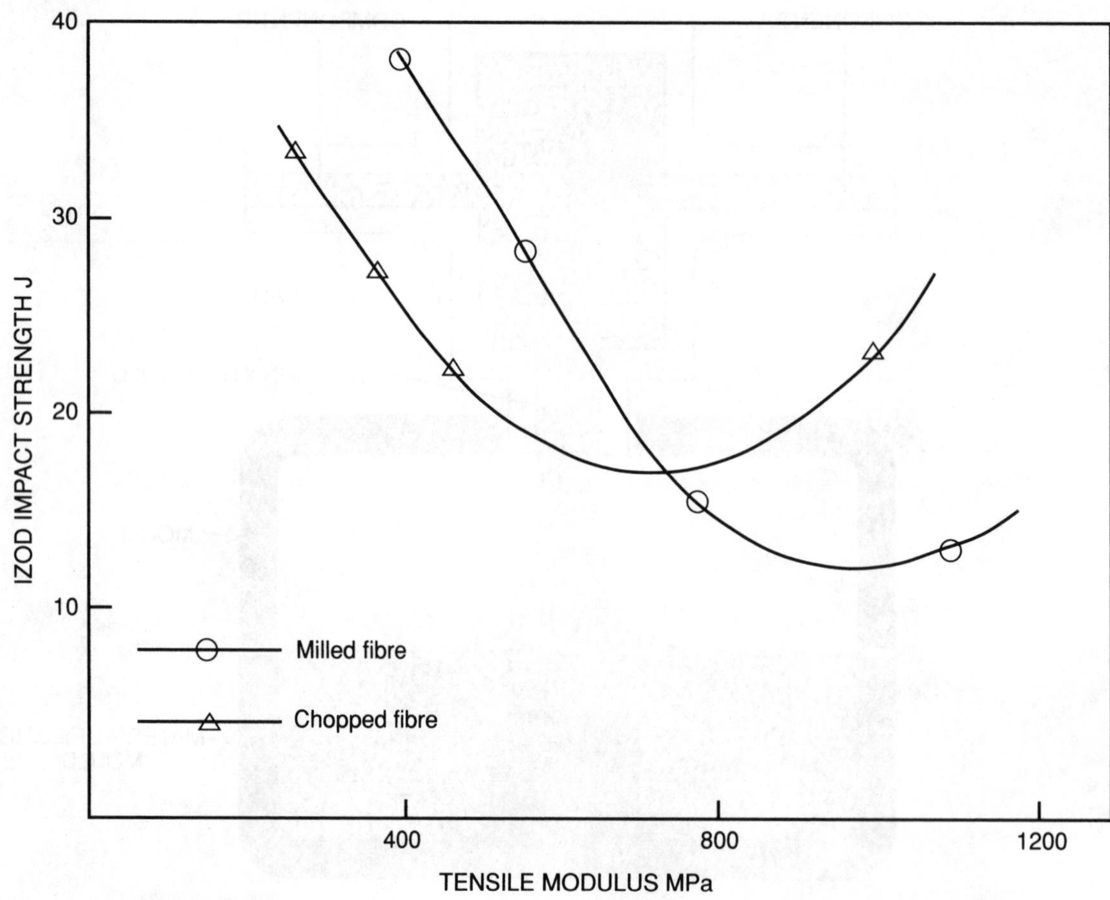

FIG 3.4.8 Variation of impact strength of polyurethane RRIM as a function of Tensile Modulus for two primary forms of glass

(source: Fibreglass Ltd)

Fig. 3.4.8

3–5

Thermoplastic elastomers

Contents

List of tables

List of figures

3.5.1 Introduction to thermoplastic elastomers

Thermoplastic elastomers (TPE) fill a gap between synthetic rubbers and flexible plastics, by combining elastomeric properties with the processability of thermoplastics. Injection moulding and extrusion offer much shorter cycle times than those used for compression or transfer moulding rubbers.

Thermoplastic elastomers can be used for applications requiring flexibility and recovery with stability of properties under moderate stress and temperature conditions.

Advantages of thermoplastic elastomers include:

- — Greater elasticity than most plastics
- — Ability to be used over wider temperature ranges than the usual flexible plastics, although not as versatile as some cured elastomer systems such as EPDM.
- — Ability to be moulded on standard plastics processing machinery.
- — Re-use of scrap and out-of-specification mouldings.

AVAILABLE TYPES

The types of thermoplastic elastomer available are listed in:
Table 3.5.1 *Basis and range of major available types of thermoplastic elastomer*

TYPICAL PROPERTIES

The more important properties of thermoplastic elastomers are listed for comparison in:
Table 3.5.2 *Typical comparative properties of thermoplastic elastomers*
Fig 3.5.1 *Comparisons of critical properties of TPEs, PVC and SBR*

Information on the extension under stress in presented in:
Fig 3.5.2 *Stress–strain properties of thermoplastic elastomers*

TABLE 3.5.1 Basis and range of major available types of thermoplastic elastomer

TPE type	*Typical properties*	*Typical applications*
Polyurethanes	Abrasion and cut resistance, load-bearing capacity. Polyester types for low temperature environments and moisture resistance; polyether for mechanical properties and oil resistance.	Wire insulation and jacketing, hose, solid tyres and roller skate wheels; small gears (for light loads) and exterior auto components. Linings, diaphragms, seals and vibration damping applications.
Olefinics (EPDM-PP)	Good weatherability and deformation resistance; low-temperature impact and flexibility; heat-ageing resistance; heat stability at high temperatures; has lowest specific gravity of four types.	Auto parts, sight shields and body-filler panels; weatherstripping hose, sporting goods and footwear. Wire insulation and jacketing (for battery booster cables, etc.) and moulded electrical products.
Styrene block copolymers	Good frictional properties, low temperature flexibility; resistance to abrasion, flex cracking and cutting; SBS has poor resistance to oxygen, ozone and weathering.	Shoe soles, exterior and interior auto parts: armrests, door panels, visors, sight shields, fender extensions, and rub strips. Also extruded seals and tubing.
Copolyesters	Resistance to high temperature; has high modulus, good elongation and high tear strength. Broad range of service temperatures; oil and chemical resistance, and good recovery from tension or compression deformation.	Exterior auto parts, solid and low-pressure pneumatic tyres, roller skate wheels. Under-the-hood auto components, low-voltage insulation, drive belting, flexible couplings and gears, mechanical moulded parts and sporting goods.

TABLE 3.5.2 Typical comparative properties of thermoplastic elastomers

	Property		*Units*	*Butyl polyethylene 5–20% glass*	*[a]Ethylene vinyl acetate*	*Polyester unfilled*
Mechanical	Tensile strength		MN/m^2	22–25	5.3–22	25–40
	Tensile Modulus 100%		MN/m^2	—	—	7–26
	Tensile Modulus 300%		MN/m^2	—	—	—
	Elongation		%	30–40	550–900	350–700
	Flexural Modulus		MN/m^2	760–1170	—	48–518
	Compression set (70–75 h)		—	—	72—103 (20–25 h)	3–60
	Izod impact, Notch		J/cm	4–>21	—	2–10
	Abrasion resistance (Taber)		mg/1000 cyc	15–16	—	3–27
	Hardness		Shore	D61–64	A91–97	A96–D72
	Resilience		%	—	—	43–62
Physical	Specific gravity		—	—	—	1.17–1.25
	Linear Thermal Expansion		10^{-5} K^{-1}	—	—	18–21
Electrical	Volume resistivity		ohm m	—	—	1.4×10^{11}–4.5×10^{13}
	Dielectric strength		10^2 kV/m	—	—	250–380
	Dielectric constant	60 Hz	—	—	—	4.35
		10^3 Hz	—	—	—	4–6
		10^6 Hz	—	—	—	3.99
	Power factor	60 Hz	—	—	—	0.007
		10^3 Hz	—	—	—	0.008–0.011
		10^6 Hz	—	—	—	0.033
	Arc tracking resistance		s	—	—	120–160
Thermal and environmental	Max. cont. service temp.		°C	—	—	>120
	Brittle temp.		°C	—	—	<–70
	Vicat softening point		°C	—	38–80	112–202
	Water absorption 24 h		%	0.02–0.04	—	0.06–0.5
Processing	Linear mould shrinkage		%	1.3–2.1	—	0.3–1.5

[a] See also Section 3.2.12.

Table 3.5.2

TABLE 3.5.2 Typical comparative properties of thermoplastic elastomers—*continued*

Polyester 5% glass-filled	*Polyester 20% glass fibre*	*Polyester 30% glass-filled*	*Polyester 40% mineral-filled*	*Polyolefin unfilled*	*Polyolefin 10% glass fibre*
24	49	50	22	4–36	7.6
—	—	—	—	4–18	—
—	—	—	—	6-11	—
300	12	5	—	230–1000	—
448	1172	2345	455	131–552	586
—	—	—	—	40–94	—
48	4.27	214	2.67	—	5.87
—	10–11	—	—	20–180	—
D58	D64	D72	D65	A54–D62	R40–50
—	—	—	—	—	—
1.23	1.34	1.61	1.54	0.84–1.02	0.96
—	—	—	—	9–54	2.7
—	—	—	—	5×10^{12} 5×10^{15}	—
—	—	—	—	120–320	—
—	—	—	—	2.2–6.3	—
—	—	—	—	2.2–	—
—	—	—	—	2.2–	—
—	—	—	—	0.0003–0.011	—
—	—	—	—	0.0006–	—
—	—	—	—	0.0011–	—
—	—	—	—	—	—
—	—	—	—	120–150	—
—	—	—	—	–50 to –80	—
—	—	—	—	40–91	—
0.19	—	0.19	—	0.03–0.07	0.04
1.0	0.4	—	—	0.1–2.0	0.6–0.8

Table 3.5.2—*continued*

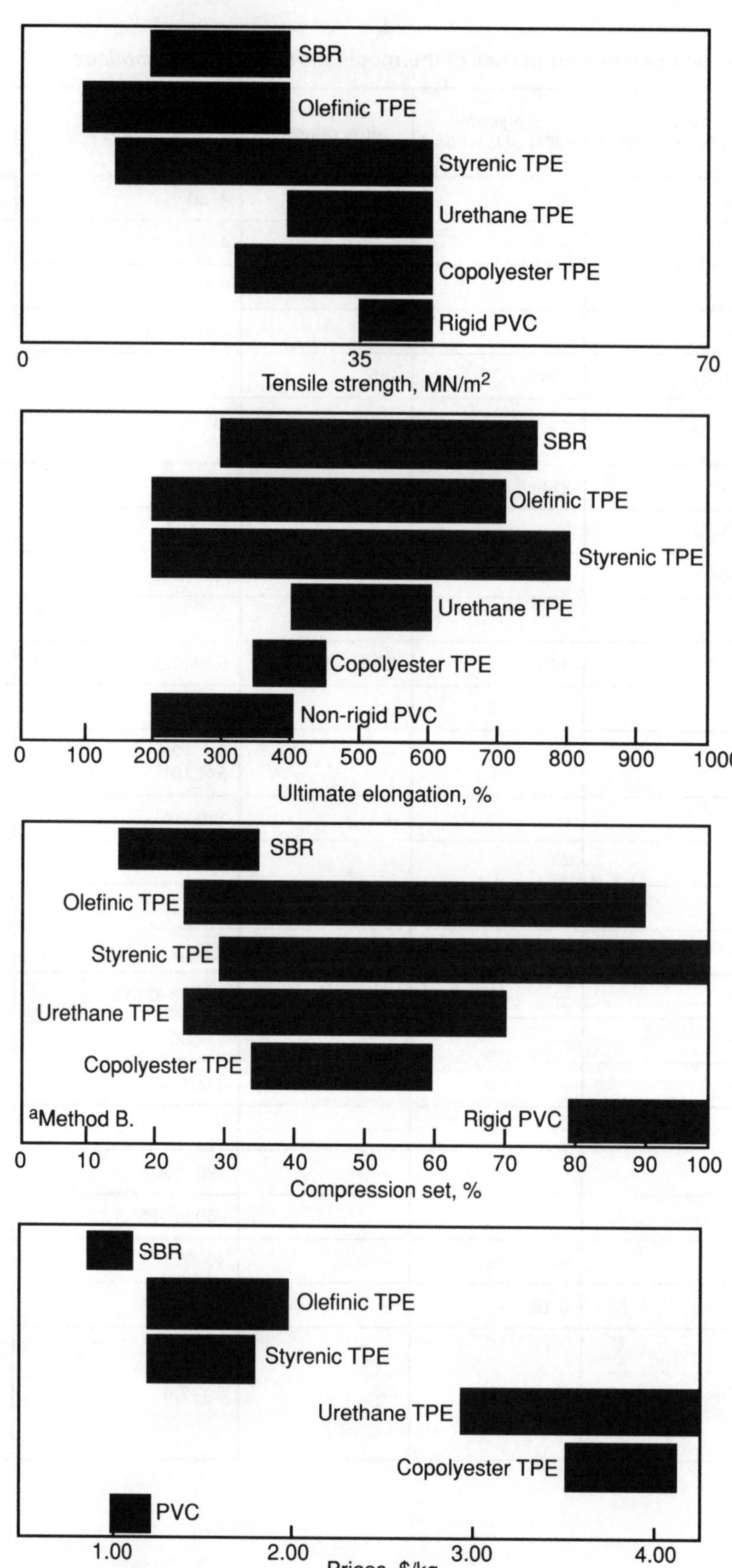

FIG 3.5.1 Comparisons of critical properties of TPEs, PVC and SBR

Fig. 3.5.1

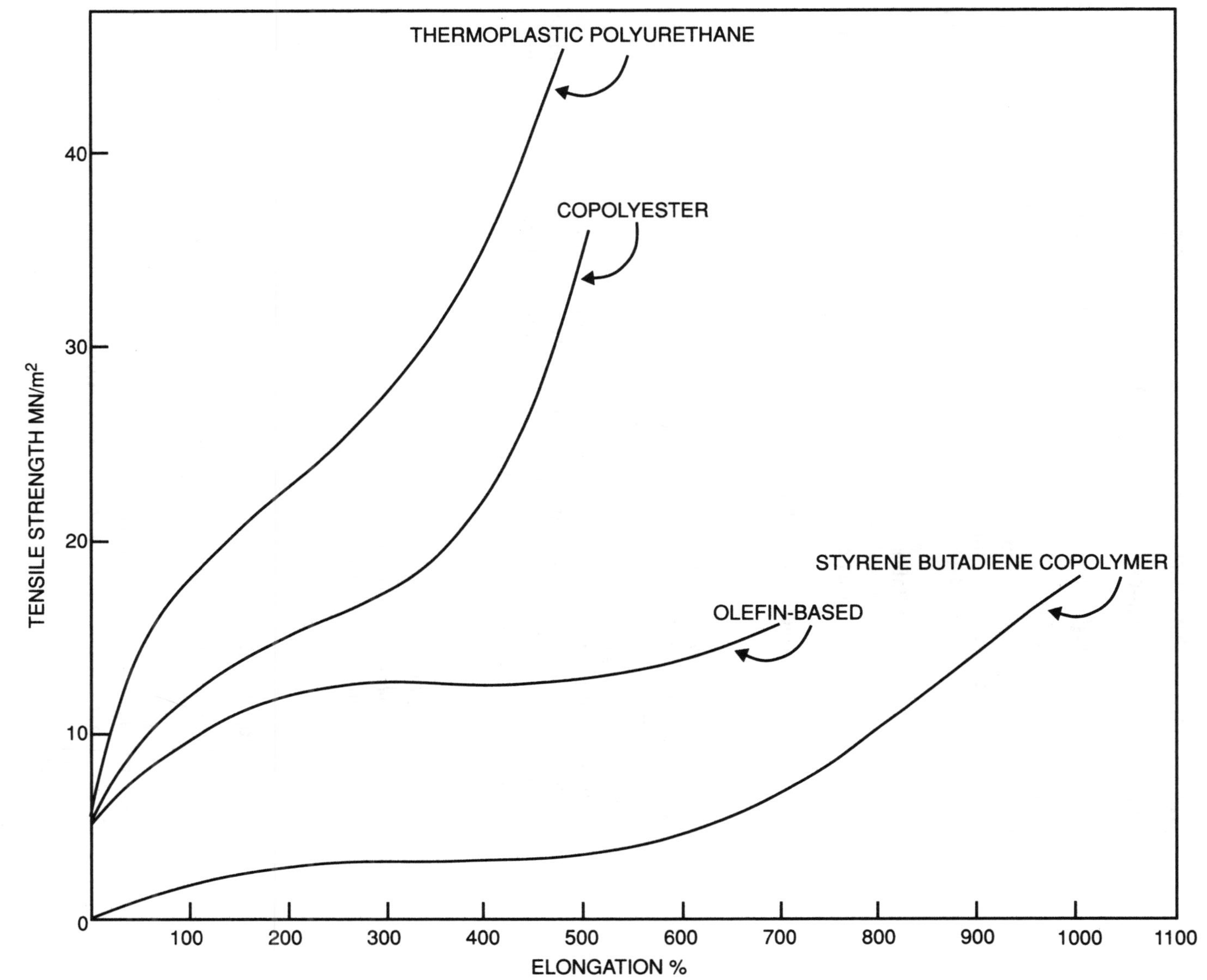

FIG 3.5.2 Stress–strain properties of thermoplastic elastomers

Fig. 3.5.2

3.5.2 Thermoplastic polyurethanes (for thermosetting polyurethanes see Section 3.4.14)

3.5.2.1 General characteristics and properties

The extreme versatility of polyurethane polymers allows manufacturers to formulate material grades for specific applications. Urethane elastomers are classed as thermoplastic, castable and millable gums. Castable and millable gums are thermosetting and are discussed in Section 3.4.14. Rigid and flexible urethane foams are discussed in Chapter 3.6.

The advantages and limitations of thermoplastic polyurethanes are listed in:
Table 3.5.3 *Comparative properties of thermoplastic polyurethanes*

The position of thermoplastic polyurethanes in the plastics hardness spectrum is indicated in:
Fig 3.5.3 *Hardness of thermoplastic polyurethanes relative to other polymers*

AVAILABLE TYPES

There are two basic types, polyester and polyether urethanes. They are supplied in the various grades in regular or irregular granule or chip form. General purpose base resins, adhesive grades, solution grades, coating grades, sheet and film, extrusion and injection moulding grade compounds are available. To prevent blocking of the granules, many grades (natural compounds particularly) either have an included antiblock additive or are dusted with, for example, talc or stearic acid or stearate, etc., depending on the use of the material. Polymer blends of the polyurethanes and compounded weather resistant grades are also available.

APPLICATIONS

The applications of thermoplastic polyurethanes are listed generally in:
Table 3.5.4 *Applications of thermoplastic polyurethanes*
and, as a function of the techniques in common use in:
Table 3.5.5 *Application of thermoplastic polyurethanes using processing techniques of the plastics industry*

TYPICAL PROPERTIES

Mechanical and some environmental properties of some grades of thermoplastic polyurethanes are listed in:
Table 3.5.6 *Mechanical and some other properties of thermoplastic polyurethanes*, and
Table 3.5.7 *Typical properties of thermoplastic polyurethanes*

Electrical and some of the properties of polyester and polyether based polyurethanes are listed in:
Table 3.5.8 *Electrical and other properties of thermoplastic polyurethanes*

3.5.2.2 Environmental resistance

WATER

Thermoplastic polyurethanes are sensitive to hot water, saturated steam and hot moist air. The water can react chemically with the polyurethane which can cause the polymer chain to break down, thus impairing the mechanical properties.

With polyester based polyurethane the deterioration in water is caused mainly by swelling, but once the materials become saturated the properties cease to change unless there is later deterioration owing to chemical or thermal attack.

However, the hydrolysis is temperature dependent and, at room temperature, products can be immersed in water for years without substantial change in mechanical properties.

OILS AND GREASES

Thermoplastic polyurethanes are resistant to pure mineral oils and greases, but technical oils which include additives may under certain circumstances be incompatible with the polyurethane due to the additives.

PETROL

Good chemical resistance to petroleum hydrocarbons provided they do no contain alcohol, otherwise alcoholysis can take place. Liquid fuels containing aromatics may swell products.

SOLVENTS

Good resistance to most solvents, but some polar solvents may cause swelling. Specialised soluble grades available for use as coatings and adhesives.

ACIDS AND ALKALIS

The thermoplastic polyurethanes have poor resistance to concentrated acids and caustic solutions.

WEATHERING

The degree of the effect of ozone, UV radiation, oxygen and moisture are temperature dependent. The resistance to ozone and oxygen is fairly good. UV radiation merely causes yellowing and does not affect the mechanical properties. However, intense UV radiation, as in the desert, requires the goods to be protected such as colouring black or adding UV stabilisers.

The chemical resistance to a number of acids, salts, solvents and other environments is listed in:
Table 3.5.9 *General environmental resistance of thermoplastic polyurethanes*

3.5.2.3 Processing

GENERAL

Thermoplastic Polyurethanes are available in many grades covering a wide spectrum of properties such as hardness, flexibility, etc., for processing by

Liquid casting
Injection moulding
Compression moulding
Transfer moulding
Extrusion
Centrifugal casting
Calendering and vacuum forming of sheet

PRETREATMENT OF GRANULES

As supplied the granules can normally be used without preliminary drying and heating. However, the granules must be kept in closed containers to prevent absorption of moisture from the air. The presence of a small amount of moisture does not disturb the injection moulding process, but excessive moisture may be absorbed if the granules are left in open containers, and this can lead to the surface of the moulding being disfigured by blisters and/or silvery streaks. Granules which are too moist should be dried at 100–110°C in an air circulation oven or a high-speed drier for 30 min–2 h depending upon the moisture content.

INJECTION MOULDING

Reciprocating screw machines with 18 : 1 – 24 : 1 L/D ratios are preferred for optimum temperature control and plasticisation. The urethanes can be moulded on any type of machine if adequate plasticising is achieved. Temperatures depend upon the compound and the part and should be kept between 170 and 250°C. Moderate screw speeds are advised. The power of the machine must be sufficient to develop high torque at low rpm, otherwise screw-stalling will occur.

After-treatment of mouldings

With some grades of thermoplastic polyurethane (but not the truly thermoplastic types) after-treatment of mouldings is necessary to obtain the optimum physical properties. Conditions of 15–20 h at 80–90°C for hardnesses below 90 Shore A and 110–120°C for hardnesses above that level are recommended.

For grades that do not require heat treatment of the cured moulding to obtain the physical properties, it is recommended that for cases where excessive strains are built into the part during moulding, an annealing condition of 2 h at 120°C be employed.

One of the key properties of the urethanes is that they need no preparation for painting and hot stamping.

Pigments should be specified for use in polyurethanes as they must possess good thermal stability and be stable to outdoor weathering to avoid discolouration of the material. Any stabilisers used should not contain stearates.

EXTRUSION

Single-stage, non-vented machines are adequate with an L/D ratio of between 20 and 24 : 1. Barrel cooling is required. Heavy duty drives (d.c. variable speed) and high power motors (e.g. 50hp for 2 in extruder) are necessary.

Mechanical sizing equipment is not recommended for the soft thermoplastic polyurethane melt.

Flat dies are used to produced sheet and film from 0.05 to 6.25mm (2 to 250 mil) thick. The coathanger manifold die has proven best for sheet. Stock temperature is higher for sheet than for profiles.

Standard pressure-type PVC dies are used for wire and cable coating, and long wind-up units are recommended to overcome 'delayed' tack effect.

Standard calendering equipment is used but horsepower requirements on a four-roll calender are substantially greater than for flexible or rigid PVC. Polished, or lightly matted, rolls are recommended to minimise sticking.

Polyurethane resins develop a natural adhesive tack which lasts for a short while after processing, and this property, coupled with the ability to 'wet' most materials, permits direct in-line coating or lamination to a wide variety of substrates without any preliminary surface treatment.

MACHINING: JOINING

Modified epoxy or acrylic adhesives are generally recommended for bonding urethanes.

In thread cutting, only coarse threads should be cut in the urethanes.

3.5.2.4 Trade names and suppliers

The suppliers and their commonly used trade names for thermoplastic polyurethanes are listed in:

Table 3.5.10 *Suppliers and trade names of thermoplastic polyurethanes*

TABLE 3.5.3 Comparative properties of thermoplastic polyurethanes

	Advantages	***Limitations***
Mechanical properties	High load-bearing capacity, elasticity and strength. Elastomeric qualities. Excellent flex life. High tear life. Polyester types more rigid than polyether (relatively flexible).	Higher compression set than thermoset polyurethanes.
Wear	High resistance to high-angle impact abrasion (hardness, elasticity, tear resistance). Polyester types recommended for abrasion resistance and toughness (superior to other elastomers).	Less successful where low-angle impact abrasion is encountered. Wear resistance falls at higher temperatures. Soft grades have higher friction coefficient than hard grades.
Environment	Good fuel and oil resistance to 80°C. General strain resistance. Low gas permeability. Relatively high moisture vapour transmission (breathability). Migration resistance. Good compatibility with wide range of synthetic polymers. Excellent resistance to radiation compared with most elastomers. Ozone resistance. Polyester types more solvent resistant, but	Polyether types recommended for humidity resistance. hydrolysis-prone.
Thermal properties	Excellent low temperature resistance to –80°C. Polyether types exhibit lower heat build-up than polyesters.	Maximum use temperature 130°C (only 80°C continuous). Thermal expansion 15–20 x steel —in design of bushings, etc. Narrow tolerances for elastomers are senseless.
Food & medicine	Finished products do not affect the taste or smell of foodstuffs.	Use of utensils consisting of cross-linked PUs restricted to (i) dry foodstuffs (ii) pH5-7 below 30°C.
Optical	Transparent grades available.	
Processing	High speed production on thermoplastics equipment. With certain grades no curing or post-heat treatment necessary for the development of optimum properties. Vacuum formability. Heat reliability. Scrap readily processed (linear polyurethanes)	Manufacture utilises toxic isocyanates.
Miscellaneous	Wide range of materials available as natural or pigmented in a range of colours. Readily foamable prepolymers.	High cost. Foams are under attack because of fire and smoke emission properties.

Table 3.5.3

TABLE 3.5.4 Applications of thermoplastic polyurethanes

Application	*Specific advantages of polyurethane*
Automotive industry	
Vehicle chassis seals Steering systems seals Gear shift mechanisms for force-transmitting bearings and shells and bushes for oscillating movements. Fuel base lines Bumpers	Good wear resistance, ability to damp vibrations and low swelling in greases. High tear strength and flexibility throughout wide temperature range. Good resistance to ozone and UV light.
General engineering	
Seals Sleeves Wipers Rolls and roll covers Tooth wheels and sprockets, gears Anti-vibration buffers Tyres	High resistance to high-angle impact abrasion. Insensitivity to oils and lubricants. Relatively high stiffness. Ability to dampen vibration. Silent running.
Tool industry	
Hammer heads Disc sanding pads	Impact resistance.
Footwear industry	
Heel tips and bars for football boots Soles of special sports footwear	Outstanding wear resistance. Good mechanical properties and low bending resistance, free from tendency to crack in hot dry climate.
Electrical engineering	
Sheathing measuring cable (extruded)	High wear resistance, impact resistance and flexibility. Good resistance to oils. Although not significant as an insulating material in high tension applications often used in electrical appliances where material performs specific technical functions and insulates sufficiently well.
Textile machinery	
Spinner centraliser bushes Loom packing seals	

Table 3.5.5

TABLE 3.5.5 Application of thermoplastic polyurethanes using processing techniques of the plastics industry

From solution	*Extrusion coating/lamination*	*Extrusion-blown film*	*Extrusion tubing*	*Calendering/hot melt coating*	*Injection moulding*
Lightweight rainwear	Apparel	Packets for lubricants, grease and oils	Hose	Flexible containers	Shoe parts, heel lifts, football cleats
Shoe adhesives	Paper coating	Covers for foam	Cable covers	Conveyor belting	Seals, gaskets O-rings
Foil coatings and adhesives	Shoe upper materials	Protective wraps		Hand-built hose	Conveyor, impeller and fan blades
Fabric laminate adhesives	Fuel containers	Irradiated foodwraps		Tarpaulins	Gear for quiet operations in film projectors, cameras, toys, vibration insulators
Leather treatments	Multilayer packaging materials				
Synthetic leather	Coating of foils, vinylidene and polyester substrates				
Topcoats	Coatings on rigid materials which are later vacuum formed				

TABLE 3.5.6 Mechanical and some other properties of thermoplastic polyurethanes

Property		Units	Polyester [a]Elastollan S								Polyester ('Copolymer polyester') [a]Elastollan C									Polyether [a]Estollan VP 1100 series				P85A
				Nominal shore hardness							Nominal hardness									Nominal hardness				
			80A	85A	88A	90A	95A	98A	60D	74D	78A	80A	85A	90A	95A	59D	60D	64D	74D	95A	54D	60D	64D	85A
Density		kg/dm³	1.21	1.22	1.22	1.23	1.23	1.24	1.25	1.26	1.18	1.19	1.19	1.21	1.22	1.24	1.24	1.24	1.24	1.13	1.14	1.17	1.17	1.10
Tensile strength (Strain rate 100mm/min)		MN/m²	55	55	55	55	55	55	50	50	35	35	40	45	45	45	45	40	35	40	35	35	35	40
Elongation at break		%	700	700	650	600	550	500	450	300	600	550	500	450	450	450	400	350	300	400	300	300	300	400
Tensile Modulus	20%	MN/m²	2	2	3	6	8	9	13	25	2.0	2.7	4	5.6	8.5	12	16	17	28	6	10	12	15	4
	100%		4	5	6	9	11	12	16	30	3.5	5	8	10	12	17	20	24	30	10	15	20	25	7
	300%		8	8	9	13	20	23	24	—	7.5	10	13	15	18	30	35	35	35	18	30	35	35	17
Graves tear strength	DIN 53507	N/mm	45	45	50	55	60	60	75	—	35	40	40	60	65	70	80	80	90	40	45	50	60	20
	DIN 53515		55	60	65	95	110	115	135	200	55	60	70	90	110	140	160	160	200	70	80	130	140	35
Abrasion loss (DIN 53516)		mm³	40	40	40	40	35	35	30	30	50	50	50	50	55	60	60	65	70	50	50	50	50	50
Compression set	RT	%	25	25	25	25	25	30	40	55	25	25	25	25	30	30	40	40	40	40	40	40	50	20
	70°C		35	35	35	45	45	45	50	60	40	40	35	40	45	50	50	55	60	50	55	55	75	35
Rebound resilience		%	35	35	32	32	32	32	32	—	44	40	35	35	35	37	38	38	41	40	41	42	44	40
Freezing temperature		°C	−40	−40	−40	−40	−40	−35	−35	−20	−45	−45	−45	−45	−45	−45	−45	−35	−25	−60	−55	−50	−50	−60
Hydrolysis resistance Tensile strength after storage in water at 80°C	8 days	MN/m²	—	—	—	—	—	—	—	—	—	—	—	—	—	—	—	—	—	—	30	30	30	—
	42 days		—	—	—	—	—	—	—	—	25[b]	25[b]	33[b]	35[b]	40[b]	40[b]	38[b]	35[b]	32[b]	30	—	—	—	25
Elongation at rupture after storage in water at 80°C	8 days	%	—	—	—	—	—	—	—	—	—	—	—	—	—	—	—	—	—	—	350	350	350	—
	42 days		—	—	—	—	—	—	—	—	650[b]	600[b]	600[b]	500[b]	500[b]	500[b]	400[b]	400[b]	300[b]	500	—	—	—	400

All data supplied by BASF/Elastogran.
[a] Tradenames of BASF/Elastogran
[b] After 21 days.

TABLE 3.5.7 Typical properties of thermoplastic polyurethanes

Property		*ASTM test*	*Units*	*Hard cast*
Tensile yield strength		D638	MN/m^2	50
Flexural strength		D790	MN/m^2	—
Elongation		—	%	400
Modulus of Elasticity in Tension		D638	MN/m^2	—
Flexural Modulus		D790	MN/m^2	22
Impact strength		D638 (Izod)	J/cm of notch	10.6 +
Hardness		—	Shore	A95
Max. continuous service temperature (no load)		—	°C	80
Deflection temperature	0.45 MN/m^2	D648	°C	—
	1.81 MN/m^2	D648	°C	40
Thermal conductivity		—	W/m per K	—
Thermal expansion		D696	$10^{-6} K^{-1}$	160
Volume resistivity		D257	ohm–m	10^{10}
Dielectric strength		D149	10^2 kV/m	160
Dielectric constant	60 Hz	D150	—	—
	10^6 Hz			
Power factor	60 Hz	D150	—	—
	10^6 Hz			
Arc resistance		D495	—	—
Specific gravity		D792	—	1.2
Flammability		D635	UL94	HB
Water absorption		D570	% in 24 h	0.2
Mould shrinkage		—	%	1.7

Table 3.5.7

TABLE 3.5.7 Typical properties of thermoplastic polyurethanes—*continued*

Hard microcellular	*Soft microcellular*	*Reinforced microcellular*	*Semi-rigid foam*	*Structural foam*
4	7	12	4	20
—	—	—	—	—
120	450	5	150	10
—	—	—	—	—
10	2	1500	10	900
10.6 +	10.6 +	9	4	10.6 +
A50	A50	A98	A70	A95
70	70	80	80	80
—	—	—	—	—
—	—	—	30	65
—	—	—	—	—
130	150	80	110	120
10^{10}	10^{10}	10^{10}	10^{10}	10^{10}
160	160	160	160	160
—	—	—	—	—
—	—	—	—	—
—	—	—	—	—
0.9	0.5	1.3	0.8	0.6
HB	HB	HB	HB	HB
10	10	0.4	0.4	0.2
1.7	1.7	0.9	1.7	0.6

Table 3.5.7—*continued*

TABLE 3.5.8 Electrical and other properties of thermoplastic polyurethanes

			Polyester based	***Polyether based***
Hardness (Shore)			A75–D73	A80–D74
Max. cont. service temp. (°C)			80	85
Brittle temp. (°C)			–40 to –82	–73 to –76
Linear thermal exp. (10^{-5} K^{-1})			11–20	10–33
Volume resistivity (ohm m)			5×10^8–5×10^{10}	2×10^9–1×10^{11}
Dielectric Strength (10^2 kV/m)			110–140	130–152
Dielectric constant	—	60 Hz	5.6–7.9	6.0
	—	10^3 Hz	5.2–7.5	5.6
	—	10^6 Hz	4.1–6.2	4.2
Power factor	—	60 Hz	0.015–0.140	0.048
	—	10^3 Hz	0.025–0.060	0.043
	—	10^6 Hz	0.050–0.100	0.075
Arc resistance (s)			—	122
Water absorption (24 h, %)			—	0.7–0.9
Specific gravity			1.15–1.28	1.05–1.18
Linear mould shrinkage (%)			0.5–1.15	1.0–1.5

Table 3.5.8

TABLE 3.5.9 General environmental resistance of thermoplastic polyurethanes

Chemical resistance	*Polyester types*		*Polyether types*	
Acids				
Formic, 20%	Poor		Poor	
Sulphuric, 20%	Fair–Good		Fair	
Sulphuric, 30%	—		Fair	
Alcohols				
Isopropyl	Poor		Fair–Poor	
Methyl	Fair		—	
BASE				
Sodium hydroxide, 20%	Fair–Poor		Fair	
Detergent				
Mr. Clean	Good		—	
Fuels				
ASTM Fuel A	Good		Good	
ASTM Fuel B	Fair–Good		Fair	
ASTM Fuel C	Fair		Fair–Poor	
Gasoline, 100 Octane	Good		Fair	
Kerosene	Good		—	
Glycols				
Ethylene	Good		Good	
Ethylene/water 50/50	Good		Good	
Propylene	Good		Good	
Propylene/water 50/50	Good		Good	
Oils				
ASTM #1	Excellent		Good	
ASTM #2	Excellent		Good	
ASTM #3	Excellent–Good		Good–Fair	
Brake Fluid Type A	Fair—Poor		Poor	
Detergent 20W	—		Good	
Non-Detergent 20W	—		Good	
Skydrol Type B	—		Dissolves	
Transmission Type A	Excellent		Good	
Plasticiser				
Didecyl adipate	Good		—	
Dioctyl adipate	Good		—	
Dioctyl phthalate	Good		Fair	
Tricresyl phosphate	—		Poor	
Salt solutions				
Calcium chloride, Saturated	Good		—	
Sodium chloride, Saturated	Good		Good	
Synthetic perspiration	Good		Good	
Solvents				
Benzene	Fair–Poor		Poor	
Carbon tetrachloride	Fair		Poor	
Cyclohexanone	Dissolves		Dissolves	
Dimethyl formamide	Dissolves		Dissolves	
Dimethyl sulfoxide	Dissolves		Dissolves	
1.4 Dioxane	Dissolves		Dissolves	
Ethylene dichloride	Poor		—	
Methyl ethyl ketone	Fair–Poor		Poor	
N-Methyl-2 Pyrrolidone	Dissolves		Dissolves	
Perchloroethylene	Fair		Poor	
Pyridine	Dissolves		Dissolves	
Tetrahydrofuran	Dissolves		Dissolves	
Toluene	Fair		—	
Trichloroethylene	Poor		Poor	
Hydrolysis resistance				
Water immersion				
23°C	Poor–Fair		Good	
70°C	Poor		Fair–Good	
Dry heat resistance				
Properties:				
At elevated temperature	Fair–Good to 130°C		Fair–Good to 80°C	
After aging at elevated temp.	Fair–Good to 100°C		Fair–Good to 100°C	
Weathering	**Unpigmented**	**Pigmented**	**Unpigmented**	**Pigmented**
Outdoor				
Arizona	Poor	Fair–Good	Poor	Good
Florida	Poor	Fair–Good	Poor	Good
Ohio	Poor	Fair–Good	Poor	Good
Accelerated				
Fade-Ometer	Poor	Fair–Good	Poor	Good
Weather-Ometer	Poor	Fair–Poor	Poor	Good

TABLE 3.5.10 Suppliers and trade names of thermoplastic polyurethanes

Supplier	*Trade name*
Mobay Chemical	Texin
K. J. Quinn & Co. Inc. (Aromatic Polyester and Polyether)	Q-Thane
Upjohn Co.	Pellethane
Hooker Chemical Co.	Rucothane
B. F. Goodrich (Polyester and Polyether)	Estate
Uniroyal Inc. (Polyester and Polyether)	Roylar
ICI	Daltomold, Daltoflex
American Cyanamid (Polyester)	Cynaprene
Dunlop Holdings	Jectothane, Microvon
Bayer AG	Desmopan, Desmoflex, Vulkolan, Bayflex, Baydur
Ohio Rubber (Polyether, Polyester, Polycaprolactone)	Orthane
BASF/Elastogran/Elastomer Products	Lupranol, Elastofoam Lupranat, Elastolit Lupraphen Elastollan Caprollan
Fabelta	Fabeltan
Freeman Chemicals Ltd	Lastane
B & T Polymers	Durelast, Hyperlast
Avalon Chemical Company	Sorane
Lankro Chemicals	Propocon-Isocon
Du Pont	Adiprene

Table 3.5.10

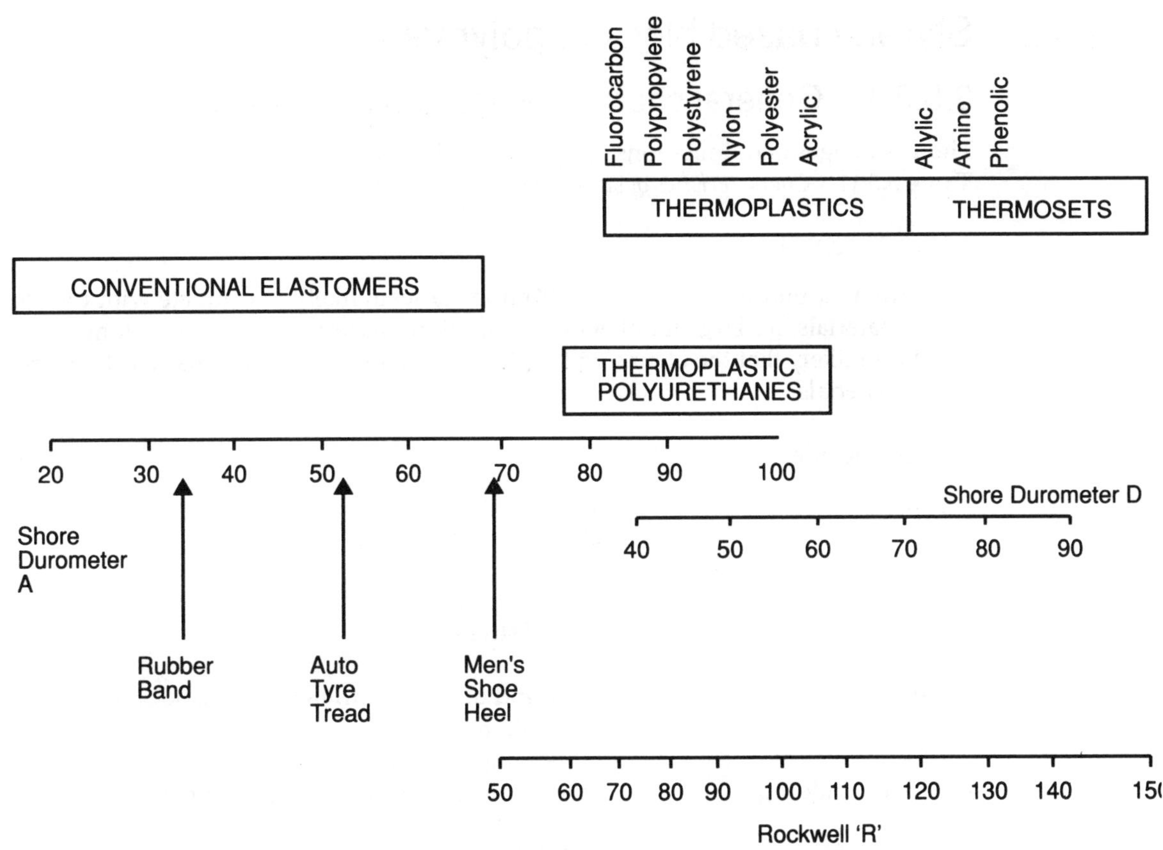

FIG 3.5.3 Hardness of thermoplastic polyurethanes relative to other polymers

Fig. 3.5.3

3.5.3 Styrene based block copolymers

3.5.3.1 General characteristics and properties

The advantages and limitations of styrene block copolymers are listed in:
Table 3.5.11 *Characteristics of styrene block copolymers*

APPLICATIONS

The saturated mid-block types (e.g. 'Kraton' E) seem likely to compete with the olefin based materials in a large number of new applications, especially in the automobile sector. It is unlikely that the olefins will capture any of the established markets for styrene-based materials.

These include:

Footwear	Unit soles Direct moulded shoes Heels Heel top pieces Sandal strips Boots
Extrusion	Hose and tubing—garden hose, elastic bands, medical, milk. Billiard table cushions.
Injection moulding	Sports equipment—hand grips, swim fins. Bottle stoppers
Plastics modification	HDPE & LDPE film for bags. Polystyrene—high impact applications. Polypropylene—impact applications.
Adhesives	Hot melt and solvent types.

TYPICAL PROPERTIES

The more important properties of one grade of styrene copolymer are listed in:
Table 3.5.12 *Typical properties of 'Kraton' G styrene/ethylene butylene/styrene block copolymers*

3.5.3.2 Trade names and suppliers

The suppliers of styrene block copolymers and their commonly used trade names are listed in:
Table 3.5.13 *Trade names and suppliers of styrene block copolymers*

TABLE 3.5.11 Characteristics of styrene block copolymers

Advantages	*Limitations*
1. Good low temperature characteristics. 2. High cohesive strength. 3. Excellent elasticity at usage temperatures. 4. Easily processed by injection, extrusion, thermoforming, blow moulding, rotational moulding. 5. Low density 6. Good clarity. 7. Wide range of properties obtainable. 8. Excellent electrical insulators (see grades). 9. Good abrasion resistance. 10. Temperature limit 50–100°C, depending on grade.	1. Styrene-butadienes have inferior weathering resistance to some other styrene block copolymers. 2. Swollen or dissolved by strong acids, chlorinated solvents, esters and ketones.

Table 3.5.12

TABLE 3.5.12 Typical properties of 'Kraton' G styrene/ethylene butylene/styrene block copolymers

Property	*Units*							
Hardness	Shore A	55A	45A	60A	90A	95A/45D	45–50D	40–45D
Tensile strength	MN/m^2	7.6	3.4	6.2	12.4	17.2	20.7	20.7
100% Modulus	MN/m^2	—	—	—	—	—	9.6	7.9
300% Modulus	MN/m^2	—	2.1	2.4	6.9	9.7	—	—
500% Modulus	MN/m^2	—	—	—	—	—	17.9	17.2
Elongation	%	—	—	—	—	—	550	600
Tear strength	kN/m	—	—	—	—	—	96	88
Flexural Modulus	MN/m^2	—	—	—	—	—	27.5	14.0
Density	gm/cm^3	0.90	1.19	1.20	1.14	1.14	0.97	0.98
Linear Expansion Coefficient	$10^{-4}/K$	—	—	—	—	—	3	3
Mould shrinkage	%	—	—	—	—	—	1.0–1.5	1.0–1.5

Data supplied by Shell.

TABLE 3.5.13 Trade names and suppliers of styrene block copolymers

Supplier	*Trade name*	*Type*	*Notes*
Anic	Europrene	Radial SBS	Licensed for Phillips
Custom Polymers Pty	Resex	SBS and modified olefins	
Dow Corning	TPSR	Copolymer of styrene and dimethyl siloxane	
Nylex Corporation	Elastrene	SBS	
Phillips Petroleum	Europrene	Radial SBS	Europe
	Solprene		US
Shell Chemical Co.	Cariflex	Styrene–butadiene	Europe
	Kraton	Isoprene–styrene	US
	Elexar	Styrene–polyolefin	

3.5.4 Olefin based block copolymers (olefin TPE)

3.5.4.1 General characteristics and properties

The advantages and limitations of olefin TPEs are listed in:
Table 3.5.14 *Characteristics of olefin based block copolymers (olefin TPEs)*

Most olefin based TPEs are physical blends of EPDM (ethylene–propylene–diene monomer) and polypropylene. The latter gives the elastomer strength and renders it mouldable. Enough EPDM is added to impart flexibility. The ratio of EPDM/PP can be varied to a certain extent to make the elastomer perform well for specific thermoplastic processes.

PP copolymers with high ethylene content provide an alternative route to stiffer olefin TPEs.

To provide a grade of olefin TPE with improved property stability characteristics over a wide temperature range with, for example, lower compression set values and comparable hardness, a small percentage of cross-linking may be introduced by a 'controlled cure process'. Such materials tend to be relatively more rubbery and softer in character than the non-cross-linked forms; they are also more difficult to process and by nature of their production method, more highly priced.

The wide range of properties which can be obtained by varying the blend ratio of the various compounds has led to some confusion in nomenclature so that olefin based thermoplastic elastomers are called thermoplastic rubbers (TPR is a registered trade name), thermoplastic polyolefin elastomers (TOPE), thermoplastic elastomers (TPE), elastomer modified PP (EMPP), or more recently thermoplastic EPDM, the latter causing great confusion when the prefix 'thermoplastic' is omitted. The term olefin TPE is used throughout the Selector.

The vulcanised or cross-linked EPDMs are not thermoplastic elastomers but 'VISTALON 3708' UNCURED exhibits both thermoplastic and elastomeric qualities. The established EPDMs have inherently good properties and it is logical to assume these will be retained in the thermoplastic version.

'SOMEL' is a thermoplastic EPDM, designed as a low cost material which can be extruded injection moulded or thermoformed. These materials have a useful service temperature range of –50–120°C with generally good resistance to many chemicals, but excluding oil.

APPLICATIONS

Olefin based elastomers fill the gap in the price-performance spectrum between general-purpose cured rubbers, plasticised PVC and EVA.
Applications include:

Automotive (60% total):	Large area grommets replacing LDPE, gaskets, tubing. Flexible bumper systems, front end, side shields competing with reaction injection moulded polyurethanes. Crash-pads replacing ABS/PVC. Steering wheel covers. Door pockets and glove boxes.
Other:	Blow moulded products requiring impact performance in acid environment. Moulded electrical plugs and fittings for use below 130°C. Low voltage cable covering and insulation, flexible cords, coaxial cable. Flooring. Golf club grips.

Premium garden hose and industrial hose.
Refrigeration gaskets.
Extruded expansion joints.
Flexible fasteners.
Footwear (especially sports shoe soles)
Fabric coatings.
Biomedical and food packaging.
Polymer blends.

Applications specifically relevant to EPDM thermoplastic elastomers include:

Industrial:	tarpaulins, hospital sheeting, reservoir linings, tubes and hoses.
Electrical:	wire and cable coatings, mouldings.
Leisure goods:	swimming pools, camping equipment, inflatables, floor mats.
Automotive:	car mats, bumper components.

TYPICAL PROPERTIES

The more important properties of olefin TPE grades are listed in:
Table 3.5.15 *Typical properties of olefin TPEs*

The flexural moduli of olefin and EPDM TPE are compared with those of other polymers and elastomers in:
Fig 3.5.4 *Comparative stiffness of olefin TPEs*

3.5.4.2 Trade names and suppliers

The suppliers of olefin TPEs and the familiar trade names of their products are listed in:
Table 3.5.16 *Trade names and suppliers of olefin TPEs*

TABLE 3.5.14 Characteristics of olefin based block copolymers (olefin TPEs)

Advantages	*Limitations*
1. Better resistance to high-temperature softening and weathering than styrene based materials. 2. Best low temperature flexibility of thermoplastic elastomers. 3. Can satisfy 30–35% of all rubber and elastomer applications. 4. 50% higher dielectric strength than PVC. Power factor and dielectric constant very similar to cross-lined polythene and polypropylene. 5. Good resistance to some acids, most bases, many organic materials, butyl alcohol, ethyl acetate, formaldehyde and nitrobenzene. 6. Low coefficient of thermal expansion.	1. Cannot accept fillers in large concentrations. 2. Cannot be compression moulded or calendered. 3. Poor flow behaviour in softer grades. 4. Swollen by aliphatic, aromatic and chlorinated solvents, turpentine and petrol.

TABLE 3.5.15 Typical properties of olefin TPEs

	Properties		Units	Unfilled grades[a]							20%[b] talc
Mechanical	Hardness		Shore	A67	A77	A88	A87	A92	A62	A62	A98 (D55)
	Tensile strength		MN/m^2	4.5	6.6	9.7	9.0	12.8	4.1	2.5	13.0
	Ultimate elongation		%	210	200	210	180	250	150	400	193
	100% Modulus		MN/m^2	3.5	5.5	8.6	8.6	12.8	3.5	2.4	—
	Compression set	20°C	%	25	30	35	30	40	—	—	—
		70°C		45	50	64	70	70	25	—	—
	Flexural Modulus		MN/m^2	10.3	18.6	69.0	55.2	241.3	24.8	18.6	—
	Bashore resilience		% rebound	50	50	45	50	45	—	—	—
	Split tear		kN/m	24.5	38.5	47.3	61.3	87.5	—	—	—
Physical	Density		kg/dm^3	0.88	0.88	0.88	0.88	0.88	0.91	0.90	—
Thermal and environmental	Abrasion resistance		g/kc	0.6	0.3	0.3	0.3	0.4	—	—	—
	Heat deflection temperature		°C	—	—	—	—	58	—	—	72

[a] Data supplied by Uniroyal.
[b] Data supplied by Chemische Werke Hüls.

TABLE 3.5.16 Trade names and suppliers of olefin TPEs

Supplier	*Trade name*	*Notes*
Bayer AG	Levaflex	
Chemische Werke Huls	Vestopren	
Copolymer Rubber	Tersyn	
DSM	Keltan TP	
Exxon/Esso/Chemical Co.	Vistaflex	
Freeman Chemicals Ltd	Larflex	
BF Goodrich	Telcar	
International Synthetic Rubber Co.	Uneprene	
Monsanto	Santoprene	
Montedison	Ovtral Dutral	
Uniroyal Inc	Uniroyal TPR	
	Thermoplastic EPDM	
Essochem Research Centre, Belgium	Vistalon 3708 uncured	
Du Pont	Nordel (Somel)	

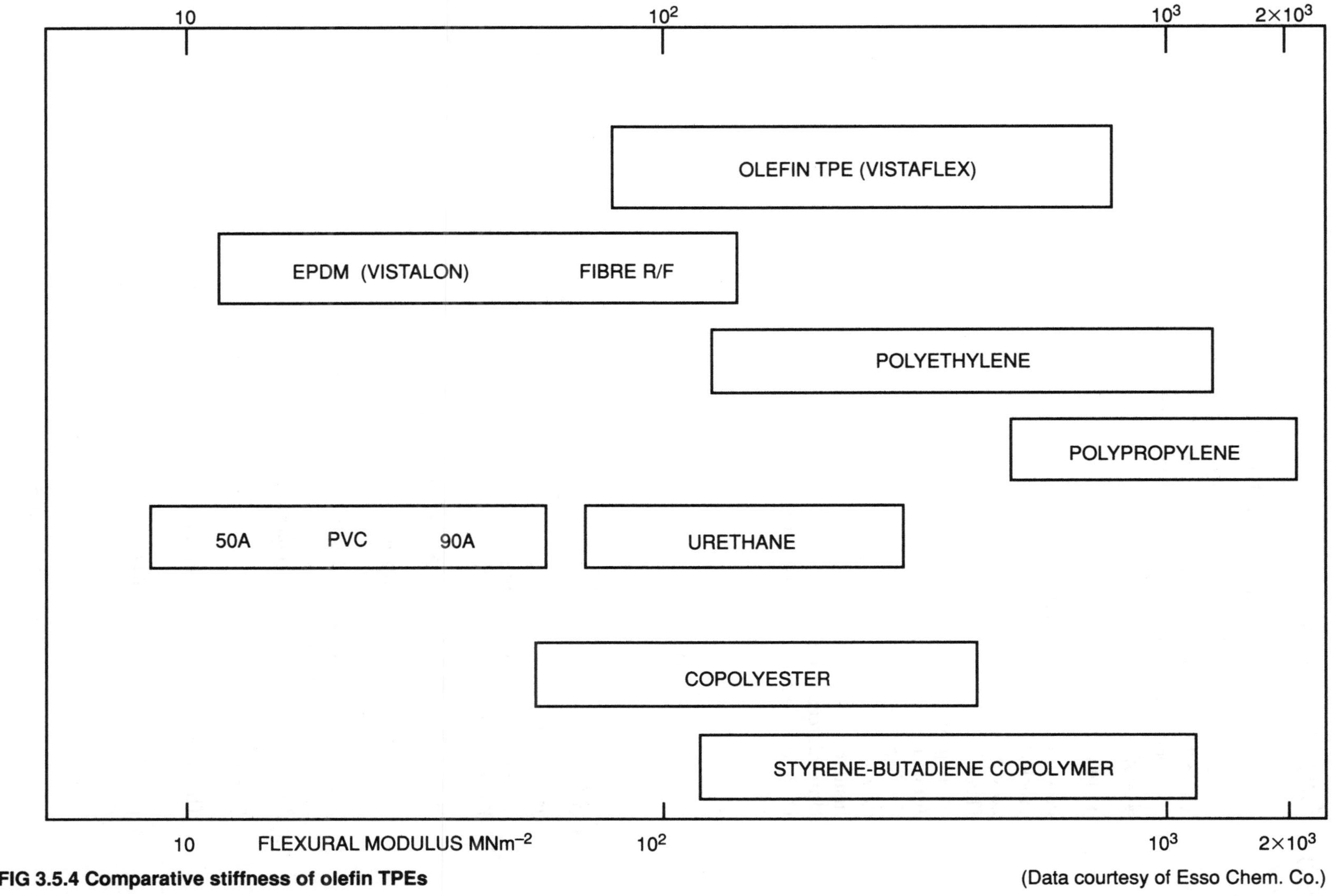

FIG 3.5.4 Comparative stiffness of olefin TPEs

Fig. 3.5.4

3.5.5 Copolyesters

3.5.5.1 General characteristics and properties

The advantages and limitations of copolyesters are listed in
Table 3.5.17 *Characteristics of copolyesters*

Copolyesters are similar to other TPEs in that they contain both hard segments for strength and soft ones for flexibility.

APPLICATIONS

Power transmission seals
Tracks for snow vehicles
Low pressure tyres
Flexible couplings
Small gears
Air compressor poppets
Cable junction covers
Covers and tubes for hydraulic hose
Insulation for spiral telephone cords
Unreinforced drive belts
Damping elements for agricultural machinery
Footballs

TYPICAL PROPERTIES

The more important properties of four different hardness grades of copolyester are listed in:
Table 3.5.18 *Typical properties of copolyester elastomer*

3.5.5.2 Mechanical properties

Thermoplastic copolyesters retain their flexibility well at low temperatures.
The creep properties of typical copolyesters are illustrated graphically in:
Fig 3.5.5 *Creep resistance of copolyesters*

3.5.5.3 Environmental resistance

The resistance of thermoplastic copolyesters to flex-cut growth at temperature extremes is given in:
Table 3.5.19 *Resistance to flex-cut growth at temperature extremes*
permeability to fuels is given in:
Table 3.5.20 *Permeability of thermoplastic copolyesters to fuel*

3.5.5.4 Trade names and suppliers

The suppliers of copolyesters and their familiar trade names are:

Supplier	*Trade name*
Akzo	Arnitel
Du Pont	Hytrel HT6 4275

TABLE 3.5.17 Characteristics of copolyesters

Advantages	*Limitations*
1. Flexibilised thermoplastics with good low and high temperature performance. 2. More stable to hydrolysis than TP polyurethanes based on aliphatic polyesters. 3. Processed by injection moulding, extrusion, thermoforming, blow moulding melt cast or flow moulding. 4. Harder than mainstream rubber compounds. 5. Useful temperature range –50–150°C. 6. Outstanding resistance to dynamic cycling. 7. Far superior to polyurethanes in load-bearing properties. 8. Outstanding resistance to flex-cut growth. 9. Good tear resistance. 10. Resistant to hydrocarbons, alcohol, ketones, esters, nitromethane, phosphate ester hydraulic fluids. Hardest grade highly resistant to permeation by isooctane/toluene fuels. 11. Good creep characteristics. 12. Resistance to permeation by non-polar molecules (excellent barrier properties towards motor fuels).	1. Glass filling, although greatly increasing tensile strength, flexural modulus and reducing mould shrinkage, decreases impact strength and severely decreases elongation. 2. Attacked by hot concentrated mineral acids and bases. 3. Soluble in phenols, cresols and some chlorinated solvents such as methylene chloride, chloroform and 1, 1, 2-trichloroethane. 4. Permeable by polar molecules.

TABLE 3.5.18 Typical properties of copolyester elastomers

	Property		Units				
	Durometer hardness classification			***40D***	***55D***	***63D***	***72D***
Mechanical	Tensile strength		MN/m^2	25.5	37.9	39.3	39.3
	Ultimate elongation		%	450	450	350	350
	25% Modulus or yield point		MN/m^2	7.6	13.8	17.2	26.2
	Stress at 25% Compression		MN/m^2	12.4	26.5	—	—
	Flexural Modulus		MN/m^2	48.3	207	345	517
	Resilience, Bashore		%	62	53	43	Not applicable
	Compression set 22 h, 70°C 25% deflection, 9.3 MN/m^2	ASTM D359A	%	60	56	Not applicable	Not applicable
		ASTM D359B	%	27	4	2	2
	Tear strength (ASTM D-624)	Die B	kN/m	110	164	185	Not applicable
		Die C	kN/m	122	158	149	Not applicable
	Resistance to flex-cut growth	Ross (pierced)	cycles to failure	$>3 \times 10^5$	$>3 \times 10^5$	2.8×10^5	Not applicable
		DeMattia (pierced)	cycles to failure	$>2 \times 10^5$	$>7 \times 10^4$	Not applicable	Not applicable
	Notched impact strength (Izod)	24°C	J/cm	>10.6 (no break)	>10.6 (no break)	>10.6 (no break)	2.1
		–40°C	J/cm	>10.6 (no break)	>10.6 (no break)	0.3	0.4
Physical	Specific gravity		—	1.17	1.20	1.22	1.25
	Linear expansion coefficient		10^{-5}/K	20	18	17	21
Thermal & Environmental	Taber abrasion CS17, 1000g		mg/1000 cyc.	3	5	8	13
	Heat distortion temperature	0.45 MN/m^2	°C	—	157	—	166
		1.8 MN/m^2	°C	—	43	—	69
	Brittleness temperature		°C	<–70	<–70	<–70	<–70
	Water absorption. 24 h		%	0.6	0.5	0.3	0.3

Data supplied by Du Pont.

Table 3.5.18

TABLE 3.5.19 Resistance to flex-cut growth at temperature extremes

	Ross-flex-pierced cycles at 5X growth	
	Room temp.	***−40°C***
Polyurethane, TP, Polyester 91A	300 000 (1X)	Immediate
Polyurethane, TD, Polyester 55D	84 000	Immediate
Polyurethane, TP, Polyester 90A	84 000–144 000	Immediate
Copolyester, 92A	300 000	12 000 (3X)
Copolyester 55D	300 000	12 000 (4X)

TABLE 3.5.20 Permeability of thermoplastic copolyesters to fuel

	Permeability ml/m² day (ASTM D-814)			
	ASTM Ref Fuel B		***ASTM Ref Fuel C***	
Sample thickness:	***0.625mm***	***1.0mm***	***0.625mm***	***1.0mm***
Copolyester (76% 4GT)	7.9	0.822	19.7	1.0
Nylon 11 plasticised	31.0	18.7	23.0	2.3
Copolyester (58% 4GT)	14.1	—	243	9.9
NBR (med-high nitrile)	401	—	955	—
CSM (43% chlorine)	880	—	2700	—
Polychloroprene	2270	—	5740	—

ASTM Ref. Fuel B = isooctane 70%, 31% toluene 30%
ASTM Ref. Fuel C = isooctane 50%, 31% toluene 50%
4GT = tetramethylterephthalate

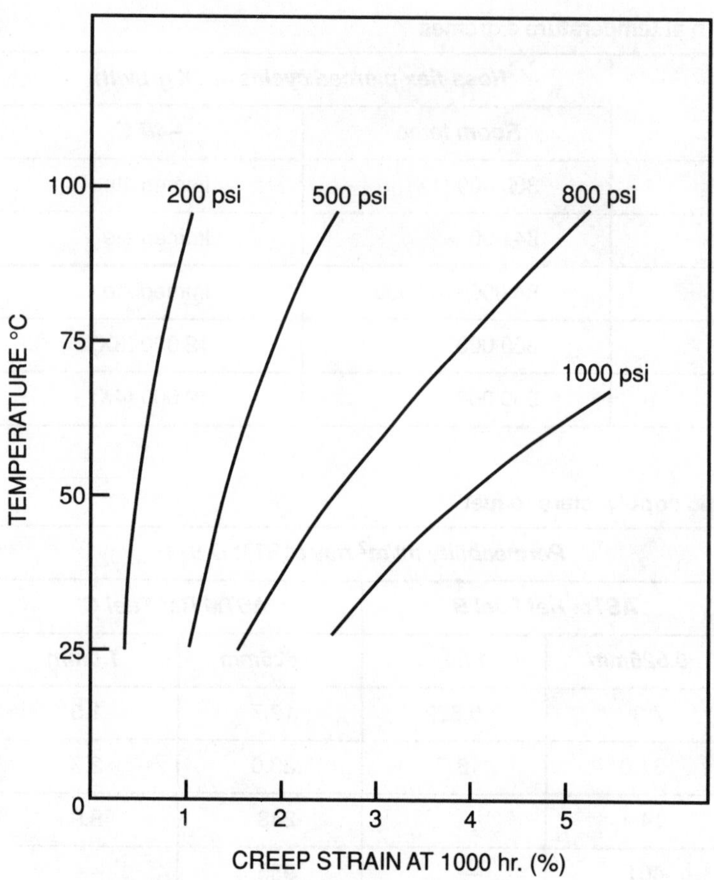

FIG 3.5.5 Creep resistance of copolyesters

Fig. 3.5.5

3.5.6 Polyamide-based thermoplastic elastomers

3.5.6.1 General characteristics and properties

As with other TPEs, the appropriate association of rigid and flexible blocks allows wide variations in properties, including hardness, flexibility, melting point, chemical resistance, density, affinity for water and antistatic behaviour.

The advantages and limitations of polyamide based thermoplastic elastomers are given in:
Table 3.5.21 *Characteristics of polyamide based thermoplastic elastomers*

APPLICATIONS

Transmission and universal joint bellows, windscreen wiper blades, pump membranes, catheters, flexible gear trains for motors and extremely thin coatings for optical fibres.

TYPICAL PROPERTIES

The more important properties of polyether block amides are listed in:
Table 3.5.22 *Typical properties of available 'Pebax' grades*

The temperature resistance of some grades is high. The moisture uptake can be varied from 1.2% to more than 100%, the latter being of significance in the textiles industry, as is the permanent antistatic nature of the hydrophilic grades.

3.5.6.2 Trade name and supplier

Atochem, trade name Pebax.

TABLE 3.5.21 Characteristics of polyamide based thermoplastic elastomers

Advantages	*Limitations*
Excellent impact toughness at low temperatures. Mechanical properties stable between –40 and +80°C. High elastic memory. Good chemical resistance. Good abrasion resistance.	High cost.

TABLE 3.5.22 Typical properties of available 'Pebax' grades

	Property	*Units*	*Grade identification*							
			5533 SNOO	*4033 SNOO*	*3533 SNOO*	*2533 SNOO*	*6312 MNOO*	*5512 MNOO*	*5562 MNOO*	*4011 RNOO*
Mechanical	Tensile break strength	MN/m^2	33	33	29	29	42	49	51	38
	Elongation (break)	%	510	620	650	680	410	530	510	530
	Tensile yield strength	MN/m^2	22	—	—	—	19	15	14	8
	Elongation (yield)	%	28	—	—	—	23	26	27	21
	Hardness	Shore	55D	40D	80A	70A	63D	55D	55D	40D
	Impact strength	kJ/m^2	No break, notched or unnotched at +20 °C or −40 °C.							
Physical	Density	kg/dm^3	1.01	1.01	1.01	1.01	1.11	1.10	1.06	1.14
	Melting point	°C	168	168	162	160	195	190	120	190
Thermal and environmental	Water absorption (24 h)	%	1.2	1.2	1.2	1.2	6.4	6.1	3.5	119
	Vicat softening point	°C	156	145	82	63	186	168	105	160
Processing	Processing	—	Injection moulding, rotational moulding, extrusion, extrusion blow moulding, spinning.				Injection moulding, rotational moulding only.			

3.5.7 Nylon 12 elastomer

3.5.7.1 General characteristics and properties

These materials are still at the developmental stage and rely upon the polymerisation of nylon 12 with proprietary components. Characteristics include high elastic memory, excellent resistance to hydrolysis and good chemical resistance.

SUGGESTED APPLICATIONS

Suggested applications include automotive bumper parts, cable sheathing, flexible couplings, tyres, seals, shoesoles, bicycle saddles.

TYPICAL PROPERTIES

Typical properties are given in:
Table 3.5.23 *Provisional properties of ELY-1256 elastomeric polyamide*

3.5.7.2 Suppliers and trade names

Supplier	*Trade name*
Emserwerke	ELY 1256
Chemische Werke-Huls	Nylon 12 Elastomer

TABLE 3.5.23 Provisional properties of ELY–1256 elastomeric polyamide

Property	*Unit*	*Value*
[a]Mould shrinkage	%	0.7–1.0
[a]Tensile yield strength	MN/m^2	20
[a]Elongation (yield)	%	45
[a]Tensile break strength	MN/m^2	31
[a]Elongation (break)	%	300
[a]Flexural strength	MN/m^2	19
[a]Elastic Modulus	MN/m^2	270
[a]Hardness	Shore D	56
Specific gravity		1.01
[b]Dielectric strength	kV/m	410×10^2
[b]Dielectric constant		5
[b]Dissipation factor		0.1

[a] Conditioned.
[b] At 50 Hz.
Data supplied by Grilon (UK) Ltd.

3–6

Foamed plastics

Contents

List of tables

3.6.1 Structural foams

3.6.1.1 General characteristics and properties

Structural foams are rigid foams comprising a low density cellular core with an integral solid skin approximating to the normal density of the base material. They are manufactured from a number of materials by a variety of techniques.

The advantages and limitations and the applications of the more important types are listed in:

Table 3.6.1 *Characteristics and applications of major structural foam materials*

The name 'structural foam' is a misnomer in relation to engineering connotations, but it is the only possible terminology that remotely describes this series of materials which are part way between solid plastics and the more conventional foams such as expanded polystyrene and non-integral skinned polyurethanes.

Density reductions of 10–40% can be achieved in most thermoplastics by the incorporation of a suitable blowing agent which can be introduced in at least four ways depending upon the material and the process.

The rigidity of a structural foam varies in proportion to its density and with the cube of its thickness, but it is very difficult to achieve significant density reductions below a thickness of 4.6 mm. With parts of equal thickness a solid plastics material is more rigid than a structural foam but the position is reversed when the components are of equal weight. This leads to advantages of material savings as a foamed article is said to require only one half to two thirds of the material weight of a similar unfoamed moulding.

The ability to mould large structural parts of complex shape in one production step leads to the acceptance of structural foams as substitutes for fabricated items made from wood, metal and other structural materials.

AVAILABLE TYPES

Structural foams are, with certain exceptions, classified into Commodity Resins and Engineering Resins as shown in:

Table 3.6.2 *Basis and range of available types of structural foam*

TYPICAL PROPERTIES

The more important properties of structural foam are listed in:

Table 3.6.3 *Typical properties of structural foams*

3.6.1.2 Processing

Structural foam can be processed in a variety of ways, many of which are specific to the manufacturer or supplier. The more important processes are listed in:

Table 3.6.4 *Processing of structural foams*

Injection moulding is one of the more important techniques for producing low weight parts.

INJECTION EQUIPMENT

Fast injection is essential for producing components in ABS foam. Special purpose structural foam injection machines and conventional in-line reciprocating screw injection machines may be used successfully. Screws having mixing heads and/or very high compression ratios are not recommended. Shut-off nozzles having a long landlength should be used.

Many grades of ABS do not need thorough pre-drying, although some may exhibit improved surface finish as a result.

TABLE 3.6.1 Characteristics and applications of major structural foam materials

Material	*Advantages*	*Limitations*	*Typical applications*
Polycarbonate	Toughest, most versatile foam. Rigidity and high strength. Dimensional stability. Temperature range −40–120°C. Good electrical properties.		Competes with metals and thermoset plastics—e.g. automotive industry. Bins and containers. Office machines.
Polyphenylene oxide	Hot water and detergent resistant. Good thermal insulator. Sound damping material. Best balance of properties at 30% density reduction.	Slightly inferior to polycarbonate foam on stiffness/weight, heat distortion temperature, modulus and tensile strength. Poorer impact strength.	FN5110 grade used for domestic appliance components (e.g. washing machines). General purpose grade used for business machine housings (FN 215).
Integral skin urethane foam (rigid) and isocyanurate	RIM process enables large dimensionally accurate parts. Isocyanurate grades have better thermal capability.	Inferior properties to above grades. Discolours without protective coating.	Structural parts. Wood replacements for decorative parts, mirror frames, chairs, etc. Ski components, tennis racquet handles. Automotive 'friendly' components. Aircraft components (mainly isocyanurate).
Polystyrene		Relatively low mechanical properties, heat resistance.	Moulding appliance and business machine housings. Automotive body components. Furniture. Window and door frames.
Polypropylene	Lowest density of any high-stiffness structural material. Good low temperature toughness. Good electrical resistivity. Can be expansion cast.	Mechanical properties and heat resistance only fair.	Truck fan shrouds. Portable kitchen and toilet parts. Handles for tools and implements. Dashboard insert panels. Extrusions to replace wood products.
ABS	Can have better strength-to-weight than metal. Good thermal and acoustic insulator.	Properties do not match polycarbonate or PPO.	Wood and metal replacement. Coextruded foam core pipe for water-well casings (replacing steel).
PBT thermoplastic polyester	High heat endurance. Good fatigue endurance. Low warpage. Good surface smoothness.		Appliance housings. Business machines. Medical equipment.

Table 3.6.1

TABLE 3.6.2 Basis and range of available types of structural foam

Classification	*Examples*	*Characteristics*
'Commodity resins'	Polystyrene Polyethylene (HD) Polypropylene ABS Nylon	Used where reasonable strength, stiffness, moisture resistance, sound absorption and weatherability are the main requirements. Not linearly elastic, hence difficult to predict stresses and deflections.
Engineering resins	Polycarbonate PPO Thermoplastic polyester	Higher stiffness and strength, which is retained at higher temperatures. Better thermal and electrical characteristics.
Polyurethane foam		Basic mechanical properties of rigid structural grades approximate to commodity resins. Generally treated separately because manufacturing process based on two component chemical reaction. Reaction injection moulding (RIM) and reinforced RIM have enabled large parts with excellent dimensional tolerance and surface finish to be produced.
Isocyanurates		Chemically distinct from polyurethanes. Superior thermal stability and flame resistance, which has prompted the development. Similar to urethanes in processing, mechanical properties and application areas, except generally higher modulus.

TABLE 3.6.3 Typical properties of structural foams

	Property		Units	'Commodity resins'			
				Polystyrene unfilled	*Polystyrene 20% glass*	*Polyethylene HD foam*	*Polypropylene HD foam*
Mechanical	Tensile strength		MN/m^2	4–14	35	8	11–30
	Tensile Elastic Modulus		GN/m^2		—	—	1
	Elongation		%	—	—	—	31 (SG = 0.65)
	Flexural strength		MN/m^2	—	—	—	25 (SG = 0.65)
	Flexural Modulus		GN/m^2	—	5.16	—	0.91 (SG = 0.65)
	Compressive strength		MN/m^2	0.46	—	8.9[c]	14
	Impact strength (Izod)		J/m notch	—	79	500	—
	Charpy impact		kJ/m^2	9 (S.G. = 0.9)	—	—	—
	Creep (300 h, 23°C, 7 MN/m^2)		%	—	—	—	—
Physical	Specific gravity[a]		—	0.16–0.9	0.84	0.56–0.83	0.56–0.81
	Thermal conductivity		W/mK	0.035–0.13	—	—	0.13
	Expansion coefficient		10^{-5} K^{-1}	—	—	7.5	7.5
Electrical	Dielectric strength		10^2 kV/m	—	—	—	—
	Dielectric constant	10^2 Hz	—	—	—	—	—
		10^6 Hz	—	1.28	—	—	—
	Dissipation factor	10^2 Hz	—	—	—	—	—
		10^6 Hz	—	0.00015	—	—	—
Thermal	Heat deflection temp. (1.8 MN/m^2)		°C	75[c]	88	110	—
	Flammability		(UL94)	—	—	—	—
Processing	Mould shrinkage		%	—	—	—	—

[a] Densities commonly used in commercial grades.
[b] Structural grades.
[c] Maximum service temperature.
[d] Parallel to flow.
[e] Perpendicular to flow.
[f] Flame retardant grade.
Lexan and Noryl are trademarks of the General Electric Co.

TABLE 3.6.3 Typical properties of structural foams—*continued*

'Commodity resins'				*'Engineering resins'*	
Polypropylene 20% glass	***ABS***	***ABS 20% glass***	***Nylon 15% glass***	***Polycarbonate (Lexan FL900)***	***Polycarbonate 10% glass***
21	14–28	50	70	36	81
—	—	—	—	1.86	—
2 (30%)	—	—	—	3–5	—
51 (30%)	—	—	—	69	—
2.8	1.8–2.1	5.2	4.5	2.26	2.76
—	23–42	—	—	35	—
185	44–96	185	170	—	470
—	—	—	—	No break	—
—	—	—	—	0.4	—
0.74	0.64–0.9	0.84	0.87	0.85–0.95	0.87
—	0.08–0.3	—	—	0.15	—
—	6.7–17.0	—	—	6.0–6.5	—
—	—	—	—	86	—
—	—	—	—	2.22	—
—	—	—	—	2.18	
—	—	—	—	0.0012	—
—	—	—	—	0.0061	—
72	68–98	100	200	126	132
—	V–0[f]	—	—	94V–0	—
—	0.5–0.8	—	—	0.5–0.8	—

Table 3.6.3—*continued*

TABLE 3.6.3 Typical properties of structural foams—*continued*

	Property	*Units*	*'Engineering resins'*			*Integral skin rigid polyurethane*[b]
			Polyphenylene oxide		*TP polyester*	
			Noryl FN215	*Noryl FN5110*		
Mechanical	Tensile strength	MN/m^2	22	32	70	20–60
	Tensile Elastic Modulus	GN/m^2	1.57	1.67	—	—
	Elongation	%	2	10	1.3	—
	Flexural strength	MN/m^2	44	57	123	—
	Flexural Modulus	GN/m^2	1.8	2.1	6.9	—
	Compressive strength	MN/m^2	34	35	78	35–103
	Impact strength (Izod)	J/m notch	—	—	185	—
	Charpy impact	kJ/m^2	19.6	19.6	—	—
	Creep (300 h, 23°C, 7 MN/m^2)	%	0.5	0.5	—	—
Physical	Specific gravity	—	0.8–0.9	0.85–0.90	1.2	0.66–1.12
	Thermal conductivity	W/mK	0.12	0.13	—	0.08
	Expansion coefficient	10^{-5} K^{-1}	7.5–8.5	7.0–8.0	—	7.2
Electrical	Dielectric strength	10^2 kV/m	85	88	61	—
	Dielectric constant, 10^2 Hz	—	2.27	—	—	—
	Dielectric constant, 10^6 Hz	—	2.18	—	—	—
	Dissipation factor, 10^2 Hz	—	0.0047	—	—	—
	Dissipation factor, 10^6 Hz	—	0.0039	—	—	—
Thermal	Heat deflection temp. (1.8 MN/m^2)	°C	83	110	171	up to 150[c]
	Flammability	(UL94)	94V–0	—	94V–0	—
Processing	Mould shrinkage	%	0.6–0.9	0.5–0.8	0.4–0.6[d] (0.6–1.0)[e]	—

[a] Densities commonly used in commercial grades.
[b] Structural grades.
[c] Maximum service temperature.
[d] Parallel to flow.
[e] Perpendicular to flow.
[f] Flame retardant grade.
Lexan and Noryl are trademarks of the General Electric Co.

Table 3.6.3—*continued*

TABLE 3.6.4 Processing of structural foams

Process type	*Description*	*Process*	*Characteristics*
Low pressure moulding	The melt is held at pressures above 14 MN/m^2 in the extruder or accumulator and is shot at this pressure. The mould is only part-filled and gas expands the plastic in the mould at 1.4–3.5 MN/m^2 pressure. As the foam contacts the mould a solid skin is formed. Collapsing gas bubbles result in 'swirl'. Aluminium moulds may be used, which are cheaper than steel and allow high production runs (up to 8×10^6).	Standard injection moulding	Chemical Blowing Agents. (CBA). Machine modified to prevent premature foaming. Limited by size and capacity to below 4.5 kg (Occasionally 9 kg).
		Union Carbide nitrogen process	Nitrogen gas injected into mould; and foams resins to fill mould. Sizes up to 55 kg or 3 × 3 m. Can also use chemical or physical blowing agents. Very common process.
		FMC process	Uses preheated moulds, permitting slower injection speeds resulting in a finished moulding with reduced material degradation. The part has time to cure and anneal instead of solidifying with the gas still pressurised; this improves surface finish and reduces internal stress. Invented by Foam Moulding Corporation.
		CBA 'fast-shot' process	Much larger shot capacity than injection machine; very high injection speed. Products have to be painted if smooth finish required.
		Expansion casting	For extra-large parts. Long cycle time, very poor surface finish.
High pressure moulding	Full shot of plastic injected into mould at 15–150 MN/m^2. After a solid skin forms the mould is enlarged to lower the pressure, permitting expansion and foaming of the plastic. Parts are 'swirl' free leading to accurate reproduction of surface detail. Sizes limited to 11 kg, smaller than most low pressure processes. Steel moulds necessary, which are also complex.	USM expanding mould process	Uses expanding moulds.
		Varitherm process	(BASF/Krauss Maffei) Mould heated to ease flow, then rapidly cooled after filling and pressurising cavity to give solid skin.
		TAD process	(Toshiba Asahi Dow) After skin has been formed selected parts are withdrawn to allow molten core to expand.
Sandwich and co-injection processes	Either the same resin, or two different resins may be employed for the resin and the core. These may have widely differing properties providing they are compatible. Surfaces with the smoothness of an injection moulded part are possible, but sizes are less than the maximum obtainable with the low pressure processes. It is possible to calculate the properties of any combination of skin and core which has been done for flexural modulus, ref. Ugine Kuhlmann (UDI-GDP Conference, Strasbourg, 1980).	Co-injection process (Battenfield)	Skin and core materials injected simultaneously. A small amount of skin material must flow before and after the core material for sealing. Moulds must be made of steel.
		ICI sandwich process	Single channel, sequential injection, expandable mould. Solid component is injected first, then foam component. Injection of solid component then fills out the space. Finally, mould expansion produces a foam core. The skin may become too thin in the gate area.
Reaction injection moulding (mainly for polyurethane)	Two liquid components are metered into a mixing chamber at relatively high pressure, then injected into the mould at atmospheric pressure. Pressure in the mould falls to between 0.2 and 0.5 MN/m^2. Part surface can be very smooth or can reproduce fine detail. The mixture is introduced at the bottom of the mould and air escapes at the top. Final foaming position of the mould is found by trial and error, so presses must be capable of tilting and rotation during trial stages. Since pressures are low, once the final position is found, the press need be no more than a self-contained mould with all the control hardware attached. Moulds are made of aluminium (except when fine details must be produced, when steel is used). This process is now so well established that integral skin polyurethanes (both rigid and semi-rigid) can be processed in shot capacities ranging from a few grams to 200 kg.		
Extrusion	The extrusion of foams can be carried out in conventional equipment processing a compression ratio of greater than 2 : 1 and L/D ratio of larger than 20 : 1. Temperature profiles rise rapidly with maximum head temperature whilst the melt temperature is determined by the polymer blowing agent decomposition temperature. Foaming occurs in the die.		

3.6.2 Non-structural foamed plastics

3.6.2.1 General characteristics and properties

Non-structural foams may be rigid or flexible. Most rigid foam is used for heat insulation. Flexible foams are used for insulation, cushioning and packaging.

The characteristics and applications of non-structural foamed plastics are listed in:
Table 3.6.5 *Characteristics of non-structural foamed plastics*

3.6.2.2 Polyurethane foams

The applications and advantages of the several grades of polyurethane foam are listed in:
Table 3.6.6 *Applications and advantageous properties of the various grades of polyurethane foams*

ADVANTAGES

Polyurethane foams offer great versatility because of the almost infinite range of properties that they can be formulated to possess, ranging from the lightweight flexible foams used for furniture cushions to the integral skin rigid foams used for structural applications.

Integral skin foams combine a solid skin with a foamed core. They are available in both rigid and flexible compositions.

LIMITATIONS

Due to the wide variations in properties possible by adjusting the formulations of polyurethane foams, there are very few properties that limit their use when compared with other cellular materials.

Early foams were extremely inflammable. The fire risk is complicated by the emission of toxic fumes, i.e. carbon monoxide and hydrogen cyanide. Flame resistant or retardant foams are available (e.g. based on polyisocyanurate) and these are acceptable under many building regulations.

Polyester based foams can suffer from degradation by hydrolysis, e.g. steam; polyether foams are more resistant in this respect. All foams suffer some degradation on exposure to sunlight. Degradation can also occur through the presence or contact with heavy metal impurities, particularly copper.

The other limitations concern risks involved in production. These involve the inhaling of dust when shaping foam slabstock, and the toxicity of the isocyanate by skin contact and inhalation of fumes.

TYPICAL PROPERTIES

Influence of chemical composition (general)

The characteristics of polyurethane foam, i.e. whether the foam is flexible, rigid or somewhere in between will depend upon the degree of cross-linking in the product. Basically this depends upon the type and functionality of the polyol, i.e. the number of reactable OH groups per molecule of polyol, and the type and functionality of the isocyanate. Toluene diisocyanate (TDI) and 4-4′ diphenylmethane diisocyanate (MDI) are the two most commonly used isocyanates. The higher the functionality, the higher the degree of cross-linking and the more rigid the product.

Density

The density is affected by the content of blowing agent, whether water or halocarbon. The density of halocarbon systems can also be affected by the catalyst level and the temperature of reaction. With moulded halocarbon blown products the density increases at the surface to give a skinned product. The effect of halocarbon content on the properties of rigid foam is detailed in:

Table 3.6.7 *Variation in properties of rigid polyether-based foams with trichlorofluoromethane content*

Flexible foams can be produced in a range of densities varying from below 32 to above 400 kg/m^3. Rigid foams can also be produced in a range of densities but the lower limit for most practical applications is around 96 kg/m^3. Self skinning foams cannot be produced at very low density.

Cell structure and cell size

The cell structure of polyurethane foams consists of a network of interlinked fibrils bridged by membranes. Where the structure is such that the membranes are ruptured this produces an 'open cell' foam. Flexible foams have a high ratio of open cells—the cell membranes may be ruptured by mechanical treatment, etc., to give 100% or near 100% open cell structure, e.g. Scott foam.

Rigid foams have a high closed cell content and foams of 95–100% closed cell content can be made.

The cell size of foams is controlled by addition of cell additives, e.g. silicones to the foam system.

Mechanical properties

The mechanical properties of flexible foams can vary greatly depending upon the polyol used. The variation of properties with density of typical flexible foams (polyester) are listed in:

Table 3.6.8 *Typical mechanical properties of flexible polyester urethane foams*

The corresponding data for rigid foams is listed in:

Table 3.6.9 *Typical mechanical properties of a rigid urethane foam*

Thermal properties

Rigid urethane foams have excellent insulation properties especially when physically (solvent) blown. Their properties with other competitive materials in:

Table 3.6.10 *Comparison of properties of various insulating materials*

Electrical properties

Rigid urethane foams are good insulators. Owing to the hydrophobic nature of the foam, electrical insulation is satisfactory even at high moisture contents. The foam can therefore be used for foam filling of electrical component housings and, as well as preventing tracking, provides protection against shock. Typical electrical properties of rigid urethane foams are listed in:

Table 3.6.11 *Typical electrical properties of rigid urethane foam panels*

Acoustical properties

Generally open cell foams give a very satisfactory absorption of sound, while closed cell foams do not. The situation is entirely different with transmission of sound as the primary factor is density, i.e. higher densities give better sound insulation. Hence urethane foams have limited suitability in this respect.

Solvent and chemical resistance properties

Generally the solvent and chemical resistance of polyurethane foams is good. They are resistant to most aqueous solutions except strong acids and alkalis that cause degradation. Solvents cause swelling to varying degrees and reduction of properties, but these are regained upon drying out of the solvent. The degree of swelling varies with the cross-link density, the higher the density (rigid foams) the less the swelling. Generally polyether foams are more resistant than polyester based foams.

Flammability

As previously mentioned polyurethane foams are inflammable and give off toxic gases when burning. This flammability increases with decreasing density due to the increase in surface area and air inclusion within cells. Modification to the chemical composition can be made to reduce the fire hazard and toxic gas emission.

ECONOMICS AND PROCESSING

The raw materials used in polyurethane foam manufacture are expensive, but, as very low densities are achieved, the actual cost per unit volume can be economical. Moulding of foam, either flexible or rigid, can utilise machinery of relatively low capital cost. Due to the low pressure exerted when moulding, very cheap, simple moulds may be employed, and this makes short production runs possible and economical. Rigid foams can be made to imitate wood very effectively, reproducing grain, etc., to a high degree. Although not as strong as wood, the material is becoming competitive, due to world shortages of timber, especially in intricate design work.

AVAILABLE FORMS

Polyurethane foams are available either as slabstock which may be cut to the desired shape, or as multicomponent systems that are reacted together to produce the desired product.

Processes employed for conversion

Foam slabstock of either the flexible or rigid variety is shaped by using either band saw or hot wire techniques. As only simple shapes are possible this has lead to the recent and rapidly expanding development of moulding to form. Mostly self-skinning foams are used in this application. The two or more components as necessary are metered, mixed and injected into a closed mould of the desired shape. The foaming takes place inside the mould to fill the cavity and after a suitable time for cure the moulding is ejected.

When rigid foams are used for insulation they are often injected and let to foam inside a cavity, i.e. the gap between the inside and outside case of a refrigerator or cavity walls in buildings. Insulation foams may also be applied by spray technique directly onto a component.

Automatic metering and dosing machines are available for dispensing the polyol and isocyanate components in the required regulated proportions and mixing as appropriate.

TABLE 3.6.5 Characteristics of non-structural foamed plastics

Type	*Characteristics*	*Applications*
Cellulose acetate	Excellent strength (i.e. 1 MN/m^2). Thermally stable. Vermin and fungus resistant. Temperature range –55–180°C. Good buoyancy. Good electrical properties.	Rib structure in fabricated lightweight reinforced plastics. Core material in sandwich structures. Tank floats, life floats, buoys. X-ray and electronic applications.
Ionomer	Tough, solvent resistant. Moulded parts have good impact toughness. Low-temperature flex-strength. Thermal and moisture barrier. Good energy absorber.	Packaging. Protective padding.
Phenolic	Rigid or flexible (newer types). Low corrosion. Low flammability, low smoking. May undergo flameless combustion (punking). Max. temperature 130–150°C. Good thermal insulator. Good dimensional stability (rigid). Low expansion coefficient. Inert to organic solvents. Adheres to wood, cardboard, cement.	Packaging (for energy absorption). Dry cell batteries (to contain H_2SO_4). Acoustic panels. Thermal insulation (where fire resistance and low smoke are important). Roofing. Sandwich panels. Truck and trailer insulation.
PVC foam	Ease of application in coating and laminating. Rigid or flexible.	Upholstered goods, cushioned pads, moulded shapes, gaskets.
Cross-linked polyethylene (XLPE)	Good impact energy absorber. Inert to most chemicals. Thermal and electrical insulator. Moisture vapour barrier. Resilient but tough for weight. Less than half the volume cost of natural rubber with 5 x life in some applications.	Safety padding for automotive interiors. Gaskets and expansion joints. Anticondensation liners. Weatherproof closure strips. Floor backing.
Extruded low density polyethylene	Chemical, solvent and moisture resistance. Toughness and flexibility over wide temperature range. 7–30 times lighter than solid LDPE.	Cushioning (commercial and military). Insulation. Marine flotation devices. Gaskets (noise and water seals)—e.g. automotive tail-light housings.
Compression moulded low density polypropylene	Tough. Heat and chemical resistant. Excellent insulator and energy absorber.	Auto air conditioner housing insulation. Packaging. Toys.
Extrudable, flexible polypropylene (sheet)—uniform matrix of small closed cell gas-filled bubbles	Density 0.01 g/cm^3. Outstanding toughness and strength. Resistant to tearing and flexing. Excellent water and chemical resistance.	Packaging or products requiring protection from impact and shock.

TABLE 3.6.5 Characteristics of non-structural foamed plastics—*continued*

Type	*Characteristics*	*Applications*
Low density expandable bead polystyrene (EPS)	Density 0.016 g/cm^3. Low thermal conductivity. Available as mouldings, sheet or extruded profiles.	Hot drink cups. Panels for cold storage containers.
Extruded foam sheet and board polystyrene	Density 0.08–0.32 g/cm^3.	Egg cartons, meat trays, produced by thermoforming.
Silicone	Broadest range of service temperatures of any elastomeric foam; –55–150°C continuous. Low toxicity.	Encapsulant for electronic equipment. Penetration seal for fire walls. Cushioning, sound deadening and thermal insulation in fire sensitive applications.
Urea formaldehyde	Cold setting foam, No water absorption. No load-bearing capacity (unless additives used). Low water vapour and gas permeability. Fungus resistant. Self-extinguishing.	Acoustic insulation by injection into hollow walls. Raising sunken vessels. Containing fires in coal mines. Protection of concrete from frost.
Polyurethanes	See Table 3.6.6.	
Polyester (thermosetting)	Can be rotational moulded, transfer moulded and reaction injection moulded. Densities lower than with syntactic foams can be obtained.	Reinforced with glass fibres, used in sandwich laminate structures for the construction of boat hulls, decks and moulds, swimming pools, building panels.

Table 3.6.5—*continued*

TABLE 3.6.6 Applications and advantageous properties of the various grades of polyurethane foams

Foam type	*Application*	*Advantageous properties*
Flexible	Furniture cushions, car, aircraft seats	Lighter weight, greater strength, ease of fabrication, dimensional stability and comfort.
	Fabric lining	Adds dimensional stability to fabric, high insulation, excellent hand and drape, outstanding crease and wrinkle resistance.
	Carpet underlay	Cushioning, non-skidding, no mat down, luxurious feel.
	Mattresses	Durability; freedom from odour; non-allergenic properties; ease of cleaning; resistance to dry cleaning; solvents, oils, perspiration, etc. Lightweight.
	Automobiles, e.g. underlays, roof insulation, padding	Good insulation, crash protection.
	Air filters	Excellent filtration
	Packaging	Excellent shock protection, easy processing and shaping to fit article.
Integral skin flexible and semi-flexible mouldings	Automobile applications, e.g. safety pads, armrests, floor mats, sun visors, weather strips.	Ease of moulding in complicated shapes, no waste, excellent safety protection.
Rigid	Insulation panels. Foamed in place insulation, e.g. cavity walls, refrigerators. Spray applied insulation.	Excellent thermal insulation, excellent weathering and thermal resistance, ease of processing and application, lightweight, contributes to structural strength, low moisture pick-up, solvent resistance, excellent sound insulation.
	Sandwich construction	High strength, low weight.
	Chair shells	Can be cheaply moulded to complicated shapes; high strength, low weight.
	Flotation equipment	High closed cell content, ability to be foamed in place, low density.
Integral skin rigid	Structural parts Wood replacement for decorative parts, mirror frames, chairs.	High strength/weight ratio, freedom of design. Ease of moulding, cheap equipment, complicated designs possible.

TABLE 3.6.7 Variation in properties of rigid polyether-based foams with trichlorofluoromethane content

Trichlorofluoromethane content % of total weight	41	25	19.5	15	12	4.0
Density kg/m^3	13.3	17.65	20.8	27.3	33.7	121.9
Compressive strength at yield point N/m^2	3.95×10^4	8.27×10^4	16.1×10^4	23.7×10^4	33.4×10^4	138×10^4
Deflection at yield point %	12.7	9.0	9.0	7.5	9.0	—
% Closed cells	85–90					
Thermal conductivity W/mK	0.026	0.0231	0.0202	0.0188	0.0159	0.0209

TABLE 3.6.8 Typical mechanical properties of flexible polyester urethane foams

Foam density kg/m^3	24	32	48	64
Tensile strength N/m^2	17.2×10^4	17.0×10^4 – 21×10^4	17×10^4 – 21×10^4	21×10^4 – 27×10^4
Tensile elongation %	250–300	250–300	300–400	300–400
Tensile elongation set %	25	6–13	4–13	3–10
Tear strength kg/m	70–100	70–100	70–100	90–125
Compression set, 70°C %	10–15	2–10	2–6	2–6

TABLE 3.6.9 Typical mechanical properties of a rigid urethane foam

Foam density kg/m³	25	30	40	50
Tensile strength N/m²	15 × 10^4	28 × 10^4	50 × 10^4	73 ×10^4
Elongation at break %	9–17	8.5–16	8.0–14.5	7.5–13.5
Compressive strength N/m²	9 × 10^4	15 × 10^4	25 × 10^4	34 × 10^4
Compressive reformation %	8–11	7–9	5–8	4.5–7
Flexural strength N/m²	22 × 10^4	34 × 10^4	56 × 10^4	79 × 10^4
Modulus of Elasticity N/m² 22°C/5% RH	2.0 × 10^6– 4.3 × 10^6	3.3 × 10^6– 5.7 × 10^6	5.8 × 10^6– 8.4 × 10^6	8.2 × 10^6– 11.0 × 10^6

TABLE 3.6.10 Comparison of properties of various insulating materials

Material	***Cork***	***Polystyrene bead foam***	***Glass-fibre***	***CO_2 blown rigid polyurethane foam***	***Solvent blown rigid polyurethane foam***
Density kg/m³	104–120	19.2	8–192	32	32
Water absorption kg/m³	low	1.3-1.7	high	0.48	0.48
Compressive strength N/m²	high	8.3 × 10^4	low	33.2 × 10^4	33.2 × 10^4
Thermal conductivity W/mK	0.0375	0.0361	0.0332– 0.0433	0.0361	0.0159

TABLE 3.6.11 Typical electrical properties of a rigid urethane foam panels[a]

Test	*Test conditions*	*Value measured*	*Influential parameters*
Dielectric strength	50 Hz in air and oil disc type electrodes 55 m in diam. 0.5 k/V sec voltage increase.	75–100 kV/cm	No influence of density, cell orientation (foaming direction) and previous conditioning of the test specimens and high humidity level are noticeable within the variations effected.
Volume resistivity	At 1000 V, 1-min, value with compressed electrodes	10^{13}–10^{14} ohms cm	
Surface resistance	At 1000 V, 1-min, value with reeded-edge electrodes	10^{13}–10^{14} ohms	
Dielectric constant	50 Hz, 500 V with compressed electrodes	1.1–1.15	The higher value refers to the higher density.
Dissipation factor	50 Hz, 500 V with compressed electrodes	0.02–0.045	Coordination of the data measured with the densities and humidity levels tested (conditioning for 14 days, 80% RH, 22°C) is possible; e.g. a 50% increase in the lowest value applicable to a density of 35 kg/m³, with a density increase to 65 kg/m³, or after conditioning at the above mentioned humidity level.

[a]10 mm thick, for densities from 35 to 65 kg/m³ and 55% RH at room temperature.

Table 3.6.9, Table 3.6.10 and Table 3.6.11

3.6.3 Syntactic foams

The introduction of bubbles of glass, ceramic or plastic into a matrix (either a resin or an inorganic material) results in a product known as a syntactic foam. Glass microballoons are used to fill epoxy casting resins (SG 0.78–0.88) for use in embedding circuitry in deep-water floats, and to fill polyester resins for synthetic wood. Microballoon-filled polyurethane is used in aerospace and hydrospace as a flexible, buoyant or lightweight material.

Glass microballoons may be used in combination with blowing agents where conventional fillers (e.g. short fibres, glass beads, mineral fillers) would reduce the effectiveness of the weight reduction achieved by foaming. Such foam has a higher strength to weight ratio than the gas-blown foam containing no microballoons.

Microballoons cannot be used in conventional injection moulding equipment as the higher pressures (compared with low-pressure foam moulding) would cause the sphere to collapse.

Silica microballoons are used with epoxy resins in place of glass as lightweight, low loss plastic dielectrics (e.g. SG 0.78; dielectrical constant 2.9; loss tangent 0.01 at 10^6 Hz).

Syntactic foam is good thermal insulating material; because it resists compression and water intrusion even under high pressure, it is a good choice for marine insulation.

A typical range of its composition and serviceability is listed in:

Table 3.6.12 *Typical properties of a range of syntactic foams*

Being lightweight materials with high stiffness, rigid syntactic foams conduct sound at very high speed; this trait, combined with low damping or loss, make for a number of acoustic uses. Lenses made of syntactic foam serve to focus or collimate sonar signals. Syntactic foam windows enclose and protect hydrophones without impairing their efficiency. Elastomeric based foams are more glossy and serve as damping materials or acoustic barriers.

Elastomeric resins, such as polyurethane, produce flexible syntactic foams. These are highly resistant to hydrostatic compression and are used in aerospace applications where low density is important, and in hydrospace as a flexible buoyancy material.

Multifoam systems, which are foams containing both gas bubbles and hollow spheres, can be formulated to have a higher strength-to-weight ratio than gas-blown foam, but lower density than 'conventional' syntactic foams.

TABLE 3.6.12 Typical properties of a range of syntactic foams

		Units	Cellular foam/MS	Flexible systems		Aggregate systems				
				UG	DB	PG	FJ	RJ	FL	RG
Principal characteristics		—	Lowest density.	Do-it-yourself. Flexible.	More flexible than UG, yet resists compression.	Do-it-yourself large sections.	Low density.	High performance version of FJ.	General purpose.	High performance version of FL.
Forms available		—	Factory-made blocks only.	Do-it-yourself kits	Factory-made sheets only.	Do-it-yourself kits.	Factory-made shapes only.			
Composition	Plastic matrix	—	Epoxy	Polyurethane	Polyurethane	Polyester	Epoxy	Hydrocarbon	Epoxy	Hydrocarbon
	Glass[c] microballoons	—	None	Yes	Yes	Yes	Yes	Yes	Yes	Yes
	'Eccospheres'[c]	—	Polystyrene	None	None	EP	EP & HG	EP & HB	EP	EP
Density (nominal)		kg/dm^3	0.192	0.577	0.721	0.449–0.513	0.320–0.385	0.288–0.385	0.417–0.513	0.385–0.513
Service depth[a]		m	61	183	458	610–915	305–610	305–915	610–1525	610–1830
Service pressure[a]		MN/m^2	0.61	1.84	4.6	62–93	3.1–6.1	3.1–9.2	6.1–15.3	6.1–18.4
Crush depth[b]		m	92	305	760	915–1525	460–915	460–1525	915–2440	915–3040
Crush pressure[b]		MN/m^2	0.92	3.06	7.7	6.3–15.3	4.6–9.2	4.6–15.3	9.2–24.5	9.2–30.6

Data supplied by Emerson-Cuming.

Notes: [a] All buoyancy materials absorb some water when subjected to hydrostatic pressure. The service depths given are for reference only and should not be construed as depth ratings.

[b] Crush depth is determined in a manner similar to ASTM D2736 Method B.

[c] Glass microballoons are very small, with average diameters under 100 μm. 'Eccospheres' are much larger with typical diameters 10–15 mm; they are called 'macrospheres' and are made of fibre-reinforced plastic. Addition of these to syntactic foam reduces density and cost.

TABLE 3.6.12 Typical properties of a range of syntactic foams—*continued*

Solid syntactic foams		*High performance syntactic foams*					
EF	***1421***	***TI***	***TG***	***EX***	***EL***	***TL***	***EG***
Do-it-yourself high performance.	Deep sea encapsulant.	Can be moulded to custom shapes.	Lowest density.	High strength/ weight ratio.	General purpose to high performance.	Available in large blocks.	Can be moulded to customer shape.
Do-it-yourself kits.		Custom shapes.	Blocks only.	Blocks only.	Blocks only.	Blocks only.	Custom shapes.
Epoxy	Epoxy	Hydrocarbon	SEE TECHNICAL BULLETINS				
Yes	Yes	Yes	Yes	Yes	Yes	Yes	Yes
None	None	None	None	None	None	None	None
0.609–0.641	0.673	0.417–0.609	0.353–0.417	0.417–0.481	0.481–0.641	0.417–0.609	0.481–0.641
1830–3660	4575	2050–10975	2050–5030	2050–6860	3780–12350	2050–10975	3780–12350
18.4–36.8	46.0	20.7–110.3	20.7–50.5	20.7–69.0	37.9–124.1	20.7–110.3	37.9–124.1
3050–6100	7625	5490–13720	3090–6870	5490–8300	5490–16460	5490–13720	5490–16460
30.6–61.3	76.6	55.2–137.9	31.0–69.0	55.2–82.7	55.2–165.5	55.2–137.9	55.2–165.5
	Cost	Special quotation	Medium	Medium	Low–high	Low–high	Special quotation

Table 3.6.12—*continued*

3–7

Plastics film and sheet

Contents

List of tables

List of figures

3.7.1 General characteristics and properties

The advantages, limitations and applications of plastics film and sheet are listed by material in:
Table 3.7.1 *Characteristics of plastics film and sheet*

TYPICAL PROPERTIES

The more important properties of plastics film and sheet are listed according to material in:
Table 3.7.2 *Typical properties of plastics film and sheet*
An important criterion in the choice of plastics film is its capacity to retain moisture which may or may not be desirable. This is shown diagrammatically in:
Fig 3.7.1 *Dependence of the water permeance of various polymers on temperature*

ENVIRONMENTAL RESISTANCE

The suitability for given environment of a film made from a specific polymer can be assessed by means of:
Fig 3.7.2 *Environmental resistance of plastics films*

TABLE 3.7.1 Characteristics of plastics films and sheet

Type	*Advantages*	*Limitations*	*Typical applications*
ABS See also Section 3.2.8	High abuse resistance and rigidity with good resistance to almost all household chemicals.	For outdoor exposure protection by a lamination with acrylic is advised. Flame retardancy and compliance with regulatory requirements may be lost with the addition of pigments.	Refrigeration door and tank liners, margarine tubs, luggage, furniture, automotive and recreational.
Acrylics See also Section 3.2.5	Good weatherability. Excellent optical clarity. Can be laminated to a wide range of substrates including wood, fibreboard, metal, ABS, polycarbonate.	More expensive than PVC and polystyrene, Combustible although some sheet grades self-extinguishing. Low scratch resistance. Flexible grades not available. Crazes under long-term stress.	**Film** Surface protection of metal, wood, plastics from outdoor or UV exposure. **Sheet** Sign industry, glazing, aircraft canopies and undershields.
Cellulosics See also Section 3.2.4	**Cellulose acetate** Good scuff and grease resistance. Good electrical insulation. Moderate water–vapour transmission. Good clarity.	High gas permeability. Optical properties inferior to acrylics. High dissipation factor. Burns slowly.	**Cellulose acetate** Widely used in packaging, particularly blisters, skins and transparent rigid containers. Also electrical insulators, shields, lenses, eyeglass frames, microfilm.
	CAB is very tough and some grades offer UV resistance and weather durability.		**CAB** Printed and formed signs, skylights, etc., and as above.
	Cellulose propionate shows good properties over wide range of temperature and humidity.		**CP** Similar applications to CA and CAB.
	Cellulose triacetate is highly resistant to greases and solvents as well as moisture. Good resistance to heat and high dielectric strength.	Does not vacuum form well.	**Cellulose triacetate** Visual aids, graphic arts, photographic albums, protective folders, photographic film base, high gloss release sheet, electrical coil forms and layer insulation.
Fluoroplastics See also Section 3.2.7	**General** Excellent chemical, physical, thermal and electrical properties.	Expensive, use limited to speciality applications.	
	CTFE Lowest water vapour transmission of any transparent film. Non-flammable, compatible with liquid oxygen. Can be heat-sealed by variety of techniques, and laminated to polyethylene, polyester, nylon, PVC, paper, foil and copper. Suitable in contact with food.	Surfaces for laminating (in most cases) must be specially prepared.	Packaging: strips and blister packs for pharmaceutical products—aerospace hardware. Non-packaging: cap liners, pressure sensitive tape, electroluminescent panels, flexible cables and circuits.

Table 3.7.1

TABLE 3.7.1 Characteristics of plastics films and sheet—*continued*

	FEP Service temperature –250–+200°C. Good anti-stick properties. Does not burn. Suitable for use in contact with food.		Conveyor belt surfacing, high temperature release sheeting for reinforced plastics, heat-shrinkable industrial roll covers. High performance electrical applications, e.g. computer and aerospace electronics.
	ETFE, ECTFE/ETFE, ECTFE Good tensile strength, flex life, tear strength and cut-through resistance—better than other fluoroplastics. Radiation resistance.	Electrical properties similar to other fluoroplastics (not a limitation as such).	Demanding high temperature electrical applications—insulation of computer wire, oil-well logging cable, nuclear power cable.
	PVF Good weathering properties. Good abrasion resistance. Good colour retention. Can be laminated to plywood, PVC, hardboard, polyester, galvanised steel, metal foils.		Aircraft cabin interiors, lighting panels, wall coverings, building applications.
Nylons See also Section 3.2.1	High tensile strength, elongation, impact strength good. Good abrasion resistance. Can be sealed by all conventional methods. Can be coated with PVDC to enhance oxygen barrier properties.	Barrier and physical properties vary with moisture content.	FDA approved for packaging, principally vacuum packaging. May be combined with ionomer or polyethylene. Heat stabilised grades used in sterilising and aerospace applications where temperatures of 200°C are maintained for some hours.
Polycarbonate See also Section 3.2.13	Impact strength. Heat resistance. Dimensional stability. Transparency plus complete selection of translucent grades. Ease of fabrication—normally joined by adhesives or solvent welding.	Necessary to remove equilibrium moisture content before elevated temperature operation. Heat-sealing and ultrasonic bonding only suitable for thinner gauges.	***Glazing and construction*** Vandal-proof and burglar-proof windows, machine guards, safety helmets. Signs—higher initial material cost may be offset by ease of maintenance. Sterilised food trays, etc. Aircraft windshields and canopies.
Thermoplastic polyester (PETP) See also Section 3.2.18	Good strength, clarity and temperature performance. Low shrinkage at elevated temperature. Good long-term ageing.		Defect-free films used for magnetic tapes, photographic base, insulation, reprographic products. Flexible printed circuits, slot liners and phase insulation, wire and cable insulation, electronic capacitors (very thin gauges). Packaging and retort pouches. Protective covers for books. Release films for GRP manufacture.
Polysulphones See also Section 3.2.16	High resistance to heat and combustion. Low smoke emission. High impact resistance (sheet). High tear initiation and elongation (film). Dimensional stability. Fair to good chemical resistance. Transparent, coloured transparent, or opaque. Suitable for heat and fire safety, food service, medical service. Can be vacuum-metallised.	Expensive—limits use to speciality applications.	Electrical—integrated circuits, portable calculators. Class H pressure sensitive electrical tape. Decorative lighting panels. Aerospace/Aircraft—lightweight expanded core acoustical structural panels. Medical—sterilisable respirators, bags and containers.

Table 3.7.1—*continued*

TABLE 3.7.1 Characteristics of plastics films and sheet—*continued*

Type	*Advantages*	*Limitations*	*Typical applications*
Polyethyene See also Section 3.2.9	Large range of grades offered characterised by melt index. Major advantage is low cost versatility.		Packing and protective barrier material HDPE—co-extruded film, institutional disposable items, panelling.
Polypropylene See also Section 3.2.10	High melting point. Chemical and grease resistance. Higher stiffness than medium density polyethylene. May be orientated to improve performance characteristics—modulus of cellophane and functional to –80°C.	Poor low temperature impact strength. Poor UV resistance of non-pigmented films.	Retortable pouch and hospital packaging. Wrapping of power cables. Multiwall bag liners. Overwrapping bread. Orientated film employed for shrink films, snack packaging. Electrical applications include high voltage capacitors, pressure sensitive tapes.
Polyimide See also Section 3.3.6	Tough, flexible. Highly resistant to combustion. Composite with FEP available for applications requiring heat sealing. Good dimensional stability.	Most grades not heat sealable.	Insulation of magnet wire requiring subsequent winding. Printed circuits, magnetic tape. Laminated with copper for high performance flexible printed circuits with excellent flexural properties and fatigue life.
Polystyrene See also Section 3.2.8	***Sheet*** Sheet properties range from clear but brittle to translucent high impact. Low cost (disposability).	Not recommended for outdoor use.	***Sheet*** Food packaging; formed trays for vending machines. Medical disposable devices. Furniture components. Luggage and instrument cases.
	Oriented film and sheet Less brittle than standard sheet.	Film has high gas permeability and intermediate water vapour transmission.	Packaging fresh produce. Wallet transparent wings. Sign printing.
Polyurethane See also Section 3.5.1	Main advantage in mechanical properties. Extreme toughness. Good abrasion and impact resistance.		Textile laminations to increase wear resistance, tear resistance, strength weight. Packaging oil and engine additives (blister packs). Thicker sheets used as conveyor belting and can be laminated to heavy duty fabric.
PVC See also Section 3.2.3	Physical and chemical properties can be varied widely. Very large number of end-uses.		***Cast*** Thinnest films used for packaging (food and other). Also certain grades for medical and surface tapes. ***Extruded*** Packaging, also tinsel, wood surfacing. ***Calendered*** Sheeting for construction, agriculture, industrial products, medical and many other application areas. Coated fabrics, wall coverings and other surface finishes. Stationery supplies, flexible hoses and ducts. As food storage bags the use of PVC film has been questioned; the variety of processes and compounding ingredients available suggest this problem may be overcome.
Polybutylene See also Section 3.2.11	Excellent creep strength, puncture resistance and tear strength.		High performance applications—packaging food, including boil-in-bag.

TABLE 3.7.2 Typical properties of plastics film and sheet

Property		*ASTM Test method*	*Units*	*ABS*	*Cellulosics*	*Regenerated cellulose (cellophane)*
Tensile strength		D882	MN/m^2	35–70	28–112	50–126
Elongation		D882	%	10–50	10–100	10–50
Bursting strength (1 mil)[b]		D774	Mullen pts	—	20–85	20–65
Tearing strength, initial		D1922	N/m	—	0.175–140	19–60
Folding endurance		D2176	—	—	500–4000[a]	—
Gas permeability	CO_2	D1434	[a]	9–12	51.6–360	0.024–0.36
	N_2			0.3–0.6	1.8–36	0.030–0.096
	O_2			3–4.2	7–120	0.012–0.30
Rate of water vapour transmission	23°C	E96 (E)	g-mm	12–16	4–16	0.2–53
	38°C	E96 (E)	m^{-2} 24 hr·			
Dielectric strength		D149	10^2 kV/m	140–160	600–2000	800–1000
Dielectric constant at 10^6 Hz		D150	—	2.5–3.0	2.5–3.8	3.2[a]
Dissipation factor at 10^6 Hz		D150	—	0.01–0.06	0.01–0.06	0.015[a]
Maximum service temperature		D150	°C	90–105	80–95[b]	120
Minimum service temperature		D150	°C	—	–25 to –35	–55
Heat sealing temperature range		D759	°C	—	175–235	85–175
FDA approved grades		—	—	Yes	Yes	Yes
Methods of processing		—	—	Extrusion, calendering	Extrusion, casting	Extrusion, into bath
Forms available		—	—	Rolls, sheets	Rolls, sheets, tubes, tapes	Rolls, sheets, ribbons
Thickness range			mm	0.25–12.5	0.02–8.9	0.02–0.04
Maximum width			m	2.7	1.06–1.70	1.16–1.70
Area factor (area covered by 1 kg film 1mm thick)			m^2 × 10^6	0.965	0.76–0.87	0.67–0.71
[a]1 cm^3 cm^{-2} s^{-1} (cm Hg)$^{-1}$ ≡ 7.6 × 10^{-8} m^4 N^{-1} s^{-1} [b]1 mil ≡ 0.001 in 1 mm ≡ 40 mil			Remarks		[a] Ethyl cellulose has much better folding endurance [b] Cellulose triacetate 150–250°C	[a] at 10^3 Hz

Table 3.7.2

TABLE 3.7.2 Typical properties of plastics film and sheet—*continued*

Fluoroplastics					*Polyurethane*	*Acrylic PMMA*
PTFE	*FEP*	*PVF*	*ETFE*	*E-CTFE*		
10–32	18–21	50–125	50–56	56–70	35–85	35–60
100–350	300	115–250	300	150–250	200–700	4–75
—	11	19–70	—	—	—	14–40
—	105	174–245	—	—	61–105	59–66
—	—	Very high	—	250,000	—	1,500
—	100	0.66	15	6.6	27–99	1
—	19	0.015	1.18	0.6	2.4–7.2	1
—	45	0.18	6	1.5	4.5–20	1
—	—	—	—	—	—	—
—	0.16	1.28	0.65	0.2	16–30	0.5
170	2800	1400	1400	2000–2400	240–520	160
2.0–2.1	2.0–2.05	7.4	2.6	2.5	5.5–7.1	3–3.5
0.0002	0.0004	1.6	0.005	0.013	0.04–0.06	0.03–0.04
260	225–270	105–120	—	150–205	90	65–90
–250	–250	–70	—	–60	<–70	Good
—	280–370	205–215	270	245–260	135–185	95–175
Yes	Yes	No	—	No	—	No
Skiving, casting, extrusion	Extrusion	Extrusion	Extrusion	Extrusion	Calendering, casting, extrusion	Extrusion
Sheets, tapes, tubing	Rolls, sheets, tubes	Rolls	Rolls, sheets, tubes	Rolls, sheets, tubes	Rolls, sheets, tubes	Rolls, sheets
up to 3.18	0.013–2.4	0.013–0.1	—	0.013–2.25	>0.013	0.05–0.25
1.2	1.2	3.5	1.2	1.3	1.7	1.1–2.7
0.45	0.47	0.64–0.72	0.59	0.60	0.81–0.90	0.79–0.88

Table 3.7.2—*continued*

TABLE 3.7.2 Typical properties of plastics film and sheet—*continued*

Property		*ASTM Test method*	*Units*	*Polycarbonate*	*Polyester PETP*
Tensile strength		D882	MN/m^2	60–62	140–280
Elongation		D882	%	85–105	60–165
Bursting strength (1 mil)[b]		D774	Mullen pts	25–35 or No Brk	55–80
Tearing strength, initial		D1922	N/m	274–2012	175–525
Folding endurance		D2176	—	250–400	>100 000
Gas permeability	CO_2	D1434	[a]	63	0.9–1.5
	N_2			3	0.04–0.06
	O_2			18	0.18–0.24
Rate of water vapour transmission	23°C	E96 (E)	g-mm	—	—
	38°C	E96 (E)	m^{-2} 24 hr·	6.7	—
Dielectric strength		D149	10^2 kV/m	600	3000
Dielectric constant at 10^6 Hz		D150	—	2.93	3.0
Dissipation factor at 10^6 Hz		D150	—	0.01	0.016
Maximum service temperature		D150	°C	130	150
Minimum service temperature		D150	°C	–100	–70
Heat sealing temperature range		D759	°C	205–220	215–230
FDA approved grades		—	—	Yes	Yes
Methods of processing		—	—	Casting, extrusion	Extrusion, biaxial orientation
Forms available		—	—	Rolls, sheets	Rolls, sheets, tapes, tubes
Thickness range			mm	0.006–12.5	0.003–0.36
Maximum width			m	1.1 (cast) 1.30 (extr).	1.7–3.0
Area factor (area covered by 1 kg film 1mm thick)			m^2 × 10^6	0.83	0.71–0.72
[a]1 cm^3 cm^{-2} s^{-1} (cm Hg)$^{-1}$ ≡ 7.6 × 10^{-8} m^4 N^{-1} s^{-1} [b]1 mil ≡ 0.001 in 1 mm ≡ 40 mil			Remarks		

Table 3.7.2—*continued*

TABLE 3.7.2 Typical properties of plastics film and sheet—*continued*

Polyamides				*Polyethylene and copolymers*		
Nylon 6	***Nylon 6/6***	***Nylon 11***	***Nylon 12***	***LDPE***	***MDPE***	***HDPE***
60–125	60–85	60–80	60–85	10–20	14–25	18–43
250–550	200	250–400	290–330	100–700	50–650	10–650
Elongates	No break	60	—	10–12	—	—
175–210	—	—	175–190	11–100	—	—
250 000	Excellent	Excellent	—	Very high	High	Good
0.6–0.7	0.54	9	9.3–20	162	60–150	35
0.054	0.04	0.2	0.7–1.1	11	5–19	2.5
0.16	0.3	2	3.1–5.5	30	15–32	11
—	—	—	—	—	—	—
—	—	—	25	0.512	—	—
520–600	340	340	400	190	200	200
3.4 (dry)	3.4	3.1	3.1	2.2	2.2	2.3
—	0.004	0.03	0.0003	0.0003	0.0003	0.0005
95–205	—	95–150	75–125	85–95	105	120
–70	—	—	–75	–55	–55	–45
195–220	240–250	165–200	160–195	120–205	120–205	135–205
Yes	Yes	Yes	Pending	Yes	Yes	Yes
Extrusion	Extrusion	Extrusion	Extrusion	Extrusion	Extrusion	Extrusion
Rolls, tubes,	Rolls, tubes	Rolls, tubes, sheets	Rolls, sheets, tubes	Rolls, sheets tapes, tubes	Rolls, sheets tapes, tubes	Rolls, sheets tapes, tubes
0.013–0.8	0.013–0.03	0.025–1.3	0.013–1.5	>0.008	>0.008	>0.01
2.1	0.76	0.71	—	12.3	6.9	1.5
0.88	0.88	0.97	0.99	1.08–1.10	1.06–1.08	1.04–1.06
				Density 0.91–0.925	Density 0.976–0.940	Density 0.941–0.965

Table 3.7.2—*continued*

TABLE 3.7.2 Typical properties of plastics film and sheet—*continued*

Property		*ASTM Test method*	*Units*	*Polyetheylene and copolymers*		
				UHMWPE	*Ionomer*	*EVA*
Tensile strength		D882	MN/m^2	21–40	35	7–20
Elongation		D882	%	300	250–450	400–800
Bursting strength (1 mil)[b]		D774	Mullen pts	—	—	10–12
Tearing strength, initial		D1922	N/m	—	—	14–88
Folding endurance		D2176	—	—	Very high	Excellent
Gas permeability	CO_2	D1434	[a]	—	—	360
	N_2			—	—	24
	O_2			—	36	50
Rate of water vapour transmission	23°C	E96 (E)	g-mm	—	—	—
	38°C	E96 (E)	m^{-2} 24 hr·	—	1.6	—
Dielectric strength		D149	10^2 kV/m	520	400	180–220
Dielectric constant at 10^6 Hz		D150	—	—	2.4	2.7–2.9
Dissipation factor at 10^6 Hz		D150	—	—	<0.007	0.01–0.02
Maximum service temperature		D150	°C	120	60–70	60–80
Minimum service temperature		D150	°C	–45	<–75	Excellent
Heat sealing temperature range		D759	°C	130–205	120–180	65–150
FDA approved grades		—	—	Yes	No	Yes
Methods of processing		—	—	Skiving from moulded billets	Extrusion	Extrusion
Forms available		—	—	Rolls, tapes	Rolls, sheets, tapes, tubes	Rolls, sheets, tapes, tubes
Thickness range			mm	0.05	0.02–0.2	0.02
Maximum width			m	0.6	1.5	12.3
Area factor (area covered by 1 kg film 1mm thick)			$m^2 \times 10^6$	1.06	1.06	1.06–1.08
[a]1 $cm^3\ cm^{-2}\ s^{-1}\ (cm\ Hg)^{-1} \equiv 7.6 \times 10^{-8}\ m^4\ N^{-1}\ s^{-1}$ [b]1 mil ≡ 0.001 in 1 mm ≡ 40 mil			Remarks	Density 0.94		

Table 3.7.2—*continued*

TABLE 3.7.2 Typical properties of plastics film and sheet—*continued*

Polyimide	*Polypropylene*	*Polystyrene*	*BDS*	*Polysulphone*	*Polyether-sulphone*
175	30–280	55–85	28	60–74	70–85
70	35–1000	3–40	10–100	64–110	20–150
75	—	16–35	—	—	—
40	175–260	47–86	—	41–42	73–166
10 000	Very high	—	—	—	—
2.7	30–48[a]	54	79–140	57	24
0.36	1.2–2.9[a]	—	—	2.4	2.4
1.5	9–14.4[a]	24	35–81	13.8	5.4
2.1	—	—	—	—	—
—	0.14–0.27	2.76	—	—	—
2800	1200–4000	2000	—	3000	960
3.4	2.2	2.4–2.7	[a]2.5	3.03	3.5
0.01	<0.0002–0.0003	0.0005	0.0004–0.001	0.0034	0.006
400	135–150	80–95	—	185	205–230
–265	–20 to –50	–55 to –65	—	–70	–70
—	90–205	120–170	—	260–285	290–315
No	Yes	Yes	Yes	Yes	Pending
—	Extrusion, orientation (cast)	Orientated	—	Extrusion	Casting, extrusion
Rolls, tapes,	Rolls, sheets, tapes	Rolls, sheets	—	Rolls, sheets	Rolls, sheets
0.006–0.125	0.012–0.25	0.006–0.5	0.99	0.002–6.35	0.002–1.0
0.9	1.5	1.9	—	1.3	1.3
0.70	1.1–1.13	0.95	—	0.81	0.73
	[a] includes biaxially orientated grades. Coated grades have greatly reduced gas permeability.		[a] at 60 Hz.		

Table 3.7.2—*continued*

TABLE 3.7.2 Typical properties of plastics film and sheet—*continued*

Property		***ASTM Test method***	***Units***	***PVC and related compounds***	
				Plasticised	***Non-plasticised***
Tensile strength		D882	MN/m^2	140–160	40–160
Elongation		D882	%	100–500	3–10
Bursting strength (1 mil)[b]		D774	Mullen pts	20	30–40
Tearing strength, initial		D1922	N/m	19–85	19–85
Folding endurance		D2176	—	—	—
Gas permeability	CO_2	D1434	[a]	1.8–114	1.2–3.0
	N_2			0.6–4.2	0.06–0.6
	O_2			1.8–9.6	0.48–1.8
Rate of water vapour transmission	23°C	E96 (E)	g-mm	—	—
	38°C	E96 (E)	m^{-2} 24 hr·	2–30.0	0.35–2.0
Dielectric strength		D149	10^2 kV/m	100–400	170–520
Dielectric constant at 10^6 Hz		D150	—	3.3–4.5	2.8–3.1
Dissipation factor at 10^6 Hz		D150	—	0.04–0.14	0.006–0.017
Maximum service temperature		D150	°C	65–90	65–90
Minimum service temperature		D150	°C	–25 to –45	—
Heat sealing temperature range		D759	°C	160–185	150–215
FDA approved grades		—	—	Yes	Yes
Methods of processing		—	—	Calendering, casting, extrusion	Calendering casting, extrusion
Forms available		—	—	Rolls, sheets, tapes	Rolls, sheets, tapes
Thickness range			mm	0.01–0.25	0.02–2.03
Maximum width			m	1.3–2.0	1.3–2.1
Area factor (area covered by 1 kg film 1 mm thick)			$m^2 \times 10^6$	0.56–0.83	0.67–0.83
[a]1 $cm^3\ cm^{-2}\ s^{-1}\ (cm\ Hg)^{-1} \equiv 7.6 \times 10^{-8}\ m^4\ N^{-1}\ s^{-1}$ [b]1 mil ≡ 0.001 in 1 mm ≡ 40 mil			Remarks		

Table 3.7.2—*continued*

TABLE 3.7.2 Typical properties of plastics film and sheet—*continued*

PVC and related compounds		*PVC – nitrile rubber alloy*	*Rubber hydrochloride*	*Polybutylene*
Vinyl chloride acetate copolymer	*Vinylidene chloride/ vinyl chloride copolymer*			
18–56	56–112	18–30	25–35	35–38
3–500	30–80	250–500	200–800	200
20	25–35	—	Stretches	—
—	<0.35	—	—	200 000
—	>500 000	—	250 000	—
2.4–4.8	0.22–2.7	14.4	17–810	49
—	0.0072–0.09	0.3	—	6.5
0.9–9.0	0.048–0.42	3.0	23–135	23
—	—	—	—	0.46
0.23–1.46	—	—	—	—
100–520	1200–2000	—	—	—
2.8–4.5	3.0–4.0	—	3.0	—
0.006–0.140	0.05–0.08	—	0.006	—
85–95	145	—	85–95	—
Good	—	—	-25	—
130–180	120–150	105–130	—	—
Yes	Yes	—	—	—
Calendering, casting, extrusion	Extrusion	Casting	Casting, extrusion	Extrusion
Rolls, sheets. Non-rigid also in tapes, tubes.	Rolls	Rolls, sheets, tapes, tubes	Rolls, sheets, tubes	—
0.02–1.0	0.01–0.15	0.02–0.08	0.01–0.06	—
1.8–2.1	1.75	1.37	1.52	—
0.7	0.5–0.6	0.77–0.91	0.90	1.099
Plasticised and unplasticised grades included.				

Table 3.7.2—*continued*

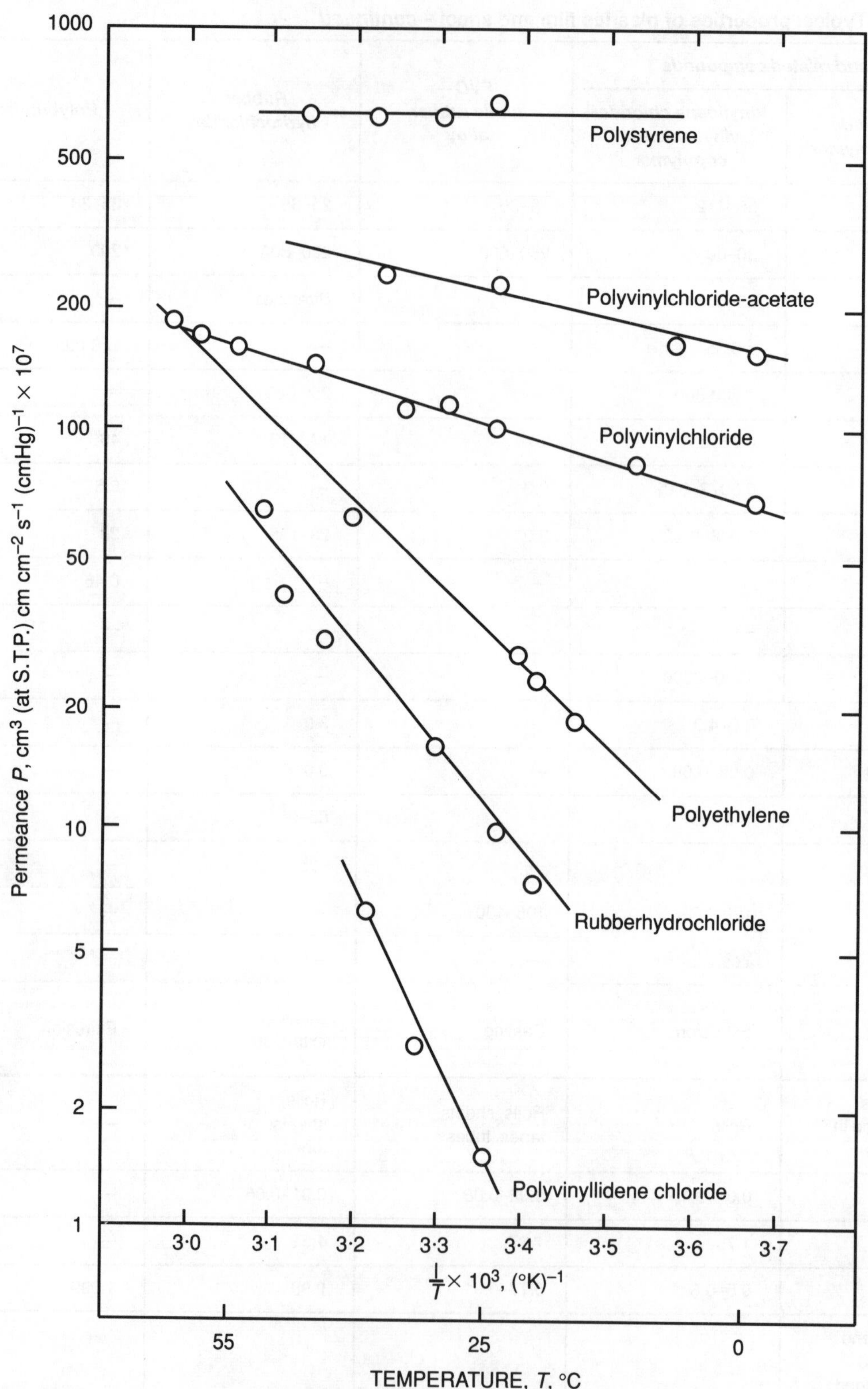

FIG 3.7.1 Dependence of the water permeance of various polymers on temperature

Fig 3.7.1

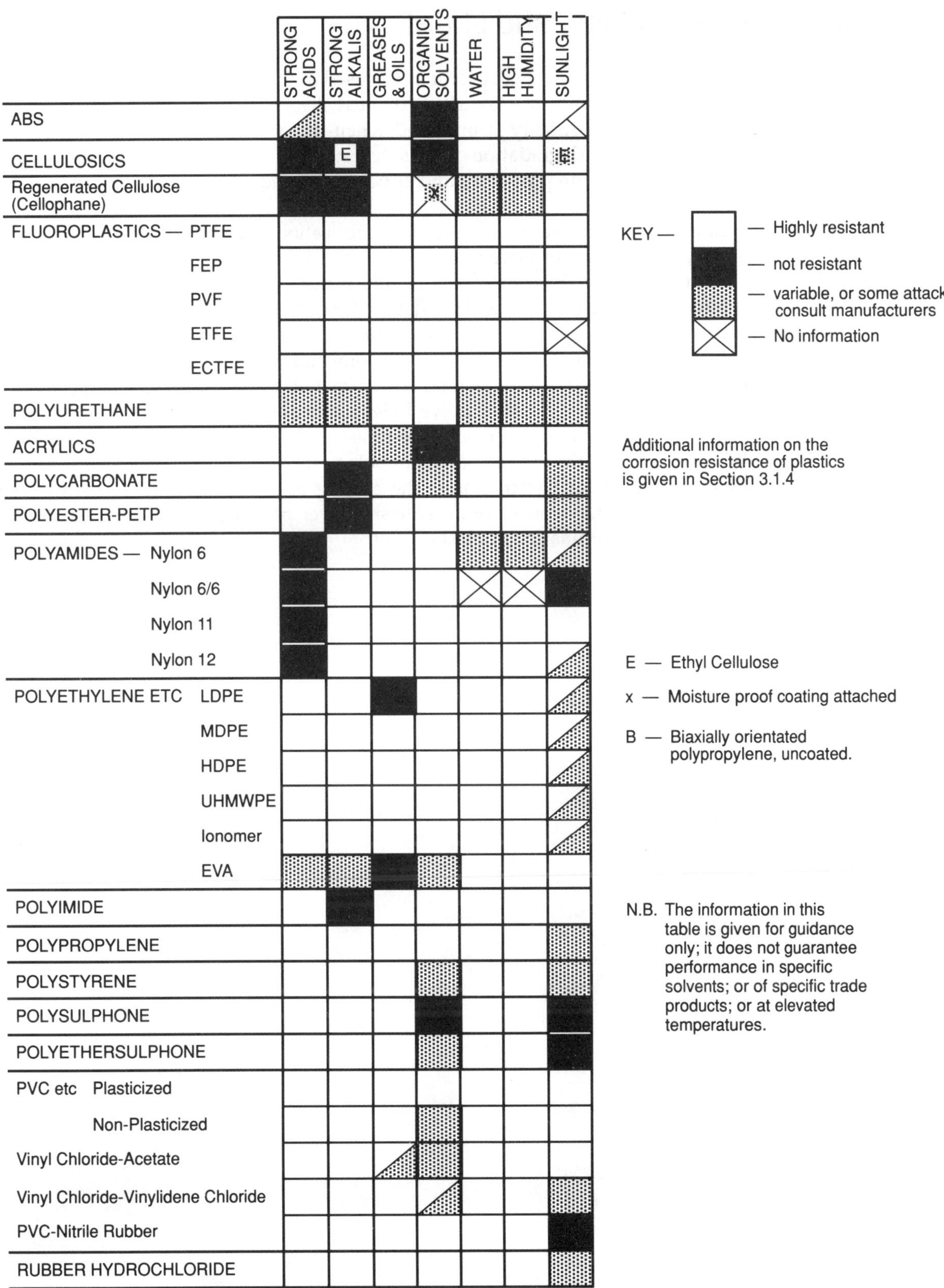

FIG 3.7.2 Environmental resistance of plastic films

3.7.2 Coextruded film and sheet

Many coextruded film and sheet products are used in the packaging industry. One example of a functional and cost-saving coextruded sheet structure is an acrylic resin on ABS. Outdoor weatherability is imparted when the acrylic resin is pigmented adequately to prevent UV degradation of the substrate ABS. By coextruding a clear acrylic over ABS, structures with properties and appearance adequate for sanitary ware can be made.

For films, one-step coextrusion can reduce fabrication costs by half compared to the cost of lamination of two or more monolayer films.

Techniques used for the coextrusion of film include multimanifold circular die for blown bubble film; feed block-standard flat die for cast film. The former method is limited normally to three-layer structures maximum; the latter can produce any number of layers, being dependent only on the layer-stacking ability built into the feedblock.

The combined properties of a layered structure tend to be additive. Characteristics combined include the low cost of polyethylene and polypropylene, the toughness of ionomer, the oxygen barrier of nylon, the moisture barrier of polyethylene, the heat sealability of EVA and the oxygen barrier of polyester.

An emerging area for coextrusion is the coating of substrates such as foam. The coextruded coating may consist of an adhesive layer plus nylon. Another use of coextruded film is for lamination of plastic sheet.

3.7.3 Trade names and suppliers

The more important suppliers of plastics film and the commonly used trade names of their products are listed in:

Table 3.7.3 *Typical trade names and suppliers of plastics films*

TABLE 3.7.3 Typical trade names and suppliers of plastics films

Supplier	***Trade name***	***Notes***
	Cellulose	
Rhone Poulenc	Cellophane	Regenerated
	PETP Polyester	
ICI	Melinex	
Rhone Poulenc	Terphane, Claryl	
	Polyester	
Du Pont	Mylar	
	Polyethylene	
BCL (British Cellophane)	Tensiltarp	
	Polyimide	
Du Pont	Kapton	
Upjohn	Polyimide	
	Polypropylene	
BIP	Bexphane	
ICI	Propafilm	
Rhone Poulenc	Pryhane	
	Polymethane	
B. F. Goodrich	Tuftane	
	PVC, PVC/VA	
Rhone Poulenc	Vinan, Vipak	PVC
	PVF	
Du Pont	Tedlar	

Table 3.7.3

3–8

Composites and laminates

Contents

List of tables

List of figures

3.8.1 Introduction to composites

The advantages and limitations of composite materials are detailed in:
Table 3.8.1 *Characteristics of composite materials*

The term 'composite' implies a material comprising more than one phase so that glass-and asbestos fibre-reinforced thermoplastic and thermosets are simple examples of composites, (as are civil engineering materials such as cement and concrete reinforced by asbestos, steel or polypropylene).

However, the term 'composite' when applied to engineering materials generally implies higher performance materials, of which the most common is directional glass reinforcement of plastic (GRP), which is dealt with in Section 3.8.2. This may be seen as a logical extension of the random glass-filled thermosets (particulary polyester SMC and DMC considered in Section 3.3.4), in which improved performance is obtained by increasing the quantity, directionality and/or length of the reinforcing fibre. Directional glass reinforcement comes in many forms and can be in one, two or three dimensions, continuous or discontinuous. Laminating is possible by layering the reinforcement to achieve desired effects. The strength of GRP is comparable to metal, but the low stiffness has led to a requirement for improvement.

Composites based on carbon, ceramic and boron fibres have been developed as high stiffness low weight materials for the aerospace industry. See Section 3.8.3.

GRP may also be stiffened by forming a composite sandwich structure which has led to the development of a further class of composites, not restricted to GRP skins, and with, in some cases, exotic materials as cores. See Section 3.8.4.

In Section 3.8.5, the properties of available industrial laminated plastics are compared and contrasted. These include, in addition to glass, laminates based on paper, cotton, asbestos and nylon fibres and textiles.

TABLE 3.8.1 Characteristics of composite materials

Advantages	*Limitations*
High (or very high for advanced composites) strength/weight and stiffness/weight ratios. Properties can be varied according to design requirements. Zero or negative expansion coefficients possible (for advanced composites). Inert to most acids and alkalis. Versatile low-cost fabrication techniques — elimination of conversion stages. Complex shapes possible. Good friction and wear characteristics (CFRP). Honeycomb materials are good insulation materials and have good resistance to fatigue and sonic fatigue. All composites have high damping capacity.	Stiffness of GRP composites low in relation to metals but can be improved by stiffening techniques (rib-stiffeners, corrugation, sandwich structures, advanced fibre reinforcement). Properties tend to be anisotropic, but this problem can be minimised by control of directional reinforcement, and should be used to advantage at the design stage. Susceptible to splitting and delamination. Cracks often propagate along fibre direction. Degraded by water and solvents. Carbon fibre based materials electrolytically positive — accelerate corrosion of adjacent materials. Very high raw materials cost (advanced composites). Residual stresses due to shrinkage during curing.

3.8.2 Glass fibre-reinforced plastic (GFRP) composites

3.8.2.1 General description

Glass fibre reinforcement of thermoplastic and thermosetting matrices has been considered in earlier chapters. In these cases the degree of reinforcement has generally been below 40% (by weight) and has been in the form of 3D random short fibres (injection moulding or extrusion compounds — although a degree of orientation may be imparted during processing). In sheet moulding compound (polyester SMC) 2D fibres are employed which may be random or oriented.

Orientation of fibres in the direction of loading is a very effective means of improving the properties of a glass-reinforced plastic. The ultimate example is 1D (unidirectional) reinforcement in which the continuous fibre is impregnated with the wet matrix and then wound continuously on a pattern or mandrel. Unfortunately, properties transverse to the direction of orientation are only about 25–30% of the longitudinal values and the nature of loading therefore must be accurately determined.

One solution to this problem is to use alternate layers of fibres orientated at angles to each other to gain strength in two directions. This is done by a filament winding or hand lay up. Another is to use woven fabrics (woven roving) or randomly orientated chopped strand mat (CSM) for 2D reinforcement. 3D reinforcement may be obtained through the use of knitted preforms.

Fibre length also has a marked effect on mechanical properties: increasing fibre length from 2.5cm to 5cm can improve strength values by 10–15%. Highest values are obtained in filament wound or pultruded components containing continuous fibres. The advantage is, however, offset by a sacrifice of mouldability, and of versatility of shape and detail.

Finally, techniques developed for increasing glass content to 70% or even higher have resulted in strength values more than double those of materials containing more conventional glass quantities. Many GFRP structures have tensile strengths comparable to those of common metals and in combination with the good processing, fabricating and shaping characteristics lead to structural applications and applications in chemical process plant (inertness at a lower cost than stainless steel). However, because a GFRP composite structure must maintain integrity at each fibre/matrix bond, a safety factor of three to four times is normally applied to the yield strength to insure against environmental and assembly factors, and uncertainties in dynamic loading.

AVAILABLE TYPES

The polymer matrices commonly used for GFRP are listed in:
Table 3.8.2 *Resinous matrices for GFRP*

3.8.2.2 Mechanical properties

For a unidirectional (1D) composite, the tensile modulus may be predicted using the law of mixtures equations:

$$E_c = E_f V_f + E_m V_m$$

E_c = Modulus of composite
E_m = Modulus of matrix
E_f = Modulus of fibres
V_f = Volume fraction of fibres
V_m = Volume faction of matrix.

2D and 3D fibre arrays can be regarded as roughly equivalent provided due allowance is made for the relative inefficiency of the fibres which are not aligned in the direction of the applied load.

The equation is modified thus:

$$E_c = kE_fV_f + E_mV_m \text{ where } k = \begin{cases} 1 \text{ for unidirectional composites} \\ 0.5 \text{ for bidirectional (woven roving)} \\ 0.375 \text{ for 2D random (CSM)} \\ 0.2 \text{ for 3D random.} \end{cases}$$

The simplest model for the transverse modulus is given by the following equation:

$$E_T = \frac{E_fE_m}{(E_mV_f + E_fV_m)}$$

although more complex models have been postulated (notation as above).

More detailed information on mechanical properties of glass fibre reinforced plastics is given in an excellent publication; *Design Data — Fibre Glass Composites*, by J A Quinn. It contains numerous graphs and formulae presented in a very clear and practical way. The publication may be obtained from the Reinforcement Division of Pilkingtons plc at St Helens, Lancashire, UK.

The properties of GFRP are compared with other materials in
Table 3.8.3 *Typical properties of GFRP compared to other materials*

The influence of proportion of glass content on tensile properties is illustrated in:
Fig 3.8.1 *Tensile strength vs glass content (by weight) of glass-reinforced polyester resin*,
and
Fig 3.8.2 *Tensile modulus vs glass content (by weight) of glass-reinforced polyester resin*

The properties of glass fibre laminates are reduced by heating and by exposure to water as shown in Table 3.8.21. These effects are time dependent and the influence of time and temperature are illustrated in:
Fig 3.8.3 *Effect of environment on flexural stress of polyester/MAT laminates*

3.8.2.3 Processing

There are a number of processing methods by which GFRP components can be fabricated. The more important of these are listed and described in:
Table 3.8.4 *Comparison of major GFRP processing methods*

ECONOMICS OF PROCESSING

Processing procedures vary in labour and capital requirements with, usually, an inverse correlation, as shown in:
Fig 3.8.4 *Capital cost and labour requirements of GFRP production processes compared*

3.8.2.4 Stiffening techniques

The tensile strengths of some GFRP composites compare favourably with those of steel and aluminium, but they have comparatively low stiffness. Stiffness is a 'composite' parameter, depending upon both design and materials properties, and techniques for enhancing stiffness of GFRP include:

(i) addition of rib stiffeners
(ii) corrugations
(iii) design with compound curvatures (highly complex for specialist applications)
(iv) selection of alternative fibre reinforcement to glass (Section 3.8.3)
(v) fabrication of sandwich panels with honeycomb or foam cores (Section 3.8.4)

Combinations of (iv) and (v) are often referred to as 'advanced composites'.

RIB STIFFENERS

Rib stiffeners are generally used for special design applications, since they are difficult to treat in general terms. Fabrication complications are involved which include the need for an extra mould, and attachment to the flat panel, which is an additional process and reliability problem. In general, sandwich panels (Section 3.8.4) are superior in load-bearing capability and stiffness to rib-stiffened panels.

Fig 3.8.5 *Typical designs of attached stiffeners for use in GFRP components*

CORRUGATIONS

Corrugated panels can provide good strength and stiffness in comparison to flat panels, but they are not as effective as sandwich panels (Section 3.8.4) and processing costs may be greater. A GFRP panel of thickness 2 cm corrugated with 7.5 cm depth to pitch ratio of 1–4 can support five times as much load and is 13 times as stiff as a flat panel of equivalent thickness.

Disadvantages of corrugated panels also include the increased use of the vertical dimension (also true of sandwich panels) and the use of about 15% more material than a flat panel. The load-bearing capability and stiffness of corrugated panels are compared to those of other GFRP fabrications in:

Fig 3.8.6 *Comparison of load -bearing capability for various GFRP structures*, and
Fig 3.8.7 *Comparison of stiffness for various GFRP structures*

TABLE 3.8.2 Resinous matrices for GFRP

Polyester	Most widely used. Low flammability, self-extinguishing. Suitable for mechanical and electrical applications. Special grades available for chemical resistance.
Epoxy	Most suitable for high performance or chemical environments.
Phenolic	Low cost. Most reinforced phenolic resins are used in the powder form and parts are produced by the transfer compression moulding technique. New laminating grades are available, suitable for hand lay, resin transfer, and possibly filament winding, pultrusion, etc. Low flammability and smoke emission increasingly attractive.
Silicone	Used with glasscloth for heat resisting electrical applications. Low moisture absorption, superior arc resistance.
Furan	High resistance to process plant environments.
Melamine	Electrical applications. High strength combined with flame and alkali resistance.
Polyimide	High strength particularly at elevated temperature.

TABLE 3.8.3 Typical properties of GFRP compared to other materials

	Glass content (wt %)	*Density (g/cm^3)*	*Tensile strength (MN/m^2)*	*Tensile Modulus (GN/m^2)*	*Compressive strength (MN/m^2)*	*Compressive Modulus (GN/m^2)*	*Flexural strength (MN/m^2)*	*Flexural Modulus (GN/m^2)*	*Coefficient of thermal expansion (10^{-6} K^{-1})*	*Heat resistance (°C)*
Unidirectional										
Wound epoxide	60–90	1.7–2.2	530–1730	28–62	310–480	—	690–1860	34–48	4–11	260
Unidirectional polyester	50–75	1.6–2.0	410–1180	21–41	210–480	—	690–1240	27–41	5–14	260
Bidirectional										
Satin weave polyester	50–70	1.6–1.9	250–400	14–25	210–280	9–17	207–450	17–23	9–11	175
Woven roving polyester	45–60	1.5–1.8	230–340	13–17	98–140	8–17	200–270	11–17	11–16	175
Random										
Preform polyester	25–50	1.4–1.6	70–170	6–12	130–160	—	70–240	—	18–33	200
Hand and spray-up polyester	25–40	1.4–1.5	63–140	6–12	130–170	6–9	140–250	5–8	22–36	175
Moulding compounds										
DMC	10–40	1.8–2.0	30–70	12–14	140–180	—	40–140	—	24–34	150–180
SMC	20–35	1.8–1.85	50–90	9	240–310	—	140–210	9–14	18–33	150–205
Nylon 6/6	30	1.34–1.42	117–200	9.6 (dry)	245	—	260	9.1 (dry)	17–36	93–150[a]
Polypropylene	30	1.12–1.13	47–103	4–7	—	—	60–140	4.2–6.5	31–38	90–120[b]
Metals										
Mild steel	—	7.8	410–480	210	410–480	—	—	—	11–14	500
Aluminium (Al Mg 2)	—	2.7	180–250	70	—	—	—	—	24	150–200
Austenitic stainless steel	—	7.9	550–1050	193–200	—	—	—	—	14–19	700

[a] Heat resistant grade.
[b] Exceptionally 140 °C.

TABLE 3.8.4 Comparison of major GFRP processing methods

Process	*Description of process*
Sheet moulding	Chopped glass rovings are fed onto a moving belt of polythene sheet to which a coat of resin has been applied. A second resin-coated sheet is fed on top of the chopped fibres and the sandwich is drawn through rollers. An additive causes the resin to thicken to a consistency at which, in due course, the polythene can be stripped off. The sheet is then cut to size ready for placing on a press for shaping.
Hand lay-up (or 'wet lay-up')	A male or female mould is used depending upon which side requires the better surface finish. A releasing agent is applied to the mould and either mat or cloth is laid onto it before applying resin either by roller or brush. A gel coat is used to give a smooth resin-rich surface.
Wet spray	As above except resin and glass fibre deposited simultaneously from a spray gun.
Hot press moulding (compression moulding)	Used for Dough (Bulk) Moulding Compound (DMC/BMC) or Sheet Moulding Compound (SMC), for high production items. Reinforcement and resin are placed in high grade male or female dies. The die is closed at a controlled speed. In minutes the moulding is cured by applied heat. Good finishes obtained on both surfaces with accurate dimensional tolerances. Expensive tooling.
Cold press moulding	Low pressure cheaper moulds (often made of GRP). Curing heat developed exothermically within resin.
Transfer moulding	A variation of compression moulding of polymers. It involves the use of a plunger and heated pot separate from the mould cavity, i.e. a three part mould assembly. Polymer and reinforcement is loaded into the separate heated pot chamber. When the material reaches the correct temperature it is transferred under pressure into the closed multi-die cavity through a sprue.
Filament winding	Used for items such as cylinders, tubes and spheres. The process consists of winding roving or tape on a rotating mandrel, the rovings having been previously resin impregnated by passage through a resin bath.
Pultrusion	Glass fibres pulled through steel dies, the resin being applied by passing the fibres through a wet bath or injecting it into a die. Profiles are broadly similar to those produced by extrusions. Curing by applying heat, usually electric strip. Lower capital costs and labour requirements than hot press moulding.
Mandrel wrapping (or 'dry lay-up')	Prepreg sheet or tape is cut to shape with the reinforcing fibre suitably orientated in the prepreg. The prepreg is rolled onto a support mandrel under slight pressure, with a degree of heating to confer tack and aid consolidation. Pressure can be applied during cure by over-wrapping with a shrink film or vacuum bag, or by the use of a split mould. After cure the mandrel is extracted mechanically. Used for fast production of tubular components.
Vacuum bag moulding	Similar to matched die processes but only a single-sided mould is used. Pressure is applied by means of a vacuum bag, which gives a less well consolidated component suitable for use with high flow resin systems for limited production runs.
Reaction injection moulding and reinforced reaction injection moulding.	Similar die arrangement to standard injection moulding. Two low molecular weight monomers are used, drawn from separate storage tanks, and they mix and polymerise in the mould itself. RRIM has reinforcement added to the monomer.
Resin injection moulding	Preforms of glass fibre are inserted in a mould, the mould is closed, and liquid resin is injected. Foam preforms are sometimes used to locate glass fibre preforms.
Vacuum assisted resin injection moulding (Vari process)	As above, but a vacuum is pulled on the mould to assist resin penetration.
Network injection moulding (NIM[R]). Trade mark of 3-D Composites Ltd, UK.	As for resin injection moulding, but patented process permits higher levels of reinforcement together with easier penetration of resin, allowing larger, more complex mouldings to be made.

Table 3.8.4

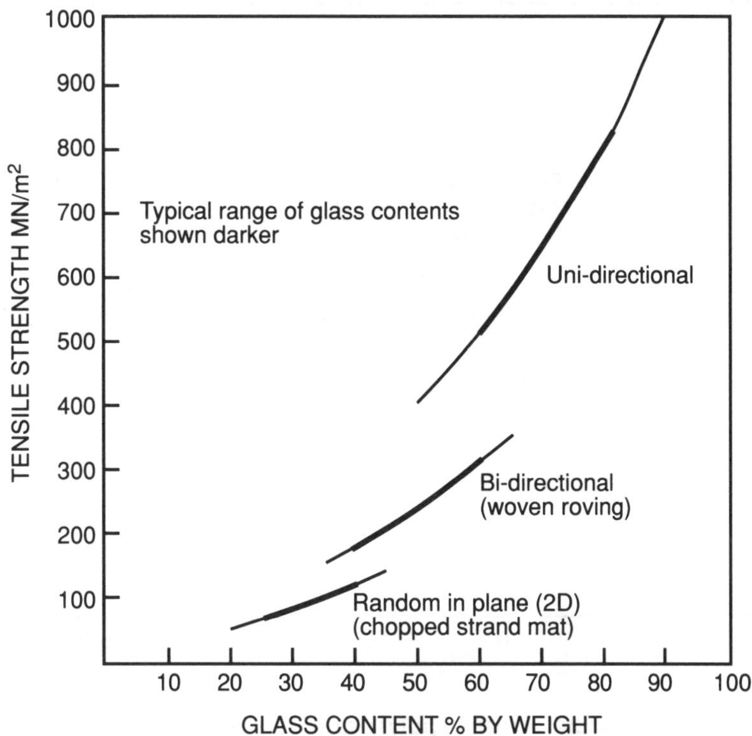

FIG 3.8.1 Tensile strength vs glass content (by weight) of glass-reinforced polyester resin. Reproduced by kind permission of J.A. Quinn & Pilkingtons from the publication referenced in the text.

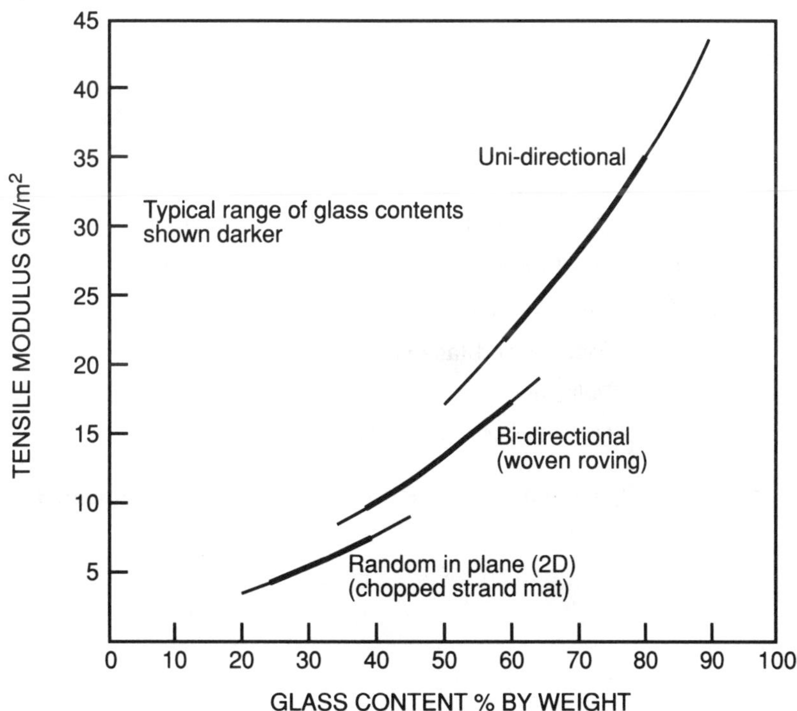

FIG 3.8.2 Tensile Modulus vs glass content (by weight) of glass-reinforced polyester resin. Reproduced by kind permission of J.A. Quinn & Pilkingtons from the publication referenced in the text.

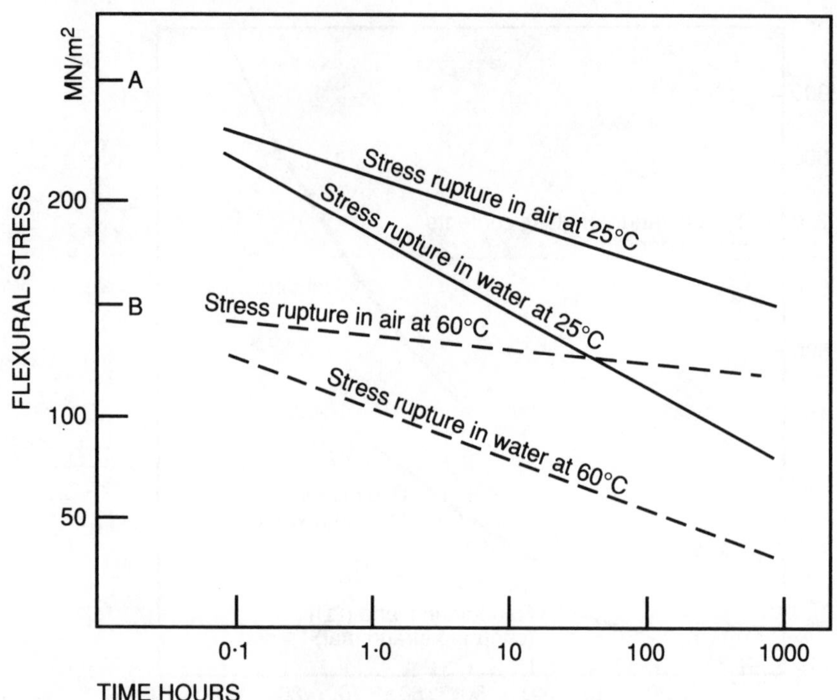

FIG 3.8.3 Effect of environment on flexural stress of polyester/MAT laminates.
A = Short-term flexural strength: 25°C.
B = Short-term flexural strength: 60°C.

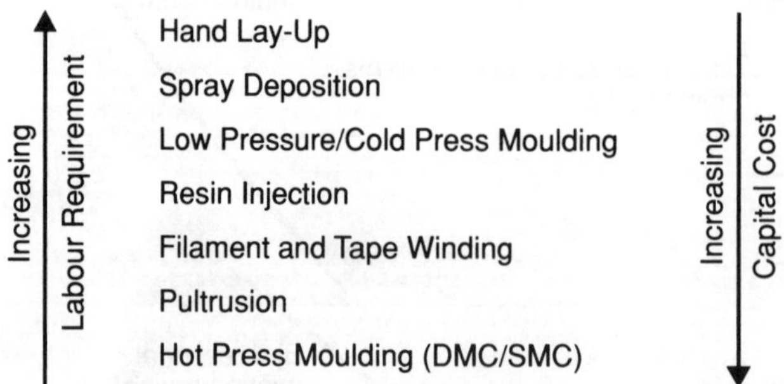

FIG 3.8.4 Capital cost and labour requirements of GFRP production processes compared.

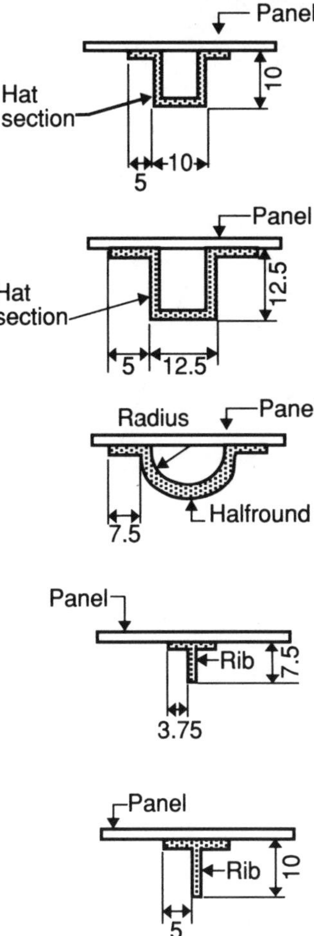

FIG 3.8.5 Typical designs of attached stiffeners for use in GFRP components

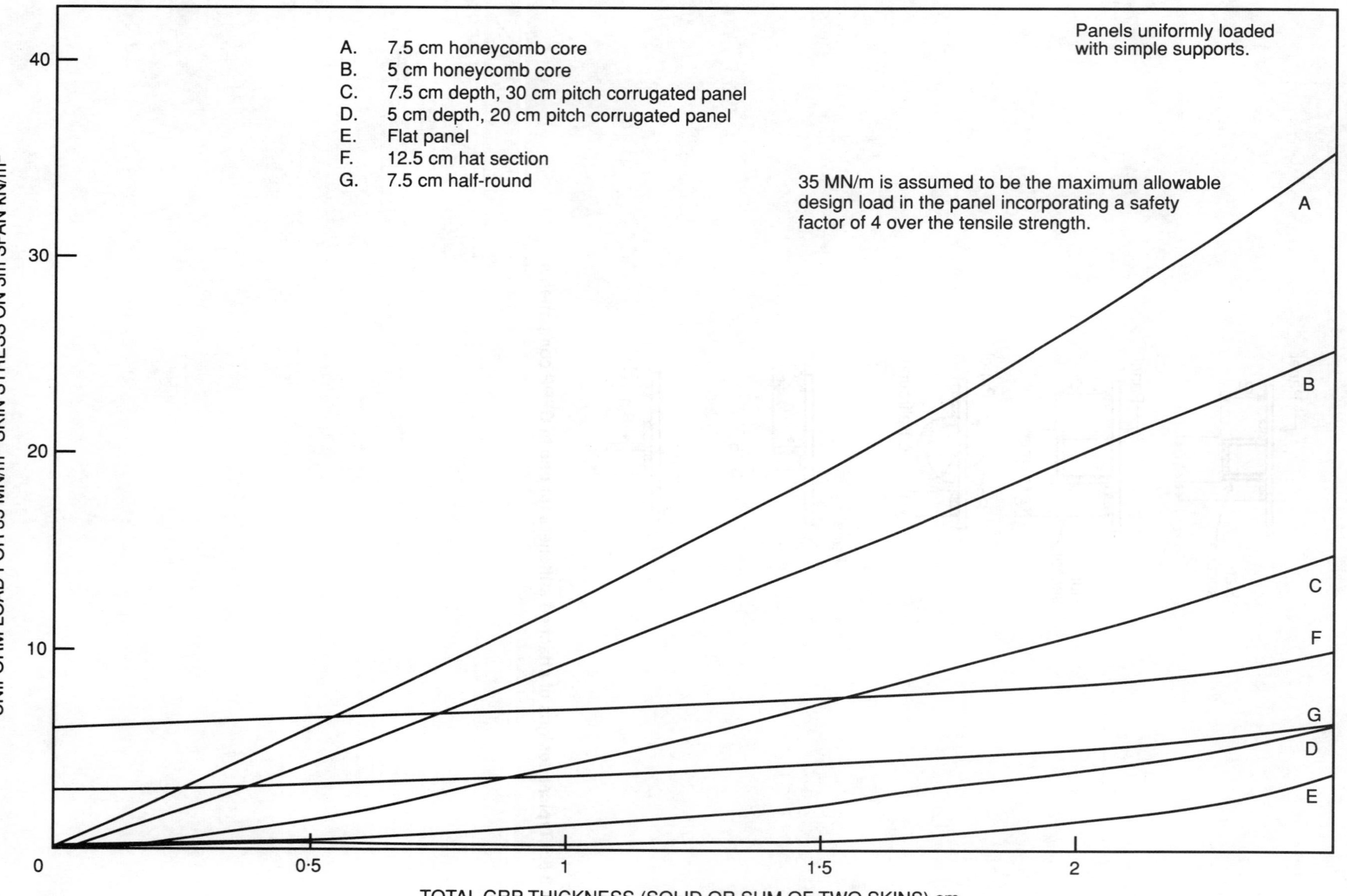

FIG 3.8.6 Comparison of load-bearing capability for various GFRP structures

Fig 3.8.6

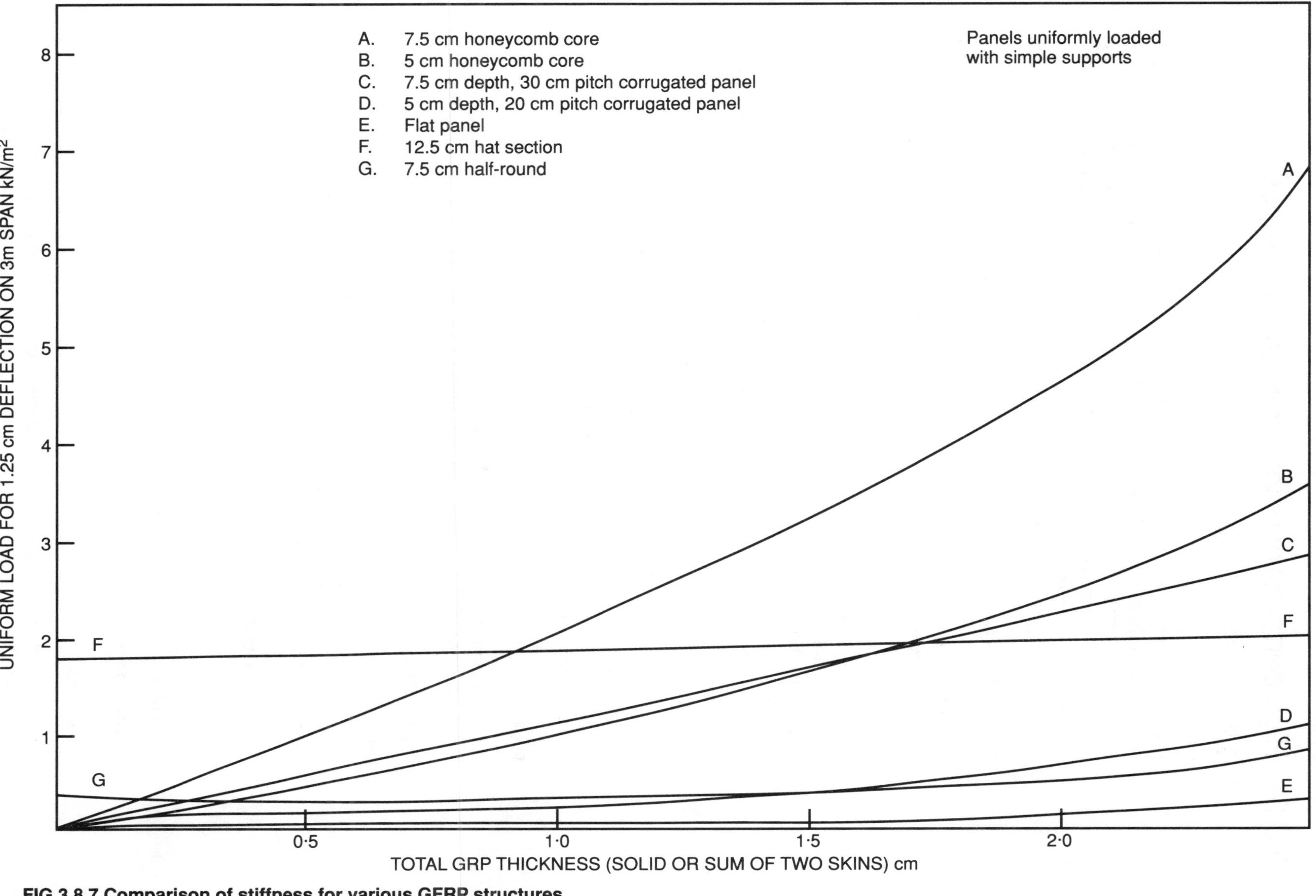

FIG 3.8.7 Comparison of stiffness for various GFRP structures

Fig 3.8.7

3.8.3 High performance fibre-reinforced composites

3.8.3.1 General description

This section deals with those composites which offer exceptionally high values for the strength-to-weight and stiffness-to-weight ratios (specific strength and specific stiffness), and is therefore limited to materials reinforced by low density, high specific strength and stiffness fibres. Carbon, boron, aramid, metal and ceramic fibres are all available in continuous form. In general the reinforcing phase usually exceeds 40% by volume of the composite and can be as high as 70%.

The advantages and limitations of high performance fibre-reinforced composites are given in:
Table 3.8.5 *Qualitative characteristics of advanced resin based composites*

The comparative characteristics of carbon fibre composites and metals are given in:
Table 3.8.6 *Comparative characteristics of carbon fibre composites and metals*

The composite materials covered in this section derive their special strength and stiffness properties from fibres individually bonded to a continuous matrix. The potentional range of materials is very large; metals, plastic and ceramics being feasible materials for both fibres and matrix. In addition fibres of the same material are capable of wide variation in terms of properties and geometry and may be incorporated in the matrix directionally, randomly or in a prescribed multidirectional fashion as single strands or woven cloth according to the requirements of the product.

Out of the wide range of possible composition the following are commercially available:

Fibre	*Matrix*
Carbon	Epoxide, Polyester, Polyimide, Carbon
Boron	Polyester, Aluminium, Epoxide, Polyimide
Aramid	Polyester, Epoxide
Alumina	Aluminium
Silicon carbide	Aluminium, Epoxide
Oriented polyethylene	Polyethylene

FIBRES

A comparison of available fibres is given in:
Table 3.8.7 *Qualitative comparison of commercially available high performance fibres*

Carbon fibre

—Continuous unimpregnated tows.
—Woven fabric. Aligned short fibre sheet. Excellent shape properties but lower mechanical properties than long fibre material.
—Chopped strand. Good drape properties. Available at lower cost if off cuts are acceptable.
—Pre-impregnated (prepregs) cut lengths or continuous rolls of sheet or tape which can be cured and bonded by heat.
—Moulding compounds. Chopped fibre in thermoplastic or thermoset resins.

Boron fibre

7.5 cm wide preimpregnated tape spools of filament.

Aramid fibre

'Kevlar' 29 High strength intermediate modulus fibre, available as yarns.
'Kevlar' 49 High strength high modulus fibre, available as yarns or rovings.

Matrices

The main matrix types currently in use or at the R & D stage for use as advanced composites are given in:
Table 3.8.8 *Main matrix types*

3.8.3.2 Typical properties

Properties such as strength, stiffness, etc., of the composite components are dependent on design, method of manufacture and the type and amount of constituent materials. A general guide to properties of available materials is given in:
Table 3.8.9 *Some typical properties of advanced composites*

Specific properties of suitable fibres are listed in:
Table 3.8.10 *Comparative properties of reinforcing fibres*
and illustrated in
Fig 3.8.8 *Comparison of specific stiffness and strength of advanced composite, ceramic and metal fibres*

The specific strengths and stiffnesses of these materials are compared with each other and with other commonly used engineering materials in:
Fig 3.8.9 *Comparison of specific tension strength and stiffness of unidirectional advanced composite laminates with metals, etc*

3.8.3.3 Processing

The main manufacturing processes used in the production of advanced composite components are listed in:
Table 3.8.11 *Advanced composites — main component manufacturing processes*

A comparison of the processing of thermoset and thermoplastic advanced composite matrices is given in:
Table 3.8.12 *Processing of thermosets vs thermoplastics*

3.8.3.4 Applications and future prospects

Although high performance fibres are comparatively expensive raw materials, when comparing composites with more conventional metal fabrication, it is frequently found that the composite material is cheaper, because of reduced machining and assembly costs and the high price of skilled labour. For example, the all composite tail fin of the A320 Airbus is assembled from 90 piece parts. The equivalent design in aluminium alloys requires over 2000 piece parts. Applications of high performance composites are listed in:
Table 3.8.13 *Current and potential applications of high performance fibre-reinforced composites*

TABLE 3.8.5 Qualitative characteristics of advanced resin based composites

Advantages	*Limitations*
High to very high, strength/weight and stiffness/weight ratios compared with conventional materials, e.g. steels and aluminiums, which give improvements in product performance even after the application of larger safety factors.	High raw material costs compared with conventional materials.
Versatile manufacturing techniques for small batch production.	Apart from filament wound or pultruded components, current manufacturing methods are labour intensive and expensive to automate.
Properties in different parts of component can be varied by design to match anticipated stressing.	Anisotropic properties can limit permissible loadings or inhibit design freedom in some applications.
Zero or negative expansion coefficients possible.	
Carbon fibre composites are inert to most acids and alkalis.	Degraded by water and solvents and in some cases UV radiation. Carbon fibre based materials are electrolytically positive which can accelerate the corrosion of adjacent materials, e.g. aluminium alloys.
Carbon fibre based materials have good friction and wear properties.	Glass fibre reinforced components can cause accelerated wear of metal counterfaces.
Excellent fatigue resistance when correctly fabricated. All composites have good damping characteristics. Honeycomb materials are good sound insulators.	Poor design and manufacture can cause voiding and delamination.

TABLE 3.8.6 Comparative characteristics of carbon fibre composites and metals

Category/property	***Composite characteristics***	***Metal characteristics***
Strength and stiffness	High in fibre direction. Low transversely and in shear.	Approximately uniform
Thermal expansion	Zero or negative longitudinally, positive transversely.	Approximately uniform, positive
Built-in stresses	Inevitable due to shrinkage in cure.	Reducible and relievable
Fracture characteristics	No inelastic ductility. Sensitive to 'secondary' stresses. Susceptible to splitting and delamination. Cracks often propagate in fibre direction. Low energy absorption without multiple failures.	Ductile Normally insensitive Cohesive Cracks random or along 'grain' Higher elastic/plastic energy
Environmental characteristics	Inert to most acids and salts. Degraded by water and solvents. Susceptible to erosion. Electrolytically positive—corrodes adjacent metal. Highly anisotropic electrical conductor: susceptible to lightning damage.	Corrosion risk Erosion resistant Corrosion between dissimilar metals Good conductors
Fabrication characteristics	Finished material produced *in situ.* Wide range of lay-ups and properties. Properties vary in nominally identical materials. Extensive non-destructive testing and inspection needed to monitor quality.	Available in final form Standard treatments, properties Low variability Limited non-destructive inspection necessary except for bonded or welded structures.
Economic	High raw material costs. Potentially low assembly costs.	Low–medium raw material costs High fabrication and assembly costs

TABLE 3.8.7 Qualitative comparison of commercially available high performance fibres

Fibre	*Advantages*	*Limitations*
Asbestos	Unique combination of heat resistance, chemical inertness, high strength and low cost.	Use presents serious health hazard, so now limited to a few military applications.
Aramid	High specific strength and good ductility. Good impact resistance.	Poor compressive strength. Limited temperature capability. Difficult to cut.
Boron	High stiffness to weight ratio. Particularly good under compression loading.	High cost. Relatively thick fibre so processing possibilities are limited.
Oxidised PAN	Fire and chemical resistance for clothing/fabric.	Limited strength and abrasion resistance.
PAN carbon	Highest combination of strength and stiffness in advanced fibres.	Brittle compared with aramid or glass fibre. Electrolytic attack on adjacent metal parts can occur.
Pitch carbon	Originally a cheap carbon fibre; now mainly used in high modulus grades for space components.	Brittle and difficult to handle. Expensive.
Ceramic	Good fire and temperature resistance.	Brittle. Expensive.
Glass	Low cost, good insulator, easily fabricated.	Limited strength and stiffness.
Polyethylene	High strength to weight ratio. Low density.	Limited temperature range. Bonding limitations. Stiffness decreases with stress applied.
Hybrid comingled yarns	New route to manufacture carbon fibre/thermoplastic components.	Limited range available. Still in early development.

© Copyright of Quo-Tec Ltd 1987.

TABLE 3.8.8 Main matrix types

Thermosets (in use)		*Thermoplastics (in R & D)*	
Types	***Trade names***	***Types***	***Trade names***
Epoxy	Shell Epon Ciba MY series	Polyarylene ether or sulphide	ICI PEEK Phillips Ryton Phillips PAS-2
BMI	Ciba Matrimid Technochemie Compimide	Amide or amideimide	Du Pont J Polymer Amoco Torlon GE Ultem Du Pont Avimid K-111
Polyimide	Rhone Poulenc Keramid	Polyimide Polysulphone	Mitsui Larc-TPI ICI Victrex Amoco Udel P1700 Amoco Radel 400
Phenolic	BP Bakelite	Polyester	Dartco Xydar

Trade names are examples only.

TABLE 3.8.9 Some typical properties of advanced composites

	Property		Units	*Epoxy, 60% Kevlar 49, u/d* [a]	*Epoxy, 60% Kevlar 49, woven fabric*	*Vinyl ester 60% carbon type II u/d*	*Vinyl ester 31% carbon II 29% glass E u/d*
	Void content		%	<2			
Mechanical	Tensile strength, 10^5 cycles		MN/m^2				
	Tensile strength	Longitudinal	MN/m^2	1400	550	1132	749
		Transverse			N/A		
	Tensile Modulus (longitudinal)		GN/m^2	90	37	115	74.6
	Compressive strength (longitudinal)		MN/m^2	260	140		
	Compress Modulus (longitudinal)		GN/m^2	76	39		
	Flexural strength (longitudinal)		MN/m^2	670	450	1240	953
	Flexural Modulus (longitudinal)		GN/m^2	80	30	101.2	69.3
	Interlaminar shear strength		MN/m^2	69	35	38	40
Physical	Specific gravity			1.38			
Thermal	Useable temperature range		°C	–40 – +120	–40 – +120		

[a] u/d – Unidirectional. All fibre contents expressed in per cent by volume.

TABLE 3.8.9 Some typical properties of advanced composites—*continued*

Vinyl ester 21% carbon II 39% glass E u/d	*Vinyl ester 16% carbon II 44% glass E u/d*	*Vinyl ester 13% carbon II 47% E-glass u/d*	*Vinyl ester 60% E-glass u/d*	*60% Carbon fibre u/d*				*Epoxy/ aromatic amine 65% woven carbon fibre*
				Low temp. epoxy		*High temp. epoxy*		
				High strength	*High stiff-ness*	*High strength*	*High stiff-ness*	
								<2
480	400	330	200					
715	686	660	780	1448	931	1724	1103	500 (warp)
					52			520 (weft)
65.2	60.2	56	39.9	127	207	141	193	72
				1138	882			
943	843	773	840	1441	1103	1379	965	860
59	50.6	46.9	34.6	113	190	122	186	69
43	42	43	40	90	94	131	83	62
				1.5	1.6	1.6	1.6	
				–196 to +74	–196 to +74	–54 to +177	–54 to +177	–130 to +180

Table 3.8.9—*continued*

TABLE 3.8.10 (Part 1) Comparative properties of reinforcing fibres

Material		***E-Glass-fibre***	***S & R Glass fibre***	***Aramid fibre***	***Boron / tungsten fibre***
Product		(Roving)	(Roving)	Kevlar 49/ Twaron HM	—
Supplier		Fibreglass	Owens Corning & Vetrotex	Du Pont & Enka	Avco
Density	(lb/in³)	0.092	0.090	0.052	0.094
	(g/cm³)	2.54	2.49	1.44	2.52
Tensile strength	(10^6 psi)	370–500	550–650	460–525	400–510
	(MPa)	2.5–3.5	3.8–4.5	3.1–3.6	2.8–3.5
Tensile modulus	(10^6 psi)	10.5	12.6	18.0	58.0
	(GPa)	72	87	124	400
Ultimate elongation	%	4.8	5.4	2–2.5	0.9
Specific strength	(10^6 ins)	4.0–5.1	6.0–7.2	8.8–10.1	4.2–5.3
	(MPa/sg)	1.0–1.4	1.5–1.8	1.9–2.2	1.1–1.3
Specific modulus	(in 10^8 ins)	1.1	1.3	3.5	6.2
	(GPa/sg)	28	35	86	153
Filament diameter	(mils, 10^{-3} ins)	0.2–0.5	0.35	0.5	4.0
	(10^{-6} m)	5–14	9	12	10
Thermal conductivity	(W/mK)	0.97	—	0.05	—
Thermal expansion	(10^{-6} /K) (0–100°C)	5.0	5.6	–2.0	2.0
Other suppliers				Tejin make a strong & less stiff grade.	

TABLE 3.8.10 (Part 2) Ceramic fibres

Material		***Boron/ carbide***	***Silicon carbide/carbon***		***Alumina***		***Boron nitride***	***Ceramic***
Product				Nicalon	FP	Saffil[a]		Nextel
Manufacturer		AVCO	AVCO	Nippon	DuPont	ICI	Carborundum	3M's
Density	(lb/in³)	0.082	0.12	0.09	0.143	0.120	0.065–0.069	0.098
	(g/cm³)	2.24	3.3	2.55	3.95	3.3	1.8–1.9	2.7
Tensile strength	(10^3 psi)	475	500	350–470	300	290	120 (ave)	250
	(MPa)	3.3	3.5	2.4–3.2	2.1	2.0	0.8	1.7
Tensile modulus	(10^6 psi)	53	62	27	55	44	30 (ave)	22
	(GPa)	365	428	186	377	300	210	155
Ultimate elongation	%	0.9	0.8	1.3	0.4	0.7	0.4	1.1

[a] Discontinuous.

Table 3.8.10 (Part 1 and Part 2)

TABLE 3.8.10 (Part 2) Ceramic fibres—*continued*

Material		*Boron/ carbide*	*Silicon carbide/carbon*		*Alumina*		*Boron nitride*	*Ceramic*
Product				Nicalon	FP	Saffil[a]		Nextel
Manufacturer		AVCO	AVCO	Nippon	DuPont	ICI	Carborundum	3M's
Specific strength	(10^6 ins)	5.8	4.5	3.9	2.1	2.4	1.8	2.5
	(MPa/sg)	1.5	1.5	1.1	0.5	0.6	0.5	0.6
Specific modulus	(10^8 ins)	6.4	5.6	3.0	3.5	3.7	4.6	22
	(GPa/sg)	160	130	73	95	91	114	57
Filament diameter mls	(10^{-3} ins)	5.6	5.6	0.4/0.6	1.0	0.2	0.24	—
	(10^{-6} m)	140	140	10–15	20	3	0.6	—
Thermal conductivity	(W/mK)	—	—	—	0.13	—	3.0	—
Thermal expansion	(10^{-6} /K)	—	0.8	—	8.5	—	—	—

[a] Discontinuous.

TABLE 3.8.10 (Part 3) Carbon fibres

Material		*Standard*	*High strength intermediate modulus*		*High modulus*		*Ultra-high modulus*
Typical product		Filkar (Toray) T300	Thornal T650/42	Toray T1000	Hercules XUHM	Hysol Apollo	Celion GY–80
Density	(gm/cm^2)	1.77	1.79	1.82	1.82	1.8	1.95
	(lb/in3)	0.065	0.065	0.066	0.067	0.066	0.070
Tensile strength	(10^6 psi)	510	700	1,000	550	500	220
	(MPa)	3.53	4.83	7.06	3.8	3.46	1.52
Tensile modulus	(10^6 psi)	33	42	42.6	62	55	80
	(GPa)	230	290	294	428	380	552
Elongation	%	1.5	1.7	2.4	0.75	0.9	0.3
Specific strength	(10^6 ins)	7.8	10.8	14.9	8.2	7.6	3.1
	(MPa/sg)	2.0	2.7	3.9	2.1	1.9	0.8
Specific modulus	(10^8 ins)	5.1	6.5	6.5	9.3	8.3	11.4
	(GPa/sg)	130	162	162	235	211	283
Filament diameter	(10^{-6} m)	7	—	5.3	—	4.5	8.4
	(10^{-3} ins)	0.28	—	0.21	—	0.18	0.33
Thermal conductivity	(W/ml)	20.8	—	—	—	—	—
Electric conductivity	μ-cm	2000	—	—	—	—	650
Thermal expansion	(10^{-6} /K)	–0.5	—	—	—	—	—

Table 3.8.10 (Part 2 and Part 3)—*continued*

TABLE 3.8.11 Advanced composites—main component manufacturing processes

Thermoset matrix	*Thermoplastic matrix*
Autoclave moulding	Hot press moulding
Hot press moulding	Hydroclave moulding
Filament winding	Filament winding
Braiding	Cold press forming
Roll wrapping	Stretch forming
Vacuum bag moulding	Pultrusion
Pultrusion	Extrusion

TABLE 3.8.12 Processing of thermosets vs thermoplastics

	Thermoset	*Themoplastic*
Tooling	Metal or composite 1 part	Metal 2 part therefore more expensive
Prepreg/impregnation	Easy	Difficult
Prepreg handling	Tack and drape make cutting harder, but lay up easier	No tack or drape
Process temperature	Low to moderate 120–200°C-epoxy; 350°C polyimide	High 330–400°C
Process time	Long	Short
Advantages	Experience of systems	Simpler formations. Easier mechanisation of process
Disadvantages	Complex formulations. Variability in resin cure	Poor solvent resistance
Stage of development	Mature	Early
Automation	Elaborate feedback needed	Appears easier
Component cost	High	Even higher now. Potential of lower cost
Process routes	Mainly autoclave moulding via prepreg. Also filament winding, press moulding, pultrusion, extrusion, braiding, roll wrapping.	Press or hydroclave moulding via prepreg. Pultrusion, extrusion, braiding, stretch forming, injection moulding

Table 3.8.11 and Table 3.8.12

TABLE 3.8.13 Current and potential applications of high performance fibre-reinforced composites

Application area	*Typical examples*	*Reasons for use*
Aerospace	Very wide range of applications including primary structure, e.g. complete tail plane on Airbus and complete wing sections on AV8-B.	High strength/weight and stiffness/weight justify cost o materials. Ability to be economically fabricated into complex shapes for low production totals.
Sports and leisure goods	Gold club shafts (CFRP). Badminton and tennis racquets (laminated wood, CFRP). Skis (CFRP/GRP). Canoes, kayaks, high performance sailing craft (carbon/glass, carbon/Kevlar, Kevlar).	Light weight High stiffness Low friction and wear Prestige value
Road transport	Critical, primary load-bearing components under evaluation (e.g. drive shaft, body spring).	Need to reduce weight to conserve energy.
General engineering	Textile machinery (CFRP).	Reduction of momentum and inertia in reciprocating and rotating components (light weight).
	Cigarette manufacture—filter tip distributor(chopped carbon fibre reinforced nylon 6/6).	Good wear properties
	Touch switches (carbon fibre-reinforced polycarbonate).	Conductivity
	Pumps, impellers, chemical plant (may be carbon or carbon/glass).	Inert
	Gauges and templates (carbon).	Low thermal expansion, light weight.
Medical applications	Surgical Implants (carbon, Kevlar, textiles).	Chemically inert Weight reduction Easy to mould cf metal. May promote healing.
Atomic plant		Low neutron capture cross-section.

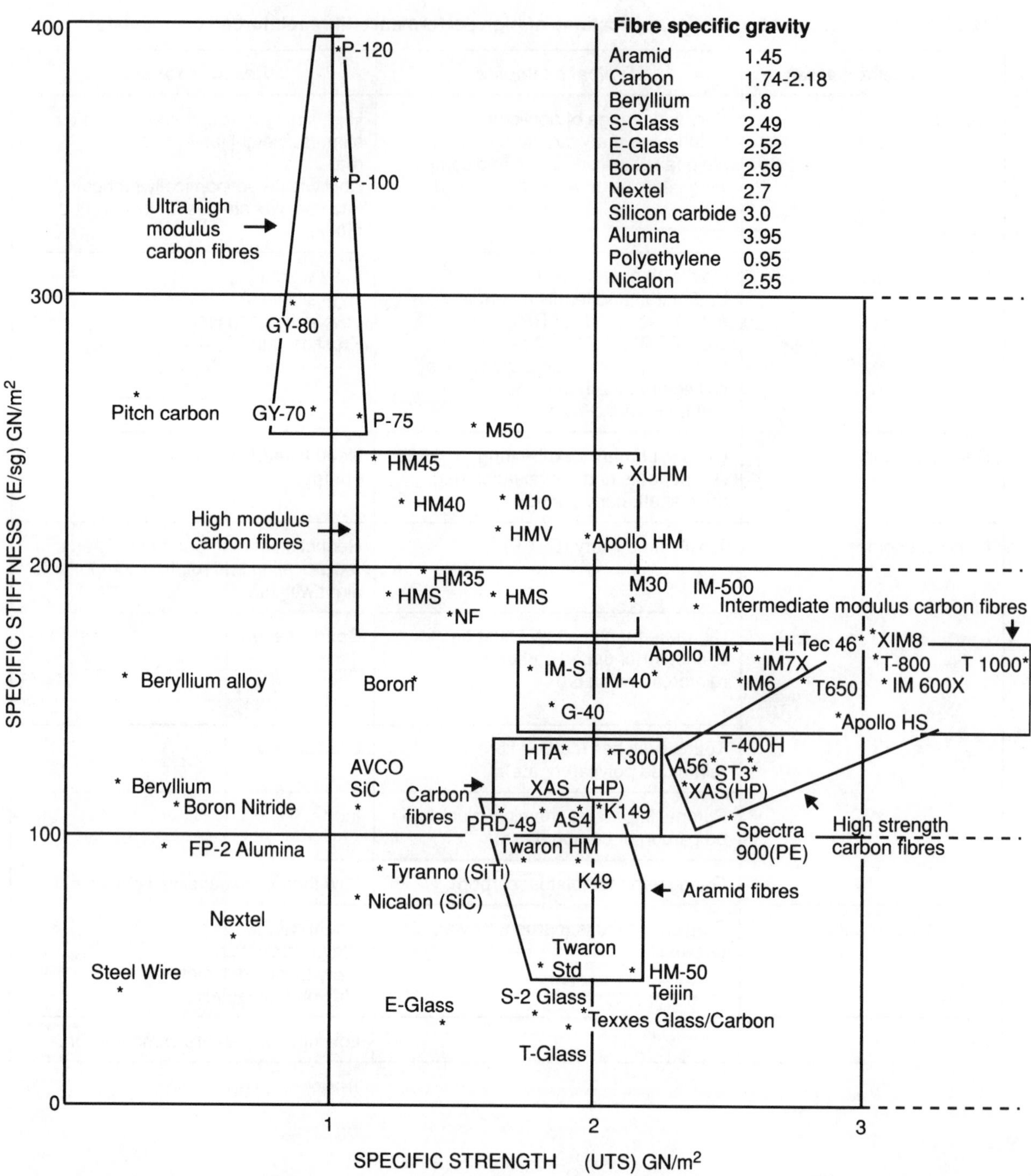

FIG 3.8.8 Comparison of specific stiffness and strength of advanced composite, ceramic and metal fibres

Fig 3.8.8

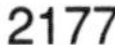

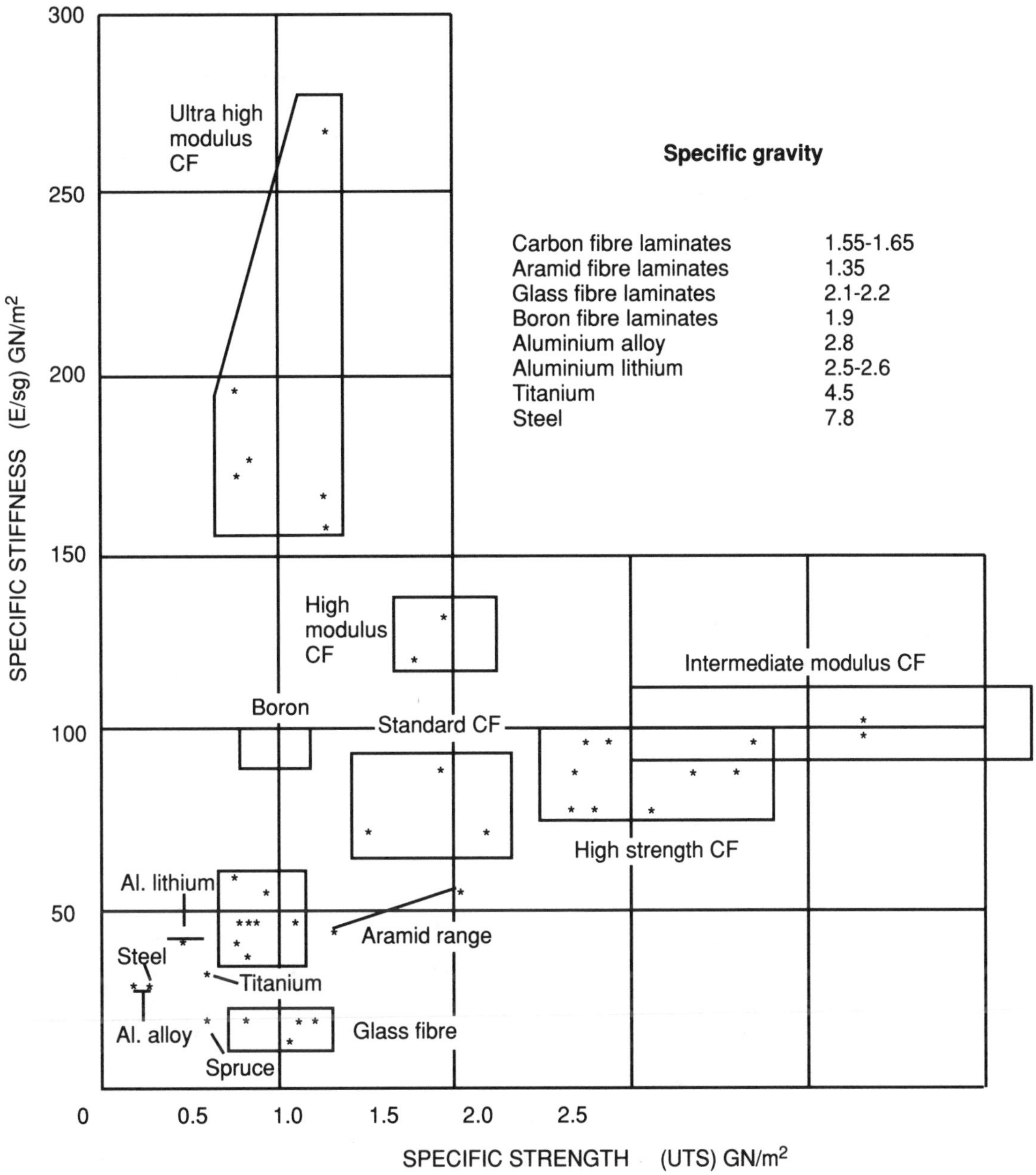
300
250
200
150
100
50
0
0.5
1.0
1.5
2.0
2.5
SPECIFIC STIFFNESS (E/sg) GN/m2
SPECIFIC STRENGTH (UTS) GN/m2
Ultra high modulus CF
High modulus CF
Boron
Standard CF
Intermediate modulus CF
High strength CF
Al. lithium
Steel
Titanium
Al. alloy
Spruce
Aramid range
Glass fibre
Specific gravity
Carbon fibre laminates 1.55-1.65
Aramid fibre laminates 1.35
Glass fibre laminates 2.1-2.2
Boron fibre laminates 1.9
Aluminium alloy 2.8
Aluminium lithium 2.5-2.6
Titanium 4.5
Steel 7.8

FIG 3.8.9 Comparison of specific tension strength and stiffness of unidirectional advanced composite laminates with metals, etc.

3.8.4 Sandwich structures

3.8.4.1 General characteristics and properties

Sandwich structures are high stiffness, lightweight structures comprising two skins or facings of high strength or modulus and a porous core. They can therefore be considered as integrated miniature I-beams.

ADVANTAGES

The following advantages of sandwich construction may be exploited in products:

Very high stiffness/weight ratio.
Reduced weight.
High resistance to fatigue, especially sonic fatigue.
Skins can be stressed well into plastic region.
Good vibration damping properties.
Can be used in noise attenuation applications.

AVAILABLE TYPES OF SKIN AND CORE

Skins

Materials which are employed as skins are as follows:

Metal — steel sheet (galvanised or plastic coated for corrosion protection).
— aluminium (may br profiled for greater rigidity).
— copper (for decoration).
— titanium alloys (for maximum strength/weight ratio and creep resistance).
Wood — plywood, chipboard, hardboard.
Plastics — decorative laminates.
— PVC
— ABS
— GRP
— CFRP (for high strength/weight ratio).

Cement (may be fibre-reinforced with asbestos, glass, polypropylene, etc.).
Foamed or lightweight concretes.
Plasterboard and reinforced gypsum.
Paper, cardboard.

All these skins can be applied either during manufacture or on assembly on site. They are compact and can be transported at maximum density. Different skins can be used on each side if required (e.g. for impermeability, corrosion resistance, etc.)

Cores

Materials available as cores are in three forms:

basically lightweight (e.g. balsa wood);
lightweight because foamed;
lightweight because honeycombed.

Materials for cores include:

Metals — steel, aluminium, titanium as hexagonal honeycomb. (Maximum shear strength, expensive.)
Wood — chipboard, plywood, strawboard, woodwool/cement.

Plastics — rigid polyurethane foam, expanded polystyrene, honeycomb, other plastics foams (excellent thermal insulation, weaker and more expensive than paper).

Others — low density porous masses of glass fibre or mineral wool with resin.
— porous cement or concrete.
— sand/resin
— paper or cardboard honeycomb (moderate insulation, good strength at minimum cost).

Various core materials are compared in:
Table 3.8.14 *Qualitative comparison of core materials*
and their properties are listed in:
Table 3.8.15 *Properties of sandwich core materials*

3.8.4.2 GFRP/paper honeycomb

While paper is used both as skin and core for honeycombs its major engineering application is for the core of sandwich structures with stronger skin material; typically GFRP. The three types of paper honeycomb core are listed, described and compared in:
Table 3.8.16 *Paper honeycomb core types*

The corrugated and hexagonal structures are illustrated in:
Fig 3.8.10 *A sandwich panel using Verticel's paper honeycomb*
Fig 3.8.11 *Panel using hexagonal honeycomb expanded core*

Shear strength, shear modulus and compressive strength are the usual structural concerns for a core material, since these are the core properties that hold the facings apart, yet make them work conjointly. Paper honeycombs, unlike foam cores, are subject to the buckling criteria of long, thin columns (Euler).

Typical properties of paper (and other core materials) are listed in:
Table 3.8.17 *Properties of paper and plastic sandwich core materials*

FABRICATION OF GFRP/PAPER HONEYCOMB

Sandwich panels are superior to the corrugated and rib stiffened panels described in Section 3.8.2 in strength and stiffness.

Paper honeycomb can usually be pressed into a wet polyester or epoxy/glass matrix to accomplish good wetting, impregnation, and bond of core to skin. Where flat panels are being made, one side can be prepared at a time, using either layup or gun techniques for making the GRP skin. A slight pressure is usually sufficient to maintain the core/skin contact during gelation and cure of the resin.

3.8.4.3 Aluminium honeycomb

Aluminium honeycomb is treated for corrosion resistance. The principal application is in core material for structural sandwich panels, exhibiting very high strength/weight ratio. The honeycomb may be skinned using resin-impregnated GRP for use as lightweight tooling and checking fixtures, for which purpose it is widely employed in the automotive industry. Related applications include drawing machine beds, trolley ways, platens. Aluminium skins are used for the manufacture of aerospace and marine construction boards.

In addition to its usefulness as a sandwich board core material, the metallic honeycomb has the ability to absorb mechanical energy by plastic deformation with constant resistance, thus providing a means of absorbing shock at constant deceleration and without rebound (e.g. emergency lift pads). It is also an efficient medium for fluid flow straightening, and for use in light collimating and heat exchanging devices.

Its aesthetic qualities have led to its use in light-diffusing ceilings and other decorative schemes. These applications require prime quality honeycomb with no 'cell wander'.

3.8.4.4 High strength polyamide ('Nomex') honeycomb

This is manufactured from high-temperature resistant polyamide in paper form, bonded and coated with specially developed high strength resin systems to produce mechanical properties necessary for structural honeycomb material. It has lower density than the metallic honeycomb, but is tough, resilient and has excellent resistance to impact and handling damage. Properties include excellent fatigue resistance, corrosion resistance, good dielectric and radar transmission properties, and is non-flammable. Principal applications include radomes, radar covers and Doppler windows, aircraft interior partitioning, flooring , furnishings and fittings.

3.8.4.5 Glasscloth honeycomb

This is manufactured from high quality flexible glasscloth. When impregnated with a suitable resin, the honeycomb acquires very high strength and stiffness. Recommended uses include radomes, fairings, hatches and applications taking advantage of the excellent chemical and water resistance.

3.8.4.6 Sandwich and co-injection processes

These processes involve the production of a foamed core and skin out of two different polymers. The basic Battenfield and ICI processes are discussed in Section 3.6.1.

TABLE 3.8.14 Qualitative comparison of core materials

Form	*Material*	*Advantages*	*Limitations*
Solid	Balsa wood	Robust. Easy to shape	Variable, affected by moisture
Foamed	PVC Klegercell		Temperature limitation
	Polymethoenylimide (Rohacell)	Controlled density	Poor fire performance
	Polyurethane	Can be foamed *in situ*	
	Melamine	Service temperature to 200°C	
Honeycomb (1) Aluminium	Pure AL strip some times epoxy coated	High strength Low weight	Can corrode next to carbon fibre reinforced plastic
(2) Nomex R	Nylon paper strip	Good strength weight ratio. Good fatigue resistance	More difficult to shape—expensive
(3) Titanium	Ti alloy strip	High temperature resistan-ce	Cost and weight

Nomex R   Trade name of Hexcel's synthetic (nylon) paper.

Table 3.8.15

TABLE 3.8.15 Properties of sandwich core materials

Core material *Trade name*	*Classification*		*Cell size ins*	*Density kg/m³ (lb/ft³)*	*Lengthwise*		*Compression strength MPa (psi)*	*Temperature resistance °C (F)*	*Notes*
					Shear strength MPa (psi)	*Shear Modulus GPa 10³ (psi)*			
End-grain balsa	Wood		—	100–250 (6–15)	0.3–20.0 (50–3000)	2.2–8 (330–1200)	1.2–3.6 (180–500)		
Rohacell R	Plastics Foams	PM1 [a]	—	30–90 (2–6)				175 (350)	
Klegecel R	Plastics Foams	PVC	—	30–50 (2–3)	0.5/1.0 (125–250)			80 (185)	
Phenolic	Plastics Foams	Phenolic	—	40–180 (3–6)	0.1/0.5 (25–125)		0.2/0.9 (22–85)	250 (5000)	Good fire resistance.
Honeycomb	Al sheets		—	30–133 (2–8)	10–5.5 (150–800)		1.0–11 (150–1600)		High mechanical strength.
Honeycomb	Nomex R		⅛	27–150 (1.8–9)	0.6–3.6 (90–520)	(3.7–17)	0.9–12.5 (130–1800)		Good fatigue strength and moisture resistance.
Honeycomb	Nomex R		3/16	90 (5.6)	2.7 (400)	80 (11,600)	10 (1,450)		Good fatigue strength and moisture resistance.
Honeycomb	Nomex R		½	128 (8)			7 (1,000)		Good fatigue strength and moisture resistance.

[a] Polymethoenylimide.

TABLE 3.8.16 Paper honeycomb core types

Classification	*Description*	*Characteristics*
Corrugated/flat-web (Verticel)	Fluted, corrugated core with flat-paper sheets between layers (Fig 3.8.10).	Maximum longitudinal shear strength and stiffness per unit weight. Lower transverse-shear and compressive strength than hexagonal cells. If long, unidirectional spans are involved this construction geometry is generally cost-effective.
Hexagonal unexpanded core	Flattened core for ease of shipping.	Greater shipping efficiency. Requires expander at the fabrication line. Difficult to use with wet, slow-curing GRP because of retraction tendencies.
Hexagonal expanded core	Cells opened and ready for bonding to skins (Fig 3.8.11).	Superior transverse-shear and compressive strength to corrugated paper honeycomb, but inferior longitudinal shear strength and stiffness per unit weight.

Table 3.8.17

TABLE 3.8.17 Properties of paper and plastic sandwich core materials

Core material	*Classification*	*Kraft paper basis weight (pounds)*	*Cell size (in)*	*Phenolic resin (%)*	*Density (kg/m^2)*	*Lengthwise shear strength (kN/m^2)*	*Lengthwise shear modulus (MN/m^2)*	*Compressive strength (kN/m^2)*
Polystyrene	Plastic	—	—	—	16–32	275–550	—	140–275
Polyurethane	Foam	—	—	—	32	250	—	110–300
Verticel	Corrugated	60 (flat), 60 (flute)	¼	20	55	1120	140	965
	paper	60 (flat), 80 (flute)	⅜	15	40	690	80	565
	honeycomb	60 (flat), 60 (flute)	½	15	23	390	47	300
Union	Hexagonal	80	½	18	36	275[a]	—	1170
Camp	paper	99	¾	11	28	165[a]	—	630
	honeycomb	99	1	11	21	110[a]	—	365
Hexcel		80	½	18	35	280[a]	30[a]	965
		99	¾	11	27	210[a]	22[a]	714
		99	1	11	19	150[a]	15[a]	440

[a] Perpendicular to ribbon direction.

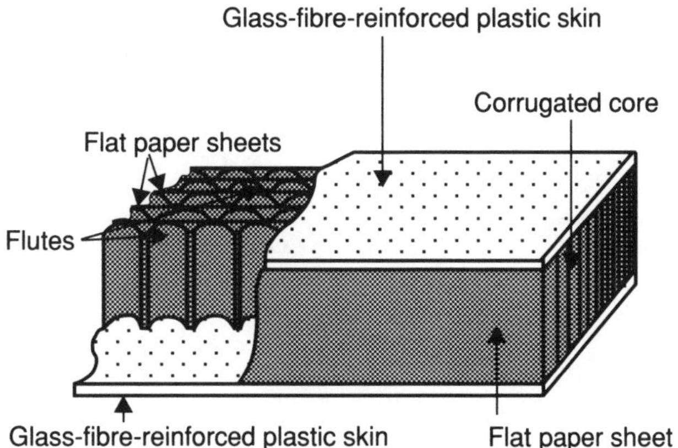

FIG 3.8.10 A sandwich panel using Verticel's paper honeycomb

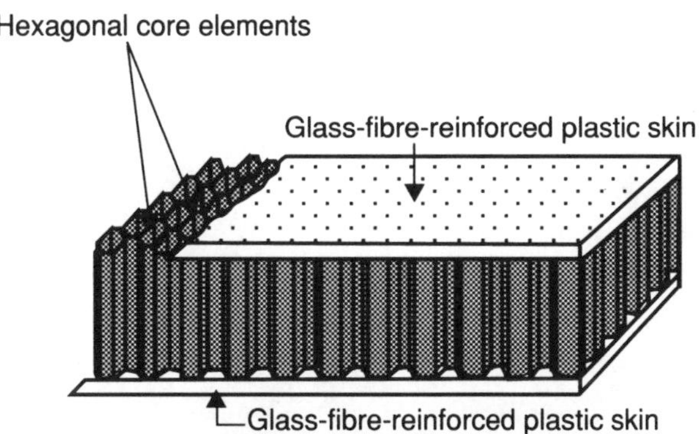

FIG 3.8.11 Panel using hexagonal honeycomb expanded core

3.8.5 Laminated plastics

3.8.5.1 General description

High-pressure rigid laminated plastics consist of superimposed layers of reinforcing materials impregnated with a resinous binder and cured under heat and pressure. The basic materials, which are available in sheet, tube, rod or moulded form can be tailored to the application by controlling the types and proportions of reinforcements and resinous binders.

The resinous binders used for plastics laminates are listed in:
Table 3.8.18 *Resins used for plastic laminates*
and the reinforcement types in:
Table 3.8.19 *Reinforcement types for plastic laminates*

APPLICATIONS

Phenolics — electrical industry, aircraft, gears and bearings (cotton-phenolics, asbestos-phenolics).
Process plant (not for alkaline environments).
Epoxy — electrical/electronic (epoxy-paper); printed circuit boards (epoxy-glass).
Melamine — marine power equipment (melamine-glass).
Silicone — high-temperature laminates requiring low dielectric loss properties.

TYPICAL PROPERTIES

The more important properties of the most commonly used plastic laminates are listed in:
Table 3.8.20 *Typical properties of plastic laminates*

3.8.5.2 Glasscloth laminates

The highest strength laminates and those most commonly used for electrical applications are glasscloth laminates.

The effect of long-term heating on the high temperature strength is shown in:
Fig 3.8.12 *Flexural strength of glasscloth laminates on heat ageing at 250°C (measured at 250°C)*

Their electrical properties and environmental resistance are listed in:
Table 3.8.21 *Properties of glasscloth-laminated boards*

TABLE 3.8.18 Resins used for plastic laminates

Resin type	***Characteristics***
Phenolic	Low cost, wide range of applications, wide experience of use. Low flammability and smoke emission.
Melamine	Electrical applications (high arc resistance), mechanical strength combined with flame and alkali resistance.
Polyester	Low flammability, self-extinguishing, suitable for mechanical and electrical applications.
Epoxy	Primarily for high humidity or chemical applications. Superior mechanical and bond strengths.
Silicone	Mainly used with glasscloth for heat resistance, electrical properties and low moisture absorption.
Polyimide	High temperature resistance.
Phenol-aralkyl	Prolonged exposure to high temperatures.
PTFE	Controlled, low dielectric constant for microwave equipment.
Vinyl ester	Corrosion resistance.
Furan	Corrosion resistance.

TABLE 3.8.19 Reinforcement types for plastic laminates

Reinforcement types	*Forms available*	*Characteristics*
Paper	Kraft	Good mechanical strength. Good dielectric strength perpendicular to laminations.
	Alpha	Good electric/electronic properties. Machinability. Dimensional stability and uniform appearance.
	Cotton linter	Greater strength than alpha combined with good moisture resistance.
Cotton	Woven fabric	Mechanical strength, increasing with weight. Lighter weights are superior for machining intricate parts.
Asbestos	Paper Mats Woven fabric	Good thermal endurance. Flame, chemical and wear resistance.
Glass fibres	Woven fabric Mats	Highest strength Superior electrical properties, maintained under high temperature and humidity. High resistance to temperature.
Nylon	Fabric	Good electrical and impact strength. Tendency to creep at elevated temperatures.
Wood	'Compreg' (Not to be confused with plywood.)	Used in electric or power applications in Europe.

TABLE 3.8.20 Typical properties of plastic laminates

		Properties	Units	Paper-phenolic	Paper-phenolic (plasticised)	Paper-epoxy	Cotton-fabric phenolic (heavy)
Dimensions		Min. thickness	mm	0.1–0.8	0.38–0.76	0.8	0.4
		Max. thickness	mm	50	9.5	6.3	5–200
Mechanical		Tensile strength, longitudinal	MN/m^2	103–138	72–86	96	62–69
		Tensile strength, transverse	MN/m^2	82–110	58–65	82	48–55
		Compressive strength, flat	MN/m^2	220–248	151–172	206	255–269
		Compressive strength, edge	MN/m^2	130–175	—	—	162–169
	(min)	Flexural strength, longitudinal	MN/m^2	93–172	83–96	138	117
		Flexural strength, transverse	MN/m^2	81–151	69–83	110	96–110
		Elastic modulus, longitudinal	GN/m^2	8.9–12.4	6.2–8.2	10.3	6.2–6.9
		(Flexure), transverse	GN/m^2	6.9–8.9	4.8–6.2	6.9	5.5–6.2
		Shear strength	MN/m^2	69–83	55–76	76	76–83
	(min)	Impact strength, flatwise	J/m	52–210	—	—	121–170
		Izod notch, edgewise	J/m	18–26	—	—	74–100
Physical		Density	g/cm^3	1.32–1.36	1.30–1.34	1.36	1.33–1.36
		Thermal expansion coefficient	10^{-5} K^{-1}	20	20	20	20
Electrical	(max)	Dielectric constant (1 MHz)	—	5.3–5.0	4.6–5.0	4.6	5.8
	(max)	Dissipation factor (1 MHz)	—	0.04–0.06	0.035–0.060	0.035	0.05–0.10
		Dielectric strength (short, 3 mm)	10^2 kV/m	185–196	167–196	180	60–145
Thermal and environmental		Water absorption (24 h, 3 mm)	%	0.95–3.3	0.55–3.0	0.5	1.6–2.5
		Max. cont. service temperature	°C	140	135–140	130	130
Remarks				Hard. Electrical properties, machinability and humidity resistance vary widely according to end-use.	More flexible than non-plasticised paper-phenolic. Tend to be slightly better electrically and can be punched. Flame retardant grades available.	Superior electricals to flexible phenolics. Excellent mechanical properties. Punchable. Flame retardant.	Tough, strong, high impact strength. Readily machinable. Some grades for electrical applications requiring mechanical strength.

Table 3.8.20

TABLE 3.8.20 Typical properties of plastic laminates—*continued*

Cotton-fabric phenolic (light)	*Asbestos paper—phenolic*	*Asbestos fabric—phenolic*	*Nylon fibre—phenolic*	*Continuous filament glass fabric—phenolic*	*Continuous filament glass cloth—silicone*	*Continuous filament glass fabric—melamine*	*Continuous filament glass fabric—epoxy*
0.15–0.38	0.8	0.8	0.4	0.15	0.15	0.15	0.10
50	50	100	25	50	50	50	50
82–90	70	82	58	160	160	255	275
58–62	55	70	55	138	127	205	240
241–255	275	260	—	345	310	480	410
162–172	117	145	—	120	96	170	240
103	90	125	70	138	158	380	380
93–96	75	110	65	125	138	240	310
6.9–7.6	16.0	11.0	4.1	10.3	9.6	17.2	18.6
5.9	9.6	9.6	3.4	8.2	8.2	13.8	15.2
79–83	62	82	95	125	117	138	130
95–148	95	190	210	345	450	635	370
53–59	32	160	160	290	400	425	290
1.33–1.35	1.72	1.72	1.15	1.65	1.68	1.90	1.82
20	15	15	80	18	10	10	9
5.8–7.0	—	—	3.9	5.5	4.2	7.1	5.2
0.05–0.10	—	—	0.030	0.030	0.003	0.017	0.025
60–145	65	20	180	240	140	140	160
1.3–1.6	0.95	2.50	0.4	2.0	0.2	0.7	0.15
130	130	155	—	180	240	140	140
Similar to heavy cotton but machines easily and recommended for fine punching, threading, close tolerance machining. Mechanical properties slightly inferior to heavy cotton.	Denser and harder than most other grades. Low moisture absorption. Resistant to heat and flame. Suitable for low voltage applications.	Stronger, higher heat resistance, more machinable than asbestos paper grade.	Electrical properties of paper, toughness of cotton. Used where high insulation properties required. Fungus resistant.	Excellent thermal endurance; mechanical strength (especially flexural, compressive, shear and impact). Very low dissipation factor.	Very good dielectric loss factor, and insulation resistance under humid conditions over wide temperature range. Excellent heat and arc resistance.	High mechanical strength and arc resistance. Excellent electric strength when wet. Flame retardant.	Excellent electrical values over wide range of humidity and temperature. Up to 50% flexural strength retained at 150 °C.

Table 3.8.20—*continued*

TABLE 3.8.21 Properties of glasscloth-laminated boards.

	Glasscloth laminates							
Property	***Phenolic***	***Special phenolic***	***Epoxy***	***Epoxy-Novolac***	***Acrlylic***	***Silicone***	***Polyimide***	***Phenol aralkyl (Xylok)***
Electrical								
Electric strength at 20°C 10^2kV/m	200	300	260	280	40	260	220	280–330
Electric-strength life at 250°C, h	144	216	700	430	0	750	144	1000–1400
Insulation resistance, MΩ	2.5×10^4	3.4×10^4	1.7×10^5	1.3×10^5	4.8×10^5	2.0×10^5	7×10^3	4.0×10^5
Comparative tracking index	195	170	180	240	115	370	180	185
Permittivity at 1 MHz								
dry	5.73	5.77	5.05	5.34	4.42	3.80	5.77	4.77
wet	5.81	5.93	5.08	5.37	4.46	3.88	5.87	4.82
Los tangent at 1 MHz								
dry	0.0122	0.0261	0.0174	0.0165	0.023	0.003	0.0218	0.0107
wet	0.0380	0.0265	0.0208	0.0182	0.024	0.008	0.0243	0.0130
Chemical resistance								
Change in weight, %, after exposure for 168 h in								
water at 100°C	1.6	—	2.0	0.6	—	0.5	1.2	0.6
10% NaOH at 90°C	destroyed	—	2.8	24.0	—	–12.5	destroyed	2.5
10% HCl at 90°C	–8.2	—	4.9	–5.0	—	–14.3	–12.5	–5.7
30% antifreeze at 90°C	9.4	—	2.0	0.9	—	1.4	0.9	0.9
engine oil at 150°C	1.4	—	0.2	0.2	—	2.3	1.3	0.1
Skydrol 500B at 100°C	1.9	—	11.6	0.0	—	Delam	0.5	0.0
transformer oil at 100°C	0.2	—	–0.5	–0.6	—	1.9	0.0	–0.2
toluene at 110°C	0.9	—	6.7	0.6	—	Delam	–0.2	–0.2
trichloroethylene at 85°C	3.3	—	13.8	1.3	—	Delam	0.2	0.0

The electric strength was measured on laminates 1.5mm thick in air between 38mm and 76-mm-diameter brass electrodes under a voltage rise of 1 kV/s. Proof testing was carried out at 11.8 MV/m for 60 s when the electric-strength life was being determined. The insulation resistance was measured according to BS 2782 and the comparative tracking index according to BS 3781. The dielectric measurements were made by Lynch's method (ERA 5183).

Table 3.8.21

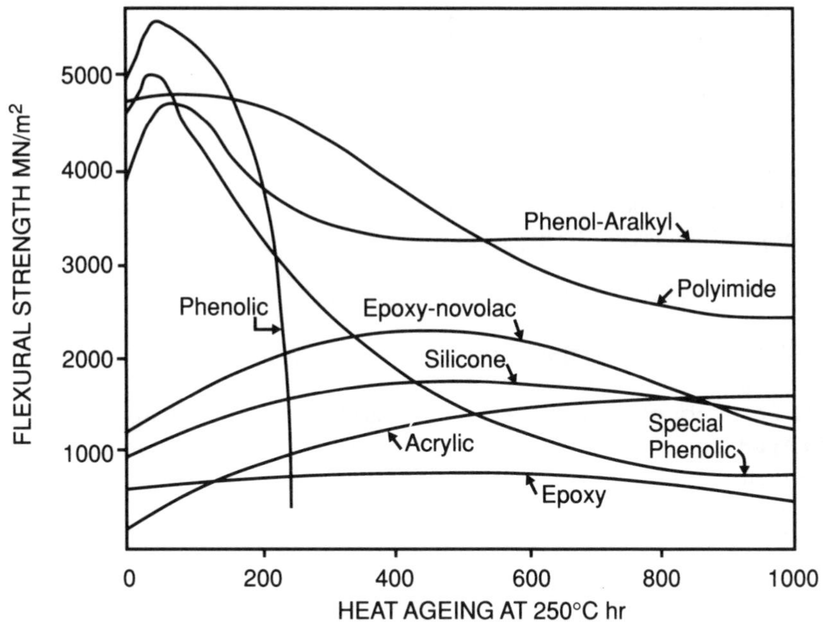

FIG 3.8.12 Flexural strength of glasscloth laminates on heat ageing at 250°C (measured at 250°C)

3.8.6 Reinforced thermoplastics

3.8.6.1 General description

The range of fibre reinforced composites has been extended to incorporate thermoplastic matrix material in addition to the previously used thermosets. The new materials may provide superior toughness, the capacity to produce hitherto unavailable designs and/or greatly increased rates of production and therefore reduced costs.

3.8.6.2 Film-stacked composites

The film-stacked composite process can produce high performance fibre-reinforced thermoplastics as alternatives to the more common epoxy or polyester laminates. Due to the flexibility of the film-stacking method, an acceptable approach where it is desired to enhance surface properties may be to substitute the outer layers of the film by one which is more appropriate to the environment under consideration.

POLYETHERSULPHONE FILM-STACKED COMPOSITES

Properties of polysulphone and polyethersulphone laminates are compared with epoxy and vinyl ester resins similarly reinforced, but by conventional methods, in:
Table 3.8.22 *Properties of polysulphone and polyethersulphone laminates compared with thermoset laminates*

Unidirectional sheet and woven fibre 'prepregs' (APC) impregnated with polyetheretherketone (PEEK) are also available commercially.

3.8.6.3 Glass filament mat-reinforced thermoplastics (GMT or GPL)

Glass filament mat-reinforced thermoplastics form a group of materials which combine simple moulding techniques of thermoplastics with a strengthening effect of filament glass fibres. Moulding—a strictly thermomechanical process—is accomplished with heat, pressure and subsequent cooling. Cycle times of about one third of glass fibre reinforced unsaturated polyester resins are claimed. Almost all available thermoplastic materials may be used as base matrix. Ribs and beads are achievable. Mechanical properties are mainly dependent on glass content. The ductility of thermoplastics provides for tough materials with high elongation at break. The new glass filament mat-reinforced thermoplastics (GMT) combine the advantages of reinforced resins—high strength, rigidity and isotropic behaviour—with the advantages of thermoplastics, achieving shape and strength via cooling in a mould. The glass content may be between 10 and 50% by weight, corresponding to 6–30% by volume. Technically most feasible is the area of 25–35% by weight.

TYPICAL PROPERTIES

The properties of GMT with a number of thermoplastic matrices are compared with those of polyester GFRP in:
Table 3.8.23 *Comparative properties of GMT*

APPLICATIONS

Applications range from lamp housings for automotive vehicles to lawn mower chassis. In Germany important developments include bumper reinforcements, motorcycle fairings, air filter housings, engine shields and brake coverings for automotive vehicles and others.

PROCESSING

The GMT processing scheme is illustrated in:
Fig 3.8.13 *GMT production principle*

3.8.6.4 Possible future applications

THE COMPOSITE BEAM-BUILDER

Towards the end of this century it is suggested that a machine may be needed to automatically produce high-volume low density structures in space. Such a machine, which manufactures aluminium beams, has been designed and developed by Grumman.

Aluminium provides a well-known structural material which is easily formed and fastened together, but it is recognised that a composite would be required to provide minimal distortion characteristics. The favoured materials are carbon/acrylic for passive structures and carbon/polyethersulphone for components operating in higher temperature regimes.

Commercial applications

Thermoformed components for aircraft: a shaft satin weave is under test at Ferranti, secondary structures for helicopters are under evaluation and prototype fire protection panels and items of internal trim are envisaged on the Fokker F29.

GEODETIC STRUCTURES

Geodetic, or Mathweb, structures are fabricated from resin-coated continuous fibres arranged in an open lattice form, the geometry of which allows fully-structured products to be designed and fabricated to meet a performance specification. Most have been fabricated to date from glass and catalytically-cured polyester, but thermoplastics have obvious potential.

Applications have been in radio masts, vehicle side-intrusion beams, structural frame members for building, trailer chassis, loading ramps, booms for dispersant spraying, underride guards for articulated trailers and specialised marine structures.

3.8.6.5 Trade names and suppliers

The suppliers of reinforced thermoplastic materials, and the commonly used trade names of their products are listed below:

Supplier	*Product*	*Trade name*
	Film-stacked composites	
Specmat (UK)	Polyethersulphone impregnated glass fibre	
ICI	PEEK impregnated industrial sheet	APC2
	Glass filament reinforced thermoplastic	
BASF		GMT, GPL

TABLE 3.8.22 Properties of polysulphone and polyethersulphone laminates compared with thermoset laminates

Resin system	Flexural strength (MPa)	Interlaminar shear strength (MPa)	Tensile strength (MPa)	Izod impact strength (kJ/m)	V_f
Vinyl ester Derakane 411–45	Mean = 1108 CV = 6.0%	Mean = 57.0 CV = 2.9%	Mean = 1085 CV = 6.0%	Unnotched= 1.54 Notched = 1.22	0.41
Cold-set epoxy Shell 162/113	Mean = 1122 CV = 8.7%	Mean = 43.8 CV = 4.6%	Mean = 858 CV = 8.9%	Unnotched= 1.84 Notched = 1.08	0.42
Hot-set epoxy Ciba MS 1778	Mean = 1151 CV = 8.3%	Mean = 46.2 CV = 6.0%	Mean = 969 CV = 7.9%	Unnotched= 2.49 Notched = 2.10	0.43
Polysulphone Union Carbide P1700	Mean = 1311 CV = 4.6%	Mean = 39.7 CV = 8.9%	Mean = 1184 CV = 3.2%	Unnotched= 2.50 Notched = 1.80	0.65
Polyethersulphone Victrex (ICI) 200P[a]	Mean = 1213 CV = 3.2%	Mean = 40.6 CV = 8.6%	Mean = 1035 CV = 5.7%	Unnotched= 2.50 Notched = 1.80	0.54

[a] Manufactured by Specmat Ltd. (UK).

TABLE 3.8.23 Comparative properties of GMT

Property	Unit	Glass filament mat-reinforced thermoplastic materials (GMT)			GFRP (Polyester)
		HDPE	Polypropylene	Nylon B	
Glass content	% by weight	23–35	25–35	25–35	30–35
Tensile strength	MN/m^2	45–65	50–70	80–100	120
Bending strength	MN/m^2	60-90	70–100	—	150–200
E-Modulus	MN/m^2	3700–5400	3900–5500	5200–7500	11 000
Linear elongation	%	2.9	3.1	4.5	3.0
Energy absorption (area under stress–strain curves)	N/mm	1.3	1.5	2.2	1.0
Specific gravity		1.14–1.23	1.08–1.17	1.18–1.35	1.45
Coefficient of thermal expansion	K^{-1}	$25 \cdot 10^{-6}$	$25 \cdot 10^{-6}$	$24 \cdot 10^{-6}$	$28 \cdot 10^{-6}$
Loss factor	10^6 Hz	0.0027	0.002	0.065	0.19/0.021
Dielectric constant	10^6 Hz	3.1	2.6	4.7	6.2/3.9
Max. temperature of use	°C	110	130	170	180

Data supplied by BASF Elastogram.

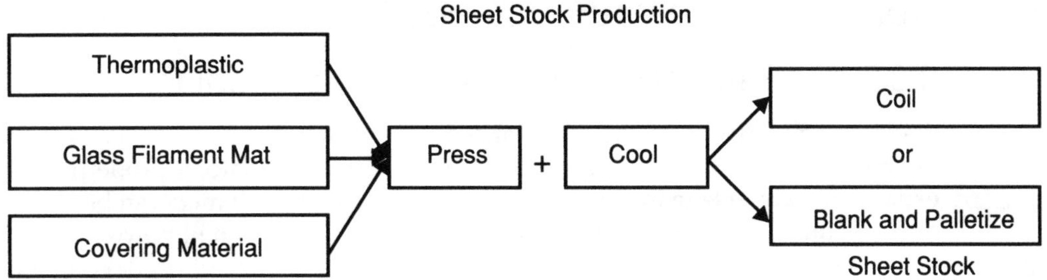

FIG 3.8.13 GMT production principle

3.8.7 Conducting polymer composites

3.8.7.1 General characteristics and properties

Polymeric materials unfilled, or enhanced with most of the conventional fillers and reinforcements, are outstanding electrical and thermal insulators, properties which are exploited to advantage in many of their uses. However, polymers can be made moderately conductive in varying degrees by the addition of carbon in powder or fibre form, or by the incorporation of metals in the forms of powder, fibre, flake or metallised glass. Aluminium tends to be the favoured metal, but iron, copper, nickel and even silver may be encountered according to end-use requirement. The degree of conductivity required of a polymer also depends upon the nature of the end-use, the most common of which are summarised below.

STATIC ELECTRICITY

This is prone to buildup on plastic parts, and can be a serious problem causing sparking and surface-dust pick-up. In some applications the problem is particularly pernicious, including wheels and rollers; conveyor belts and plates which convey dusty products where a static spark could trigger a dust explosion; sensitive instrumentation such as medical electronics, where erroneous readings or loss of data can be caused by a sudden static discharge with obvious catastrophic results.

Accumulated surface charges can be dissipated by conduction within or along the material surface, and by gradual decay to the air. Resistivity, the reciprocal of conductivity, is a measure of the electrical charge conducted through and along the material surface. So materials having a low resistivity will exhibit good anti-static properties.

Metallic conductors have surface resistivities which range from 10^{-6} to 10^{-8} ohm-m. Polymeric insulators have surface resistivities of 10^{14}–10^{16} ohm. Bridging this conductance gap are the carbon or metal-filled polymer systems. These may be classified according to surface resistivity. 'Antistats' used primarily in packaging in the form of sheets, films and elastomeric foam are in the range 10^{5}–10^{12} ohms. 'Semiconductive' composites in the range 10^{4}–10^{7} ohm represent the greatest area of current activity including part storage bins, housings and structural applications.

The resistivity covered by conductive polymer systems is shown diagrammatically in:
Fig 3.8.14 *Approximate ranges of surface resistivity of engineering materials*

Their thermal conductivities are compared with those of other engineering materials in:
Fig 3.8.15 *Thermal conductivity of engineering materials*

EMI ATTENUATION

Stray electromagnetic interference (EMI) must be guarded against in digital and analogue logic circuits for computers or control systems. The sources of EMI are frequently receptors as well, emitting EMI at a number of frequencies and susceptible at a number of frequencies.

Equipment may be protected using an enclosure which shields out extraneous EMI and keeps in any generated by the equipment. These shields have historically been metal, the housing or cabinet itself.

Conductive polymers combine the shielding aspects of metal with the inherent advantages of the plastic moulding process. EMI attenuation properties depend on both electrical conduction and *energy-reflection properties*. Metal-filled and carbon fibre-reinforced plastics perform well in these modes. This third category of polymers is truly 'conductive' with resistivities 10^{2} – 10^{-2} ohm.

The effectiveness of specific conducting polymers in reducing EMI over a range of frequencies are shown in:
Fig 3.8.16 *EMI attenuation properties of conducting polymers*

OTHER APPLICATIONS

Conducting polymers are also used in resistive heaters and heat sinks, the latter utilising thermal, as opposed to electrical, conductivity properties. They may also be used to provide conductive ground paths, to bond or 'solder' at low temperatures to carry out field or emergency repairs, or to fill gaps with conductive materials where this is required.

AVAILABLE FILLERS

Conduction properties of a polymer depend upon the number of particle contacts and the average interparticle distance. These parameters, in turn, are a function of particle shape, concentration and dispersion.

Metallic silver is probably the most effective addition to promote conductivity.

Carbon blacks have long been used to impart antistatic and conductive properties to plastics and elastomers for wire and cable.

Carbon fibres provide a graphite conductor of different geometry and surface chemistry. Although significantly more expensive than carbon powder, their mechanical properties exceed those obtainable with glass fibres and they are therefore valuable where a combination of specific strength or stiffness and conductivity is needed.

Carbon powder composites are more conductive than carbon fibre composites at equal loading, but the resistivity of carbon fibre composites varies less markedly with carbon content. Thus use of carbon powder composites is limited to electrical applications in which a wide range of conductivity values is allowable. At loadings greater than 15–20% volume of carbon fibres, electrical conductivity increases minimally with further addition of fibres.

Aluminium fibres give the highest conductivity at the lowest loading of the commercially available systems. The fibres are available in lengths up to 18mm and in aspect ratios of up to 300. Because of their energy reflectivity as well as electrical conductivity they are in widespread use for EMI shielding. However, while the highest aspect ratio fibres are well suited to compression moulding, attempts to use them in injection mouldings can lead to lower conductivity, attributed to curling and breakage of the fibres in high-shear processing. Aluminium flakes are therefore available for injection mouldings.

Finally, metallised glasses are available (Hexcel Corp—trade name Thorstrand) in chopped form, roving or mat and give conductivity at relatively low concentrations. They are the most economic fibre; the glass is generally 12–20μm in diameter, and the aluminium coating 2.5μm in thickness. However, they are brittle and prone to break up during injection moulding. Their effectiveness to date seems confined to compression mouldings and woven prepregs.

The influence of filler type and quantity on resistivity is shown in:

Fig 3.8.17 *Effect on volume resistivity of filler type and quantity for injection moulding*

3.8.7.2 Suppliers and trade names

Suppliers and trade names for conductive polymer systems are:

Supplier	*Trade Names*	*Notes*
Emerson Cuming	Eccoshield	Sealants, gaskets, tapes and grease
	Eccobond	Adhesives
	Eccocoat	Coatings

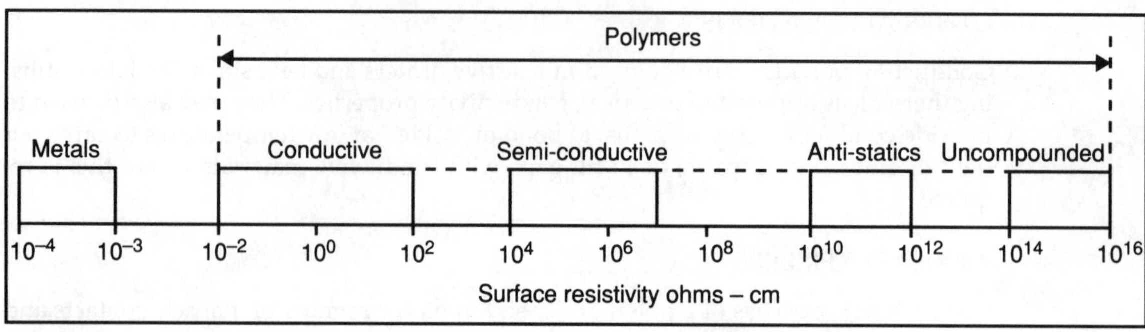

FIG 3.8.14 Approximate ranges of surface resistivity of engineering materials

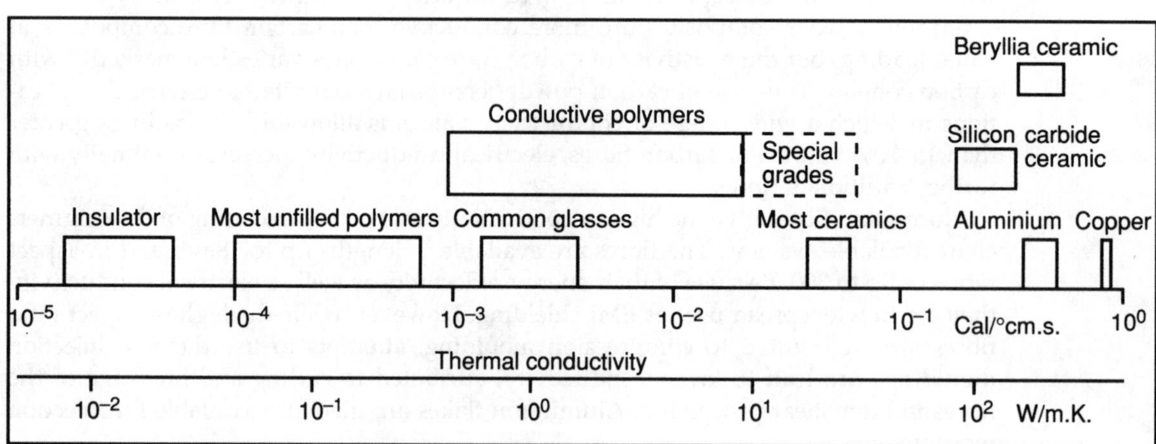

FIG 3.8.15 Thermal conductivity of engineering materials

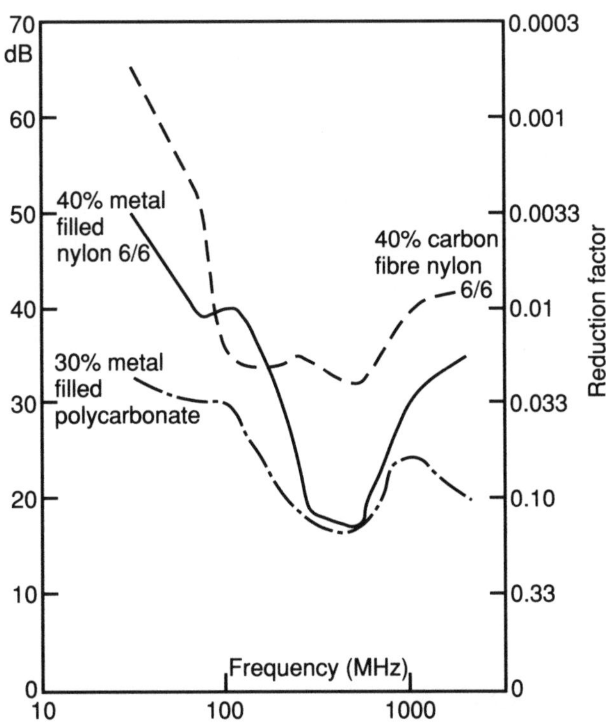

FIG 3.8.16 EMI attenuation properties of conducting polymers (courtesy LNP)

Fig 3.8.16

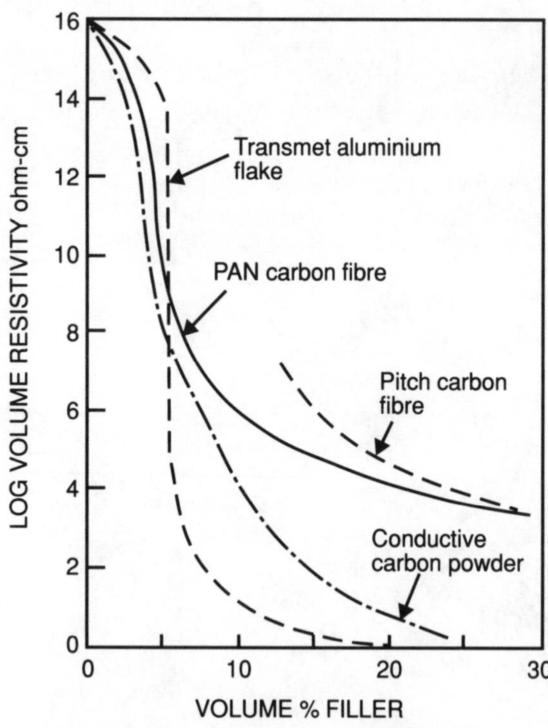

FIG 3.8.17 Effect on volume resistivity of filler type and quantity for injection moulding

Fig 3.8.17

3.8.8 Asbestos products and potential substitutes

3.8.8.1 Characteristics

Asbestos is a naturally occurring fibrous silicate which can be fabricated into a wide variety of forms, including textiles, paper and board. It is used because of its fibre characteristics, as a reinforcement material in cement and resinous matrices.

The three significant forms of asbestos are:

CHRYSOTILE (WHITE)

This is the most commonly found asbestos mineral and accounts for about 96% of total annual production. It is characterised by relatively long, flexible fibres, and resistance to degradation by alkaline attack. Principal sources of chrysotile are Canada and the USSR, although little fibre is imported into the UK from the latter.

AMOSITE (GREY-BROWN)

This has superior resistance to chrysotile and its principal use in the UK is for the manufacture of insulating boards. The fibres have a harsh, brittle acicular texture.

CROCIDOLITE (BLUE)

This fibre has a variable texture, between flexible and brittle, depending on source. Although this mineral has greater fibre strength and is more resistant to acid attack than both the above types and can be used in conjunction with chrysotile in the manufacture of asbestos-cement products, its importation into the UK has been effectively banned on health grounds since 1970. A few imported products may however, be made from crocidolite (blue asbestos) and it is to be found in the form of spray applied insulation in buildings constructed before stringent restrictions were imposed upon its use in 1970.

Amosite and crocidolite are both members of the group of asbestos fibres known as amphiboles, and their principal source is Southern Africa. The other two commercial forms of asbestos, the amphiboles, anthophyllite and tremolite, are little used in the UK Anthophyllite is used in the USA in cement and plastics products.

LEGISLATION CONCERNING THE USE OF ASBESTOS AND ASBESTOS PRODUCTS

In many countries throughout the world stringent precautions have been introduced concerning the use of asbestos and acceptable environmental exposure levels. Frequently, as in the UK, crocidolite asbestos has been effectively banned and in some countries, notably Sweden, Denmark and Japan, legislation is even more stringent. In considering the applications of asbestos therefore, it is necessary to consider what alternatives are available.

3.8.8.2 Properties

The numerous applications of asbestos are a consequence of its desirable physical and chemical properties, combined with a low material cost. It is this unique combination that makes the replacement of asbestos very difficult in many applications.

Some of the properties of asbestos are summarised together with the comparable properties of some of the synthetic fibre materials that have been suggested as replacements for asbestos in some applications, in:

Table 3.8.24 *Properties of materials commonly considered as asbestos substitutes*, and
Table 3.8.25 *Costs of materials commonly considered as asbestos substitutes*

Some of these properties warrant further comment.

THERMAL PROPERTIES

The most widely known property of asbestos is its heat and fire resistance, although this resistance does not go as far as is popularly believed. Asbestos cannot be classed as refractory, although normally its properties are sufficient to withstand superheated steam and other elevated temperature industrial environments. Degradation of the crystal structure of asbestos and major loss of strength occur at temperatures in the range 300–500°C. However, useful performance can be obtained at higher temperatures than this; specified working temperatures for some asbestos products may be as high as 600°C.

Further, the solid decomposition products are inert and of low thermal conductivity, providing additional protection to the remaining fibres, and maintaining structural integrity. It has been shown that, in some cases, asbestos can maintain its integrity at temperatures up to 1700°C.

MECHANICAL PROPERTIES

Values quoted for the strength of asbestos fibres are very high. However values quoted may not tell the whole story. The measurements of strength are inevitably derived from testing in controlled laboratory conditions, and the values obtained may not be representative. Discussion with suppliers of asbestos has suggested that the reliable strength of chrysotile fibres, as produced and used commercially, is no higher than approximately 700 MN/m^2.

OTHER PROPERTIES

Various other properties make asbestos a valuable material. For instance, its resistance to chemical and biological attack is valuable in applications involving hostile environments, and in achieving a useful service life.

The friction and wear characteristics of chysotile and its thermal decomposition product forsterite, a non-fibrous silicate, make chrysotile a widely used material in such applications as friction clutches, brake linings and bearings.

The high aspect ratio of asbestos fibres makes them useful as a mechanical reinforcement in both polymer and cement-base products.

3.8.8.3 Applications

The applications of asbestos, and the materials that are now, and may in future, be employed in substitution for it are listed in:

Table 3.8.26 *Major applications of asbestos*

3.8.8.4 Thermal insulation and dry packings

Heat resistant fibres, which can be made into textiles, extend over a range of service temperatures from 200°C to above 1200°C. A heat resistant textile must withstand extreme conditions such as welding sparks, molten metal splash and bare flame hazards. They may be worn as personal protective garments. Other applications of heat resistant textiles include: safety curtains in theatres; welding screens; fire blankets; flange seals on oven and boiler doors; thermal insulation on pipe work, boilers etc. (not exclusively textiles); conveyor belting; flue curtains; hot gas filtration; fire barriers in electrical switchgear and for wrapping field and armature coils in large power generators; heavy-duty friction materials; heat resisting, rigid-reinforced plastic laminates; and packings and jointings. Some of these applications are considered in separate sections.

Asbestos textiles, contrary to their popular image, occupy a position near the middle of this range, with a service temperature of about 600°C. Aluminised asbestos cloth is the only fabric amongst the heat resistant textiles which can be safely used for fully protective fire-fighting garments where contact with bare flame is likely. In situations involving molten metal splashes and welding sparks, asbestos cloth can be used at temperatures well above 1500°C.

Asbestos cloths have good drape characteristics and are widely used in safety curtains or screens in many situations. They also crease and fold easily, so that when they are applied as thermal or electrical insulation barriers in industrial machinery, boilers and power station equipment, they can form to tight corners and uneven surfaces without difficulty.

Below asbestos in terms of temperature performance stand the synthetic organic fibres and glass fibre; above it stand the inorganic refractory fibres. Of the many substitutes, most are expensive. Of those which can withstand comparable or higher temperatures than asbestos, only glass and the cheapest of the aluminosilicate ceramic fibres come close to being competitive with asbestos.

Glass textiles, like asbestos, lie in the middle of the temperature range of heat resistant fabrics. Continuous filament glass fibre (A, E and C glass) in the form of textiles and fabrics has overtaken the manufacture of asbestos based textiles and is used in many traditional applications of the latter. It now competes more with synthetic organic fibres than with asbestos.

While the protective clothing garments of all reputable manufacturers conform to British Standard Specifications (BS 3119 and BS 5438), the appropriate tests are conducted in the laboratory using small flames, even though it is recognised that scaling effects can be very important in determining behaviour. Work carried out under the auspices of the CEGB has led to the CEGB National Safety Code of Practice; GS:EII.10, Materials and Equipment for Personal Protection against Flames and Heat.

Glass, asbestos and ceramic fibre textiles are also extensively used as protective outer coverings on other, non-load-bearing forms of insulation, which nowadays is likely to be one of the glassy man-made mineral fibres (MMMF). MMMF is a generic term used to describe the amorphous glassy fibres made from molten blast furnace slag or other readily fusible slags, natural rocks and minerals, basalt, diabase, olivine and borosilicate or calcium aluminium silicate glass. These are frequently referred to collectively as mineral wools.

Generally, these fibres have much lower strengths than those of asbestos. The surfaces of fibres are sensitive to rubbing which creates surface defects, which in turn reduce the mechanical properties of the fibre. Furthermore, all the MMMF materials lose strength from about 300°C upwards and soften in the range 650–700°C. Conversely, asbestos may change structurally at these temperatures but does not melt below 1000°C.

The refractory fibres, primarily aluminosilicates and higher performance derivatives, have a temperature limitation in continuous use of 1260°C or higher. This means that they are primarily intended for applications where asbestos is no longer useful. Examples would be high temperature insulation above 700°C, heat shields and insulation seals. Refractory fibres may also be used in furnace linings, where they are claimed to be more efficient and more economical than refractory bricks.

The ceramic fibres developed a reputation as alternatives to asbestos at a time when the incentive to replace the material was high enough to accept their high cost and a number of drawbacks, such as poor abrasion resistance and resistance to flexing. With the development of many materials based on glass fibres, it is probably no longer true to treat the ceramic fibres as primary asbestos substitutes.

The temperatures up to which the thermal insulators described may be employed are illustrated diagrammatically in:

Fig 3.8.18 *Maximum temperature capability of thermal insulants*

3.8.8.5 Reinforced cement

Alternative fibres which may be considered for inclusion in fibre reinforced cement (FRC) provide adequate reinforcement at much lower loadings, or 'furnishes', than required of asbestos fibres. But while the longer fractions of asbestos make a major contribution to reinforcement in asbestos-cement, the shorter fractions of the fibre have an important function as a retention carrier for cement particles in the manufacturing process.

The most frequently quoted and realistic contenders as direct substitutes for asbestos in FRC are glass fibres and fibrillated polypropylene. However, both have deficiencies in comparison with asbestos fibre and in both cases there is a limit to the proportion of these materials than can be successfully incorporated into the cement matrix.

The overall progress of glass reinforced cement (GRC) as an asbestos-cement substitute is still disappointing, though it has been used for over a decade. GRC products are currently being made in the UK, West Germany, Japan and Australia. The properties are similar to those of cements based on fibrillated polypropylene—high impact strength, good flexural strength and pseudo-ductility. The range of products based on Cem-FIL include: continuously-produced medium density flat sheet (Northwich Board), promenade tiles, low pressure pipes, roof slates and fittings and corrugated sheet properties for cladding and roofing.

Basically, there are three potential methods of producing substitute products for asbestos-cement using glass fibres:

(a) Application of established glass fibre reinforced cement processes such as direct spray, premix casting or press moulding. These have been developed since 1969 when Cem-FIL alkaline resistant fibres were first introduced.
(b) Direct substitution of glass fibre for asbestos in traditional asbestos cement processes: Hatschek, Magnani, Mazza, etc.
(c) Development of purpose-built, high throughput machinery.

CELLULOSE FIBRE

Cellulose fibre is light, has a moderate strength and similar capacity to carry cement particles to that of asbestos; it is a likely candidate to replace asbestos. However, because it is combustible and may be degraded at quite low temperatures, it can only be a partial substitute for asbestos in asbestos-cement sheet products (as in Japan) or otherwise as a fibrous reinforcement in conjunction with mineral wools and certain minerals, in order to densify the product and make it more fire-proof.

High quality cellulose fibre has more in its favour as a replacement for asbestos than other types of potential reinforcement in cement. Inexpensive and abundant, cellulose fibre can have good reinforcing properties, depending upon how it is used. Finnish wisakraft pulp is an exceptionally good, specially refined type, in which lignin, believed to have an inhibiting effect on the curing of the cement, is removed by bleaching.

SYNTHETIC ORGANIC FIBRES; PVOH

Polyvinyl alcohol (also known as PVOH—or sometimes wrongly as PVA, which can be confused with polyvinyl acetate) fibres are tough, easily processable and durable. First developed in 1939, there has been a revival of interest in this fibre. The chemical and saline resistance is demonstrated by their use in Japanese fishing nets. They have good tensile strength and are compatible with cement. Compared with other organic fibres, PVOH fibres have a relatively high modulus, except under prolonged high temperature exposure, when polyester fibres are superior. Even so, on a weight-for-weight basis, although having about half the density, PVOH fibres are much more expensive than glass.

STEEL FIBRES

Steel fibres offer a low reinforcing value and difficulty of dispersion. There is some service experience of both chopped and continuous steel wire in load-bearing situations.

The largest application area by far has been for airport pavements. Other included factory floors, hydraulic structures (where resistance to cavitation erosion by high velocity water flow is a prime factor) and shotcrete for mine or tunnel linings.

The fibres are 2–3 times more expensive than asbestos but usually only 1–2% are incorporated. The system suffers from the same drawback as the Cem-FIL process for GRC in that the drainage characteristics of the mix demand specially developed production techniques. Thus a high capital outlay is involved.

Nevertheless, the attitudes of both producers and users have changed considerably over the last two decades. The characteristics and properties are now quite well established, although not always fully understood by designers encountering the material for the first time. The material has also changed in detail: no longer do the fibres constitute a convenient means of disposing of large volumes of what were essentially reject materials from the steel industry. Instead, it is recognised that the fibres should be 'designed' to fulfil specific requirements; smooth, straight, uniform fibres have virtually disappeared, replaced by fibres with rough surface, with enlarged or hooked ends, to improve their resistance to pull-out from the matrix. These features also permit fibres to be shortened, facilitating better mixing and more uniform fibre dispersion in freshly-mixed concrete. The fabrication of fibres in bundles held together with water-soluble glue helps to accomplish the same objective. To improve durability, corrosion-resistant carbon steel and stainless steel fibres are now available as alternatives to plain carbon steel.

3.8.8.6 Building boards

A comparison of properties of asbestos-containing and asbestos-free building boards is given in:

Table 3.8.27 *Properties of some asbestos-based (now withdrawn) and asbestos-free flat sheet and building boards*

Assessment of possible alternative fibres is given in:

Table 3.8.28 *Assessment of fibres as possible replacements for asbestos in building boards.*

3.8.8.7 Seals and gaskets

Seals fall into two categories-

(a) Static seals, where there is little or no movement between the adjacent surfaces;
(b) Dynamic seals, where sealing takes place between surfaces that have relative motion. The most common examples are seals between housings and rotating and reciprocating shafts, rods, etc.

Gaskets are probably the most common form of static seal, but other popular types, notably the O-ring and its variants, are widely used for static and dynamic applications. Other devices, such as bellows and gaiters, accommodate larger relative movements, but are generally regarded as flexible static seals.

In some countries, import regulations and compulsory marking for products containing asbestos have been in force for some considerable time. As long as there is no general prohibition for use of seals and gaskets containing asbestos, the substitution over time for asbestos in this area will depend largely upon the price/performance ratio of the alternatives described.

The technical properties of jointing materials based on substitutes for asbestos, as described in the literature, require qualification in terms of both mechanical performance and mechanical application. The tolerances of glass or aramid fibre-based jointings to high stress levels at elevated operating temperatures fall short of those of composite asbestos fibre (CAF) jointings and these alternative products may only be used with confidence when they are applied under carefully controlled seating stress conditions and at thicknesses less than those traditionally used with CAF jointings. This situation has invoked more precise mechanical standards for flange facings, bolts and above all for the handling and fitting of alternative jointing products.

Properties of materials used in packing and jointings are given in Table 3.8.29:

Table 3.8.29 *Properties of basic materials used in packings and jointings*

3.8.8.8 Millboard

Millboard is one of the most versatile asbestos materials used in industry. Features of millboard which contribute to this versatility are: ease of cutting or punching to size or shape; useful thermal insulation properties; the ability to be impregnated with bonding agents and cements; the ability to be wet-moulded; compressibility.

Millboard finds applications wherever a low cost, relatively soft, low density board material with modest mechanical properties, high heat resistance and good thermal and electrical insulation characteristics are required. Examples include the fabrication of rollers for the transport of hot materials in the steel and glass industries, as formers for wire-wound electrical resistances, flange gaskets for joints in ducting and trunking used for high volume/low pressure gas transport, cylinder head gaskets, insulating linings to minimise heat losses from ovens and moulds, as plugs and stoppers for molten metal containers and, in resin impregnated form, for clutch facings and for low voltage electrical insulation at temperatures up to 180°C.

Table 3.8.30 *Properties of non-asbestos millboard. TBA industrial products*

For many applications, millboards made from aluminosilicate fibres or mineral wool fibres, bonded with a high temperature inorganic (silica) binder can provide a direct replacement, although they tend to be more expensive. They are available in thicknesses up to 50 mm and are made by a suction method from an aqueous slurry of fibres.

For less stringent applications, cellulosic fibres have been proposed as a base for these products: for example, a combination of bleached sulphite pulp, diatomaceous earth, barytes with a cement binder; or cellulose fibre with mica and ball clay and a starch binder. Such materials will have limited thermal properties, but with resin impregnation will suit a number of jointing and gasket uses.

Many indirect substitution possibilities exist by consideration of the individual application area. For example, for thermal insulation applications, if the thermal and mechanical conditions be not too severe, one of the several types of mineral fibre block and slab products should be an effective substitution.

3.8.8.9 Dry rubbing bearings

A significant application of asbestos composite materials has been in plain rolling bearings. In this application, a matrix of thermosetting phenolic resin is used to impregnate asbestos cloth or yarn. The principal advantage of these materials is that, although they can be lubricated with oil or grease and, in some cases, are supplied impregnated with up to 7% mineral oil, they are also able to function effectively without lubrication, or lubricated by in-situ process fluids or sea-water.

REINFORCED THERMOSETS AS DRY RUBBING BEARINGS

Asbestos composite bearings belong to the reinforced thermoset family of plain bearing materials. Common reinforcements are textiles, such as cotton, polyester, asbestos and other fibres. Solid lubricants are often incorporated, either into the surface layers or as a uniform dispersion. Reinforced thermosets provide a higher stiffness bearing than is possible with thermoplastics and can usually tolerate short periods of high temperature operation. On the other hand, wear rates are usually higher than that of filled PTFE. Liquid lubrication is usually beneficial. Reinforced thermosets find application in journal and thrust bearings, or for duties involving contamination by abrasives and/or water.

Among the other members if this family which do not contain asbestos are:

(a) polyester bonded textile laminates with molybdenum sulphide (MoS_2) or graphite;
(b) cellulose-fabric based phenolic laminates with uniformly distributed PTFE or graphite.
(c) various proprietary bearing materials incorporating unspecified organic fibres.

The members of this family have broadly similar physical and chemical properties. However, the maximum operating temperature of asbestos-reinforced composites is generally higher; up to 175°C; compared with 100–130°C for other composites. Also the coefficient of linear expansion is lower.

For applications in which such substitutes are not acceptable, alternatives have to be sought from other families of plain rubbing materials.

OTHER CLASSES OF PLAIN BEARING MATERIALS

Unfilled polymers

Unfilled polymers can provide an inexpensive bearing with good resistance to abrasion and tolerance of misalignment in, for example, simple bushes and thrust washers, such as might be found in consumer durables. Operation at high loads and temperatures is rarely possible, though it is well known that special high temperature polymers have now been developed. Plastic-based materials have comparatively high coefficients of thermal expansion and, as a consequence, bearing clearance tends to decrease with rise in temperature. Some types are also prone to swell in water.

Filled polymers

The addition of a filler to a polymer improves mechanical properties, but abrasion resistance may be reduced. Other fillers include carbon fibres. Liquid lubricants, such as mineral oil, may also be incorporated. Filled polymers are used for various general applications in, for example, consumer durables and automobiles.

High temperature polymers

These polymers retain their mechanical and tribological properties at elevated temperatures (up to about 250°C). They are available in either unfilled form or with the addition of fillers, such as graphite and carbon fibres. They are, however, currently more expensive than other polymers.

Thin layer materials

In thin layer materials, the backing may consist of a metal, a filament wound fibre/thermoset or a textile-reinforced thermoset. All current types use a PTFE-based surface layer. Metal backings will offer a coefficient of expansion similar to metal housings into which the bearing may be fitted, but may require protection in corrosive environments. Metal backings can also be used to dissipate heat.

Thin layer bearings are generally not recommended for abrasive conditions and are rather less tolerant of misalignment than other types. They are used extensively as pivot bearings and slideways and in control linkages, notably in the aerospace industry.

Filled PTFE

There are more types of filled PTFE available than of any other composite bearing material. Common fillers include glass, graphite, minerals and metals, such as bronze. Although these are suitable for a wide operating temperature range and have good chemical resistance, it is not always appreciated that the presence of a liquid may impair performance. Neither are they suitable for high specific loads, though high speeds may be possible. Filled PTFE is used extensively for compressor piston rings and machine tool slideways, sometimes at cryogenic temperatures.

Solid lubricant impregnation metals

The solid lubricant, commonly graphite, is either uniformly distributed throughout the metal matrix or is incorporated as discrete inserts.

These materials have good electrical and thermal conductivity and a coefficient of thermal expansion that is similar to the housing. They can be used at relatively high loads and temperatures and are tolerant of contamination by abrasives and fluids. They find application in particular in the mining, metallurgical and offshore industries.

Many uses where impregnated metals are technically and economically satisfactory could also be adequately covered by the family of oil-impregnated porous bearings.

Carbon graphites

Carbon/graphites can be used at relatively high temperatures (limited in air by oxidation) and high sliding speeds, but the materials are brittle and have a low tensile strength which limits the allowable bearing pressure. Resin impregnation improves the strength and wear resistance, but at the expense of the high temperature capability.

Carbon/graphites have good thermal and electrical conductivities (although care is required when electrolytes are likely to come into contact with any metallic components of the bearing), but the low coefficient of thermal expansion necessitates shrink fitting of the bearing in the housing. Carbon/graphites are not usually recommended for abrasive conditions and tolerance to misalignment is poor. Also, applications in vacuum or dry gases are not recommended, unless the material has been impregnated with a suitable additive.

Corrosion resistant counterface materials, such as stainless steel, are recommended and a hardness exceeding 400 VPH (as provided by stellite and ceramic coatings) is often preferred to minimise abrasive wear.

The forms of dry rubbing bearings that are readily available are given in:

Table 3.8.31 *Dry rubbing bearings; forms readily available*

3.8.8.10 Reinforced composites

Excluding such composite products as are dealt with in other sections (friction materials, bearings, pipes, etc), there remains a wide range of asbestos/polymer composites in use across the engineering spectrum. These are usually made with a thermoset matrix, although composites using thermoplastic matrices, such as nylon, polypropylene and PVC, are in fairly widespread use. Typical applications are for small machine parts, usually made by injection moulding, impact moulding, single press-action sheet moulding, extrusion or hand lay-up methods. Parts include vehicle distributor caps, fans, fan shrouds, small casings and other similar products. In these applications, the fibres are randomly aligned in three dimensions.

The main argument for economical reinforcement of plastic composite materials revolves around asbestos and glass fibres. Up to the 1970s, little advantage was gained by replacing asbestos with glass fibre. Since then, the situation has changed, with the development of finer chopped strand glass fibres which are less susceptible to degradation in compounding and with the introduction of coupled glass fibres, which give greatly improved physical properties to reinforced plastics.

In the Americas and Western Europe, asbestos has been partly displaced in reinforced plastics by other materials, mainly glass fibres. Over the past 10 years, glass fibre reinforcement has almost entirely replaced asbestos in phenolic moulding, though the residual Western consumption, added to consumption in Eastern Europe and the Far East, suggests that a substantial amount of asbestos may still be used in this application area.

Generally, the maximum continuous operating temperature for a glass-reinforced composite is comparable to that of the corresponding asbestos material, although with glass, mechanical properties may be retained at higher temperatures for longer periods of time.

The Ferobestos range of materials shown in Table 3.8.32 comprises sheets, rods and tubes made from asbestos cloths, felt or paper, bonded with phenolic or silicone thermosetting resins. The main characteristics are a high degree of wear resistance, plus thermal stability and corrosion resistance. Much of the production is in the form of fully machined components, which include bearing bushes, bearing washers, machine tool slides, rotor blades for compressors and heavy steel rolling mill bearings.

Table 3.8.32 *Comparison of properties of asbestos and asbestos-free laminates and composites*

3.8.8.11 Friction materials

Various other materials, including some fibres, have been used to substitute for asbestos which, historically, have all had their drawbacks. Most well-known manufacturers of friction materials in the UK have had asbestos-free products on the market for up to 5 years. This is an area where the companies concerned are generally unwilling to divulge the nature of the substitutes used, though to a certain extent their efforts are revealed in patents.

A considerable volume of work has been carried out on the use of steel and glass fibres.

Table 3.8.33 *Advantages and disadvantages of some asbestos alternatives in friction products*

TABLE 3.8.24 Properties of materials commonly considered as asbestos substitutes

Heat resistance	*Reinforcing strength*	*Chemical resistance*
Good (above 400°C) Mineral wool Refractory fibres All minerals Steel fibre (asbestos)	***Good*** Aramid fibre Carbon fibre PVA Glass fibre Steel fibre (asbestos)	***Good*** Polybenzimidazole (PBI) PAN PTFE PVA PP Carbon fibre Refractory fibre, except in alkalis Most minerals (asbestos, except in acids)
Moderate (200–400°C) Aramid fibre PVA PTFE PBI PAN Carbon Glass fibre	***Moderate*** Cellulosic fibres PP PAN Refractory fibre Mineral wools	***Moderate*** Aramid fibre Steel fibre
Poor Cellulosic fibres PP	***Poor*** All minerals PTFE	***Poor*** Cellulosic fibres

TABLE 3.8.25 Costs of materials commonly considered as asbestos substitutes

Less than £1/kg	*£1–5/kg*	*£5–10/kg*	*Over £10/kg*
Mineral wool Glass wool Cellulose pulp Vegetable fibre Perlite Diatomite Mica Vermiculite Talc Wollastonite Attapulgite, Sepiolite (Asbestos, grades 4, 5, 6, 7)	Continuous Filament glass Alkali-resistant (AR) glass Polypropylene (PP) Polyvinyl alcohol (PVA) Aluminosilicates Steel fibres (Asbestos grade 3)	Continuous Filament glass Polytetrafluoroethylene (PTFE)	Aramid fibre Pitch carbon fibre Polyacrylonitrile (PAN) carbon fibre Alumina fibre Silica fibre

TABLE 3.8.26 Major applications of asbestos

Asbestos product		***Properties***	***Application areas***	***Potential substitutes***
Asbestos textiles	Cloth Webbing	Incombustibility Resistance to corrosion	Fireproof clothing, safety curtains. Thermal insulation material	Aramid (Kevlar/Nomex) fibres (aluminised) Special wood blends Glass/ceramic fibre composites
	Tape Paper	Electrical insulation (iron-free) Thermal insulation	Heat resistant electrical insulation (not suitable for high voltage or high frequency insulation).	Ceramic fibre cloth, tape or sleeving with or without glass or metal inserts. High temperature plastics (e.g. polyimide, polyethersulphone). Polymer films.
Jointing and sealing materials, gaskets		Mechanical strength Stiffness Resilience Heat resistance: in rubber matrix to maintain mechanical strength and integrity above normal flow temperature—*c.* 150°C	Static sealing in industrial and marine applications. Low-cost gaskets for automotive applications. Lubricated packings.	Unreinforced rubber sheet Cork/rubber. Impregnated cellulose or jute fibres. Machined or stamped sheet metal. Materials incorporating 'Rockwool' fibre (under development) in rubber matrix. PTFE, graphite, aramid fibres
Millboard		High temperature resistance Easy formability—can be cut, wet moulded or cemented Inexpensive	Rollers for conveying steel or glass during processing and heat treatment. Furnace linings. Many miscellaneous applications.	Millboards made from aluminosilicate fibres Mineral fibre blocks (depending on temperature) Generally applications are too diverse to discuss collectively.
Insulating boards		Asbestos board bonded with an inorganic (calcium silicate) bond Suitable to *c.* 600°C	Fire protection structural panels for civil engineering and marine use.	Many proprietary types of asbestos—free board containing mineral wools, mica, cellulose fibre, etc.
Asbestos-cement	Pipes	Chemical resistance, good formability	Sewage and drainage	GRP, rigid PVC, HDPE, cast iron (small bores), concrete (large bores), depending on performance and price constraints.
	Sheet		Building and roofing panels	Glass-reinforced cement GRC, GRAC Cement containing 'Rockwool' fibres + cellulose Polypropylene-reinforced cement (inferior impact strength) Metal sheets. See text.
High density A/C product	'Sindanyo'[a]	Strength, thermal and electrical insulation	Various.	Non-asbestos grades available as part of present range
Reinforced plastics (also includes asbestos felt and flock, impregnated with thermosetting resins and pressure moulded into intricate shapes for heat protection)	Friction materials	Complex stiffening, strengthening and high temperature function	Brake and clutch linings.	In phenolic binders: steel, glass, mineral, wool aluminosilicate (all have drawbacks). Sintered metals, silicon nitride, carbon/carbon composites (high performance, not suitable for cars). Vermiculite reinforced phenolic. Kevlar composites (recent development).
	Bearings	Asbestos-reinforced phenolic Low modulus of elasticity (misalignment less serious) of metals Tolerance of line and spot loading Corrosion resistance	Dry rubbing bearings for high load, low sliding speed applications (e.g. marine bearings, lock gates, railway bearings).	See Vol. 1, Chapter 1.5. Polyester—bonded textile laminates with MoS_2/graphite Cellulose—fabric based phenolic laminates with uniformly distributed PTFE/graphite. Polyimides. Woven and resin-bonded PTFE fibre. Graphite or PTFE impregnated metals. (All the above can involve a cost or performance penalty).
	General engineering	High temperature stiffness Corrosion resistance Mouldability, uniform shrinkage	Polypropylene, styrenes, thermosets.	Some applications can be satisfied by glass-reinforcement, mineral fibre reinforcement, or combination. Exotic reinforcements e.g. quartz, graphite, high-silica glass.
General insulation	Bulk	Moderate resistance to high temperatures	Various linings and seals.	Mineral wools. Ceramic fibres, Microtherm. Generally alternatives can be selected to give better temperature resistance, but mechanical properties and corrosion resistance may be inferior.

[a] Registered trade mark of TBA.

Table 3.8.26

TABLE 3.8.27 Properties of some asbestos-based (now withdrawn) and asbestos-free flat sheet and building boards

Board	*Description*	*Source*	*Density (kg/m³)*	*Flexural strength (N/mm²)*	*Strain to failure (%)*	*Impact strength (kJ/m²)*	*Young's modulus (N/m²)*	*Thermal shrinkage 4 h at 100°C (%)*	*Fire rating*
Asbestolux	Asbestos based building board for walls and ceilings	①[b]							
Marinite	As above For marine environments	①[b]							
LDR Turnasbestos	Asbestos-based building boards	②[b] ②[b]							
Pical Promat	Asbestos-based building boards	③[b] ③[b]							
Masterclad[a]	Cellulose-based building boards	① G	1200	17	0.6	2.2	5500	6.0	Class 0
Masterboard	Cellulose-based building boards	① G	875	10	0.6	2.0	3600	4.0	
TAC Board[a]	PVOII-based insulating board	① F	1200	15		2.5			Class 0
TAC Board SP[a]	PVOII-based insulating board	② F	1700	26		3.5			Class 0
Tacfire	Glass-fibre based building board	② F	875	7.5					
Northwich Board[a]	Glass-fibre based building board	⑥	1700						Class 0
CP Board	Wood-chip/cement board	⑤ F	1250	9–13					Class 0 up to 2.5h
Promatect II	Cellulose-reinforced building board								
Vermiculux	Vermiculite based insulating board	① F	450	2	0.2	0.6	1200	3.3	4h
Vicuclad	Vermiculite based insulating board	④ F							4h
Supalux	Cellulose-reinforced building board	① F	875	9	0.4	2.2	3600	1.6	
Monolux	Cellulose-reinforced building board	① F	640	5	0.2	0.7	2600	2.0	
Limpet	Cellulose and Glass insulating board	② F							

[a] NB—The distinction between asbestos-cement substitutes and insulating boards has become progressively blurred.
[b] Withdrawn.
Source—① Cape Boards and Panels. ② TAC Engineering Materials. ③ Eternit. ④ William Kenyon. ⑤ CP Boards Ltd. ⑥ Northwich
F—Substitutes for fire resistance. G.—Substitutes for general purpose applications.

TABLE 3.8.28 Assesment of fibres as possible replacements for asbestos in building boards

Fibre	*Dispersability in water*	*Film-forming ability*	*Alkali resistance*	*Autoclave temperature resistance*	*Flexural reinforcement*	*Toughness contribution*
Glass and ceramic						
E-glass	••	•	•	•••	•	•
Cem-FIL	••	•	•	•••	•	•
Mineral fibres	•••	•	•	•••	•	•
Kao-wool	•	•	•	•••	•	•
Carbon fibres						
High modulus	•	•	•••	•••	•••	••••
Low modulus	••	••	•••	•••	••	•••
Plastic						
Polypropylene	••	•	•••	•	•	•
Rayon	••	•	•••	•	•	••
Polyester	••	•	•	•	•	•
Nylon	••	•	•••	•	•	•
Kevlar	••	••	•••	•••	•••	•••
Cellulose						
Wood pulp	•••	•••	•••	•••	•••	•••
Mechanical pulp	•••	•••	•••	•••	••	•
Newsprint	••	•••	•••	•••	••	•
Metal						
Steel	•	•	•••	•••	•	•

• poor •• reasonable ••• good •••• very good.

Table 3.8.28

TABLE 3.8.29 Properties of basic materials used in packings and jointings

Material	***Service temperature (°C)***	***Chemical resistance, pH range***	***Advantages and limitations***
Asbestos, plain	540	3–14	Heat and pressure resistant, low price. — Abrasive, heat insulator, low service life.
Asbestos/graphite Asbestos/PTFE Glass fibres	425 250 450	3–14 } 3–14 } 5–11	Better than plain asbestos in packings. Good general-purpose jointing. — Limited chemical resistance.
Aramid fibres	250	3–11	Strong fibre, long service life. — Difficult processing characteristics, limited chemical resistance.
Graphite/PTFE	250	1–14	Non-abrasive, good heat dissipation, excellent chemical resistance, long service life — High price.
PTFE	250	1–14	Low friction, excellent chemical resistance. — Thermal expansion, low shaft speed in gland packings, high price.
Carbon	650 (inert conditions e.g. steam)	1–14	Good heat dissipation, high shaft speed in packings, excellent chemical resistance. — Brittle, no resistance to oxidising agents, high price.
Graphite	650 (inert conditions e.g. steam) 3000 (non-oxidising conditions)	1–14	Good heat dissipation, very high shaft speed in gland packings, excellent chemical resistance. — Brittle, not resistant to oxidising agents, high price.
Vegetable fibres, cork composites	100	5–10	Inexpensive. — Limited applications, not chemical-resistant, for low pressure use.

Note: pH values: 1 very strong acids.
7 neutral.
14 Very strong alkalis.

TABLE 3.8.30 Properties of non-asbestos millboards. TBA industrial products

Properties		*Units*	*Mineral wool/inert fillers to simulate asbestos millboard*			*High quality shot-free non-asbestos millboard*			*Fume-/smoke-free millboard for IIT Insultation*		
Density		kg/m³	950			875			950		
Tensile strength	20°C	MN/m²	3			4			2.6		
	500°C	MN/m²	1.2			1.8			—		
	800°C	MN/m²	0.6			3.5			—		
Flexural strength		MN/m²	6			7			4.8		
Compression at 21 MPa		%	30–40			35–45			50.3		
Ignition loss		%	15			15			7		
Moisture content		%	3			3			3		
Thermal conductivity	(50–400°C)	W//m per K	0.12–0.15			0.11–0.15			—		
			L	W	T	L	W	T	L	W	T
Shrinkage	500°C	%	−0.04	−0.25	−3.50	−0.60	−0.78	+9.6	—	—	—
	800°C	%	−0.52	−0.50	−1.80	−0.38	−0.45	−2.2	—	—	—
	1000°C	%	−0.60	−0.56	−4.89	−0.89	−0.25	−1.9	—	—	—

L = length.
W = width.
T = thickness.

Table 3.8.30

TABLE 3.8.31 Dry rubbing bearings; forms readily available

Material group	*Finished bearings*				*Stock shapes*				*Moulding material*
	Bush	*Thrust washer*	*Spherical*	*Rod end*	*Bar*	*Tube*	*Plate*	*Strip*	
A. Unfilled polymers	★	★			•	•	•	•	•
B. Filled polymers	★	★			•	•	•	•	•
C. High temperature polymers	★	★			•	•	•		•
D. Thin layer materials	•	•	•	•				•	
E. Filled PTFE	★	★			•	•		•	•
F. Reinforced thermosets	•	•			•	•	•		
G. Solid lubricant impregnated metals	•	•	•				•		
H. Carbon/graphites	★	★			•	•	•		

• Generally available. ★ Available with large batch orders.

Table 3.8.32

TABLE 3.8.32 Comparison of properties of asbestos and asbestos-free laminates and composites

Product range	*Grade*	*Description*	*Density (kg/dm³)*	*Shear strength (MPa)*	*Flexural strength (MPa)*	*Ultimate crushing strength (MPa)*	*Water absorption (%)*	*Electric strength (air, 90°C) (kV)*	*Surface breakdown (air, 90°C) (kV)*	*Arc resistance (s)*	*Thermal conductivity (W/m K)*	*Coefficient of thermal expansion (°C⁻¹ × 10⁵)*	*Max. cont. service temp. (°C)*	
SINDANYO	Natural CS51	Manufactured from asbestos fibres and cement formed into fully compressed boards under high pressure. Good heat and arc resisting characteristics and good anti-tracking properties. Treatments include surface oil impregnation (CS76), silicone resin surface treatment (CS85), silicone resin impregnation (CS86).	1.95[a]	34–41[a]	34–45[a]	124–151[a]	Can be 15	~0.8–1.9	>10	—	0.66	8	350	High temperature arc and heat resistant insulation; induction furnaces, handling molten metals, handling hot glass, controlling and extinguishing electric arcs.
	L11	Cement/cellulose-synthetic fibre composite.	1.42	10	8	59	24.7	3.11	>10	240	0.37	5.31	200	General purpose material available in sheets or machined components.
	L21	Similar composition as above, but denser and mechanically stronger.	1.84	16	19	104	15.7	3.31	>10	240	0.34	1.87	200	For mechanical or electrical use.
	M31	Cement-based composite incorporating cellulose and carbon fibres.	1.81	14	17	71	15.4	—	—	—	1.04	2.45	350	Not for electrical use. For demanding applications where maximum temperatures of 350°C are required. Available in sheet form or machined components.
	1141	Does not contain fibrous material. Maximum operating temperature must be reached carefully to avoid shattering.	2.38	26	22	119	2.0	2.7	>10	—	1.26	6.37	650	Available only as machined components for both mechanical and electrical usage.

TABLE 3.8.32 Comparison of properties of asbestos and asbestos-free laminates and composites—*continued*

FEROBESTOS	LA	Phenolic resin with heavy weave asbestos cloth base. Rods.	1.6 to 1.8 depending on grade	80	100	—	—	—	—	—	Hot face (120°C) 4–9.2	11–17 depending on grade	175	Established uses include bearings, bushings, machine tool slides, wearing pads and plates, sealing strips and vanes for pumps and compressors. See under Feroform for non-asbestos alternatives.
	LA3	As above with graphite inclusions. Tubes.		75	90	—	—	—	—	—			175	
	LA33	As LA3.		70	90	—	—	—	—	—	Cold face (10°C) 0.4–0.93 depending on grade		175	
	LA4	As for LA, but with a silicone resin. Rods and tubes.		60	35	—	—	—	—	—			350	
	EC	Asbestos felt base. Sheets.		100	200	—	—	—	—	—			175	
FEROFORM	F21	Phenolic resin/organic fibre designed to complement the Ferobestos LA grades.	1.4	50	90	—	—	—	—	—	—	—	140	Low duty applications on bearings or wearing surfaces. Washing machines, and industrial machinery are areas currently under evaluation. Available as sheet, rod, tube or machined parts.
	F31	Medium temperature phenolic resin and synthetic fibre material.	1.4	77	181	—	—	—	—	—	—	—	175	For medium duty rotor vane applications where Ferobestos LA and EC grades have previously been used. Currently under evaluation for high capacity vacuum and compressor units. Available only as machined parts.
	F61	Combination of silicone resin and specific glass cloth to complement Ferobestos LA4 and LA6.	1.6–1.7	66	142	—	—	—	—	—	—	—	300	Aimed at the aerospace industry and available only as machined parts.

[a] Direct substitutes for Ferobestos come under the trade name Feroform and there are at present five grades on offer: D11—resin-based graphite material for high load/low speed applications; F21—high modulus organic fibre reinforced; F31—resin-bound speciality paper-based; F61—high temperature glass reinforced; C11—high chemical resistance.

Table 3.8.32—*continued*

TABLE 3.8.32 Comparison of properties of asbestos and asbestos-free laminates and composites—*continued*

Product range	Grade	Description	Density (kg/dm^3)	Shear strength (MPa)	Flexural strength (MPa)	Ultimate crushing strength (MPa)	Water absorption (%)	Electric strength (air, 90°C) (kV)	Surface breakdown (air, 90°C) (kV)	Arc resistance (s)	Thermal conductivity (W/m K)	Coefficient of thermal expansion ($°C^{-1} \times 10^5$)	Max. cont. service temp. (°C)	
FEROFORM (continued)	C11	Phenolic resin matrix incorporating a unique glass reinforcement specifically for use in chemically hostile environments. Readily machinable.	1.75	110	152	—	—	—	—	—	—	—	175	Baffles and agitators in mixing tanks. Available in sheet, rod, tube and machined components. Will resist most organic chemicals and acids to 175°C, but not strong oxidising conditions, caustic alkalis (above cold and dilute), conc. sulphuric acid above 90°C and hydrofluoric acid.
	D11	Modified phenolic resin with selected additives intended for dry running applications.	1.87	—	72	—	—	—	—	—	—	7.4	175	Pumps with dry running rotors are being evaluated. Available only as machined parts.
FEROGLAS	—	A range of polyester glass mat laminates is available which have successfully replaced asbestos-based products as electrical insulators in applications up to and including Class F specification.	—	—	—	—	—	—	—	—	—	—	—	—

TABLE 3.8.33 Advantages and disadvantages of some asbestos alternatives in friction products

	Advantages	*Problems*
Kevlar	Strength, modulus, density thermal stability, wear	Mixing, machining
Steel	Strength, modulus, density thermal stability	Density, corrosion, low cold friction
Glass-fibre	Strength, modulus	Melt causes fade, low wear resistance, aggressive wear
Synthetic fibres	Strength	Melt causes fade
Cellulose fibres	Infusible	Low strength, modulus low char
Carbon fibres	Strength, modulus, thermal stability, infusible	Fibre form lost in mixing—price
Asbestos	Strength, modulus, thermal stability, infusible, wear	Environmental and health problems

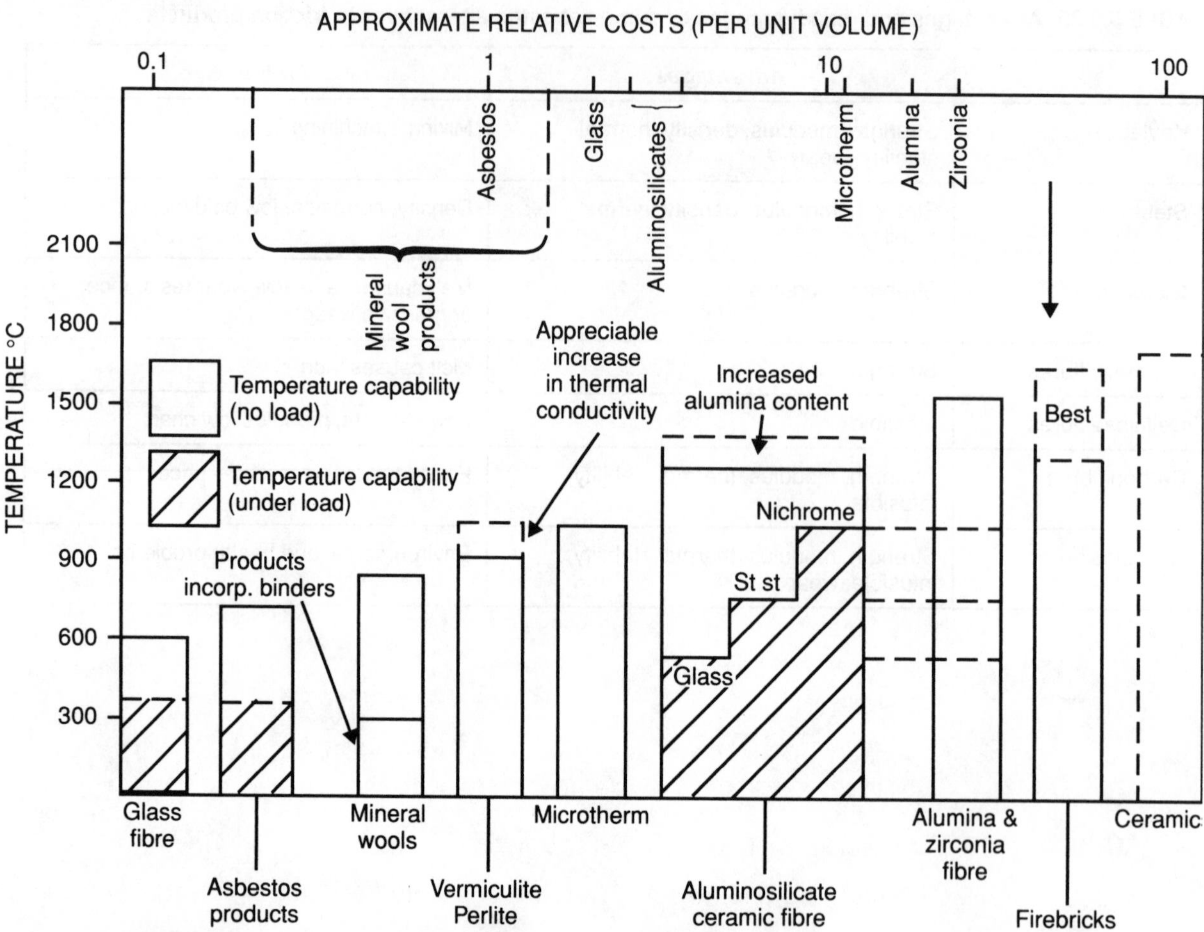

FIG 3.8.18 Maximum temperature capability of thermal insulants

Fig 3.8.18

3.8.9 Flexible laminated plastics (including flexible printed circuits)

Flexible laminates consist of combinations of plastic films (see Chapter 3.7), metal foils and fabrics, bonded together with adhesives, by fusion or by coextrusion.

MECHANICAL PROPERTIES

When designing a flexible laminate to meet a certain strength requirement, it is imperative to consider the modulus of each component and the strain region in which the laminate must operate.

The strength of each component at that elongation of interest can be added to estimate a total useful strength.

ENVIRONMENTAL PROPERTIES

Considering the effects of temperature, humidity and chemicals, the flexible laminate is usually only as good as the weakest layer. These effects are, at best, only delayed by a more impervious layer. On the other hand, radiation effects, such as sunlight, can be effectively resisted by shielding with resistant layers.

APPLICATIONS

Applications include low-permeability membranes (plastic film/metal foil), flexible printed circuitry, e.g. polyester or polyimide film/copper foil/epoxy (usually matrix/glass cloth), weatherseals and helicopter blade liners.

Flexible printed circuits

A most important application of flexible laminated plastics is in the production of flexible printed circuits. The materials used in the production of these are listed in:

Table 3.8.34 *Characteristics of substrate materials for flexible printed circuits*

TABLE 3.8.34 Characteristics of substrate materials for flexible printed circuits

Material	*Advantages*	*Limitations*
PETP Polyester film	Primary established material. Excellent mechanical and electrical properties. Maximum temperature of 130°C.	Melting points limits assembly and soldering options; crimped tags or pressure connectors generally used. Not readily adapted for flexible multilayer systems.
Polyimide film/thermosetting adhesive	Excellent mechanical and electrical properties. Maximum temperature of 125°C but solderable. Compatible with conventional laminates and used for interconnection packages and multilayers.	More expensive than polyester.
Polyimide film	Temperature capability exceeds 400°C. Self-extinguishing. Suitable for precision applications.	Very expensive, bearing in mind that the covercoat must be made of the same material.
Glass-epoxy	Solderable. Adaptable to multilayering. Properties, being similar to rigid PCBs, are well-established.	Susceptible to cracking around small radius bends: low tear resistance and moisture trap. Not easily available.
Fluoroplastic FEP	Excellent environmental resistance. Solderable.	Covercoat laminating operation causes dimensional stability problems.

3.8.10 Metal–plastic laminates

Metal–plastic laminates, like sandwich structures (Section 3.8.4), function like I-beams, in that the metal facings support bending stresses while the plastic core supports shear stresses between facings. They provide weight savings in 'stiffness limited applications'.

Because of the low strength of the plastic core and the thinness of the metal facings, the laminates should be considered only for components subject primarily to bending loads. Tensile and compressive loads would have to be quite low.

Because of the thin facings and relatively soft core, the laminates may be susceptible to localised denting from stone impact. Therefore, exterior body parts may require thicker facings and stronger, more rigid, cores.

Formability of the laminates is governed largely by the formability of the metal facings, provided the plastic core is at least as formable.

Joining methods for assembling the laminates are generally limited to mechanical fastening and adhesive bonding. Conventional resistance and fusion welding are largely ruled out by the plastic core and the thinness of the metal facings.

Cost of steel-faced laminates is approximately equivalent to that of cold rolled steel, while that of aluminium faced laminates is lower than that of aluminium, as illustrated in:

Fig 3.8.19 *Relative weight and cost of metal–plastic laminates*

Scrap also must be considered. Even if laminate scrap could be remelted in steelmaking furnaces, its market value would be considerably less than that of conventional sheet scrap because of its large plastic content. Separate recycling of the steel and plastic is conceivable, but would require considerable laminate production to warrant development.

Flexible sheets of lead–plastic laminate could be used for containers, cable sheathing, etc., where the corrosion resistance and impermeability of lead combined with lightness and formability of plastics can be employed in harness. The main area of application is likely to be shielding against high energy radiation using composites of lead with PVC, polyethylene or polypropylene.

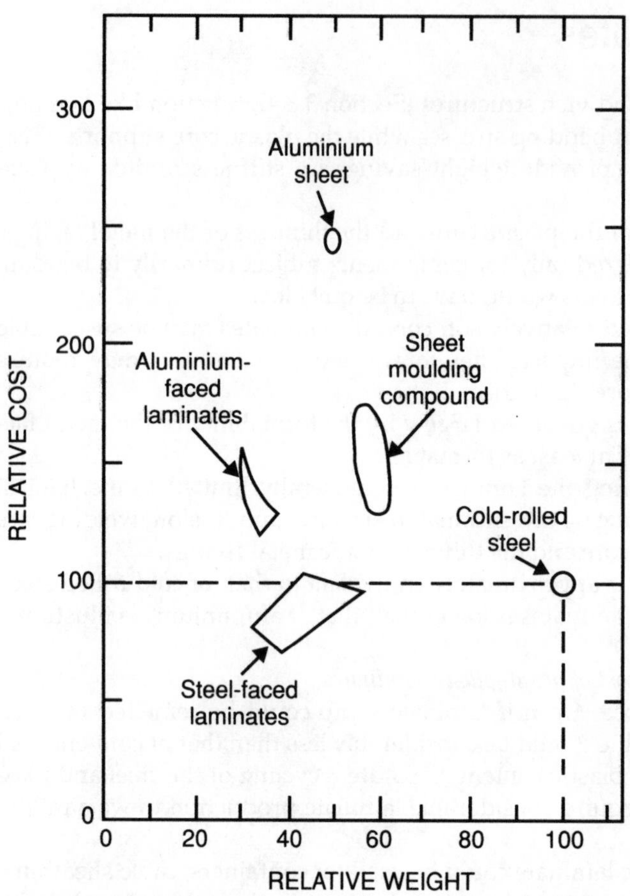

FIG 3.8.19 Relative weight and cost of metal–plastic laminates
(source: General Motors)

Fig 3.8.19

Index

Vols 1–3

Note: Tables are indicated by **bold page numbers**, and Figures by *italic page numbers*.
For longer Tables and Figures, the first page of the Table/Figure is given; please look through for the particular material/property

A

C

D

G

H

I

J

K

L

O

P

S

T

U

X

Y

Z